Deformation and Flow of Polymeric Materials

Helmut Münstedt · Friedrich Rudolf Schwarzl

Deformation and Flow of Polymeric Materials

Springer

Helmut Münstedt
Friedrich Rudolf Schwarzl
Lehrstuhl für Polymerwerkstoffe
Friedrich-Alexander-Universität
Erlangen-Nürnberg
Erlangen
Germany

ISBN 978-3-642-55408-7 ISBN 978-3-642-55409-4 (eBook)
DOI 10.1007/978-3-642-55409-4
Springer Heidelberg New York Dordrecht London

Library of Congress Control Number: 2014940158

Printed on acid-free paper

Springer is part of Springer Science+Business Media (www.springer.com)

Preface

During the last few decades, the use of polymeric materials has grown so that they form the most important class of materials if counted by volume and it can be regarded as certain that their growth rates will remain high in the future. While at the beginning of the era of polymers Chemistry was the predominant science, Engineering and Physics have become more and more requisite for the successful development of polymeric materials. Engineering, as the efficient production of materials of good quality and their highly sophisticated processing have developed into key factors of economic success; and Physics, as material properties have become decisive for specific applications and the continuing innovation and improvement of products. These requirements cannot be fulfilled without a fundamental knowledge of polymeric materials and a profound understanding of the relations between the molecular structure of polymers and their physical and processing properties.

These developments and the increasing number of joint uses of various materials in highly sophisticated technical products led to the foundation of special institutes devoted to materials science at some universities about 40 years ago. Besides classical materials like metals, glass, and ceramics, polymers became a central point of academic teaching and research. This book has its origin in lectures for students in the field of polymeric materials within the Department of Materials Science and Processing at the Friedrich-Alexander-University Erlangen-Nürnberg. Its main intention is to teach the basics about polymers necessary for everybody working with these materials. One part is based on the textbook "Polymermechanik," which appeared in 1990, the other stems from more recent lectures.

The book follows two main guidelines. One is a quantitative description of the molecular structure, though this has sometimes had to be simplified due to its complex nature. The other presents relationships between molecular quantities and material properties, which cover the solid and the molten state. The temperature is the key external parameter according to which the mechanical behavior is comprehensively discussed. Nevertheless, this work has to be regarded neither as a complete text book on the mechanics of polymer materials nor on their rheology. Rather, it is meant to discuss these fields from a common viewpoint, which encompasses the transitions between the solid and molten states.

"Tools" are described insofar as they are believed to be necessary for a deeper understanding of the results presented. They comprise measuring devices as well as mathematical formalisms required for dealing with large deformations.

Much space is devoted to the linear theory of viscoelasticity as it is coherent in itself and allows quantitative insights into relations between properties and molecular structure. Theories covering the nonlinear viscoelastic regime which dominates processing are not yet in a state of development enabling quantitative descriptions of properties and processes with a generality comparable to the linear behavior. This has to be said, although many theories have been published since 1990, the year of the appearance of "Polymermechanik." Nevertheless, an overview of constitutive equations based on various models is given and some of their predictions are compared with experimental findings.

In many parts of the book, results of investigations performed at the Institute of Polymer Materials are presented. They are examples of how research and established knowledge can become complementary parts of teaching. Regarding the references, long lists, which could easily be obtained today from various electronic bibliographies, have deliberately been avoided. Instead, a selection of original literature is cited, which opens the door to deeper information for those who are interested in more details.

The book is written from our experience of teaching the knowledge on polymeric materials, which we think to be useful for people interested and engaged in this class of materials. We hope it will be helpful for students to consolidate and broaden their knowledge, to researchers in the field of polymers at various institutions, and even to those working in industry, whenever they would like to get some fundamental questions answered, which arise from dealing with polymers.

As it is obvious from the reference lists, the originality and actuality of the results presented in this book are based, to a high degree, on the research of doctoral students under the guidance of the authors of this book. Particularly appreciated are the contributions from the theses of:

Dr. Dietmar Auhl
Dr. Hans-Jürgen Grieß
Dr. Daniela Hertel
Dr. Jens Hepperle
Dr. Claus Gabriel
Dr. Ute Maria Kessner
Dr. Stefan Kurzbeck
Dr. Julia Maria Resch
Dr. Martin Schwetz
Dr. Florian Stadler
Dr. Jens Stange
Dr. Thomas Steffl
Dr. Erik Wassner
Dr. Friedrich Wolff
Dr. Robert Greiner
Dr. Joachim Kaschta
Dr. Günther Link
Dr. Walter Pfandl
Dr. Michael Wolf
Dr. Franz Zahradnik

H. Münstedt thanks Dr. Ute Maria Kessner, Dr. Martin Schwetz, and Dr. Friedrich Wolff for reading parts of the manuscript, Dr. Florian Stadler for fruitful discussions and M.Sc. Ute Zeitler for her support in preparing computer-based versions of many of the figures.

F. R. Schwarzl wishes to remember the late Dr. ir. J. Heijboer with whom many fruitful discussions in the field of secondary relaxation mechanisms of polymers were held. He wishes also to thank Prof. Dr. H. Janeschitz-Kriegl for a life-long friendship and scientific cooperation in Graz, Delft, Linz, and Erlangen, Prof. Dr. ir. L. C. E. Struik for the very stimulating discussions concerning the solid state of polymers in Delft and Prof. Dr. M. H. Wagner, Technical University Berlin, for the introduction to his theories on polymer rheology.

Erlangen, March 2014

Helmut Münstedt
Friedrich Rudolf Schwarzl

Contents

Chapter 1
Introduction

1.1 General Aspects of Polymeric Materials

Polymers consist of macromolecules composed of a great number of basic units called monomers, which are connected to each other in an identical way. Polymers that are ubiquitous in nature and can be regarded as the basis of life are called *natural polymers*. Examples are proteins and polysaccharides. Proteins are the basic elements of various nucleic acids or natural silk, for example, polysaccharides of cellulose, starch, or chitin. Cellulose is the main component of the extremely versatile material wood and starch the base of many kinds of food. Another example is natural rubber, a cis-isoprene (cf. Table 4.4) which has found an important technical application in tyres.

Synthetic polymers are products of modern times. Polyvinylchloride was synthesized on a laboratory scale in 1913, but not industrially produced before the 1930s, the time when polystyrene and polymethylmethacrylate made their first steps in applications. The breakthrough of synthetic polymers started in the 1950s due to the invention of the Ziegler–Natta catalysts which made an economic production of polyethylenes and polypropylenes possible. Ziegler and Natta received the Nobel Prize for their inventions in 1963. Another key for the overwhelming success of polymers as a new class of materials was the abundance of cheap monomers based on improved and scaled-up petrochemical processes.

Synthetic polymers can be divided into materials that are able to be shaped and those that can only be applied together with a substrate. The latter comprise dispersions, varnishes, paints, and functional polymers. The shapeable polymers are split into *duromers*, *elastomers*, and *thermoplastics* (cf. Sect. 4.1), whereby thermoplastics have the largest market share.

Synthetic polymeric materials can be made in a great variety of modifications as becomes evident from the schematics in Fig. 1.1. Not all thinkable combinations of chemical moieties arranged in various structures can be synthesized, however, and even of those feasible to make, only a few show properties interesting for applications. Nevertheless, Fig. 1.1 gives an idea of the wide scope of products obtainable by polymerization.

H. Münstedt and F. R. Schwarzl, *Deformation and Flow of Polymeric Materials*,
DOI: 10.1007/978-3-642-55409-4_1, © Springer-Verlag Berlin Heidelberg 2014

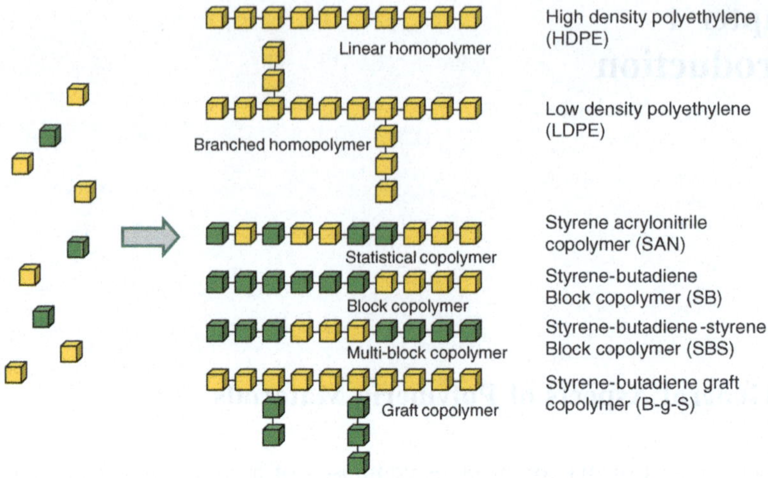

Fig. 1.1 Schematic examples of molecular architectures of synthetic polymers

Linear polymers synthesized from one special monomer, like high density polyethylene, polypropylene, polyvinylchloride, and polystyrene, for example, form the largest stake in the market. Besides the chemical nature of the monomer the chain lengths and their distributions can be varied. As will be discussed later, these changes in molar mass and molar mass distribution have a decisive influence on properties of polymeric materials. Additionally, branches can be introduced. As their length, concentration, and distribution along the main chain can be changed, a great variation in the architecture and following from that the tailoring of properties become possible. Low density polyethylene is an example for such a branched product with a broad field of applications.

The number of products becomes still more diversified if two different monomers are copolymerized. As sketched in Fig. 1.1 this can be done in various ways. The monomers can statistically be distributed along the chain or arranged in blocks. These species are named *statistical or block copolymers,* respectively. Composing a main chain of blocks of different lengths or generating branches with one species of monomer by grafting, open up a field of macromolecules with great chemical and structural varieties. Styrene is suitable for copolymerization with acrylonitrile or butadiene resulting in engineering copolymers consisting of statistically arranged styrene and acrylonitrile (SAN) or styrene-butadiene blocks (SBS), respectively. Once again one can imagine how manifold the chemical and structural compositions of the molecules could in principle be.

Copolymerization plays an important role for polyethylenes, too. Olefinic monomers up to eight carbon atoms (octene) are copolymerized with ethylene and yield short-chain branches. These materials are called linear low density polyethylenes (cf. Sect. 4.1). Copolymerization of ethylene and polar monomers like vinylacetate result in grades with specific adhesion properties.

These few examples demonstrate the diversity of polymeric materials and their potential to get tailored for special applications. Therefore, it is not surprising that polymers have substituted metals and natural materials like wood in many applications. Indeed, they have reached a very important economic and technical position within only 50 years and the industries related to them play an important role in many countries today.

In parallel to the development of synthetic polymers on a laboratory scale, scientific questions regarding their structure and physical properties have come up. Three highlights among the many steps in research should be mentioned. In 1920, H. Staudinger published his idea on the macromolecular structure of polymers [1]. His scientific achievements were honored in 1953 by the Nobel Prize. In the 1930s, Guth and Mark [2] and Kuhn [3] postulated the coil structure of macromolecules in solution, and starting around 1940 the statistical mechanics of coiled molecules by Flory added an important step to a deeper understanding of the behavior of macromolecules [4–6]. Flory was awarded the Nobel Prize in 1974. During the 1960s and 1970s a lot of experimental data was collected regarding properties of polymers in their solid and rubber-elastic states which were complemented by investigations in the melt later on. These results are the basis for a better understanding of many of the application and processing properties of polymeric materials. Furthermore, the progress in the analytical techniques of polymers made it possible to correlate properties with their molecular structure. This knowledge was and is the basis of specific developments of polymeric materials.

This book tries to collect findings and expertise about the deformation and flow of polymeric materials under the special aspects of how external parameters and here, in particular, temperature and time, and internal parameters based on the molecular structure influence this behavior.

1.2 Nomenclature

The generic names of polymers have been standardized by a Working Party of the International Union of Pure and Applied Chemistry (IUPAC) [7]. Particularly, for polymeric materials of high technical relevance acronyms are used that can be found in [8]. For example, for polypropylene the acronym is PP, polyetheretherketone is designated as PEEK. The generic name of some polymers, the chemical structure of their repeat units, and their acronyms are presented in Tables 4.2–4.4. Besides this standardized nomenclature, different trade names for products of the same chemical structure are common. That goes back to the intention of companies to assign brand names to their products. As norms do not exist, only in rare cases the chemical structure or information on some basic properties can be derived from the trade name.

1.3 General Classification of Polymeric Materials

Synthetic polymers can be classified according to different aspects. One is the chemical group they belong to. *Polyolefins*, *polyamides*, and *polyaryletherketones* are examples. Another classification is related to the kind of chemical reaction like *polymerization* (e.g., PE, PP, PS, PVC), *polycondensation* (e.g., PA, PET), or *polyaddition* (e.g., PUR). The products belonging to the acronyms are listed in Tables 4.2–4.4. More details on the different polymerization methods can be found in [9], for example. A third classification aspect is related to the dynamic-mechanical behavior as a function of temperature (cf. Sects. 4.3–4.4). Thermoplastics, which are either totally amorphous or semicrystalline, can be melted and favorably processed in this state, elastomers are characterized by a wide-meshed network, and duromers by a high density of chemical crosslinks.

A very application-related classification is based on the *continuous use temperature* T_{cu} the determination of which is standardized [10]. It defines the heat stability of various polymer materials with respect to a chosen property like the yield stress, for example. T_{cu} is of interest in technical applications at higher temperatures. It is essentially determined by the chemical composition of the macromolecular chain. In Fig. 1.2 some polymeric materials are presented and ordered according to T_{cu}. The different polyethylenes LDPE, LLDPE, and HDPE (cf. Table 4.7), polystyrene (PS), and polyvinylchloride (PVC) belong to the group of the so-called *standard polymers* with the largest market share. The next category are the so-called *technical or engineering polymers*. They are used in applications more demanding with respect to temperature. Some so-called *high temperature thermoplastics* can be applied up to 260 °C for longer periods of time without loosing too much of their properties. The products belonging to most of the acronyms can be found in Tables 4.2–4.4. PAEK designates the polyaryletherketones which polyetheretherketone (PEEK) is a part of. LCP stands for liquid

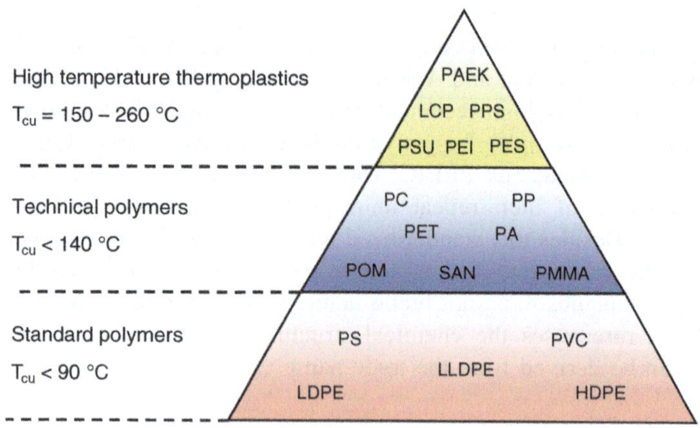

Fig. 1.2 Polymeric materials classified according to their continuous use temperature T_{cu}

crystalline polymers, PSU for the amorphous polysulphone, and PEI for the amorphous polyetherimide. Particularly, in case of the semicrystalline PEEK and PPS the excellent temperature stability is coupled with a pronounced resistance to a lot of chemicals and aggressive media. For PEEK and PPS, these properties lead to the designation as high performance products which compete with metals in some applications. However, due to their demanding prices, high temperature resistant polymers comprise a relatively small market till now.

References

1. Staudinger H (1920) Ber d Deut Chem Ges 53:1073
2. Guth E, Mark H (1934) Monatsh Chem 65:93
3. Kuhn W (1934) Kolloid Z 68:2
4. Flory PJ (1941) J Chem Phys 9:660
5. Flory PJ (1949) J Chem Phys 17:303
6. Flory PJ (1969) Statistical mechanics of chain molecules. Interscience Publisher, New York
7. http://www.chem.qmul.ac.uk/iupac/rssop
8. DIN EN ISO 1043-1
9. Elias HG (1984) Macromolecules. Plenum Press, New York
10. DIN EN ISO 2578

Chapter 2
Physical Structure of Macromolecules

In this chapter, the structure of single macromolecules is discussed, together with the possibilities for their internal motion. This leads to the concept of the shape and extension of macromolecules as calculated by random walk statistics, in theta solution, in bulk and in the melt. The chapter further contains the description of molar mass and molar mass distribution.

2.1 Structure and Brownian Motion of Macromolecules

Let us consider the spatial shape of a molecule consisting of five consecutive carbon atoms, i.e., of a pentane or a substituted pentane. The structural formula of pentane

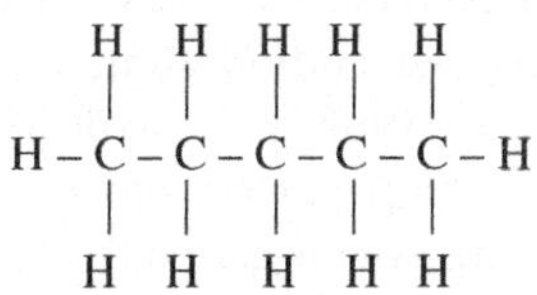

suggests an incorrect picture of the spatial shape of the molecule. In reality, the four bonds of a carbon atom can never be situated in the same plane; neither will the single bonds between successive carbon atoms form parts of the same straight line. The four single bonds of a carbon atom build up a regular tetraeder, as illustrated in part a of Fig. 2.1.

With the carbon atom in its center, the bonds point to the vertices of the tetraeder. The spatial angle between each pair of bonds has the same fixed value of $\vartheta = 109.47°$ ($\cos\vartheta = -1/3$). In Fig. 2.1, this structure is further illustrated by two models of the methane molecule. The "*Buechi-model*" places emphasis on the direction of the bonds, which are indicated by small tubes. This model allows the illustration of the change of the spatial shape of a molecule by the rotation around single carbon-carbon bonds (the *conformational changes of the molecule*). The

H. Münstedt and F. R. Schwarzl, *Deformation and Flow of Polymeric Materials*,
DOI: 10.1007/978-3-642-55409-4_2, © Springer-Verlag Berlin Heidelberg 2014

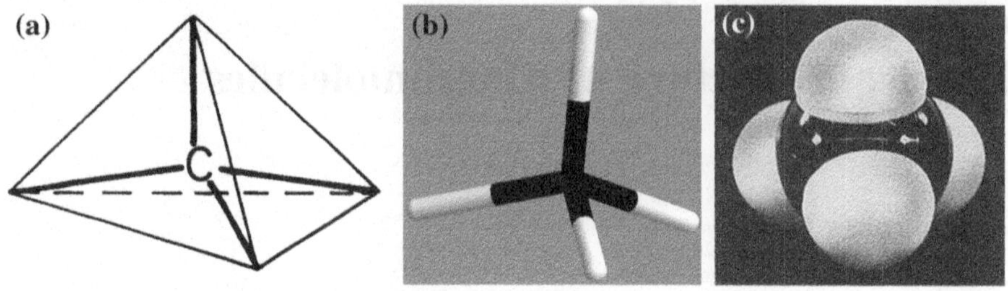

Fig. 2.1 **a** The directions of the four single bonds extending from a carbon atom **b** Buechi-model of the methane molecule **c** Stuart-model of the methane molecule. Reprinted from [1]

"*Stuart-model*" visualizes the occupation of space of a molecule by its outer electron shells. The dimensions of Stuart models are mostly in proportion to the real molecule (scale $1{:}10^8$).

Next, we discuss the spatial shape and the conformations of the pentane molecule as illustrated in Fig. 2.2. The direction of the first carbon-carbon bond (C_1–C_2) is assumed to be fixed. On rotation around this bond, the three further bonds starting from atom C_2 will move on the surface of a cone with a generating line, including the angle ϑ with the direction of the first bond. The next carbon atom C_3 will then be situated anywhere on the basic circle of this cone. Fixing the position of C_3 for the moment, the fourth carbon atom C_4 will move on the basic circle of a cone with a generating line including the angle ϑ with the direction (C_2–C_3), and so on.

If all rotations around the bonds C_1–C_2, C_2–C_3, C_3–C_4 can take place independently of each other, the last carbon atom C_5 can reach a large number of spatial positions, even if the position of C_1 and the direction of (C_1–C_2) are kept fixed. The change in shape of the pentane molecule occurring in this manner looks similar to the twisting of a worm and is called the "*micro-Brownian motion*" of the macromolecule.

The micro-Brownian motion of the macromolecule is the conformational change in shape of the molecule originating from the rotations around the single C–C bonds of the main chain.

By the continuous micro-Brownian motion originating from the thermally induced kicks of the environment, a macromolecule will pass through all its conformations permitted by the positions of its neighbors. Most conformations will lead to the shape of a coil. A few conformations will adapt the shape of a stretched ellipsoid, only one will lead to the *completely stretched conformation*. Consequently, we expect the macromolecule averaged over time to assume the *shape of*

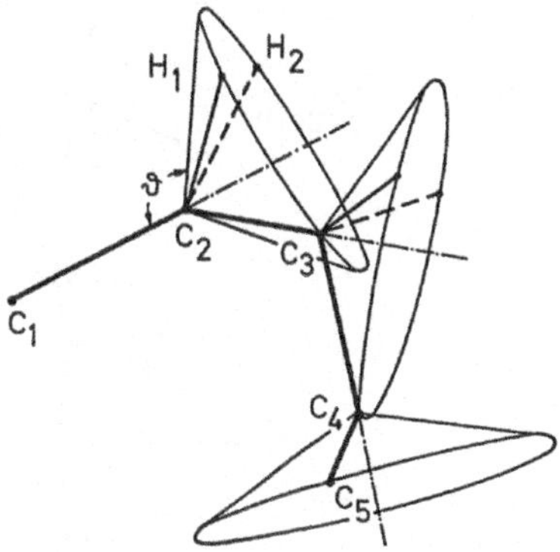

Fig. 2.2 Conformations of the pentane molecule

a coil, whose details will continuously change with time. Description and calculation of these coils will be handled in Sect. 2.3.

The *completely stretched conformation* is defined by the condition that the distance between the beginning and the end of the molecule is maximal. In Fig. 2.3, the carbon chain with single bonds is shown in its completely stretched conformation in the representation of a Buechi-model.

In this conformation, all bonds of the main chain are situated in one single (horizontal) plane in the shape of a "zick-zack". The other two bonds of the carbon atoms emerge from this plane, one into the direction upwards and one into the direction downwards.

The completely stretched conformation is well suited to introduce the concept of *tacticity* of *vinyl-polymers*. The class of vinyl-polymers is characterized by a structural unit of the following type

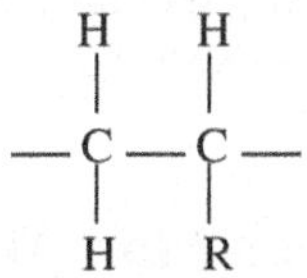

with R being an arbitrary side-group different from H. The spatial placement of the side groups R plays an important role for the properties of the vinyl-polymer. This is illustrated for the case of poly(propylene) PP ($R = \mathrm{Me} \equiv CH_3$) in Fig. 2.4.

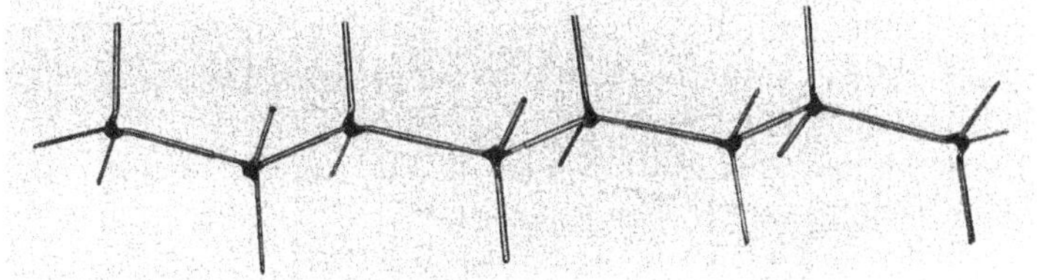

Fig. 2.3 Buechi-model of a carbon chain in the completely stretched conformation

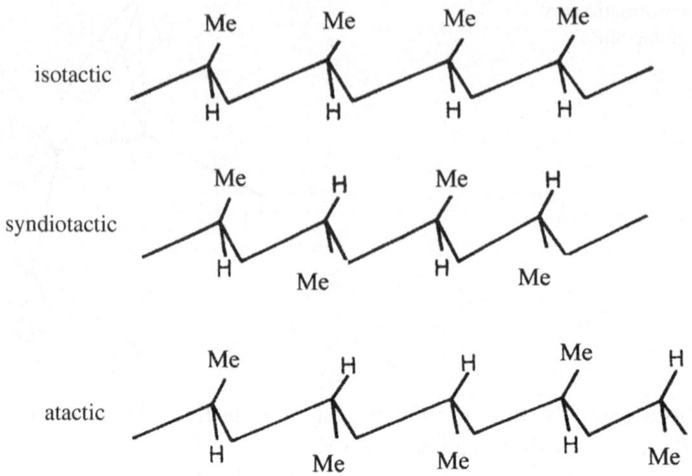

Fig. 2.4 Illustrating the tacticity of poly(propylene)

Consider the completely stretched conformation of the PP–chain. If all methyl-groups are situated on the same side of the principle plane, the polymer is called *isotactic*. If the methyl-groups are alternatively situated above and below the main plane, the polymer is called *syndiotactic*. If the position of the methyl-groups is irregular, the polymer is called *atactic*. Isotactic and syndiotactic polymers tend to *crystallization*, because their molecules fit better into the crystal lattice. Stereo-specific polymers are obtained by special coordination polymerization methods with mixed catalysts (Ziegler and Natta).

2.2 Molar Mass and Molar Mass Distribution

Next, we deal with the molar mass of the macromolecules, which is a very important characteristic property of polymers. The definition of the *molar mass* (sometimes also called molecular weight) is reasonable only for *uncrosslinked* polymers. For *uncrosslinked, not branched homopolymers*, the molar mass characterizes the structure of the molecule, apart from its tacticity. For *uncrosslinked, branched homopolymers*, this information has to be supplemented by the number, length and the topology of the branches.

We start with the definition of *Avogadro's number*. This number does not have a physical dimension, but a unit is designated to it, called mol^{-1}, meaning “per mol”.

$$N_A = 6.0225 \cdot 10^{23} \mathrm{mol}^{-1} \tag{2.1}$$

The mol is used as a unit of the amount of a material. The totality of N_A identical items is designated as one mol of those items. Let m be the mass of a molecule (in kg), then its molar mass is defined as

$$M = N_A \cdot m \tag{2.2}$$

The unit of the molar mass is kg/mol. The number N_A is defined in such a way that the molar mass of the nuclide ^{12}C becomes exactly 0.012 kg/mol = 12 g/mol.

Consider the molecule of a non-*crosslinked, unbranched homopolymer*:

$$A_B + A + A + \cdots\cdots + A + A_E$$

The moieties A_B and A_E of the beginning and the end of the molecule respectively, will generally differ from its center part, which consists of P repetitions of one single group A. P is called the *degree of polymerization*; the repeating unit A is called the *structural unit*. If M_g, M_B, and M_E are the molar masses of the structural unit, the beginning and the end moiety of the molecule, respectively, the molar mass of the complete molecule is

$$M = M_B + P \cdot M_g + M_E \approx P \cdot M_g \tag{2.3}$$

The last part of this equation is valid approximately at sufficiently large values of P. Some examples are:

poly(vinylchloride) *PVC* 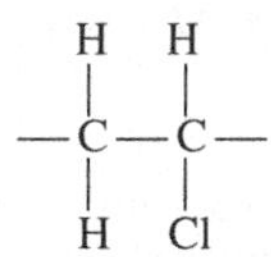 $M_g \approx 2 \cdot 12 + 3 \cdot 1 + 1 \cdot 35 = 62$ g/mol

poly(ethylene) *PE* 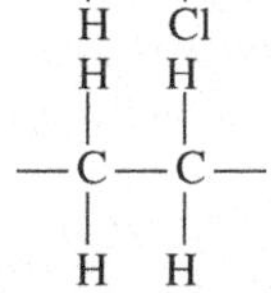 $M_g \approx 2 \cdot 12 + 4 \cdot 1 = 28$ g/mol

As P has to be an integer, the molar mass can only assume a finite number of discrete values, which are, apart from the small term $M_B + M_E$, multiples of M_g.

We describe the distribution of the molar mass by either *counting* or *weighing* the molecules with a certain molar mass and arrive at a scheme like that of Table 2.1, in which the k classes are arranged in increasing order of their degree of polymerization, i.e., $P_i < P_{i+1}$ for i = 1 to k − 1. Let n_i be the number of molecules with molar mass M_i and let us use the abbreviation:

$$m_i = n_i M_i \tag{2.4}$$

Table 2.1 Discrete values for the distribution of the molar mass arranged in k classes

Number of molecules	Degree of polymerization	Molar mass	Mass of the fraction
n_1	P_1	$M_1 \approx P_1 M_g$	$W_1 = n_1 M_1/N_A$
n_2	P_2	$M_2 \approx P_2 M_g$	$W_2 = n_2 M_2/N_A$
.....			
n_k	P_k	$M_k \approx P_k M_g$	$W_k = n_k M_k/N_A$

The *mass* of the *fraction i*, W_i, is given by the equation

$$W_i \approx n_i M_i / N_A = m_i / N_A \tag{2.5}$$

We further define the total number n_O of the molecules considered, by

$$n_0 = \sum_{i=1}^{k} n_i \tag{2.6}$$

and the *number distribution function* of the molar mass, $f(M_i)$, as the *relative number* of molecules with a molar mass equal to M_i.

$$f(M_i) = n_i / n_0 \tag{2.7}$$

The number distribution function obeys the normalization condition

$$\sum_{i=1}^{k} f(M_i) = 1 \tag{2.8}$$

and constitutes a *discrete distribution function*, i.e., it is not defined for values $M \neq M_i$. In addition, we define the *mass distribution function* of the molar mass, $h(Mi)$, as the *relative mass* of molecules with a molar mass equal to M_i:

$$h(M_i) = \frac{W_i}{\sum_{i=1}^{k} W_i} = \frac{n_i M_i}{\sum_{i=1}^{k} n_i M_i} = \frac{m_i}{\sum_{i=1}^{k} m_i} \tag{2.9}$$

which, of course, is also normalized

$$\sum_{i=1}^{k} h(M_i) = 1 \tag{2.10}$$

and also constitutes a discrete distribution function.

For mathematical descriptions, it is often necessary to change from the *discrete distribution function* $f(M_i)$ to a *continuous distribution function*, which is defined for all values of the molar mass. This is achieved in two steps:
Step 1: We extend the range of definition of the distribution function to all values of the molar mass. Around each point M_i of the M-axis, an interval $[B_i\ B_{i+1})$ is created, which starts from the midpoint between M_{i-1} and M_i, $B_i = (M_{i-1} + M_i)/2$ and ends at the midpoint between M_i and M_{i+1}, $B_{i+1} = (M_i + M_{i+1})/2$. For the first interval we choose the bounds $B_1 = M_1 - M_g$ and B_2, for the last interval the bounds B_k and $B_{k+1} = M_k + M_g$. The *widths* b_i of those intervals are

$$\begin{aligned} &b_i = B_{i+1} - B_i = (M_{i+1} - M_{i-1})/2 \qquad &&\text{for} \quad i = 2 \text{ to } k-1 \\ &b_1 = M_g + (M_2 - M_1)/2 &&\text{and} \quad b_k = M_g + (M_k - M_{k-1})/2 \end{aligned} \tag{2.11}$$

As $f(M)$ is defined as the relative number of molecules with a molar mass within the interval $[B_i\ B_{i+1})$, we put $f(M) = f(M_i)$ for $B_i \leq M < B_{i+1}$. $f(M)$ is the probability to find a molecule within the interval $[B_i\ B_{i+1})$. If the molar masses would be equally distributed within this interval, $f(M)dM/b_i$ would be the probability to find a molecule with a molar mass between M and $M + dM$ within this interval. Consequently, we define a new distribution function

$$g(M) = f(M_i)/b_i \quad \mathit{for} \quad B_i \leq M < B_{i+1} \tag{2.12}$$

$g(M)$ is a step function defined for all values of M with discontinuities at the points B_i. $g(M)dM$ is the probability to find a molecule with a molar mass between M and $M + dM$ in the interval $[B_i\ B_{i+1})$.
Step 2: Generally, the number of classes k will be sufficiently large to replace this step function by a continuous function $g(M)$ which will be approximately equal to $f(M_i)/b_i$ at the points $M = B_i$ and is smooth in between. To distinguish between $f(M)$ and $g(M)$, we will call the latter the *number distribution density* of the *molar mass*. The function $g(M)dM$ describes the *relative number* of molecules with a molar mass between M and $M + dM$. It will occur in all integral representations of the molar mass. Its condition of normalization reads

$$\int_0^\infty g(M)dM = \sum_{i=1}^{k} \int_{B_i}^{B_{i+1}} g(M)dM \cong \sum_i b_i g(M_i) = \sum_i f(M_i) = 1 \tag{2.13}$$

Besides the number distribution density, an other distribution function is introduced, the *mass distribution density*, which describes the *relative mass* of the molecules with a molar mass between M and $M + dM$:

$$w(M) = M \cdot g(M)/M_n \tag{2.14}$$

with

$$M_n = \frac{\sum n_i M_i}{\sum n_i} = \frac{\sum m_i}{\sum (m_i/M_i)} = \frac{\sum m_i}{\sum n_i} \tag{2.15}$$

The denominator M_n is introduced to normalize the mass distribution density by the condition:

$$\int_0^\infty w(M)dM = \frac{1}{M_n}\int_0^\infty Mg(M)dM = 1 \tag{2.16}$$

or

$$M_\mathrm{n} = \int_0^\infty Mg(M)dM \tag{2.15'}$$

Apart from one case, the usual methods for the determination of the molar mass, do not yield the entire distribution function, but only certain averages of it. The most important are

M_n *number average* of the molar mass
M_w *weight average* of the molar mass
M_z *z-average* of the molar mass
M_v *viscosity average* of the molar mass

The *number average* M_n is the mean value of M calculated by using the number distribution of the molar mass and was defined in Eqs. (2.15) and (2.15′). The *weight average* M_w is the mean value of M calculated by using the mass distribution of the molar mass and is defined by

$$M_\mathrm{w} = \frac{\sum m_i M_i}{\sum m_i} = \frac{\sum n_i M_i^2}{\sum n_i M_i} \tag{2.17}$$

and

$$M_\mathrm{w} = \int_0^\infty M \cdot w(M)dM = \frac{1}{M_n} \cdot \int_0^\infty M^2 g(M)dM \tag{2.17'}$$

The second part of Eq. (2.17′) is found by multiplying Eq. (2.14) with M and integrating over M. The *z-average* M_z is the ratio of the number average of M^3 to the number average of M^2:

$$M_z = \frac{\sum n_i M_i^3}{\sum n_i M_i^2} = \frac{\sum m_i M_i^2}{\sum m_i M_i} \tag{2.18}$$

and

$$M_z = \int_0^\infty M^3 \cdot g(M)dM \Big/ \int_0^\infty M^2 \cdot g(M)dM \tag{2.18′}$$

Finally, the ath power of the *viscosity average* M_v is defined as the mass average of M^a. a is a number between 0 and 1, whose value depends on the choice of the system (polymer—solvent) and on the temperature (Compare also Sect. 2.4).

$$M_v = \left[\frac{\sum m_i M_i^a}{\sum m_i}\right]^{1/a} = \left[\frac{\sum n_i M_i^{1+a}}{\sum n_i M_i}\right]^{1/a} \tag{2.19}$$

and

$$M_v = \left[\int_0^\infty M^a \cdot w(M)dM\right]^{1/a} \tag{2.19′}$$

If and only if, the polymer is *monodisperse*, ($n_1 = n_2 = n_{i-1} = n_{i+1} = \ldots = n_k = 0$; $n_i = n_0$), all averages of the molar mass will be equal and equal to M_i. For a *polydisperse* polymer the following equation always holds

$$M_n < M_v \leq M_w < M_z \tag{2.20}$$

M_v is found between M_n and M_w, generally closer to M_w. The equality $M_v = M_w$ only holds in the (rare) special case $a = 1$. Equation (2.20) is not only an experimental fact, but a mathematical consequence of the definitions of the various averages.

The ratios M_w/M_n and M_z/M_w increase with increasing width of the molar mass distribution. Hence,

$$U \equiv \frac{M_w}{M_n} - 1 \geq 0 \tag{2.21}$$

and

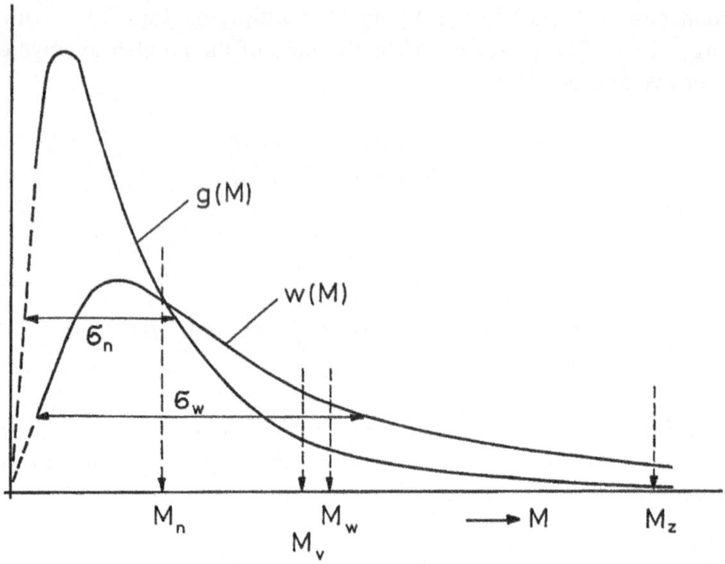

Fig. 2.5 Distribution densities of the molar mass of a technical polymer and the positions of its averages

$$\tilde{U} \equiv \frac{M_z}{M_w} - 1 \geq 0 \tag{2.22}$$

are introduced as measures for the width of the molar mass distribution. U is called the *nonuniformity* of *Schulz*, $U + 1 = M_w/M_n$ the *polydispersity factor*.

Figure 2.5 shows a schematic drawing of the *number distribution density* $g(M)$, the *mass distribution density* $w(M)$ and the positions of the molar mass averages for a technical polymer. Plotted over a linear scale of the molar mass as abscissa, those distribution densities are mostly strongly asymmetric curves with a long tail at the high molar masses. Their values at low molar masses are often difficult to determine, and are, therefore, indicated by dashed lines. According to Eq. (2.14) $w(M)$ is shifted to higher molar masses with respect to $g(M)$. Both curves intersect at the abscissa $M = M_n$. The line of gravity of the density function $g(M)$ passes through M_n, the line of gravity of the density function $w(M)$ passes through M_w. The standard deviations of the curves $g(M)$ and $w(M)$ are indicated by σ_n and σ_w. The maximum of the curve $g(M)$ is situated left to the maximum of the curve $w(M)$ and both are situated left to the value of M_n.

The averages may be interpreted as statistical parameters of the distribution densities $g(M)$ and $w(M)$. If the stochastic quantity M is distributed with the density $g(M)$, its mean value is equal to M_n and its variance is given by

Table 2.2 Parameters of the molar mass distribution of some technical polymers (M in kg/mol)

Polymer	M_n	M_w	M_z	M_w/M_n	M_z/M_w	Lit.
Poly(styrene) *PS N* 7000	182	385	771	2.12	2.00	[2]
Poly(styrene) Styron 666	120	250	–	2.08	–	[3]
Poly(styrene) *PS* 158 K	120	278	–	2.30	–	[4]
Poly(vinyl chloride) Solvic	38.6	84	154	2.18	1.83	[5]
Poly(methyl methacrylate) Plexiglas 7 N	60	95		1.6		[6]
Poly(ethylene) HDPE HE 6914	17.6	284	2430	16	8.6	[7]
Poly(ethylene) LDPE Lupolen 1840 H	15.5	258	2740	16.6	10.6	[7]
Poly(propylene) Moplen HP 501 N	42.3	254	–	6.0	–	[8]
Poly(carbonate) PC1	15	30	–	2.0	–	[4]

$$\sigma_n^2 = \int_0^\infty (M - M_n)^2 g(M) dM = M_w M_n - M_n^2 = U M_n^2 \tag{2.23}$$

If the stochastic quantity M is distributed with the density $w(M)$, its mean value is equal to M_w and its variance is given by the equation

$$\sigma_w^2 = \int_0^\infty (M - M_w)^2 w(M) dM = M_z M_w - M_w^2 = \tilde{U} M_w^2 \tag{2.24}$$

σ_n and σ_w are the *standard deviations* of the distribution densities $g(M)$ and $w(M)$, respectively. The quotient of standard deviation to mean value is known in statistics as the *variation coefficient* c and constitutes a measure for the relative width of the distribution density. From (2.23) and (2.24) we obtain the statistical meaning of the polydispersities as the corresponding squares of the variation coefficients

$$U = \frac{\sigma_n^2}{M_n^2} = c_n^2 \tag{2.25}$$

and

$$\tilde{U} = \frac{\sigma_w^2}{M_w^2} = c_w^2 \tag{2.26}$$

Equations (2.23) and (2.24) also prove the general validity of Eq. (2.20). The left-hand side of Eq. (2.23) is nonnegative, as has to be the right-hand side, proving that $M_w \geq M_n$.

The parameters of the molar mass distribution of some technical polymers are summarized in Table 2.2. All distributions are rather broad. Values for M_w/M_n between 1.5 and 3 are quite usual, while *PE* shows even values as high as 17. In cases, where U and $\tilde{U}$ could be measured, U was found to be larger than $\tilde{U}$.

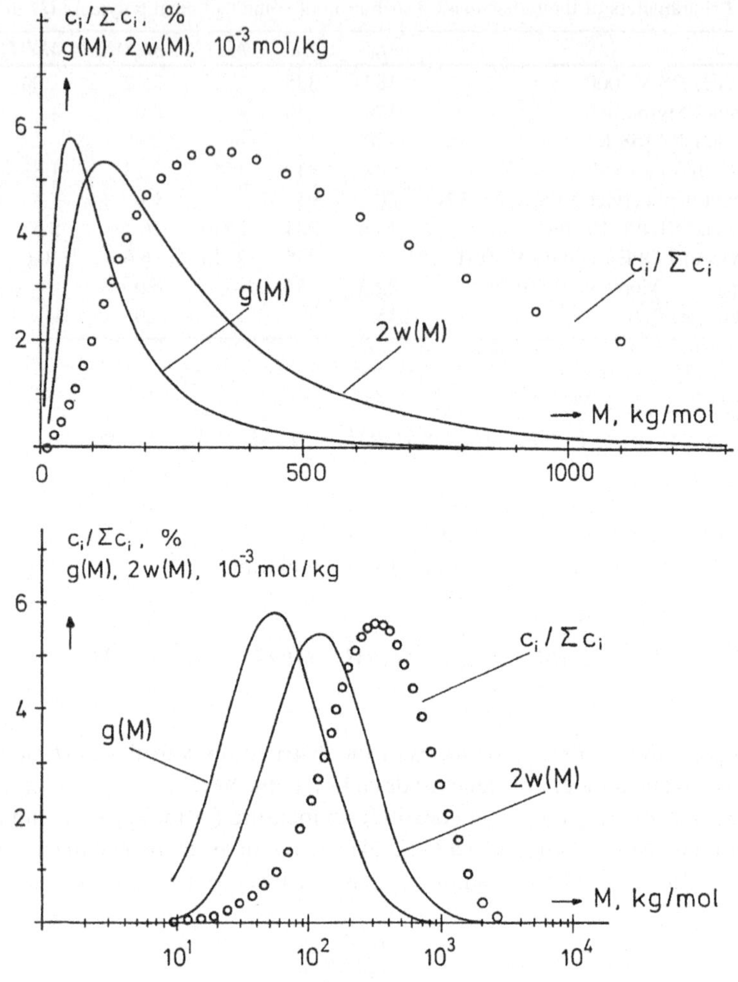

Fig. 2.6 Molar mass distributions of the technical poly(styrene) *PS N* 7000 after Wolf [2]

Figure 2.6 shows the distribution functions of the molar mass of poly(styrene) *PS N* 7000 as determined by Wolf [2] by means of *gel permeation chromatography*. The distribution functions are linearly plotted in the upper part of the figure versus a linear scale for the molar mass, and in the lower part of the figure versus a logarithmic scale for the molar mass. Note that the distribution functions are all very asymmetric when plotted versus a linear *M*-scale, but nearly symmetric, when plotted versus a logarithmic *M*-scale. The measuring points represent the concentrations c_i of the fractions, corresponding to the molar masses M_i. From these data it is possible to calculate the number and mass densities of the molar mass, as will be explained in Sect. 3.4.

Table 2.3 Parameters of the molar mass distribution of anionic poly(styrenes) (M in kg/mol)

Polymer	M_n osm. pressure	M_w light scattering	M_w/M_n	M_w/M_n from *GPC* [9]	$\sqrt{U}$, %
S1	1600	1800	1.12	–	–
S2	773	867	1.12	1.14	37
S3	392	411	1.05	1.13	36
S4	164	173	1.05	1.12	35
S5	96.2	98.2	1.02	1.10	32
S6	49	51	1.04	1.10	32
S7	19.7	19.8	1.01	1.09	30
S8	9.7	10.3	1.06	1.10	32
S9	4.6	5.0	1.09	1.10	32

For scientific investigations it is often desirable to investigate polymers with *approximately monodisperse molar mass distributions*. These may be prepared by either *fractionation* of technical polymers or by *special control* of the *polymerization process*. The company Pressure Chemicals Co, Pittsburgh, USA, has sold *anionic poly(styrenes)* with narrow molar mass distributions, which have been often used for scientific investigations. Unfortunately, those polymers are rather expensive and, therefore, only small portions of them could be investigated. Table 2.3 shows a summary of the molecular data of some of these materials. Parameter values in the 2nd and 3rd column are taken from the producer and were determined by means of osmotic pressure and light scattering. The ratio M_w/M_n calculated from these values is very sensitive to small experimental errors and, therefore, not very reliable. A better estimate of those ratios is obtained from *GPC*-measurements. In the last colomn, $\sqrt{U}$ is listed, which describes, according to Eq. (2.25), the variation coefficient of the number distribution density $g(M)$.

2.3 The Random Walk Problem in Three Dimensions

As to be seen from Fig. 2.2, even a short olefin chain with five carbon atoms (pentane) shows an enormous mobility, if the micro Brownian motion is fully developed. If the first carbon atom is fixed, and if full rotation around the carbon-carbon bonds can take place, the last carbon atom may occupy a very large number of spatial positions within a sphere with a radius equal to the end point distance of the completely stretched conformation. To answer the question on the shape of such a chain as a time and group average, the *random walk model* is investigated as a mathematical description of the real chain. This model is also called *random flight model* as the steps are assumed to occur in all three dimensions of the space (Fig. 2.7).

Starting from the origin of a Cartesian coordinate system, n steps of equal lengths b are performed in succession, the direction of each step being at random and independent of the directions of the previous steps. We ask for the probability

Fig. 2.7 The random walk (flight) problem

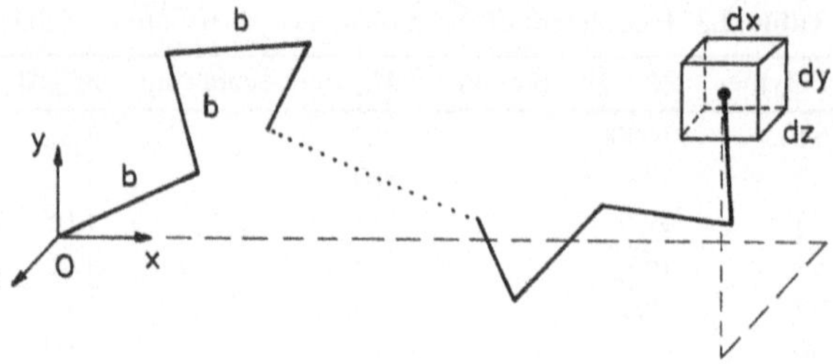

$w(x, y, z)\ dxdydz$ to arrive after the last step within a volume element with the edges dx, dy, dz around the point x, y, z. As the direction of the first step was assumed to be arbitrary, the problem has spherical symmetry and the probability function $w(x, y, z)$ can depend only on the distance r between the *starting* and *endpoint* of the random walk.

$$w(x, y, z) = w(r) \quad \text{with} \quad r^2 = x^2 + y^2 + z^2 \tag{2.27}$$

The *completely stretched conformation* results in the *stretched length*, r_{max}, which corresponds to the product of the number and the length of the steps of the random walk

$$r_{max} = n \cdot b \tag{2.28}$$

As the maximum length cannot be exceeded, we have

$$w(r) = 0 \quad \text{for} \quad r > r_{max} \tag{2.29}$$

The random walk problem can be solved exactly for all values of n and r. For the derivation of the somewhat complicated solution see the excellent comprehensive presentations of Flory [10] and Yamakawa [11]. For most applications, however, it is sufficient to use a simple approximation of the exact solution, which is valid under the following conditions (Flory [10]):

(1) $n \gg 1$ (P is large); $n > 10$ in practice
(2) We only consider conformations, which are not too close to the completely stretched one $r < r_{max}$ in practice $r < 0.8\ r_{max}$

Under these assumptions, the probability function of the random walk problem is given by the *Gaussian approximation*

$$w(r) \cong \left(\frac{3}{2\pi n b^2}\right)^{3/2} e^{-3r^2/2nb^2} \tag{2.30}$$

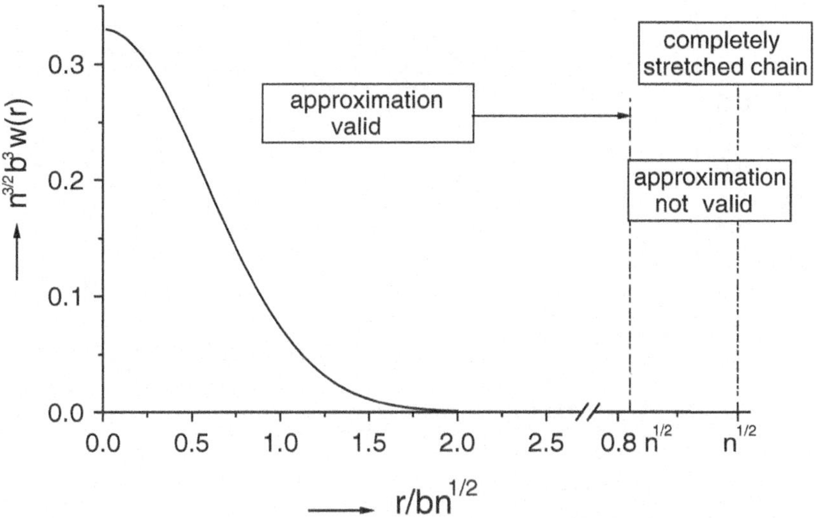

Fig. 2.8 The Gaussian approximation for the probability density of the end point position of the random walk chain

This function shows a maximum at the position $r = 0$ and decreases rapidly to zero with increasing r. It does not fulfill, however, the condition (2.29), showing that Eq. (2.30) cannot be valid for large values of r. The probability function, $w(r)$, multiplied with $b^3n^{3/2}$ is represented as function of $(r/bn^{1/2})$ in Fig. 2.8. In this representation the curve does not depend any longer on the values of n or b explicitly. The function $b^3n^{3/2}w(r)$ shows a maximum of 0.33 at the origin, a turning point at the abscissa $1/\sqrt{3} = 0.58$ and decreases to 22 % of its maximum value at the abscissa 1, to 0.2 % of its maximum value at the abscissa 2. In the figure, the *regions of validity* [from $r = 0$ to $r/(bn^{1/2}) = 0.8n^{1/2}$] and *invalidity* [from $r/(bn^{1/2}) = 0.8n^{1/2}$ to $r = \infty$] of the approximation, are also indicated.

For most questions in polymer physics the conditions for the validity of the Gaussian approximation will be fulfilled, as n will be of the order of magnitude of P, and as completely stretched conformations do not play a significant role in most problems. *Exceptions* to this rule are

The behavior of *very stiff polymers*
The behavior of polymers with a *very low degree of polymerization*
The behavior of rubbers *at large deformations* and near rupture
The behavior of materials with *frozen stresses* at high degrees of orientation

Equation (2.30) describes the probability density of the position of the endpoint of the random walk chain in space, but not the *probability* of the *occurrence of chains* with a certain *end to end distance*, independent of their orientation. The latter is called the *probability density of the end to end distance $W(r)$*. $W(r)dr$ is the probability to find a chain with an end to end distance between r and $r + dr$,

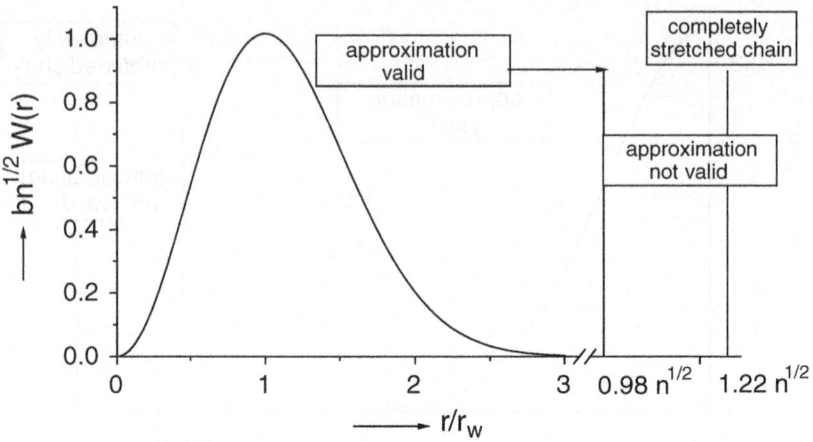

Fig. 2.9 Gaussian approximation for the end to end distance of the random walk chain

irrespectively of the orientation of the vector connecting its beginning and endpoint. To calculate *W*(*r*), we integrate the probability *w*(*r*) over the space bound by the spherical shells with radius *r* and $r + dr$ around the origin. This is achieved by introducing spherical coordinates and leads to the result

$$W(r) = 4\pi r^2 w(r) \tag{2.31}$$

Inserting the Gaussian approximation, we find for the probability

$$W(r) = \sqrt{\frac{24}{\pi}} \cdot \frac{1}{b\sqrt{n}} \cdot \left(\frac{r}{r_w}\right)^2 e^{-(r/r_w)^2} \tag{2.32}$$

with the abbreviation

$$r_w = \sqrt{2/3} \cdot b\sqrt{n} \tag{2.33}$$

The product $bn^{1/2} \cdot W(r)$ is shown as a function of r/r_w in Fig. 2.9.
The distribution of the end to end distance shows a maximum at the value $r = r_w$, meaning that r_w is the *most probable end to end distance* of the chain. The function *W*(*r*) is normalized, because

$$\int_0^\infty W(r)dr = 1 \tag{2.34}$$

as to be seen, by introducing $\xi = r/r_w$ as a new integration variable and solving the integral by partial integration.

The most important parameter of the chain statistics is not r_w but the average of the square of the end to end distance, called the *mean square distance*:

$$\boxed{\langle r^2 \rangle \equiv \int_0^\infty r^2 W(r) dr = \tfrac{3}{2} r_w^2 = nb^2} \tag{2.35}$$

Equation (2.35) is obtained by the substitution of $\xi = r/r_w$ and partial integration. The mean square distance is proportional to the number of steps in the random walk n. The quantity $\sqrt{\langle r^2 \rangle}$, which represents a *measure* for the *mean radius* of the polymer coil, increases proportional to the square root of the number of steps (square root of molar mass or degree of polymerization). Using (2.35), the two distribution densities of the random walk problem may be rewritten as

$$w(r) = \left(\frac{3}{2\pi \langle r^2 \rangle} \right)^{3/2} e^{-3r^2/2\langle r^2 \rangle} \tag{2.36}$$

and

$$W(r) = 3 \cdot \sqrt{\frac{6}{\pi \langle r^2 \rangle}} \cdot \frac{r^2}{\langle r^2 \rangle} \cdot e^{-3r^2/2\langle r^2 \rangle} \tag{2.37}$$

Similarly, the *averages* of all *positive integer powers* of the *end to end distance* may be calculated by repeated partial integration

$$\langle r^{2k} \rangle = \frac{1 \cdot 3 \cdot 5 \cdots (2k+1)}{3^k} \langle r^2 \rangle^k = \frac{1 \cdot 3 \cdot 5 \cdots (2k+1)}{2^k} r_w^{2k} \tag{2.38}$$

$$\langle r^{2k-1} \rangle = \sqrt{\frac{6}{\pi}} \frac{2 \cdot 4 \cdot 6 \cdots (2k)}{3^k} \langle r^2 \rangle^{(2k-1)/2} = \frac{2}{\sqrt{\pi}} \cdot \frac{2 \cdot 4 \cdot 6 \cdots (2k)}{2^k} r_w^{2k-1} \tag{2.39}$$

These formulae are valid for all positive integer values of k and for $k = 0$.

Equation (2.35) may be derived in another, more illustrative way, which has been used frequently by Flory [10]. Let $\underline{r}_1$, $\underline{r}_2, \ldots, \underline{r}_n$ be the connecting vectors representing the n steps of the random walk (cf. Fig. 2.10). Each of those vectors has the same length b, but an arbitrary direction in space.

The *end to end vector* of the random walk is the *sum* of all the *connecting vectors*

$$\underline{r} = \underline{r}_1 + \underline{r}_2 + \underline{r}_3 + \cdots + \underline{r}_n \tag{2.40}$$

Its absolute value, r, which equals the end to end distance, is found as the square root of the scalar product of the vector $\underline{r}$ with itself:

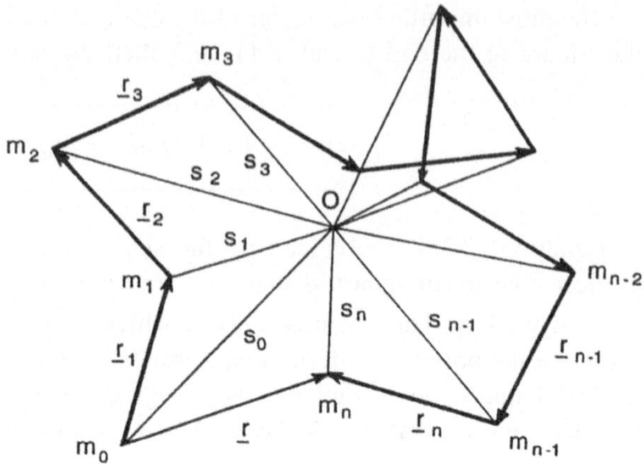

Fig. 2.10 Definition of the end to end distance as the sum of the connecting vectors and the definition of the radius of gyration of the random walk (flight) chain

$$\underline{r} \cdot \underline{r} = \sum_{i=1,k=1}^{n} (\underline{r}_i \cdot \underline{r}_k) \tag{2.41}$$

The double sum over i and k may be separated into a sum with equal indices and a sum with different indices:

$$r^2 = \sum_{i=1,k=1}^{n} (\underline{r}_i \cdot \underline{r}_k) = \sum_{i=1}^{n} (\underline{r}_i \cdot \underline{r}_i) + \sum_{\substack{i=1,k=1\\ i \neq k}}^{n} (\underline{r}_i \cdot \underline{r}_k) = nb^2 + \sum_{\substack{i=1,k=1\\ i \neq k}}^{n} (\underline{r}_i \cdot \underline{r}_k) \tag{2.42}$$

This equation is valid for each chain of n steps of equal length, independently of the law specifying the direction of the successive steps.

The average of the second sum will vanish for the random walk chain, as the angle between two different connecting vectors is arbitrary and the angle $\pi - \vartheta$ occurs with equal probability as the angle ϑ . In the first case, the scalar product of the two vectors will contribute with $-b^2 \cos \vartheta$ to the sum, in the second case with $b^2 \cos \vartheta$, and, on the average, both terms will cancel each other, i.e.,

$$\langle (\underline{r}_i \cdot \underline{r}_k) \rangle = 0 \quad \text{for} \quad i \neq k \tag{2.43}$$

Inserting this into Eq. (2.42), leads again to Eq. (2.35) for the random walk chain. Though both equations are identical, their requirements are different. We used the Gaussian approximation for deriving (2.35), but the weaker restriction of random walk only, now.

Besides the end to end distance, a further measure is often introduced to characterize the coil dimensions of a polymer, the so called *radius of gyration s*. This is explained referring to Fig. 2.10. The random walk chain of n steps contains $n + 1$ points (the starting and end points of the vectors), indicated by m_0, m_1, m_2

Fig. 2.11 End to end distance r and radius of gyration s of the random walk chain

... m_n. Assuming that the mass of the chain is uniformly distributed over these $(n + 1)$ segments, the center of gravity 0 of the segments may be calculated as well as the distances s_i (i = 0 to n) of the segments from it. The mean value of the squares of the distances yields the *square* of the *radius of gyration*

$$s^2 = \frac{1}{n+1}\sum_{i=0}^{n} s_i^2 \tag{2.44}$$

For the random walk chain, it is possible to calculate the expectation value of the square of the radius of gyration (for a derivation see e.g., [10, 11]):

$$\langle s^2 \rangle = \frac{1}{6}\langle r^2 \rangle \tag{2.45}$$

The radius of gyration may be considered as a *measure* for the *dimension* of the *coil*. From Eq. (2.45), it is seen, however, that the radius of gyration is smaller than half of the end to end distance. The reason for that is the fact that in the random walk chain the concentration of segments is much higher in the center of the coil, than in its outer regions. This is indicated in Fig. 2.11.

2.4 Macromolecules in Solution

Molar masses are determined in dilute solution. Therefore, solubility is an essential requirement for the determination of molar masses of polymers. The methods most frequently used are presented in Table 2.4:

In order to understand why the determination of molar masses is to be performed in dilute solution, we consider the shape of the polymer in solution. Micro-Brownian motion is excited in solution, therefore macromolecules which are not extremely stiff will assume the shape of coils with the radius h depending on

molar mass
interaction with the solvent
internal mobility of the macromolecule

Table 2.4 Methods of determination of molar masses and their range of applicability

Method	Average obtained	Range of applicability, kg/mol
Osmotic pressure	M_n	5–1000
Viscometry	M_v	0.1–1000
Light scattering	M_w	0.1–1000
Ultracentrifuge	M_z	10–2000
Gel permeation chromatography	$c(M)$	0.1–1000

A large number of solvent molecules, however, is included within one macromolecular coil. Therefore d_m, *the average density of the macromolecule in the coil*

$$d_m = \frac{\text{mass of the macromolecule}}{\text{volume of the coil}} \tag{2.46}$$

will be small. The properties of the isolated coils can be determined in dilute solution only, as the coils should not penetrate, not even touch, each other. Therefore, *the weight concentration of the polymer in the solution*

$$c = \frac{\text{mass of the polymer}}{\text{volume of the solution}} \tag{2.47}$$

should be chosen adequately small, viz., $c \approx 1$–50 g/l. On the other side, the dilution should not be too extreme, as the properties of the solution should significantly differ from those of the pure solvent to be able to measure their difference. This requirement often constitutes an experimental difficulty.

For a better understanding of the properties of a polymer solution, some basic concepts of its thermodynamic properties are to be discussed, especially the question of the characterization of the quality of the solvent. We consider solution or mixing at constant temperature T and constant pressure p, i.e., an isothermal-isobaric dissolving process. The thermodynamic potential under these circumstances is the *free enthalpy*

$$G = U + pV - TS \tag{2.48}$$

where U designates the internal energy, p the pressure, V the volume, T the absolute temperature and S the entropy. Under isothermal-isobaric conditions, a system tends to achieve a state with minimum free enthalpy.

We solve n_2 mol of the polymer into n_1 mol of the solvent and designate

$(G)_1$ the free enthalpy of the solvent before the process of solution
$(G)_2$ the free enthalpy of the polymer before the process of solution
$(G)_L$ the free enthalpy of the system after the process of solution

The *free enthalpy of mixing* G_m is the difference of the enthalpies after and before the dissolving process and may be decomposed into its energetic part U_m, its entropic part TS_m and its volumetric part pV_m

Table 2.5 Quality of the solvent and heat balance of the solution process

	Quality of the solvent	Heat balance of the dissolving process
$U_m < 0$	Very good	Exothermal
$U_m = 0$	Good	Athermal
$U_m > 0$	Moderate to poor	Endothermal

$$G_m \equiv (G)_L - (G)_1 - (G)_2 = U_m + pV_m - TS_m \tag{2.49}$$

S_m is the always positive *mixing entropy*, V_m the change in volume during the mixing process (which may be neglected for condensed phases) and U_m the *mixing energy*, which may be positive or negative. The dissolving process is classified according to the sign of U_m (Table 2.5).

For the characterization of the properties of the solution we need an *intensive* quantity (which depends on T, p, and the composition of the solution, but not on its mass). We describe the composition of the solution by the ratio of the *mole numbers*

$$z = n_2/n_1 \tag{2.50}$$

or by their *mole fractions*

$$x_1 = \frac{n_1}{n_1 + n_2} = \frac{1}{1+z} \tag{2.51}$$

and

$$x_2 = \frac{n_2}{n_1 + n_2} = \frac{z}{1+z} \tag{2.52}$$

It is shown in thermodynamics that every *extensive* quantity may be written as sum of the corresponding *intensive* quantities multiplied with their mole numbers

$$G_m(T,p,n_1,n_2) = n_1\mu_1(T,p,z) + n_2\mu_2(T,p,z) \tag{2.53}$$

μ_1 and μ_2 are *intensive quantities* and may be calculated by partial differentiation of the extensive quantity G_m with respect to their respective mole numbers:

$$\mu_1(T,p,z) = \left(\frac{\partial G_m}{\partial n_1}\right)_{n_2,T,p} \tag{2.54}$$

and

$$\mu_2(T,p,z) = \left(\frac{\partial G_m}{\partial n_2}\right)_{n_1,T,p} \tag{2.55}$$

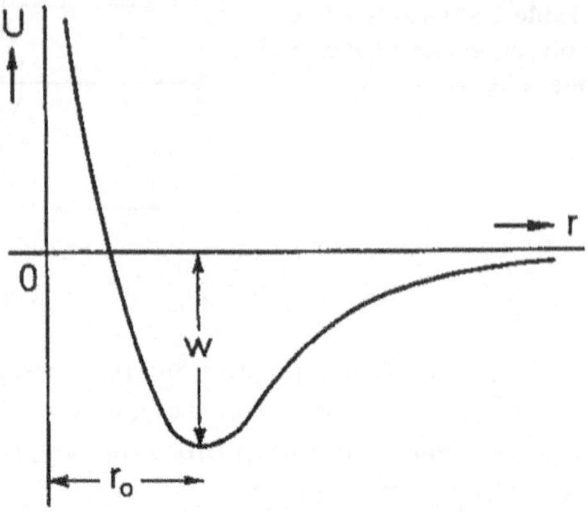

Fig. 2.12 Potential energy of two (parts) of molecules versus their distance

μ_1 is called the *chemical potential of the solvent* (or the *free enthalpy of dilution*), μ_2 the *chemical potential of the solute* (or the *free enthalpy of solution*). A further intensive quantity may be used for the characterization of the solution, called the *average molar free enthalpy of mixing* and is defined by the equation:

$$\overline{G}_{\mathrm{m}}(T,p,z) \equiv \frac{G_{\mathrm{m}}(T,p,n_1,n_2)}{n_1+n_2} = x_1\mu_1(T,p,z) + x_2\mu_2(T,p,z) \tag{2.56}$$

We turn to the molecular interpretation of the mixing energy. The energy of interaction of two solvent molecules or parts of two macromolecules as a function of their distance r is sketched in Fig. 2.12.

At the distance r_0 both molecular parts are in equilibrium, at smaller distance ($r < r_0$) *strong repelling forces* occur due to the interaction of the electron shells, at larger distance *weak attractive forces* occur. The depth of the energy minimum w at $r = r_0$ depends on the chemical composition of both partners involved (note that w is always negative!).

Solvent molecule—solvent molecule	w_{11}
Structural unit of the polymer—structural unit of the polymer	w_{22}
Structural unit of the polymer—solvent molecule	w_{12}

During the process of dissolving the contact of two solvent molecules is broken up, as well as the contact of two parts of macromolecules, and two new contacts of parts of the macromolecule and a solvent molecule are created. The change in energy due to this process will be

$$w = 2w_{12} - w_{11} - w_{22} \tag{2.57}$$

As the sign of U_{m} is the same as that of w, three cases can be regarded:

1. $w < 0$ and $U_{\mathrm{m}} < 0$: *very good solvent*; The process of dissolving is driven energetically, too, and is exothermal; the macromolecule is soluble at all temperatures, the solvent molecules are driven into the coil, which is inflated, i.e., its radius h becomes large.

2. $w = 0$ and $U_m = 0$: *athermal solvent*; the process of dissolving is driven by entropy changes, only; the macromolecule is soluble at all temperatures, the coil is inflated.
3. $w > 0$ and $U_m > 0$: *poor solvent*; the process of dissolving is a competition between the entropy effect, driving the molecule into solution, and the energy effect, driving the molecule to precipitate. The process of dissolving is endothermal, the solubility of the polymer will increase with increasing temperature. The solvent molecules are squeezed out of the coil, the radius of the coil h will become small.

Flory [12] and Huggins [13] have proposed the following simple expression as an approximation for the free enthalpy of mixing

$$G_m = RT(n_1 \ln v_1 + n_2 \ln v_2 + \chi n_1 v_2) \tag{2.58}$$

Here, v_1 und v_2 are the *volume concentrations* of *solvent* and *polymer*,

$$v_1 = \frac{V_1 n_1}{V_1 n_1 + V_2 n_2} \tag{2.59}$$

and

$$v_2 = \frac{V_2 n_2}{V_1 n_1 + V_2 n_2} \tag{2.60}$$

with V_1 and V_2 being their molar volumes. R is the *gas constant* and k the *Boltzmann* constant

$$R = k \cdot N_A = 8.3143 \text{ J/K mol} \tag{2.61}$$

$$k = 1.3805 \cdot 10^{-23} \text{ J/K} \tag{2.62}$$

χ is the "*Flory-Huggins parameter*", which characterizes the interaction between polymer and solvent. It consists of a contribution, χ_0, to the mixing entropy and a contribution, which is proportional to the mixing energy w

$$\chi = \chi_0 + \beta \frac{w}{kT} \tag{2.63}$$

χ_0 is positive and β is a coordination number of the order of unity. If Eq. (2.58) is applied to the description of the free enthalpy of polymer solutions, χ is experimentally found not to be constant but to be slightly depending on T, M, and the concentration v_2. Using (2.54), (2.58), (2.59), and (2.60), we find after some elementary calculations, the following expression for the free enthalpy of dilution

$$\mu_1 = RT\left(\ln v_1 + \left(1 - \frac{V_1}{V_2}\right) v_2 + \chi_1 v_2^2\right) \tag{2.64}$$

Table 2.6 Solvents, precipitants, and Theta-solutions of some polymers

Polymer	Solvent (T, °C)[a]	Theta-solution (T, °C)	Precipitants
PE-HD	Hydrocarbon (>80) Hydrogenated Hydrocarbon (>80)	n-hexane (133)	All organic solvents At room temperature
PE-LD	Hydrocarbon (>60) Hydrogenated Hydrocarbon (>60)	Same as for PE-HD	Same as for PE-HD
PP atactic	Hydrocarbon Hydrogenated Hydrocarbon	i-amylacetate (34) Cyclohexanone (92)	Polar organic solvents
PA 6.6	Phenols Formic acid	2.3 mol KCl in 90 % Formic acid (25)	Hydrocarbons Chloroform
PIB	Hydrocarbons Tetrahydrofuran (THF)	Benzene (22.8)	Butanol Methanol
PMMA	Benzene Xylene Chloroform	Butanon/isopropanol 50/50 (23) m-xylene (24)	Hexene Cyclohexane Methanol
PS	Chloroform Benzene	Cyclohexane (34)	Saturated hydrocarbons alcohols
PVC	THF Cyclohexanone	THF/H_2O 89/11 (30)	Hydrocarbons Alcohols
PC	Benzene Chloroform	Chloroform/ethanol 74.5/25.5 (18)	Aliphatic hydrocarbons Acetone

[a] No indication of a temperature: soluble at room temperature

χ_1 is a new *Flory-Huggins parameter*, related to χ by the equation

$$\chi_1 = \chi + \frac{n_1}{v_2}\left(\frac{\partial \chi}{\partial n_1}\right)_{n_2,T,p} = \chi - v_1\left(\frac{\partial \chi}{\partial v_2}\right)_{T,p} \tag{2.65}$$

If χ is independent of v_2, we have simply $\chi_1 = \chi$. As in poor solvents the solubility increases with temperature, a temperature will be found, at which the influence of the occupied volume of the polymer molecules is just compensated by the influence of the polymer-solvent interaction. This temperature is called the "*Theta-temperature*" and plays an important role in polymer physics. At the Θ-temperature the shape of polymer molecules may be calculated by statistics without taking into account the influence of the occupied volume or the polymer-solvent interaction. This solution is called a *Θ-solution.* The concept of the occupied volume and the Θ-temperature will be explained in more detail in the Sect. 3.1.

For a very large number of polymers, extensive tables have been given of solvents and precipitants by Bloch [14] and of Θ-solutions by Elias [15]. A few examples of solvents, precipitants and Θ-solutions for some polymers are given in Table 2.6.

Experimental methods to determine thermodynamic properties of polymer solutions and Θ-temperatures will be described in the next chapter. The reader can find extensive tables for the Flory-Huggins parameters for many polymers, solvents and temperatures in [16].

2.5 Statistical Shape of Linear Macromolecules in Θ-Solution

The linear macromolecule in theta-solution may not be directly identified with the random walk chain. Though the influence of the excluded volume and the interaction with the solvent molecules may be neglected in a theta-solution, the detailed molecular structure of the molecule in question determines the degrees of freedom for the internal motions and has to be considered in each case separately. The presence of fixed valence bonds and the partially hindered rotation around these bonds enlarge the coil dimensions. Nevertheless, it is possible in many cases to calculate the mean square end to end distance. Most of these calculations have been performed and presented by Flory [10].

The simplest case is the *valence bond chain with free rotation*. Consider a linear not branched chain of a vinyl polymer, for which all bonds of the main chain are single carbon-carbon bonds. The molecule consists of $n_0 + 1$ carbon atoms, connected by n_0 bond vectors $\underline{r}_1, \underline{r}_2, \ldots \underline{r}_{n_0}$. Let the length of each bond vector be b_0, its bond angle ϑ. Apart from movements of side groups, the shape of the main chain (its *conformation*) is fixed, if the $n_0 - 1$ rotational angles around the main bonds, $\varphi_1, \varphi_2, \ldots \varphi_{n_0-1}$ are fixed. The definition of such an angle is illustrated in Fig. 2.13 for the first two bond vectors and for the first rotational angle φ_1.

The zero points of the rotational angles are chosen in such a way, that the set $\varphi_1 = \varphi_2 = \ldots \varphi_{n_0-1} = 0$ corresponds to the completely stretched conformation. The fixed angle between successive valence bonds has an effect on the value of $\langle r^2 \rangle$. Equation (2.42) remains valid, but the average values of the second term on the right-hand side of this equation do not longer vanish. The averages of the scalar products of two different bond vectors may be evaluated as shown by Flory [10] in a very elegant manner. For the scalar product of two successive bond vectors one finds

$$(\underline{r}_1 \cdot \underline{r}_2) = b_0^2 \cos(\pi - \vartheta) = -b_0^2 \cos \vartheta \tag{2.66}$$

The scalar product $(\underline{r}_1 \cdot \underline{r}_3)$depends on the value of the second rotational angle φ_2

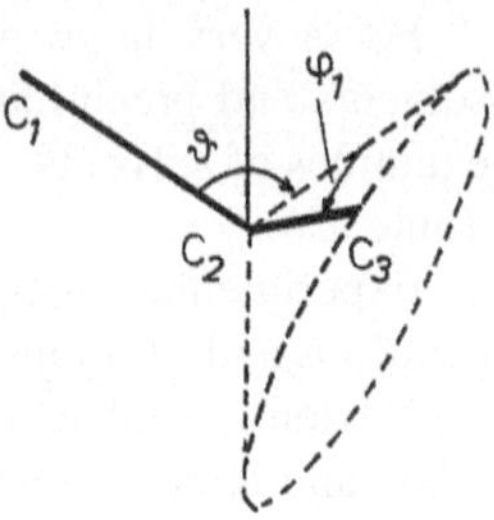

Fig. 2.13 The definition of the rotational angle φ_1

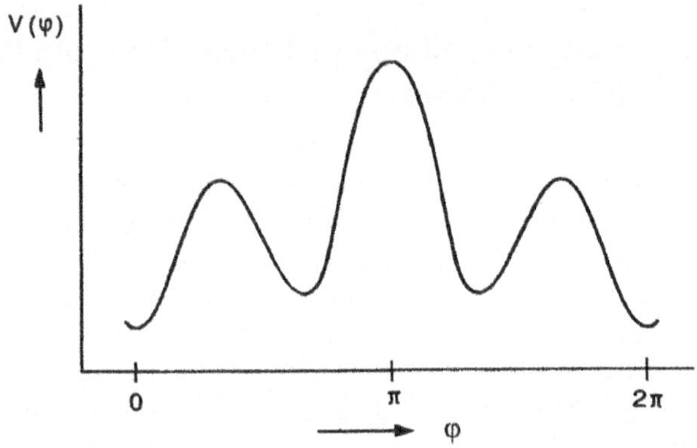

Fig. 2.14 Potential energy as a function of the rotational angle

$$(\underline{r}_1 \cdot \underline{r}_3) = b_0^2 \left(\cos^2 \vartheta + \sin^2 \vartheta \cos \varphi_2\right) \tag{2.67}$$

As all values of φ_2 are equally probable (free rotation), we have to integrate over all values of φ_2 and to divide by 2π in forming the average $\langle(\underline{r}_1 \cdot \underline{r}_3)\rangle$. In similar way all terms of the double sum of Eq. (2.42) may be calculated with the result

$$\left\langle r^2 \right\rangle_\Theta = n_0 b_0^2 \cdot \frac{1 - \cos \vartheta}{1 + \cos \vartheta} \tag{2.68}$$

For vinyl polymers $\cos \vartheta = -1/3$ and the quotient in Eq. (2.68) becomes 2.

The *valence bond chain with hindered rotation.*

If the rotation around valence bonds is hindered by the presence of side groups, the angles $\varphi_1, \varphi_2, \ldots \varphi_{n_0-1}$ are no longer equally probable.

If only side groups of neighbored C–atoms hinder each other, the hindering effect may be described by the potential energy of a part of the chain as a function of the rotational angle $V = V(\varphi)$ as shown in Fig. 2.14.

To calculate the expectation value of the scalar products at the right-hand side of Eq. (2.42), each term has to be weighed with the probability of occurrence of the corresponding rotational angle. Instead of (2.68), one obtains [10]

$$\langle r^2 \rangle_\Theta = n_0 b_0^2 \cdot \frac{1 - \cos\vartheta}{1 + \cos\vartheta} \cdot \frac{1 - \langle \cos\varphi \rangle}{1 + \langle \cos\varphi \rangle} \tag{2.69}$$

The expectation value $\langle \cos\varphi \rangle$ has to be calculated using the Boltzmann factor (notice the symmetry of V with respect to $\varphi = \pi$!):

$$\langle \cos\varphi \rangle = \int_0^\pi \cos\varphi \, e^{-V(\varphi)/kT} d\varphi \Bigg/ \int_0^\pi e^{-V(\varphi)/kT} d\varphi \tag{2.70}$$

More complicated macromolecules

If the macromolecule is more complicated, e.g., if short parts of the main chain are not movable, or if the interactions extend further than between neighbored side groups, a calculation of the end to end distance is still possible. As long as the interactions occur only between side groups of carbon atoms which are not too far apart from each other in the completely stretched conformation, a result of the following form is obtained [10]

$$\langle r^2 \rangle_\Theta = n_0 \cdot b_0^2 \cdot C \tag{2.71}$$

where b_0 is the distance of the singular C–C bond, n_0 the number of bonds in the principal chain and C is a constant depending on the detailed structure of the molecule. In many cases, it is possible to calculate C [16]; in very complicated cases, C may be determined experimentally by osmotic pressure or light scattering measurements (Compare Table 3.2 for examples of the values of C).

The *statistical equivalent random walk chain*

So far, we have considered the values of the average of the mean square distance $\langle r^2 \rangle_\Theta$, only. If this quantity may be described by Eq. (2.71), being proportional to the number of bonds, n_0, the probability of the end point position will be *Gaussian*. Inserting the values found in Eqs. (2.6), (2.68) or (2.69) into (2.36) and (2.37) the *probability densities* of the *end point position* and the *end to end distance* remains valid for linear macromolecules in Θ-solution.

To each linear macromolecule in theta-solution a *statistical equivalent random walk chain* with n steps of length b may be attributed, which behaves equal with regard to their distribution in space. Their completely stretched lengths and their mean square distances, as well correspond to each other.

The *length* and *number* of *steps* of the *statistical equivalent random walk chain* may be defined by the equations:

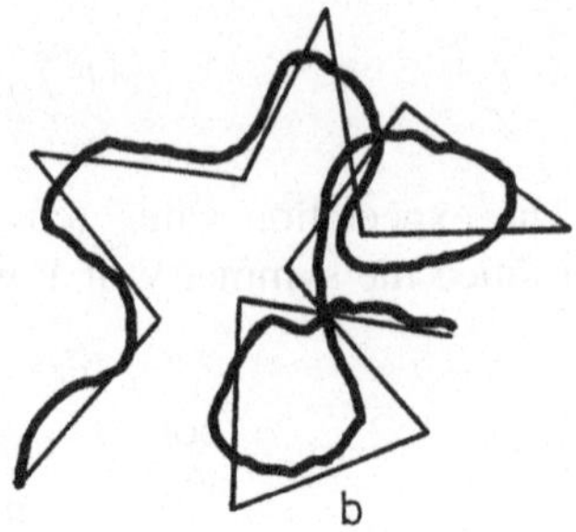

Fig. 2.15 The macromolecule and its statistical equivalent random walk chain

$$n \cdot b^2 = \langle r^2 \rangle_\Theta = n_0 \cdot b_0^2 \cdot C \tag{2.72}$$

and

$$n \cdot b = r_{\max} \tag{2.73}$$

Or vice versa:

$$b = \langle r^2 \rangle_\Theta \big/ r_{\max} = n_0 b_0^2 C \big/ r_{\max} \tag{2.74}$$

and

$$n = (r_{\max})^2 \big/ \langle r^2 \rangle_\Theta \tag{2.75}$$

b is called *Kuhn's statistical segment length*; $r_{\max}$ is the completely stretched length of the molecule, which may be calculated from its structural formula. The manner, in which the statistical equivalent random walk chain may follow the shape of the real chain, has been indicated in Fig. 2.15.

For the special case of a *vinyl polymer* we have two C–C bonds in one structural unit, the bond length $b_0 = 0.154$ nm, and the bond angle $\vartheta = 109.47°$ ($\cos\vartheta = -1/3$) and therefore

$$r_{\max} = b_0 n_0 \cos((\pi - \vartheta)/2) = \sqrt{\frac{2}{3}}\, b_0 n_0 = \sqrt{\frac{2}{3}}\, 2b_0 \cdot P = \sqrt{\frac{8}{3}}\, b_0 \cdot \frac{M}{M_g} \tag{2.76}$$

with $\sqrt{8/3}b_0 = 0.251$ nm. Inserting Eq. (2.76) into Eqs. (2.74), (2.72), and (2.45), yields the following expressions for *Kuhn's statistical segment*, the *mean square end to end distance* and the *mean square radius of gyration* in *theta solution*:

$$b = \sqrt{\frac{3}{2}C} \cdot b_0 \tag{2.77}$$

$$\boxed{\langle r^2 \rangle_\Theta = \alpha_\Theta^2 \cdot P} \tag{2.78}$$

$$\boxed{\langle s^2 \rangle_\Theta = \beta_\Theta^2 \cdot P} \tag{2.79}$$

with

$$\alpha_\Theta = b_0 \cdot \sqrt{2 \cdot C} \tag{2.80}$$

and

$$\beta_\Theta = \alpha_\Theta / \sqrt{6} = b_0 \cdot \sqrt{C/3} \tag{2.81}$$

Some values for C, b/b_0, and β_Θ / b_0 will be given in Table 3.3.

2.6 Statistical Shape of Macromolecules in Good Solvents

In good solvents the interaction between macromolecules and solvent molecules has to be considered. This interaction influences the shape of the macromolecule and the statistical distribution of its segments. The mathematical problem becomes much more complicated and is still unsolved (cf. [10]). However, it should be mentioned, that the coils will be widened in good solvents, i.e.,

$$\langle r^2 \rangle > \langle r^2 \rangle_\Theta \quad \text{and} \quad \langle s^2 \rangle > \langle s^2 \rangle_\Theta$$

Neither Eq. (2.45) nor Eq. (2.71) remain valid in this case. Instead, the *mean square values* of the *end to end distance* and of the *radius of gyration* increase more than proportional to the degree of polymerization P:

$$\langle r^2 \rangle = \alpha^2 \cdot P^{1+\varepsilon} \tag{2.82}$$

and

$$\langle s^2 \rangle = \beta^2 \cdot P^{1+\rho} \tag{2.83}$$

with ε and ρ being positive numbers. Though power laws of the form of Eqs. (2.82) and (2.83) have not been derived theoretically up to now, their validity can be checked experimentally. Using gel permeation chromatography (*GPC*) coupled with multi-angle laser light scattering (*MALLS*) it is possible to investigate

Table 2.7 Parameter values of the relation (2.83) between the mean square radius of gyration and the degree of polymerization for the linear vinyl polymers poly(propylene) and poly(ethylene) in the good solvent trichlorbenzene (TCB)

Polymer, solvent, T °C			β, 10^{-10} m	ρ	Range for P, kg/mol	Lit.
PP	TCB	140	1.52	0.17	$2.5\ 10^3 < P < 10^5$	[8, 17]
PE	TCB	140	1.66	0.16	$10^3 < P < 2\ 10^4$	[8, 17]

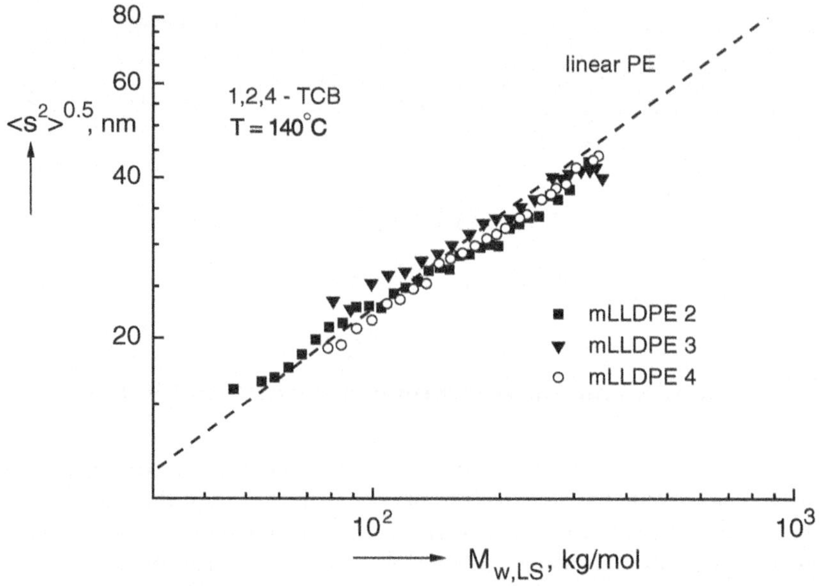

Fig. 2.16 Square root of the mean square radius of gyration versus the absolute weight average molar mass $M_{w,LS}$ determined by GPC-MALLS for three linear *PE*. The *dashed line* represents the square root of Eq. (2.83) with the parameter values given in Table 2.7. Reproduced from [7]

the relation between $\langle s^2 \rangle$ and M_w for polymer solutions in good solvents. For some linear vinyl polymers, power laws of the form of Eq. (2.83) have been found to be valid, as summarized in Table 2.7.

Two examples of measurements of this kind, taken from [8] are given in Figs. 2.16 and 2.17.

Figure 2.16 shows the square root of the mean square radius of gyration versus the weight average molar mass measured by light scattering for *GPC*-fractions of three linear low density *PE* in trichlorobenzene (TCB) at 140 °C, [8]. The broken line represents the square root of Eq. (2.83) with the parameter values given in Table 2.7. The three polymers, designated as mLLDPE 2–4 are metallocene linear low density poly(ethylenes), which are assumed to contain no long chain branches. They have similar molar mass distributions and similar values of M_w, but slightly different comonomer composition. Their values of $\langle s^2 \rangle^{1/2}$ lie on or closely to the

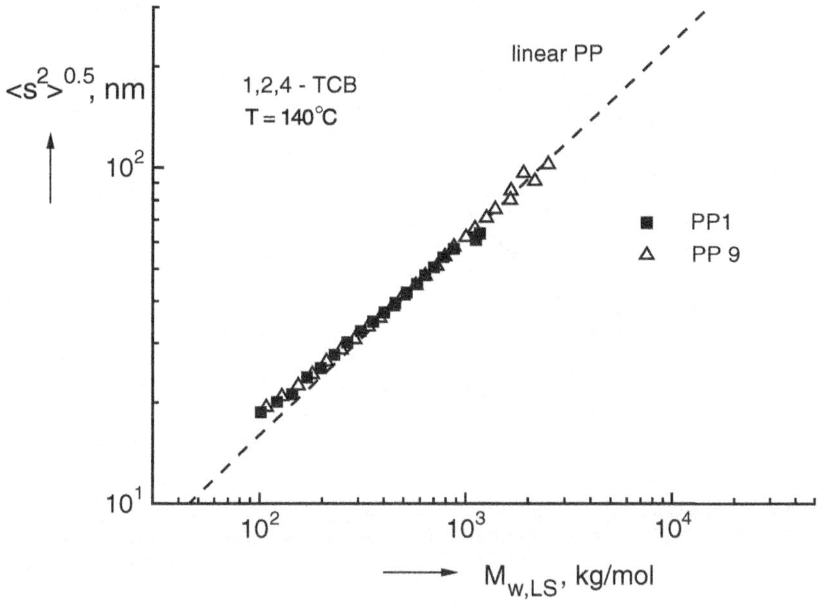

Fig. 2.17 Square root of the mean square radius of gyration versus the absolute weight average molar mass $M_{w,LS}$ determined by GPC-MALLS for two linear *PP*. The *dashed line* represents the square root of Eq. (2.83) with the parameter values given in Table 2.7. Reproduced from [7]

(dashed) reference line for *linear PE*. This result confirms, that not branched polymers show a power law relation between $\langle s^2 \rangle^{1/2}$ and M_w. It is well known that this is no longer true, if the polymer contains long chain branches. Moreover, $\langle s^2 \rangle^{1/2}$ is then found to be smaller than the radius of gyration of a linear molecule with the same M_w (c.f. Sect. 2.7).

Figure 2.17 shows the square root of the mean radius of gyration versus the weight average of the molar mass M_w for two linear *PP* in TBC at 140 °C, designated as *PP* 1 and *PP* 9 [8]. The dashed line represents the square root of Eq. (2.83) with the parameter values given in Table 2.7. The polymers are assumed to be linear with similar molar mass distributions but different weight average of the molar mass measured by light scattering. Again, the values of $\langle s^2 \rangle^{1/2}$ lie closely to the (dashed) reference line for *linear PP*, which differs slightly from the reference line for *linear PE.*

2.7 Analysis of Branched Macromolecules

As any branching within a polyolefin chain means a change of the mobility of the carbon atom it is attached to, the melt-state C^{13} *nuclear magnetic resonance*, in principle, is a suitable method to detect branches. Its elegance lies in the fact that a

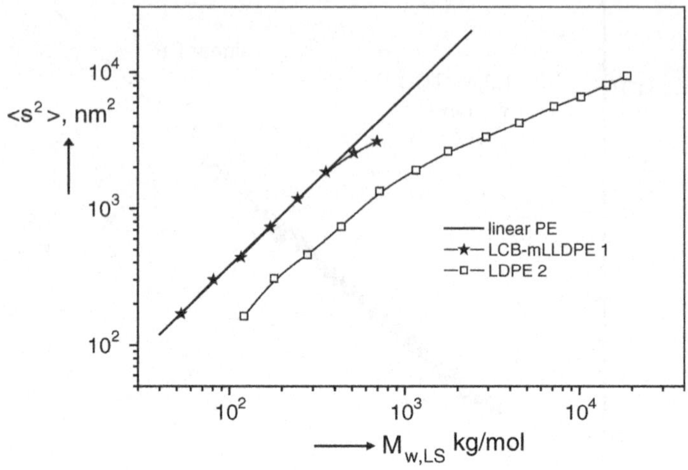

Fig. 2.18 Mean square radius of gyration $\langle s^2 \rangle$ as a function of the absolute weight average molar mass $M_{w,LS}$ measured by light scattering for polyethylenes of various branching structures [7]

quantitative picture of the branching sites can be obtained without any model assumptions. But according to [18] the analytical power of this technique is restricted insofar as it is able to merely discriminate between attached molecules up to 6 carbon atoms. Longer molecules which are typical of long-chain branches do not give particular signals. This physical fact has two consequences. A reliable information on short-chain branching can be obtained, only, if any long-chain branches are absent and, the other way round, conclusions with respect to long-chain branching can be drawn from NMR-measurements under the precondition that no short-chain branches longer than 6 C–atoms are present. Particularly the last requirement is very difficult to fulfill in the case of technically produced materials. It is well established, for example, that low density polyethylenes may contain a respectable number of short-chain besides long-chain branches.

Similar arguments hold for results from infrared-spectroscopy which is easy to perform. In the case of polyethylenes the number of CH_3-groups obtained by this method can belong to short-chain or long-chain branches as well.

Therefore, the gel-permeation chromatography coupled with multi-angle laser-light scattering (*GPC-MALLS*) as sketched in Sect. 3.4 is a very valuable method to get an insight into the long-chain branching structure of a polymer. This potential is demonstrated by Fig. 2.18 which displays in a double-logarithmic plot the mean squared radius of gyration of a low density polyethylene (LDPE) which is well known for its long-chain branching as a function of the absolute value of the weight average molar mass determined by light scattering in comparison to the corresponding curve for a linear polyethylene. It is obvious that the coil dimensions of the LDPE are significantly smaller than those of the linear counterpart. This finding can be explained by the assumption that the branches are able to at least partly fill the free space between the segments of the backbone of the molecules. According to Fig. 2.18 the differences of the coil dimensions between linear and long-chain branched molecules become the more pronounced the higher the molar mass indicating an increasing effect of branching.

A distinctly different result is found from Fig. 2.18 for the long-chain branched metallocene-catalyzed linear low density polyethylene (LCB-mLLDPE 1). Only at higher molar masses a deviation from the straight line of the linear material occurs. This finding can be interpreted in a way that the existence of long-chain branches is restricted to the longer molecules and the shorter ones behave linear although they contain short-chain branches according to their chemical structure. As a side effect this result demonstrates that the influence of comonomers on the coil dimension cannot be detected by *GPC-MALLS* at least for the compositions commonly found with LLDPE.

For a more quantitative discussion of the differences in coil size between linear and branched molecules the so-called *branching index*

$$g = \frac{\langle s^2 \rangle_{\text{br}}}{\langle s^2 \rangle_{\text{lin}}} \tag{2.84}$$

is defined. The index "br" denotes the mean square radius of gyration of the branched, the index "lin" that of the linear species taken at the same molar mass.

Although clearly defined and very descriptive there have been only very few theories up to now which try to correlate the branching index with the branching structure [19, 20]. For example, for monodisperse star polymers the conformation of which can be described by the theory of undisturbed chain statistics the relationship

$$g = (3f - 2)/f^2 \tag{2.85}$$

between g and the functionality f of a branching point is given by Zimm and Stockmayer in [19].

Furthermore, in [19] the following relationship between g and the number of statistically distributed trifunctional branching centers m per molecule is derived for monodisperse polymers:

$$g \approx \left[(1 + m/7)^{0.5} + 4m/9\pi \right]^{-0.5} \tag{2.86}$$

From m, the number of branching points per monomer unit M_g can be assessed as

$$\lambda = mM_g/M \tag{2.87}$$

with M being the molar mass. Under the assumption of equal lengths of the segments between the branching points their molar mass M_s follows as

$$M_s = M/(2m+1) \tag{2.88}$$

The prerequisite of monodispersity for the application of the Zimm-Stockmayer theory is fulfilled for an evaluation of the *GPC-MALLS* measurements as the fractions obtained are very narrowly distributed. The requirement of trifunctionality is very difficult to assess, however, as the branching structure depends on the polymerization conditions. Therefore, Eqs. (2.85) to (2.88) are of a very limited value for the description of real long-chain branched polymers, but in some cases they may be helpful for a qualitative comparison of various products.

2.8 Size of Macromolecules in the Glassy and Molten State

The large importance of the Θ-solution is due to the fact that shape and extension of linear macromolecules may be experimentally determined and calculated by means of Gaussian statistics. Moreover, the coil dimensions of linear macromolecules in the glassy state and in the melt are similar to those in Θ-solution. This has been assumed since long [10], but the experimental proof was conducted by Fischer and Dettenmaier in 1978 [21]. Their experiment is based on the strong difference in neutron-scattering between molecules containing hydrogen or deuterium atoms. If for a small part of the molecules present in the solid polymer hydrogen atoms are replaced by deuterium, it is possible to detect these molecules by small-angle neutron scattering similar to the characterization of ordinary macromolecules in dilute solution by means of light scattering. In both cases the radius of gyration of the coil is obtained.

For linear vinyl polymers in Θ-solution the relation (2.89) between the radius of gyration and the degree of polymerization

$$\langle s^2 \rangle = \beta_\Theta^2 \cdot P \tag{2.89}$$

holds with β_Θ being a constant only weakly depending on temperature. This relation remains valid with the same value of the constant β for macromolecular coils in the solid and molten state as proven by Fischer and Dettenmaier [21]. The values from their publication are given in Table 2.8. The coefficients β as determined by light scattering or viscometry in Θ-solution are compared with results of

Table 2.8 $\beta = \sqrt{\langle s^2 \rangle / P}$ for polymers in the solid and molten state and in Θ-solution

Polymer	Range of molar mass in kg/mol	β in 10^{-10} m solid or molten	β in 10^{-10} m in Θ-solution
PMMA	6–1000	2.56	2.20
PS	20–1000	2.81	2.76
SI	200	2.28	2.24
PVC	70–200	3.24	2.93
PE	10–80	2.33	2.28
	60–400	2.44	2.28

neutron scattering in the solid and molten state of various polymers in a wide range of molar masses. The agreement of the results in Θ-solution and in the bulk is very good.

References

1. Schwarzl FR (1990) Polymer-Mechanik, Springer Berlin Heidelberg
2. Wolf M (1986) Diploma thesis. University Erlangen-Nürnberg, Erlangen
3. Jungnickel JL, Weiss FT (1961) J Polymer Sci 49:437
4. Hepperle J (2003) Doctoral thesis, University Erlangen-Nürnberg. Shaker Verlag, Aachen. ISBN 3-8322-1228-0
5. Strazielle C, Benoit H (1971) Pure Appl Chem. 26:451
6. Triebel C (2012) Doctoral thesis. University Erlangen-Nürnberg, Erlangen. Verlag Dr. Huf, München, ISBN 978-3-8439-0293-9
7. Gabriel C (2001) Doctoral thesis, University Erlangen-Nürnberg. Shaker Verlag, Aachen. ISBN 3-8265-8909-2
8. Resch JA (2010) Doctoral thesis. University Erlangen-Nürnberg, Erlangen
9. Onogi S, Masuda T, Kitagawa K (1972) Makromol 3:109
10. Flory PJ (1969) Statistical mechanics of chain molecules. Wiley, New York
11. Yamakawa H (1971) Modern theory of polymer solutions. Harper and Row, New York
12. Flory PJ (1941) J Chem Phys 9:660, (1942). J Chem Phys 10:51
13. Huggins ML (1941) J Chem Phys 9:440, (1942). J Phys Chem 46:151
14. Bloch DR (1999) Solvents and non-solvents for polymers. In: Brandrup J, Immergut EM, Grulke EA (eds) Polymer Handbook, 4th edn. Wiley, New York
15. Elias HG (1999) Theta—solvents. In: Brandrup J, Immergut EM, Grulke EA (eds) Polymer Handbook, 4th edn. Wiley, New York
16. Schuld N, Wolf BA (1999) Polymer–solvent interaction parameters. In: Brandrup J, Immergut EM, Grulke EA (eds) Polymer handbook, 4th edn. Wiley, New York
17. Sun T, Brant P, Chance RR, Graessly WW (2001) Macromol 34(19):6812
18. Klimke K, Parkinson M, Piel C, Kaminsky W, Spieß HW, Wilhelm M (2006) Macromol Chem Phys 207:382
19. Zimm BH, Stockmayer WH (1949) J Chem Phys 17:1301
20. Scholte TG (1986) In: Dawkinss JV (ed) Developments in Polymer Characterisation—4, Elsevier London, New York
21. Fischer EW, Dettenmaier M (1978) J Non Cryst Solids 31:181

Chapter 3
Experimental Methods to Determine Molecular Quantities

In this chapter, experimental methods are described which are used to determine the molar mass and the interaction between macromolecules and solvent molecules.

3.1 Osmometry

Measurements of osmotic pressures are performed with an apparatus, whose principle is sketched in Fig. 3.1. The cell with the solution is separated by a semi-permeable diaphragm from the vessel, which contains the pure solvent. The diaphragm is permeable for the solvent molecules, but not permeable for the much larger polymer molecules. After some time, a constant pressure difference is reached between both reservoirs which is called the *osmotic pressure* π.

The occurrence of the osmotic pressure may be explained either by molecular or thermodynamic reasoning.

(I) The osmotic pressure results from the impact of the polymer molecules bouncing against the walls of the inner cell.
(II) The solution tries to get diluted until the decrease in free enthalpy is balanced by the repulsing pressure difference.

Ad I. If we consider the osmotic pressure as a result of the impact of the polymer molecules on the walls of the inner cell, we may expect in analogy to the law for ideal gasses

$$pV = nRT \tag{3.1}$$

the equation

$$\pi V = n_2 RT \tag{3.2}$$

with V being the volume of the inner cell and n_2 the *mole number of the solute*. Introducing the *mass concentration* c of the *polymer in the solution*

H. Münstedt and F. R. Schwarzl, *Deformation and Flow of Polymeric Materials*,
DOI: 10.1007/978-3-642-55409-4_3,

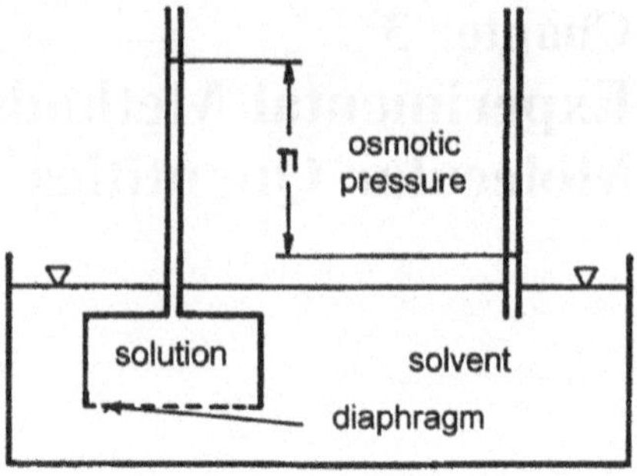

Fig. 3.1 Determination of the osmotic pressure

$$c = n_2 M/V \tag{3.3}$$

into Eq. (3.2), we find *Van't Hoff's law* for the osmotic pressure

$$\pi = RTc/M \tag{3.4}$$

For polydisperse polymers, M in Eq. (3.4) has to be replaced by the number average molar mass M_n:

$$\pi = RTc/M_n \tag{3.5}$$

The proof rests upon the following arguments: The partial pressure due to the species with the molar mass M_i is given by $\pi_i = n_i R\, T/V$ and the sum of all partial pressures forms the osmotic pressure

$$\pi = (RT/V) \cdot \sum_i n_i \tag{3.6}$$

For the species i Eq. (3.3) reads $c_i = n_i M_i/V$ and for the mass concentration of all polymer molecules

$$c = \sum_i c_i = \frac{1}{V} \sum_i n_i M_i \tag{3.7}$$

Inserting 1/V from Eq. (3.7) into Eq. (3.6) and using Eq. (2.15), yields Eq. (3.5).

Equations (3.4) or (3.5) may be applied to solutions of low-molar mass materials, but fail for polymer solutions. This is shown with reference to Fig. 3.2, in which the ratio between osmotic pressure and mass concentration is plotted versus the concentration c for solutions of fractions of poly(isobutylene) in two solvents after Flory [1].

Instead of being constant, the ratio π/c depends significantly on the polymer concentration c. Therefore, an expansion after powers of the concentration c should be used

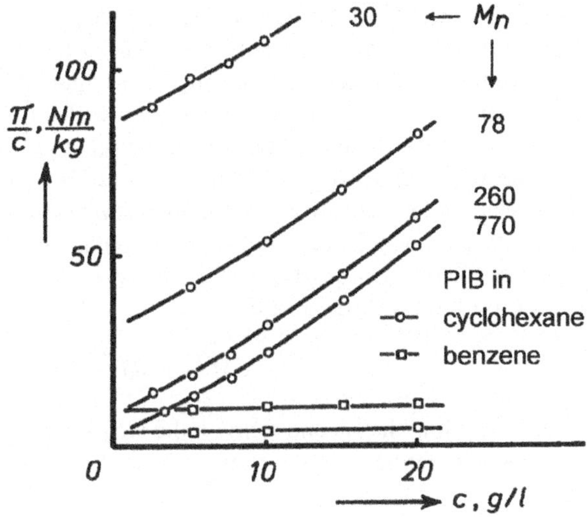

Fig. 3.2 Values of π/c versus c for solutions of poly(isobutylene) in two solvents at the temperature of $T = 24$ °C after Flory. Reproduced from [1] by permission from American Chemical Society

$$\pi = RT\left[c/M_n + Bc^2 + Cc^3 + \cdots\right] \tag{3.8}$$

to describe the course of the osmotic pressure with polymer concentration. B is called the *second virial coefficient*, C the *third virial coefficient* etc. To determine the molar mass, the ratio π/c is plotted versus c and, from the intersection with the abscissa at $c = 0$, a measure of the reciprocal value of the molar mass M for monodisperse polymers or for the reciprocal value of the number average molar mass M_n for polydisperse polymers is obtained.

The influence of the higher powers of c in Eq. (3.8) manifests itself in an increase of the ratio π/c with increasing c. In the good solvent (cyclohexane) the ratio π/c strongly increases with increasing c, showing that the second virial coefficient B is large and positive in this case. In the Θ-solvent (benzene) the ratio π/c remains independent of c in the region of smaller concentrations, showing that in this case $B = 0$. Hence, the value of the second virial coefficient constitutes a measure for the quality of the solution.

The second virial coefficient can be interpreted by the *excluded volume concept*, applied to a real gas. While for an ideal gas Eq. (3.1) is valid, for a real gas the influence of the *excluded volume b* is taken into account leading to the equation

$$p(V - b) = nRT \tag{3.9}$$

The excluded volume is the volume, which is not for disposal for the molecule under consideration. The excluded volume, however, will consist of two terms

$$b = b_1 + b_2 \text{ with } b_1 > 0 \text{ and } b_2 > 0 \text{ or } b_2 < 0 \tag{3.10}$$

b_1 is the volume occupied by the other molecules of the real gas and is therefore always positive. The sign of b_2 will depend on the interaction of the gas molecules with each other. If the gas molecules attract each other, they will stay longer in the neighborhood of other gas molecules than in the neighborhood of the walls, and the volume for disposal will appear to be larger, therefore b_2 will be negative. If the gas molecules repulse each other, they will stay longer in the neighborhood of the walls, than in the neighborhood of the other gas molecules, and the volume for disposal will appear to be smaller, therefore, b_2 will be positive.

A relation between the *excluded volume* b and the *second virial coefficient* B for solutions of monodisperse polymers is obtained as follows: We replace in the equation for the real gas Eq. (3.9) p by π, n by n_2 and V by the value following from Eq. (3.4), with the result:

$$\pi M/c = RT + \pi b/n_2$$

In this equation, the second term on the right-hand side is small compared with the first one. Therefore, we may replace π in the second term by the approximation

$$\pi \cong RTc/M + 0(c^2)$$

and obtain

$$\pi = RT \cdot \left[\frac{c}{M} + \frac{c^2}{M^2} \cdot \frac{b}{n_2} + \cdots\right] \tag{3.11}$$

From a comparison of Eq. (3.11) with Eq. (3.8), in which M_n is substituted by M one gets:

$$B = \frac{b}{n_2 M^2} \tag{3.12}$$

or with

$$n_2 = cV/M \qquad \frac{b}{V} = c \cdot M \cdot B \tag{3.13}$$

A very important special case is the *Θ-solution*, for which the effects of the attractive forces between the coils and those of their occupied volume just compensate each other, i.e.,

$$b_1 = -b_2 \quad b = 0 \quad B = 0 \tag{3.14}$$

is valid for the Θ-solution.

Similarly to the decomposition of the excluded volume into a contribution from the occupied volume b_1, and a contribution from the energetic interaction b_2, it is possible to decompose the *second virial coefficient* into an *enthalpy term* B_H and an *entropy term* B_S. As shown in thermodynamics (see, e.g., [2]), we have

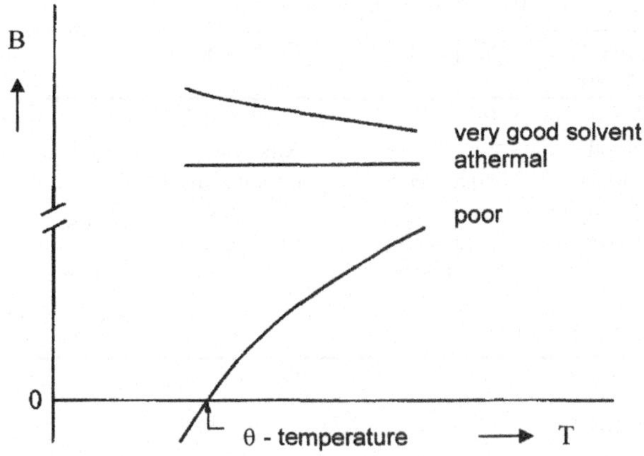

Fig. 3.3 Temperature dependence of the second virial coefficient for solvents of different quality

$$B = B_H + B_S \tag{3.15}$$

with

$$B_H = -T\left(\frac{\partial B}{\partial T}\right) \tag{3.16}$$

and

$$B_S = B + T\left(\frac{\partial B}{\partial T}\right) \tag{3.17}$$

B_H describes the energetic interaction between the coils and B_S the influence of their occupied volume. As a consequence, in a *very good solution* B_H will be positive, and therefore *($\partial B/\partial T$)* negative; in an *athermal solution B* will be independent of the temperature and therefore $B = B_s$; in a *moderate or poor solution* B_H will be negative, and therefore ($\partial B/\partial T$) positive; at the *Θ-temperature* we have $B_H = -B_S$. The temperature dependence for these three cases is presented schematically in Fig. 3.3.

Ad II: In thermodynamics [2] it is shown that the *osmotic pressure* is related directly to the *free enthalpy of dilution* by means of the equation

$$\pi = -\mu_1/V_1 \tag{3.18}$$

If the osmotic pressure has been measured as a function of the polymer concentration c, the dependence of the free enthalpy of dilution may be calculated. From $\pi = \pi(c)$ it follows $\mu_1 = \mu_1(v_2)$, when the mass concentration c is replaced by the volume concentration d_2 by means of

$$c = d_2 \cdot v_2 \tag{3.19}$$

Table 3.1 Properties of dilute solutions

Quality of the solution	Shape of the coil	Heat exchange	B	$\partial B/\partial T$	B_H	U_m, w	χ_1
Very good	Inflated	Exothermal	Large > 0	< 0	> 0	< 0	0.0–0.4
Good		Athermal	Large > 0	$= 0$	$= 0$	$= 0$	0.40–0.45
Moderate to poor	Squeezed	Endothermal	Small > 0	> 0	< 0	> 0	0.45–0 0.5
Θ-solution	Undisturbed statistical shape	Endothermal	0		> 0	$= -B_S$	> 0 0.50
Precipitation			<0				$\chi_{1k} > 0.5$

with

$$d_2 = M/V_2 \tag{3.20}$$

d_2 is the *density of the polymer in the solution* ($\cong$ density of the polymer in the unsolved state). The following relation between the *Flory–Huggins parameter* χ_1 and the *second virial coefficient* may be derived by inserting Eq. (2.61) into Eq. (3.18), developing the result into powers of c and v_2, respectively, and comparing the coefficients of c^2:

$$B = \frac{1}{V_1 d_2^2} \cdot \left[\frac{1}{2} - \chi_1\right] \tag{3.21}$$

and

$$B_H = \frac{1}{V_1 d_2^2} \cdot T\left(\frac{\partial \chi_1}{\partial T}\right) \tag{3.22}$$

Large positive values of B (good solutions) correspond to small positive values of χ_1; for the Θ-solution we have $B = 0$ and $\chi_1 = 1/2$. An overview of the properties of dilute solutions is given in Table 3.1. χ_{1k} is the value of the Flory–Huggins parameter, at which precipitation occurs.

Some experimental results for B are given in the Fig. 3.4. Figure 3.4 shows the temperature dependence of the second virial coefficient of solutions of poly(vinyl chloride) of different molar masses in toluene after Schulz et al. [3]. In this very good solution, the second virial coefficient shows large positive values, which weakly increase with temperature.

A completely different behavior was reported for solutions in bad or moderate solvents by Schulz et al. [4]. They investigated the temperature dependence of the second virial coefficient of solutions of PMMA in six bad solvents. Values for B were found which were ten times smaller than for good solvents and partially negative. For low temperatures B was negative and increased strongly with increasing temperature. For each system, a temperature was observed, at which the second virial coefficient vanished, the *Θ-temperature*.

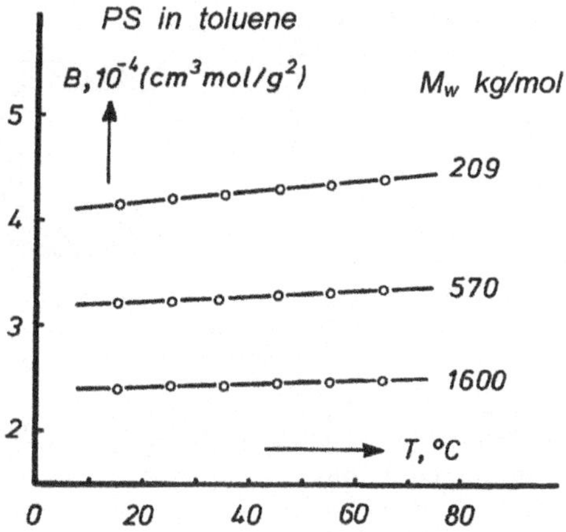

Fig. 3.4 Second virial coefficient of PS of various molar masses in toluene versus temperature, after Schulz et al. Reproduced from [3] by permission from the American Chemical Society

As already mentioned in Sect. 2.4, the Θ-solution is of large significance for theoretical considerations and molecular interpretations in polymer physics. Normally, two effects will disturb the simple calculation of the shape of the macromolecular chain structure. First, other parts of the considered chain end or other chains may occupy the place, where the chain end could otherwise find a place in the diffusion process. This restriction is called the occupied volume effect as discussed before. Second, in bad solutions, the interaction between parts of the macromolecule and solvent molecules may squeeze the solvent molecules out of the macromolecular coils, bringing them closer together. In the Θ-solution both effects just compensate each other, leaving the end of the macromolecule the freedom to diffuse undisturbed to each place in its vicinity.

In the Θ-solution the effects of the attractive forces between macromolecular coils and those of their occupied volume just compensate each other, and therefore, the undisturbed chain statistics may be applied to the macromolecules.

3.2 Viscometry

The measurement of the viscosity of dilute macromolecular solutions is a convenient and often used method to determine molar masses of polymers. For the definition of the viscosity of liquids we consider the laminar shear flow between two parallel flat plates of large extension, one being fixed, the other being moved with the constant velocity V in the x-direction (cf. Fig. 3.5).

If the liquid sticks to the walls, a flow field with a constant velocity gradient will develop. Its velocity components are:

Fig. 3.5 Simple shear flow

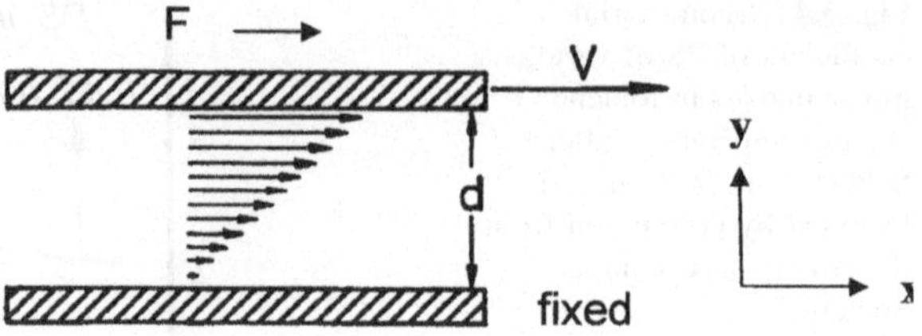

$$v_x = q \cdot y \quad v_y = v_z = 0 \quad q = V/d \tag{3.23}$$

We define the (*velocity*) *gradient* q, the *shear stress* σ, and the (*shear*) *viscosity* η by the equations

$$q = \frac{dv_x}{dy} \tag{3.24}$$

$$\sigma = F/A \tag{3.25}$$

$$\eta = \sigma/q \tag{3.26}$$

d is the distance between the two plates, A the surface of the upper plate moved by the shearing force F. η is not necessarily constant, but may depend on the magnitude of the gradient q. If η is independent of q, the liquid is called *Newtonian*, if not, it is called *non-Newtonian*. For example, water is a Newtonian fluid with a viscosity of $\eta \approx 10^{-3}$ Pas at 20 °C. Examples of non-Newtonian liquids are polymer solutions and polymer melts. The unit of the velocity gradient is s^{-1}, the unit of the viscosity is *Pas*.

Sometimes, the velocity gradient is also called the *shear rate* $\dot{\gamma}$. As velocity gradient and shear rate are two different concepts, defined by different experiments, we prefer for the present to use the designation velocity gradient. In a later stage (in Sect. 8.7) we return to the question of the relationship between both concepts and prove their identity. But this has to be postponed until the discussion of large deformations.

The measurement of the viscosity of low-viscous liquids (10^{-3}–10 Pas) is usually performed by a capillary viscometer. The fluid is forced by a pressure difference Δp through a circular capillary with the radius R and the length L. The law of *Hagen–Poiseuille*,[1] valid for Newtonian liquids, predicts for the amount of liquid Q passing through the capillary during the time t

$$\frac{Q}{t} = \frac{\pi R^4}{8L} \cdot \frac{\Delta p}{\eta} \tag{3.27}$$

[1] For derivation of this formula see for instance [5] or Sect. 15.1 of this book.

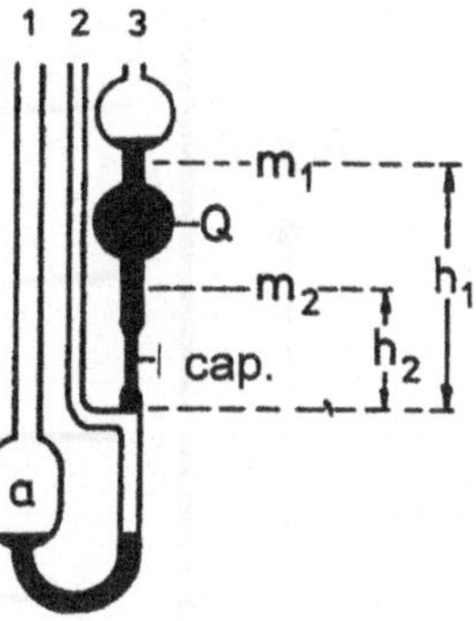

Fig. 3.6 The Ubbelohde viscometer

For the measurement of the viscosity of dilute polymer solutions, the *Ubbelohde-viscometer* is often used (cf. Fig. 3.6).

The *Ubbelohde-viscometer* is a 3-armed vessel, whose arms—designated as 1, 2, 3—may be closed on the upper end. For the measurement one proceeds as follows:

(1) Filling of the store vessel *a*
(2) Closing of 2 and sucking up the liquid through the capillary 3 up to a level above the mark m_1
(3) Closing of 3 and opening of 2, which results into separating the liquid at the joint of 2 and 3. A hanging level is created in 3.
(4) Opening of 3 and measuring the time necessary for the liquid to sink from the mark m_1 at the level h_1 to the mark m_2 at the level h_2.

With g as the acceleration of gravity and ρ the density of the liquid, the driving pressure during the measurement becomes

$$\Delta p = (h_1 + h_2) \cdot g\rho/2 \tag{3.28}$$

Inserting Eq. (3.28) into Eq. (3.27) yields for the viscosity

$$\eta = A\rho \cdot t \tag{3.29}$$

with the specific constant A of the instrument:

$$A = \frac{\pi R^4 (h_1 + h_2) g}{16 L Q} \tag{3.30}$$

which is determined in practice by calibration.

Polymer solutions are non-Newtonian fluids and may show a flow behavior as sketched in Fig. 3.7.

For small values of the gradient, a constant value of the viscosity is reached, the *Newtonian viscosity* η_0. Then, the viscosity strongly decreases with increasing gradient. For the pure solvent, the lowest value of the viscosity, the *solvent*

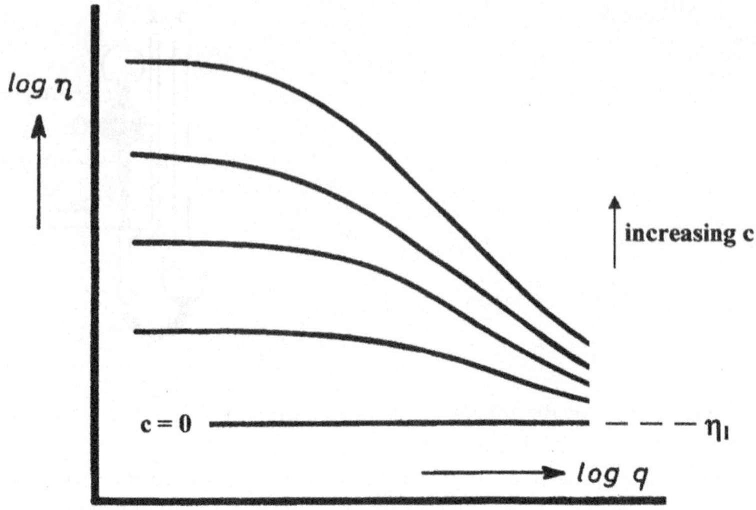

Fig. 3.7 Non-Newtonian flow behavior of polymer solutions, schematically

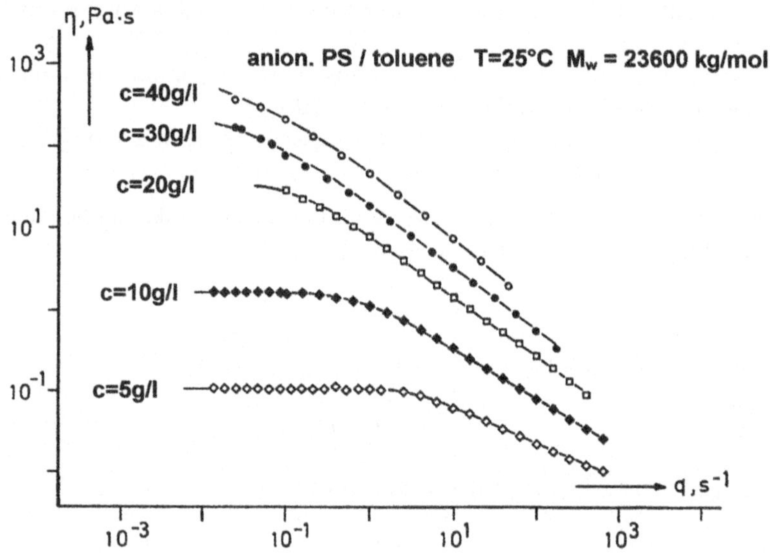

Fig. 3.8 Viscosity functions of solutions of an anionic poly(styrene) with a very high-molar mass in toluene for various polymer concentrations after Kulicke and Kniewske [6]

viscosity η_l is found, which is independent of the gradient. The ratio η/η_l strongly increases with increasing polymer concentration.

The non-Newtonian viscosity of solutions of anionic poly(styrene) in toluene [6], is shown in the Figs. 3.8 and 3.9. The polymers used in this study were

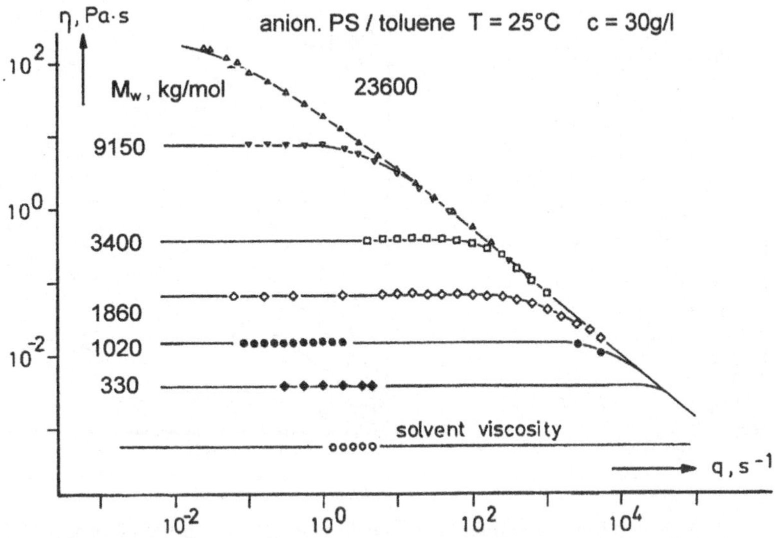

Fig. 3.9 Viscosity functions of 3 wt. % solutions of anionic poly(styrenes) of different molar masses in toluene, after Kulicke and Kniewske [6]

commercial products from Pressure Chemical Company (Pittsburgh, USA) and Polymer Laboratories Ltd. (Church Stretton, UK). Figure 3.8 presents the influence of the polymer concentration on the viscosity of solutions of a poly(styrene) with a very high molar mass. The Newtonian region of the viscosity curve, η_0, can be observed only at small values of the gradient, and only for solutions with low-polymer concentrations. As to be seen from Fig. 3.8, the transition from the Newtonian to the non-Newtonian region shifts to higher gradients with decreasing concentration of the polymer.

Figure 3.9 demonstrates the influence of the molar mass of the solute on the viscosity of solutions with a concentration of 30 g/l. The lower the molar mass, the more extended is the Newtonian region and the lower the value of the Newtonian viscosity η_0.

For the determination of the molar mass of the solute, the limit of the viscosity of the solution for small shear rates, the Newtonian viscosity of the solution, is required. The latter may be developed into a power series of the polymer concentration:

$$\eta_0 = \eta_l\{1 + \alpha\, c + \beta\, c^2 + \ldots\} \tag{3.31}$$

where η_l denotes the solvent viscosity. The coefficient of the first power in this series characterizes the property of the isolated ($c \to 0$), undeformed ($q \to 0$) macromolecular coil. This coefficient is called *viscosity number, intrinsic viscosity* or *Staudinger index* and is designated by *[η]*. The ratio

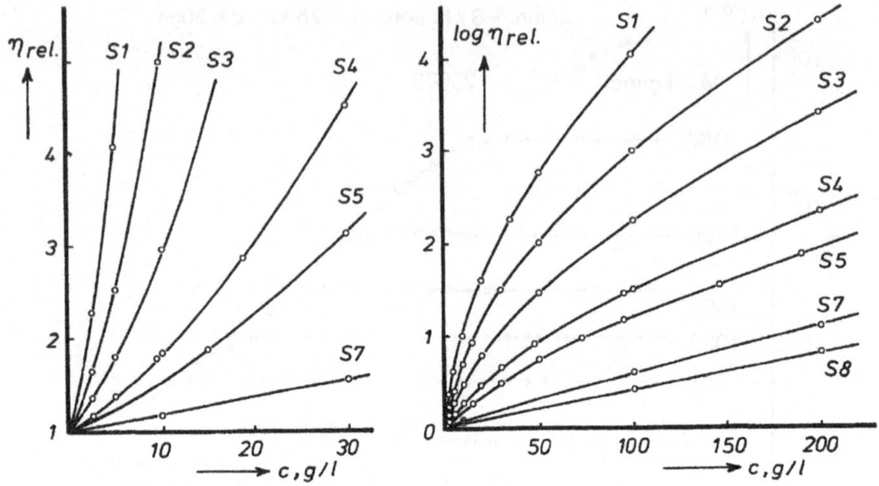

Fig. 3.10 Relative viscosity for solutions of anionic poly(styrene) in bromo-benzene at 25 °C after Daum [7]. The anionic PS used are those listed in Table 2.3

$$\eta_{\text{rel}} = \frac{\eta_0}{\eta_l} \tag{3.32}$$

is called the *relative viscosity* and the expression

$$\eta_{\text{spec}} = \frac{\eta_0}{\eta_l} - 1 \tag{3.33}$$

the *specific viscosity*, though both quantities do not have the dimension of a viscosity, but are pure numbers. The viscosity number—which is not a pure number, but has the dimension of an inverse mass concentration with the unity $m^3/kg = l/g$,—may be determined by one of the following limiting expressions

$$[\eta] = \lim_{c \to 0} \left[\frac{\eta_{\text{spec}}}{c} \right] \tag{3.34}$$

or

$$[\eta] = \lim_{c \to 0} \left[\frac{1}{c} \ln \eta_{\text{rel}} \right] \tag{3.35}$$

An example for the concentration dependence of solutions of anionic poly(styrene's) in bromo-benzene at 25 °C is shown in Fig. 3.10 [7].

The left-hand side of the figure shows the relative viscosity as a function of the mass concentration, the right-hand side displays the same values, but in a

Fig. 3.11 Illustration of the convergence of the both limits Eqs. (3.34) and (3.35). Measurements on S3 in bromo-benzene after Daum [7]. α is the coefficient of c in Eq. (3.31)

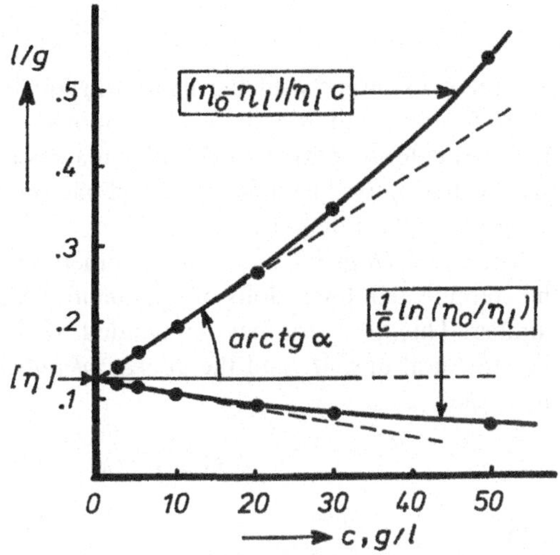

logarithmic scale for η_{rel} over a larger range of concentrations. The convergence of the two limits in Eqs. (3.34) and (3.35) at $c \to 0$ is illustrated in Fig. 3.11, showing η_{spec}/c and $(ln\ \eta_{\mathrm{rel}})/c$ versus c for solutions of the anionic polystyrene S3. The quantity η_{spec}/c becomes smaller with c and converges to the limit $[\eta\ ]$ for $c \to 0$. The function $(ln\ \eta_{\mathrm{rel}})/c$ grows with decreasing c and attains the same limit $[\eta]$ for $c \to 0$. Because of the inequality

$$\frac{1}{c}\ln \eta_{\mathrm{rel}} < [\eta] < \frac{\eta_{\mathrm{spec}}}{c} \tag{3.36}$$

the two functions represent upper and lower bounds for the error in the extrapolation to the value at $c \to 0$.

The large importance of the *Staudinger index* originates from the existence of a simple relation with the molar mass, which is experimentally found for linear macromolecules:

$$\boxed{[\eta] = K \cdot M^{a}} \tag{3.37}$$

K and a are constants, which depend on the combination chosen for the solute and solvent and on the temperature. This power law is called the *Mark–Houwink equation*. It is valid in the form of Eq. (3.37) for nearly monodisperse polymers. For polymers with a broader distribution of the molar mass, M has to be replaced by the viscosity average of the molar mass, M_v, as defined in Eq. (2.19) or (2.19′). The proof of this statement is based on the assumption, that the contributions of the various fractions to the increment of the specific viscosity behave additive, and was given by Flory [8]. The constant a is found to be between

$$0.5 \leq a \leq 1 \tag{3.38}$$

and becomes larger with the increasing quality of the solvent. For *Θ-solutions*, $a = 0.5$ is obtained. K and a are determined experimentally by fractionation of the polymer, determination of the molar masses of the fractions by osmometry or by light scattering (see next section), and measurement of the Staudinger index of the solutions of the fractions.

The *Mark-Houwink equation* is sometimes inconvenient to handle because of the complicated dimensions of the quantity K in cases for which a is an irrational number. This difficulty may be avoided by the introduction of the molar mass of the structural unit M_g and the *degree of polymerization* P into Eq. (3.37), which then reads

$$[\eta] = L \cdot (M/M_g)^a = L \cdot P^a \tag{3.39}$$

with

$$L = K \cdot M_g^a \tag{3.40}$$

for linear, monodisperse polymers and

$$[\eta] = L \cdot (M_v/M_g)^a = L \cdot P_v^a \tag{3.41}$$

for linear, polydisperse polymers. Some examples for values of K, L, and a are summarized in Table 3.2. For each polymer, at least two different solvents were chosen, one being a Θ-solvent, the other being a good solvent. Very comprehensive tables of K and a–values have been given by Kurata and Tsunashima [9].

In Fig. 3.12 the viscosity numbers are shown as a function of the molar mass in a double-logarithmic plot for fractions of poly(styrene) in three different solvents [29]. For a clearer representation of the accuracy of the measurements, different scales for $[\eta]$ were chosen. The three curves have been shifted relatively to each other in the y-direction by a factor 10. This picture demonstrates that Eq. (3.37) is valid only for molar masses, which exceed a certain value (here $M > 30$ kg/mol). Above this value the exponent in Eq. (3.37) increases with increasing quality of the solvent (cf. the different slopes for the three solvents).

Figure 3.13 shows the Staudinger index as a function of the molar mass for PS in five different solvents, over a still broader range of the molar mass than in Fig. 3.12. For the matter of clarity, the measuring points have been deleted in this picture and replaced by average lines.

Above a certain value of the molar mass, the curves fan out as straight lines with double logarithmic slopes between 0.5 (for the Θ-solvent cyclohexane) and 0.73 (for chloroform).

Next, we consider an illustrative interpretation of the Mark–Houwink equation. Einstein [31] derived the following simple equation for the viscosity of a dispersion of hard spheres in a fluid:

Table 3.2 Parameters of the Mark-Houwink-equation for some combinations of solvent and polymer

Polymer solvent	T, °C	$10^3 \cdot K^*$	a	L, 10^{-4} l/g	Range of molar masses, kg/mol	Lit.
Poly(styrene) in						
Chloroform	25	1.74	0.73	3.3	70–3730	[10]
Benzene	20	1.78	0.72	3.5	6–5200	[11]
Toluene	25	2.00	0.69	4.2	10–1040	[12]
Methylethylketone	20	2.15	0.58	5.8	10–10000	[12]
Cyclohexane	34.5	2.68	0.50	8.6	30–2000	[13]
Poly(methyl methacrylate)						
Chloroform	20	1.22	0.80	1.9	80–2000	[14]
Benzene	25	1.05	0.76	1.8	20–7400	[15]
Toluene	25	1.10	0.73	2.1	40–3300	[16]
Acetone	25	0.82	0.70	1.5	20–7800	[15]
4-heptanone	33.8	1.52	0.50	4.8	10–1720	[17]
Poly(isobutene)						
Cyclohexane	30	3.24	0.69	4.4	40–710	[18]
Benzene	24	3.38	0.50	8.0	180–1880	[18]
Poly(vinylchloride)						
Tetrahydrofuran (THF)	25	3.24	0.77	3.9	20–300	[19]
Benzyl alcohol	155.4	4.93	0.50	12.3	40–350	[20]
Poly(carbonate)						
THF	25	5.02	0.766	19.3	10–270	[21]
Butyl-benzyl-ether	170	6.64	0.50	33.5	40–310	[22]
Poly(ethylene) low pres						
Tetralin	130	7.63	0.725	5.73	40–500	[23]
Diphenyl	130	9.55	0.50	16.0	57–270	[24]
Poly(propylene) atac						
Decalin	135	2.76	0.80	2.2	20–620	[25]
Phenyl ether	153	3.79	0.50	7.8	37–210	[26]
Poly(amide 6.6)						
M-cresole	25	16.2	0.61	65.5	14–50	[27]
90 % formic acid + 2.3 mol KCl	25	8.00	0.50	38.0	14–50	[28]

* The unit for K has been chosen in such a way, that $[\eta]$ is obtained in l/g, if M is inserted in kg/mol

$$\eta_0 = \eta_l(1 + 2.5 \cdot \varphi) \tag{3.42}$$

where η_0 is the viscosity of the dispersion, η_l the viscosity of the pure fluid and φ the volume concentration of the spheres in the dispersion. Assuming that the solution of the macromolecular coils behaves just as a dispersion of hard spheres with radius h, Eq. (3.42) can be compared with the linear part of Eq. (3.31), i.e.,

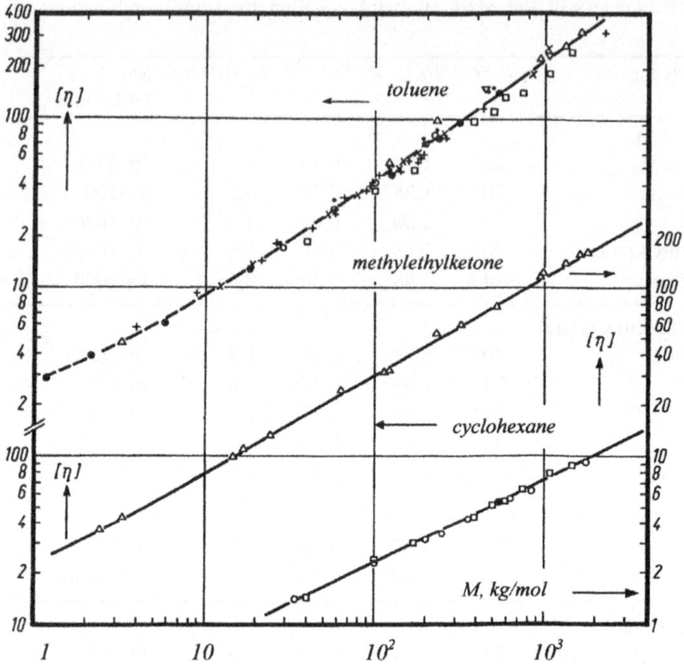

Fig. 3.12 Staudinger index, $[\eta]$, in cm^3/g, as function of the molar mass, M, in kg/mol for solutions of poly(styrene) in three solvents after Meyerhoff [29]

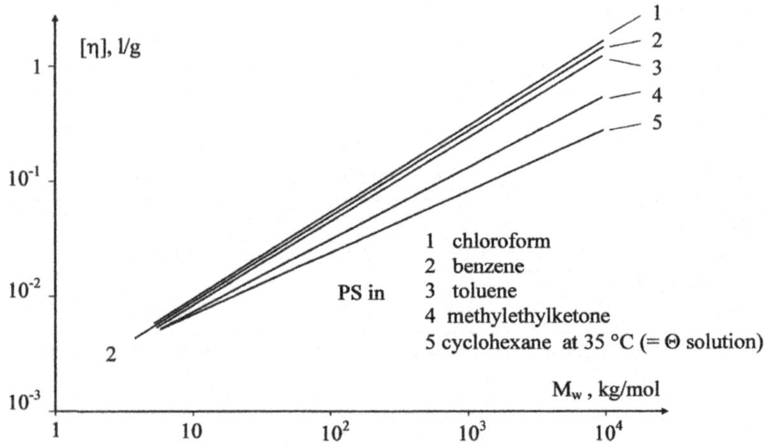

Fig. 3.13 Staudinger index, $[\eta]$, in l/g as function of the molar mass, M_w, in kg/mol for solutions of poly(styrene) in five different solvents after Hoffmann. Reproduced from [30] by permission from Hüthig and Wepf, Basel

$$\eta_0 \cong \eta_l(1 + [\eta] \cdot c)$$

this comparison yields

$$[\eta] = 2.5\frac{\varphi}{c} = \frac{2.5}{d_m} \tag{3.43}$$

where d_m is the *average density of the macromolecule in the coil* . The latter is equal to the ratio of the mass of one macromolecule to the volume of one coil

$$d_m = \frac{c}{\varphi} = \frac{M/N_A}{4\pi h^3/3} \tag{3.44}$$

and therefore

$$[\eta] = \frac{10\pi}{3} N_A \frac{h^3}{M} \tag{3.45}$$

From the Staudinger index an estimate for the coil radius in dilute solution is obtained. We insert for the radius of the coil, h, the square root of the mean square of the radius of gyration $h = \langle s^2 \rangle^{1/2}$ and use Eq. (2.45) with the result

$$[\eta] = \frac{10\pi}{18\sqrt{6}} N_A \frac{\langle r^2 \rangle^{3/2}}{M} = \Phi' \cdot \frac{\langle r^2 \rangle^{3/2}}{M}$$

The evaluation of the numerical factor yields $\Phi' \approx 4.3 \times 10^{23}\ \text{mol}^{-1}$. The consideration above is very simplified, as the macromolecular coils are not hard spheres, but are partially penetrated by the solvent and are flexible.

In two famous publications, Flory [32] and Fox and Flory [33] have calculated the resistance of the single segments of a pearl string model in the flow field and arrived at a similar result, but with a different numerical factor

$$\boxed{[\eta] = \Phi \cdot \frac{\langle r^2 \rangle^{3/2}}{M} \quad \text{with } \Phi \cong 2.6 \times 10^{23}\ mol^{-1}} \tag{3.46}$$

This *equation by Fox and Flory* yields a correlation between the *viscosity number*, the *end to end mean square distance* and the *molar mass*. Though it was originally derived under the assumption of a random walk chain, Flory later argued, that it could also be valid for solutions of good quality [8]. If $\langle r^2 \rangle^{3/2}$ is resubstituted by $\langle s^2 \rangle^{3/2}$ via Eq. (2.45), Eq. (3.46) could even describe the Staudinger index for branched polymers.

For the Θ-solution the *Mark–Houwink equation* becomes

$$\boxed{[\eta]_\Theta = K_\Theta M^{1/2} = L_\Theta P^{1/2}} \tag{3.47}$$

with

$$L_\Theta = K_\Theta \cdot M_g^{1/2} \tag{3.48}$$

and the *Fox and Flory equation* reduces to

$$\boxed{[\eta]_\Theta = K_\Theta M^{1/2} = \frac{\Phi}{M}\left\langle r^2\right\rangle_\Theta^{3/2}} \tag{3.49}$$

From the right side of this equation, we obtain the following expressions for the *mean square end to end distance* and the *mean square radius of gyration*

$$\left\langle r^2\right\rangle_\Theta = \left(\frac{K_\Theta}{\Phi}\right)^{2/3} M = \alpha_\Theta^2 P \tag{3.50}$$

with

$$\alpha_\Theta = \sqrt{M_g} \cdot \left(\frac{K_\Theta}{\Phi}\right)^{1/3} \tag{3.51}$$

and

$$\left\langle s^2\right\rangle_\Theta = \beta_\Theta^2 P \tag{3.52}$$

with

$$\beta_\Theta = \sqrt{\frac{M_g}{6}} \cdot \left(\frac{K_\Theta}{\Phi}\right)^{1/3} \tag{3.53}$$

We calculate br_{max}/P from Eq. (2.74) and substitute $\left\langle r^2\right\rangle_\Theta$ from Eq. (3.50) to obtain an expression for *Kuhn's statistical segment, b*.

$$b\frac{r_{max}}{P} = M_g\left(\frac{K_\Theta}{\Phi}\right)^{2/3} \tag{3.54}$$

The ratio (r_{max}/P) can be calculated from the geometry of the molecule, the ratio K_Θ/Φ can be measured over a wide range of molar masses. Finally, from Eqs. (3.51) and (2.80), we obtain expressions for the constants K_Θ and L_Θ of the *Mark-Houwink equation* for *linear vinyl polymers*

Table 3.3 The parameters K_Θ and L_Θ of the Mark-Houwink equation and the value of the molecular parameters C, b/b_0 and β/b_0 for some polymers in Θ-solution. The values for C and K_Θ have been taken from Flory [34]. Values for L_Θ, b and β_Θ were calculated using Eqs. (3.53), (3.56) and (3.57). ("at" stands for "atactic")

Polymer	Solvent	T, °C	K_Θ^*	L_Θ^*	C	b/b_0	β_Θ/b_0
Poly(ethylene)	Dodecanol-1	138	97.1	16.3	6.6	8.1	1.49
	Diphenylmethane	142	99.6	16.7	6.7	8.3	1.50
Poly(styrene) at	Cyclohexane	34.8	25.9	8.4	10.2	12	1.84
	Diethyl malonate	35.9	24.3	7.9	9.8	12	1.80
Poly(propylene) at	Cyclohexane	92.0	54.4	11.2	6.7	8.3	1.50
	Diphenyl ether	153	37.9	7.8	5.3	6.5	1.33
Poly(isobutylene)	Benzene	24.0	33.8	8.0	6.5	8.0	1.48
Poly(vinyl acetate)	i-pentanon hexane	25.0	27.8	8.2	8.8	11	1.71
Poly(methyl methacrylate) at	Various solvents	4–70	15.2	4.8	6.9	8.5	1.52
Poly(dimethyl siloxane)	Butanone	20.0	24.7	6.7	6.2	–	1.44
Poly(vinylchloride)	Benzyl alcohol	155	49.3	12.3	9.4	10.5	1.77
Poly(carbonate)	Butyl-benzyl-ether	170	66.4	33.5	–	–	3.93
Poly(amide 6.6)	90 % HCOOH + 2.3 mol KCl	25.0	80.0	38.0	6.8	–	3.62

* K_Θ in 10^{-4} m^3 $\text{mol}^{1/2}$ $\text{kg}^{-3/2}$, L_Θ in 10^{-4} l/g, $b_0 = 1.54 \times 10^{-10}$ m, $\alpha_\Theta = \sqrt{6}\beta_\Theta$

$$K_\Theta = \Phi \frac{\alpha^3}{M_g^{3/2}} = \Phi b_0^3 \left(\frac{2C}{M_g}\right)^{3/2} \tag{3.55}$$

and

$$L_\Theta = \Phi \frac{\alpha^3}{M_g} = \Phi (2C)^{3/2} b_0^3 / M_g \tag{3.56}$$

and, from Eq. (3.54) with Eq. (2.76), for b, *Kuhn's statistical segment*

$$\frac{b}{b_0} = \sqrt{\frac{3}{8}\frac{M_g}{b_0^2}} \left(\frac{K_\Theta}{\Phi}\right)^{2/3} \tag{3.57}$$

We conclude, that the combination of the unperturbated chain statistics with the *Fox-Flory equation* results in

- The correct expression for the Mark-Houwink equation for *theta-solutions*,
- expressions for b, α, and β from the results of viscosity measurements,
- and, for vinyl polymers, expressions for K_Θ and L_Θ in terms of the molecular parameter C.

By the measurement of the viscosity numbers of linear polymers with known molar mass in theta-solution, it is possible to determine the molecular parameters b, C, α, and β. Some examples are given in Table 3.3, taken from Flory [34] and completed with the calculation of the L_Θ-, β_Θ- and b-values. Comprehensive tables of K_Θ, C, b, have been published by Kurata and Tsunashima [9].

3.3 Light Scattering

The principle of the measurement of light scattering is sketched in Fig. 3.14. A laser beam with the wave length λ_0 travels through a semi-permeable mirror, which splits it into two branches. One part of the light falls on a reference photo cell, the other part reaches the cuvette filled with the polymer solution. The light scattered at the various angles ϑ is measured by a second photo cell. [Modern instruments use a larger number of photocells mounted at fixed angles (*multiangle laser-light scattering* (MALLS)).]

Let $i(\vartheta, c)$ be the intensity of the light scattered by the solution with the concentration c under the angle ϑ, $i(\vartheta, 0)$ the intensity of the light scattered into the same direction by the pure solvent and i_0 the intensity of the incident beam. The intensity of the light scattered under the angle ϑ, minus the intensity of the light scattered by the pure solvent into the same direction, divided by the intensity of the incident beam is proportional to the scattering volume V of the cuvette and inverse proportional to the square of the distance r between cuvette and photo cell.

$$\frac{i(\vartheta, c) - i(\vartheta, 0)}{i_0} = \frac{V}{r^2} R(\vartheta, c, M, h/\lambda, B)$$

The scattering volume of the solution depends on the scattering angle ϑ

$$V = V_0 / \sin \vartheta$$

with V_0 being the scattering volume at the scattering angle $\vartheta = \pi/2$. Therefore, the measurement yields

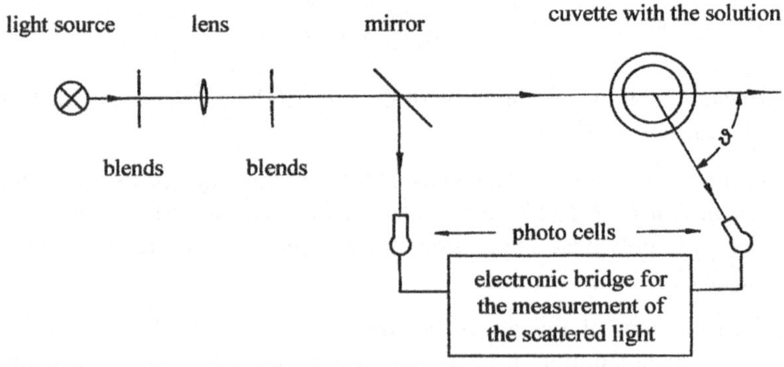

Fig. 3.14 Schematic arrangement for the measurement of light scattering

$$\frac{i(\vartheta,c)-i(\vartheta,0)}{i_0}=\frac{V_0}{r^2\sin\vartheta}R(\vartheta,c,M,h/\lambda,B) \tag{3.58}$$

The quantity R is called *Rayleigh's ratio.* It depends on the scattering angle ϑ, the wave length of the light in the solution λ, and the properties of the solution as the *polymer concentration c*, the *molar mass of the polymer M*, the *radius* of the *macromolecular coils h*,[2] and the *second virial coefficient B*. R has the dimension of a reciprocal length, and is determined as a function of the scattering angle ϑ and of the *concentration c* of the *solution*.

We consider first the most simple case, that the dimensions are small in comparison to the wave length of the light (in the solution) and that the solution is very diluted:

$$2h \ll \lambda \textit{ highly diluted; incident light unpolarized}$$

Due to the condition $2\,h \ll \lambda$, the entire coil may be considered as one single dipole which is excited by the incident light wave. The dipole oscillates with the frequency of the incident light in the direction of its polarization. It emits light of the same frequency with an intensity depending on the angle between the direction of the dipole and the direction of the irradiation. After summation over all directions of dipoles and over all coils in the scattering volume one finds for unpolarized light (for a derivation of this formula see [35] and also [36])

$$R(\vartheta,c,M)=\frac{1+\cos^2\vartheta}{2}KcM \tag{3.59}$$

with

$$K=\frac{4\pi^2n_1^2}{N_A\lambda^4}\cdot\left(\frac{dn}{dc}\right)^2 \tag{3.60}$$

The quantities occurring in this equation are

n_1	The refractive index of the solvent
n	The refractive index of the solution
dn/dc	The refractive index increment of the solution
$\lambda = \lambda_0/n$	The wave length of the light in the solution
λ_0	The wave length of the light in the vacuum
N_A	Avogadro's number

---

[2] h is used as an abbreviation for the square root of the mean square radius of gyration $h=\langle s^2\rangle^{1/2}$.

The molar mass is obtained by an extrapolation of the measured data to the scattering angle $\vartheta = 0$ and to the concentration $c = 0$. For a monodisperse polymer one gets from Eq. (3.59)

$$\lim_{\substack{\vartheta \to 0 \\ c \to 0}} \frac{R(\vartheta)}{c} = K \cdot M \tag{3.62}$$

and for a polydisperse polymer

$$\lim_{\substack{\vartheta \to 0 \\ c \to 0}} \frac{R(\vartheta)}{c} = K \cdot M_w \tag{3.63}$$

The proof that the molar mass in Eq. (3.63) is the weight average molar mass M_w is based on the additivity of the contributions to $R(\vartheta, c, M)$ of coils with different molar masses.

Next, we consider the case of a moderate dilution, for which the coil dimensions are still small in comparison to the wave length:

$$2h \ll \lambda : \textit{moderate dilution}.$$

At higher concentrations, the light waves scattered from the single coils do not longer behave additive, but interfere with each other. Consequently, the intensity of the scattered light increases less than proportional to the concentration. A theoretical treatment [35] yields instead of Eq. (3.59):

$$\frac{1 + \cos^2 \vartheta}{2} K \frac{c}{R(\vartheta, c, M, B)} = \frac{1}{M_w} + 2Bc + \cdots \tag{3.64}$$

with B being the second virial coefficient.

In order to determine B and M_w, the data are presented in the so-called *Zimm-diagram* (cf. Fig. 3.15). The quantity $(1 + \cos^2 \vartheta)Kc/2R(\vartheta)$ is plotted against an abscissa, which is composed additively of the values of $\sin^2 \vartheta/2$ and $\kappa \cdot c$. The proportionality factor κ may be chosen arbitrarily. It has the dimension of *l/g* and is introduced only to separate measuring points, which belong to different concentrations but to equal values of the scattering angle or vice versa.

Points indicated by circles are obtained directly from the measurement, those indicated by crosses are obtained by extrapolations to either $c \to 0$ or $\vartheta \to 0$. The intersection of the two extrapolated lines lies on the ordinate and corresponds to the value $1/M_w$. From the slope of the line with constant scattering angle $\vartheta = 0$, the *second virial coefficient B* is obtained, from the slope of the line with the concentration $c = 0$, a measure for the *coil dimension* follows according to Eq. (3.68).

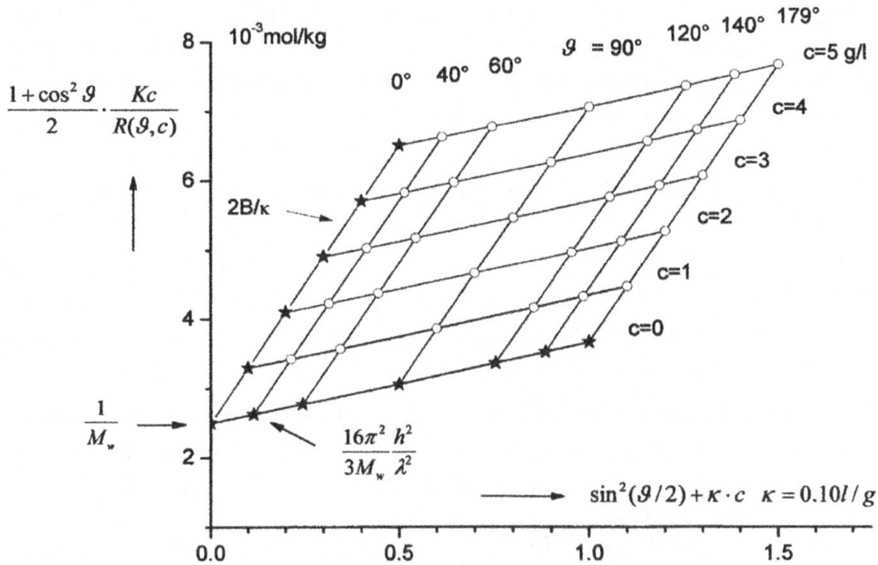

Fig. 3.15 Zimm diagram for the determination of M_w, B, and h^2/λ^2, schematically. The diagram was constructed, using Eq. (3.65) and the parameter values $M_w = 400$ kg/mol, $B = 4.10^{-4}$ S.I. units and $h/\lambda = 0.3$

In the case of a moderately diluted solution, for which the coil dimensions are no longer small in comparison to the wave length, the method yields direct information on the magnitude of the macromolecular coils:

$2h \cong \lambda$; *moderate dilution.*

If the dimensions of the coils become comparable with the wave length ($2\,h \approx 0.1\,\lambda$), it is not allowed any longer, to add the intensities of the light scattered from different parts of the macromolecule. Vibrations of the dipoles from different parts of the molecule show phase differences. This leads to interferences, which magnify the light scattered into the direction of the incidence of the original beam, and attenuate the light scattered into the backward direction. While the scattering curve is symmetrical with respect to the position $\vartheta = 90°$ for small coils, it becomes asymmetrical for larger ones. This *dissymmetry* forms a measure for the coil dimensions. Instead of Eq. (3.64), the following relation holds for large coils in moderately diluted solutions [35]:

$$\frac{1+\cos^2\vartheta}{2}\frac{Kc}{R(\vartheta, c, M, h/\lambda, B)} = \frac{1}{M_w}\cdot\frac{1}{J(\vartheta)} + 2Bc \qquad (3.65)$$

The function $J(\vartheta)$ depends on the shape and the size of the scattering particles. $J(\vartheta) = 1$ for small particles and $J(0) = 1$ for particles of all sizes. $J(\vartheta)$ has been calculated for rigid spheres, for rigid rods and for macromolecular coils.[3]

The result for coils is:

$$J(\vartheta) = \frac{2}{u^2} \cdot [e^{-u} - 1 + u] \tag{3.66}$$

with

$$u = 16\pi^2 \cdot \frac{h^2}{\lambda^2} \sin^2(\vartheta/2) \tag{3.66a}$$

$h^2 = \langle s^2 \rangle$ is the *mean square radius of gyration*, which may directly be determined from the initial slope of the curve $(1 + \cos^2 \vartheta)Kc/2R(\vartheta)$ against $\sin^2(\vartheta/2)$ for $c = 0$, as by developing $J(\vartheta)$ into a power series of u, the slope is found as:

$$d\left\{\frac{1 + \cos^2 \vartheta}{2} \frac{Kc}{R(\vartheta)}\right\} \Big/ d\left(\sin^2(\vartheta/2)\right) = \frac{16\pi^2}{3} \frac{1}{M_w} \frac{h^2}{\lambda^2} \tag{3.67}$$

This slope may easily be obtained from the extrapolated line $c \to 0$ of the *Zimm diagram*, which thus offers a measure for the coil dimensions, viz., the mean square radius of gyration.

A further possibility for the measurement of the coil dimensions is the *dissymmetry* of the *scattering curve*. One determines the ratio of the intensity of the light scattered at the scattering angles 45 and 135° in the limit of $c \to 0$. This ratio is called the *dissymmetry factor* z and is easily calculated from Eqs. (3.65) and (3.66).

$$z = \frac{R(\vartheta = 45°, c \to 0)}{R(\vartheta = 135°, c \to 0)} = \frac{J(\vartheta = 45°)}{J(\vartheta = 135°)} = 1 + \frac{8\sqrt{2}\pi^2}{3} \frac{h^2}{\lambda^2} + O\left[\frac{h^4}{\lambda^4}\right] + \cdots \tag{3.68}$$

The dissymmetry factor z, calculated as a function of the ratio of the mean square radius of gyration to the square of the wave length, is presented in Fig. 3.16

The dissymmetry factor increases linearly with $(h/\lambda)^2$ at small values of the abscissa, with a slope of 37.2 and exceeds unity by more than 30 % already for values of $h/\lambda = 0.1$. As examples, in Figs. 3.17 and 3.18 *Zimm diagrams* of a technical poly(styrene) in a *good solvent* and in a *theta solution* are shown.

In a good solvent, B has high-positive values. Therefore, the line obtained by extrapolation to $\vartheta \to 0$ with the slope B increases stronger with increasing abscissa

[3] Rigid spheres and macromolecular coils scatter differently, as spheres are uniformly filled with scattering matter, while coils show the density distribution of the random walk chain.

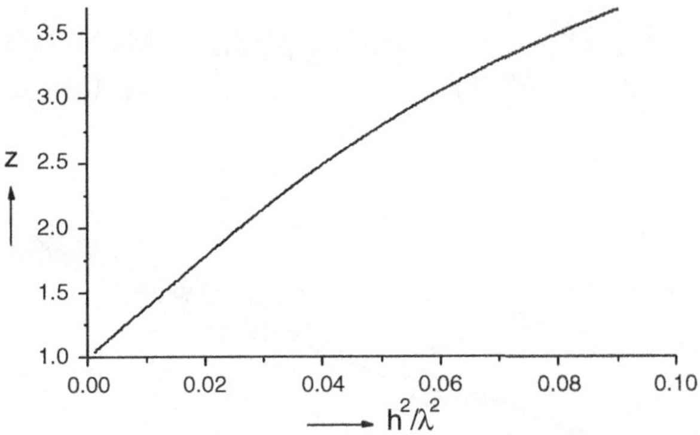

Fig. 3.16 Dissymmetry factor z as a function of the relative coil dimension h/λ

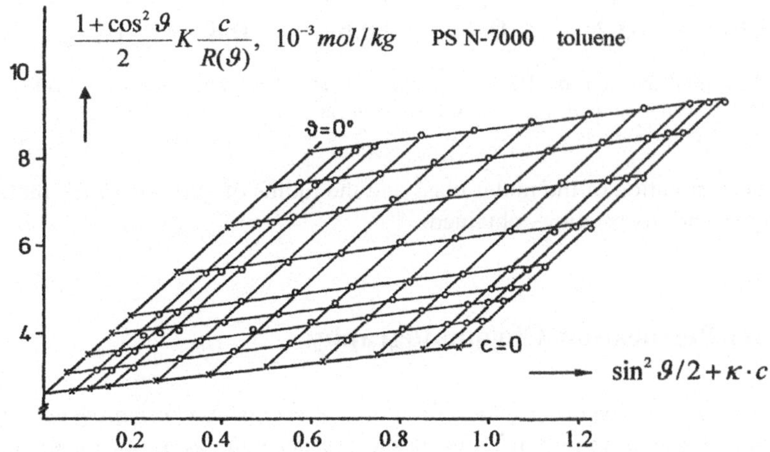

Fig. 3.17 Zimm diagram of PS in toluene at 20 °C after Jolk [37]. Measuring points at the scattering angles $\vartheta = 30$, 37.5, 45, 60, 75, 90, 105, 120, 135, 142, and 150°, and at the concentrations c = 0.53, 1.0, 1.56, 1.94, 2.98, 4.14, 5.06, and 6.0 g/l and $\kappa = 0.1$ l/g

than the line extrapolated to $c \to 0$. The first line forms the left boundary of the rhombus of the Zimm diagram, the latter its lower boundary.

In a theta solution, the line extrapolated to $\vartheta \to 0$ with the slope $B \cong 0$ forms the lower boundary of the rhombus of the diagram, the line extrapolated to $c \to 0$, its left boundary.

In conclusion it may be stated, that light scattering is one of the most powerful methods of molecular characterization. Besides the measurement of M_w and B, it is the only method which yields direct information about the mean square radius of gyration. Furthermore, light scattering in combination with gel permeation chromatography is a very important tool of molecular characterization, as it allows a

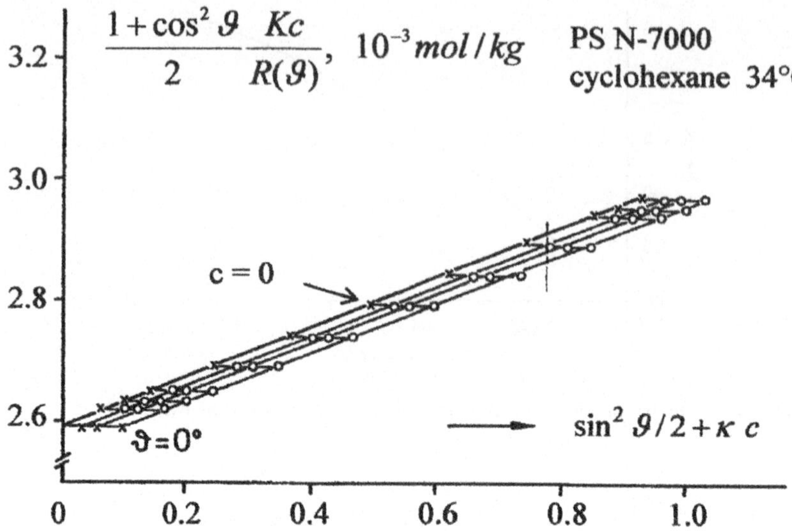

Fig. 3.18 Zimm diagram of PS in cyclohexane at 34 °C after Jolk [37] (Θ-solution). $\kappa = 0.05l/g$. Scattering angles as in Fig. 3.18, concentrations c = 0.66, 1.18, 2.0 g/l

direct determination of the molar mass and the radius of gyration of the molecules within the various fractions obtained.

3.4 Gel Permeation Chromatography

Gel permeation chromatography (GPC) or *size exclusion chromatography* (SEC) has attained a great importance during the last decennia as a tool for the characterization of polymers. By means of GPC it is possible to determine not only various averages of the molar mass distribution, but also the shape of the distribution functions, which is of special interest in dealing with bi- or multimodal distributions, and cannot be obtained by other methods.

A schematic setup for gel permeation chromatography is given in Fig. 3.19. The solvent is pumped from a container through a sampling valve, and then further through a number of separating columns into the detector. The purpose of the sampling valve is to inject at a certain time $\Delta v = 0.1$ ml of the polymer solution into of the solvent which is continuously pumped through the columns at a constant flow rate.

While the solution passes the separating columns, it is *fractionated.* The different columns are filled with gels of different pore sizes. Frequently used are styrene-divinylbenzene gels. Polymer coils which fit into the pores are retained there for longer times, while larger coils pass the column faster. By this process larger macromolecules arrive earlier at the end of the last column and are measured first

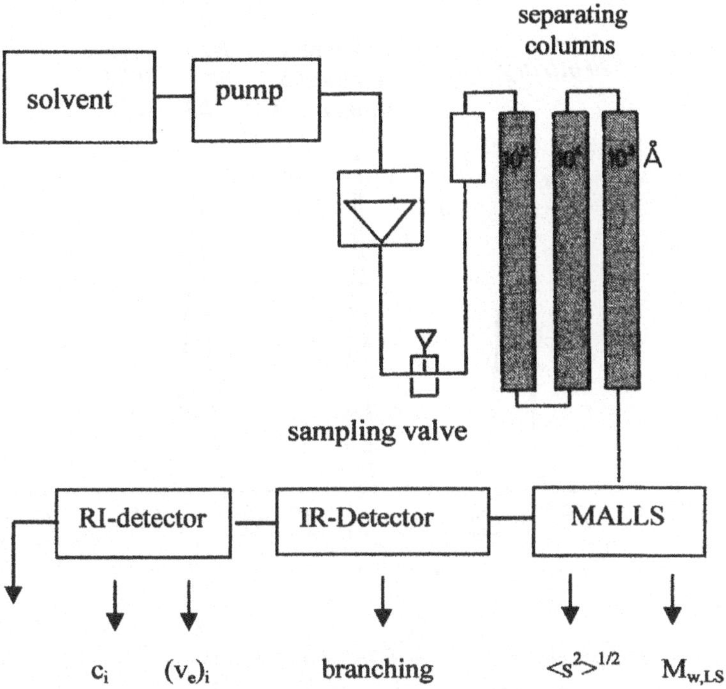

Fig. 3.19 Schematic setup for gel permeation chromatography. As solvent THF is frequently used. The numbers on the columns designate the average pore sizes 10^{-3}, 10^{-4} and 10^{-5} cm

by the differential refractometer or other probes. In the refractometer the polymer concentration of the solution is measured as a function of the elution volume by comparing its refractive index with that of the pure solvent.

The *polymer concentration* c_i, as a function of time or the *elution volume* $(v_e)_i$, respectively, already represents a good qualitative picture of the *molar mass distribution*. In order to arrive at quantitative results, it is necessary either to determine the absolute molar mass of the fractions by means of laser light scattering or to calibrate the chromatogram using nearly monodisperse samples.

As an example for the latter method, Fig. 3.20 shows the *gel permeation chromatograms* of three polystyrene standards with known narrow molar mass distributions. The elution volumes at the concentration maxima are assigned to the corresponding molar masses. In this way the calibration curve shown in Fig. 3.21 was obtained.

The result of the chromatograms of ten PS-standards with molar masses between 1 and 2500 kg/mol is shown in the *calibration curve* in Fig. 3.21.

The closed circles represent the ten PS-standards, the open circle is ascribed to the styrene monomer. The relation between the logarithm of the molar mass (in kg/mol) and the elution volume v_e (in ml) may be described by the following *calibration curve*

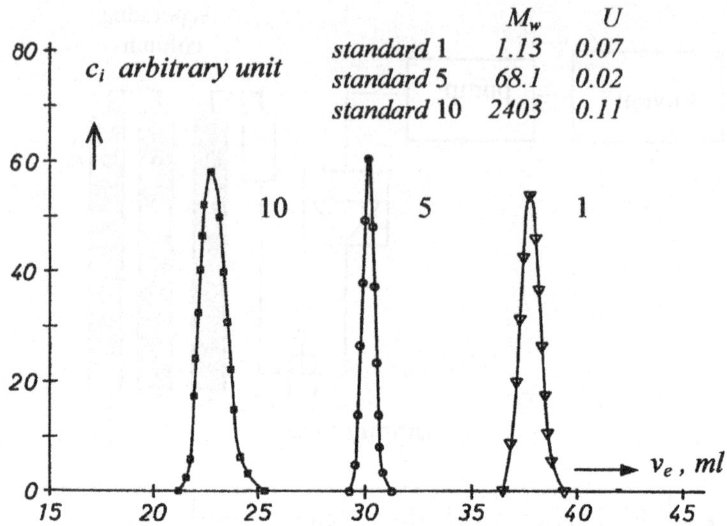

Fig. 3.20 Gel permeation chromatogram of three anionic poly(styrenes) after Wolf [38]

$$\log(M/(kg/mol)) = a + bv_e + cv_e^2 + dv_e^3 + ev_e^4 + \cdots \tag{3.69}$$

with

$$a = 50.6 \quad b = -5.85 \quad c = 0.273 \quad d = -5.76 \cdot 10^{-3} \quad e = 4.47 \cdot 10^{-5}$$

In a very coarse approximation, the calibration curve of Fig. 3.21 may be replaced by a straight line, meaning that only the first two terms of Eq. (3.69) are retained:

$$\log(M/(kg/mol)) \approx a + bv_e \tag{3.69a}$$

The concentration c_i of the ith fraction is measured at the elution volume $(v_e)_i$, to which the molar mass M_i is assigned by the calibration function Eq. (3.69). The measurement is performed every 12 s, corresponding to an increase of the elution volume by $\Delta v = 0.1$ ml. The *width* β_i of the *interval* of the *i*th *fraction* with respect to the elution volume is (in analogy to the derivation of Eq. (2.11)) the same for all fractions, viz.,

$$\beta_i = \left((v_e)_{i+1} - (v_e)_{i-1}\right)/2 = \Delta v \tag{3.70}$$

Assuming that the fractionating columns separate ideally, i.e., that the fraction i contains only molecules of the molar mass M_i, the *mass fraction* W_i of the molecules with a molar mass M_i in the solution is the product of the concentration

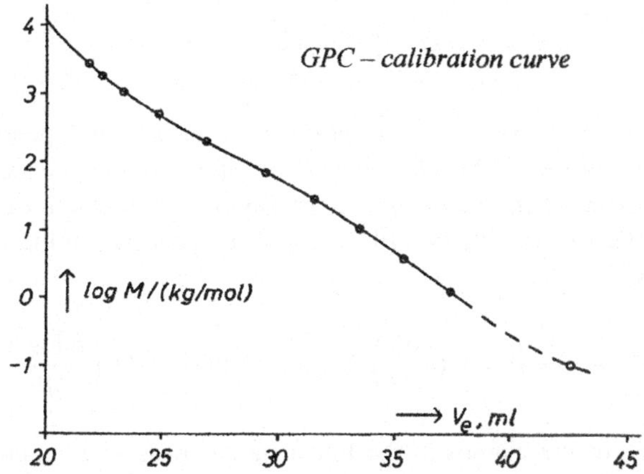

Fig. 3.21 Calibration curve of the GPC-setup after Wolf [38]. Circles indicate calibration points, the drawn line shows the calibration curve according to Eq. (3.69)

c_i measured at the elution volume $(v_e)_i$ and the width β_i of the interval around this elution volume, i.e.,

$$W_i = c_i\beta_i = c_i\Delta v \tag{3.71}$$

Due to Eq. (2.5) this should be equal to

$$W_i = n_i M_i / N_A$$

from which one obtains

$$m_i = n_i M_i = c_i N_A \Delta v \tag{3.72}$$

We insert Eq. (3.72) into the definitions of the *number average* Eq. (2.15), the *weight average* Eq. (2.17), the *z-average* Eq. (2.18) and the *viscosity average* Eq. (2.19) of the molar mass and find the corresponding expressions in terms of the measured *concentrations* c_i:

$$M_n = \frac{\sum n_i M_i}{\sum n_i} = \frac{\sum c_i}{\sum (c_i/M_i)} \tag{3.73}$$

$$M_w = \frac{\sum n_i M_i^2}{\sum n_i M_i} = \frac{\sum c_i M_i}{\sum c_i} \tag{3.74}$$

$$M_z = \frac{\sum c_i M_i^2}{\sum c_i M_i} \tag{3.75}$$

and

$$M_v = \left[\frac{\sum c_i M_i^a}{\sum c_i}\right]^{1/a} \tag{3.76}$$

Though the *width* β_i of the *interval of the* ith *fraction* is independent of i, when measured on the scale of the *elution volume*, it is *not independent* of i, when measured on a scale of the *molar mass*. By Taylor development of Eq. (3.70) we find for the *width* b_i of the interval of the ith *GPC-fraction* (on the scale related to the molar mass)

$$b_i = \frac{(v_e)_{i+1} - (v_e)_{i-1}}{2} = -\left(\frac{dv_e}{dM}\right)_i M_g = -0.43 M_g \Big/ M_i \left(\frac{d \log M}{dv_e}\right)_i \tag{3.77}$$

whereas the size of the classes in the linear molar mass scale is chosen as M_g.[4]

The derivative occurring in this equation follows from Eq. (3.69) as

$$\frac{d \log(M/(kg/mol))}{dv_e} = b + 2cv_e + 3dv_e^2 + 4ev_e^3 + \cdots \tag{3.78}$$

or from Eq. (3.69a) as

$$\frac{d \log(M/(kg/mol))}{dv_e} \approx b \tag{3.78a}$$

Inserting the latter into Eq. (3.77), we arrive at

$$b_i \approx -\frac{0.43 M_g}{b M_i} \tag{3.79}$$

For the calculation of the *mass and number densities* of the *molar mass distribution* we have to use the definitions Eqs. (2.7) and (2.9) of the mass and number distribution functions of the molar mass and to insert for n_i and m_i the values following from Eq. (3.72). We find for the *number distribution function*

$$f(M_i) = \frac{n_i}{\sum n_i} = \frac{c_i/M_i}{\sum (c_i/M_i)} \tag{3.80}$$

and for the *mass distribution function*

$$h(M_i) = \frac{m_i}{\sum m_i} = \frac{c_i}{\sum c_i} \tag{3.81}$$

[4] If the size were chosen as a fixed multiple of M_g, this would only multiply the width with a constant factor, independent of i, and would not change the following argument. The minus sign is required, because log M is a decreasing function of v_e and b_i should be positive.

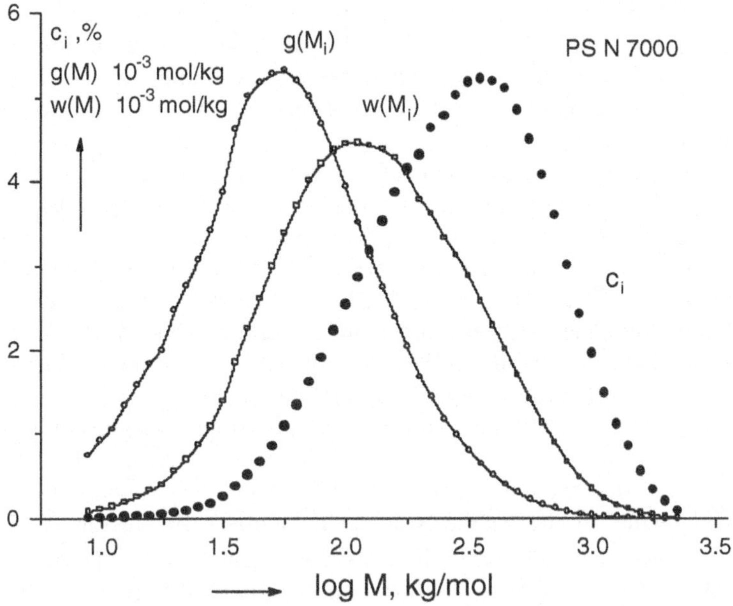

Fig. 3.22 Concentration c_i, mass distribution density $w(M)$ and number distribution density $g(M)$ of a technical poly(styrene), PS N 7000, after Wolf [38]

The *number* and *mass distribution densities* are found by multiplying Eqs. (3.80) or (3.81) with the ratio b_i/β_i and renormalizing the distribution densities

$$M_g g(M_i) = \frac{c_i/M_i(d\log M/dv)_i}{\sum_i c_i/M_i(d\log M/dv)_i} \cong \frac{c_i/M_i^2}{\sum_i c_i/M_i^2} \tag{3.82}$$

and

$$M_g w(M_i) = \frac{c_i/(d\log M/dv)_i}{\sum_i c_i/(d\log M/dv)_i} \cong \frac{c_i/M_i}{\sum (c_i/M_i)} \tag{3.83}$$

Figure 3.22 shows the *elution chromatogram* of a technical poly(styrene) for a calibrated logarithmic molar mass scale as determined by Wolf [38].

Measured c_i-values have been indicated by filled circles. From these, the values of single points of the curve of the mass distribution density (open squares) have been calculated by means of Eq. (3.83) and the values of single points of the curve of the number distribution density (open circles) have been calculated by means of Eq. (3.82). Note, that the c-values represent a discrete distribution function, while $w(M)$ and $g(M)$ are continuous distribution functions, (distribution densities). This also explains the difference of the units for these quantities.

Comparing the formulae (3.81), (3.83), and (3.82) explains, why the curves in Fig. 3.22 for the concentrations of the GPC-fractions, for the mass distribution density and for the number distribution density are similar over a logarithmic molar mass scale, but, respectively, shifted into the direction of smaller molar masses each.

It is seen, that the c_i-data are not sufficient to determine the complete curve of $g(M)$. A significant part of this density remains undetermined due to the lack of c-data at low-molar masses. Moreover, in the region, where $g(M)$ is to be calculated, the corresponding c-values are very low, and therefore show a considerable experimental error. This makes the calculation of the number distribution density from GPC-measurements rather tedious and uncertain. Though, if Eqs. (3.73) or (3.74) are used to calculate M_n and M_w from the GPC-data shown in Fig. 3.22, the correct values of 188 kg/mol and 386 kg/mol are obtained.

Recently, the GPC-equipment has been completed by a very useful tool which has enlarged its significance considerably. It has been coupled with a *multiangle laser light scattering* apparatus (*MALLS*), which offers the possibility to determine the absolute value of the weight average molar mass, $M_{w,\mathrm{LS}}$ for each fraction separately, as well as the square root of its mean square radius of gyration $\langle s^2\rangle^{0.5}$ evaluating the Zimm-diagram. This substitutes, partially the tedious calibration of the elution volume. Moreover, by separating a polymer in well defined fractions, and measuring $\langle s^2\rangle^{0.5}$ and $M_{w,\mathrm{LS}}$ simultaneously, the relationship between the *mean square radius of gyration* and the *molar mass* can directly be determined even in the case of non-theta solvents, where calculations are not available. The method also works for branched polymers.

In the case that a MALLS equipment is not available and for special polymers where well-defined narrowly distributed samples for setting up a calibration curve are difficult to obtain, the so-called *universal calibration* is applied making use of commercial polystyrene standards (cf. Fig. 3.21).

This method is based on the physical fact that the molecules are separated by the gel of the GPC-columns according to their hydrodynamic volume. Use is made of the *Mark-Houwink equation*

$$[\eta] = KM^a \tag{3.37}$$

with $[\eta]$ being the *intrinsic viscosity* and a and K quantities depending on the polymer, its solvent and the temperature of the viscosity measurement. Specific values of the Mark-Houwink equation are listed in Table 3.2, for example.

According to Eq. (3.45), $[\eta]$ is proportional to the coil volume and inversely proportional to the molar mass M. Therefore, at the same elution time or elution volume, respectively, the following relation holds between two species 1 and 2

$$[\eta]_1 M_1 = [\eta]_2 M_2$$

or with Eq. (3.37)

$$K_1 M_1^{a_1+1} = K_2 M_2^{a_2+1}$$

and therefore

$$M_1 = (K_1/K_2)^{1/(a_1+1)} M_2^{(a_2+1)/(a_1+1)} \tag{3.84}$$

For example, for the elution volume v_1 of species 1 with an unknown molar mass distribution the corresponding molar mass M_2 of polystyrene can be determined according to the calibration curve of Fig. 3.21. The molar mass M_1 of the unknown species follows then from Eq. (3.84) making use of the adequate parameters a_1, a_2, K_1, and K_2. In this way calibration curves for polymers based on the well established one of polystyrene can be obtained.

References

1. Flory PJ (1941) J Amer Chem Soc 65:372
2. Münster A (1953) Statistische Thermodynamik hochpolymerer Lösungen in Stuart HA Die Physik der Hochpolymeren Bd.2. Springer, Berlin
3. Schulz GV, Baumann H, Darskus R (1966) J Phys Chem B 70:3647
4. Schulz GV, Inagaki H, Kirste R (1960) Z F Phys Chemie 24:390
5. Joos G (1977) Lehrbuch der theor. Physik. Akad. Verlagsges, Wiesbaden
6. Kulicke WM, Kniewske R (1984) Rheol Acta 23:75
7. Daum U (1970) Centraal laboratorium TNO, Delft, private communication
8. Flory PJ (1953) Principles of polymer chemistry. Cornell University Press, New York
9. Kurata M, Tsunashima J (1999) Viscosity-molecular weight relationships and unperturbated dimensions of linear chain molecules. In: Brundrup J, Immergut EM, Grulke EA (eds) Polymer handbook, 4th edn. Wiley, New York
10. Bawn CEH, Freeman RFJ, Kamaliddin AR (1950) Trans Farad Soc 46:1107
11. Meyerhoff G (1955) Z Phys Chem 4:355
12. Meyerhoff G, Apelt B (1979) Macromolecules 12:968, (1980) 13:657
13. Inagaki H, Suzuki H, Fujii M, Matsuo T (1966) J Phys Chem 70:1718
14. Meyerhoff G, Schulz GV (1951) Makromol Chem 7:294
15. Cantow HJ, Schulz GV (1954) Z Physik Chem 2:117
16. Chinai SN, Matlak JD, Resnik AL, Samuels RJ (1955) J Pol Sci. 17:391
17. Fox TG (1962) Polymer 3:111
18. Kriegbaum WR, Flory PJ (1953) J Pol Sci 11:37
19. Freeman WR, Manning PP (1964) J Pol Sci A2:2017
20. Sato M, Koshiish Y, Asahina M (1963) J Pol Sci B1:233
21. Schulz GV, Hornnbach A (1959) Makromol Chem 29:93
22. Berry GC, Nomura H, Mayhan KD (1967) J Pol Sci A2(5):1
23. Kaufmann HS, Walsh EK (1957) J Pol Sci A2(26):124
24. Williamson GR, Cervenka A (1974) Eur Polym J 10:295
25. Kinsinger JB, Hughes RE (1959) J Phys Chem 63:2002
26. Kinsinger JB, Hughes RE (1963) J Phys Chem 67:1922
27. Burke JJ, Orfino TA (1969) J Pol Sci A2(7):1
28. Elias HG, Schumacher R (1964) Makromol Chem 76:23
29. Meyerhoff G (1961) Fortschr. Hochpol. Forschung 3:59
30. Hoffmann M (1957) Makromol Chemie 24(222):245

31. Einstein A (1906) Ann Phys 19:289; (1911) Ann Phys 34:591
32. Flory PJ (1949) J Chem Phys 17:303
33. Fox T, Flory PJ (1950) J Polymer Sci 5:745
34. Flory PJ (1969) Statistical mechanics of chain molecules. Wiley, New York
35. Elias HG (1984) Macromolecules 1. Plenum Press, New York
36. Casassa EF (1999) Particle scattering factors in Rayleigh scattering. In: Brundrup J, Immergut EM, Grulke EA (eds) Polymer handbook, 4th edn. Wiley, New York
37. Jolk W (1982) Diploma thesis, University Erlangen-Nürnberg
38. Wolf M (1982) Diploma thesis, University Erlangen-Nürnberg

Chapter 4
Structure and States of Polymers

In this chapter, a short description of the molecular structure of polymeric substances and their states at various temperatures is given. Polymers may be classified according to their application or to their molecular structure. Mechanical properties in the various structural states are addressed.

4.1 Classification of Polymeric Materials

Polymer materials are either totally amorphous or semicrystalline species in the temperature range of their applications. The morphological structures are of great importance for the end-use properties. A special feature in comparison to other materials is that amorphous and semicrystalline polymeric materials as well can be cross-linked up to various degrees. If in addition one considers the great number of variations with respect to chemical composition and molecular structure the immense scale of polymer species becomes evident. The two most important features for the understanding of the structural states of polymers are *crystallinity* and *cross-linking*. Based on these two properties a classification into the five basic types of polymers, summarized in Table 4.1, can be made.

Amorphous and semicrystalline uncross-linked polymers are used as plastics. Because of the absence of chemical cross-links they can be molten and may easily be processed in this state by extrusion, injection molding or fiber spinning. Both types of polymers are soluble in organic solvents. The amorphous materials can be dissolved in a great number of ingredients. The more difficult dissolution of the semicrystalline species has advantages and disadvantages. On the one hand they are more resistant against certain environmental influences than their amorphous counterparts, on the other they have to be handled at temperatures well above room temperatures if solutions are needed as in the cases of molecular characterization or solution spinning. Both species can repeatedly be molten and solidified by thermal treatments, and are therefore called *thermoplastics*.

Amorphous, weakly cross-linked polymers are used as *elastomers*. Because of the presence of chemical cross-links, they cannot be molten. In contact with

H. Münstedt and F. R. Schwarzl, *Deformation and Flow of Polymeric Materials*,
DOI: 10.1007/978-3-642-55409-4_4,

Table 4.1 Basic types of polymers, their processing, and some properties [1]

Structure	Melting behavior solubility	Processing	Type
Amorphous uncross-linked	Meltable easy to dissolve	Extrusion injection molding	Thermoplastic
Semicrystalline uncross-linked	Meltable difficult to dissolve	Extrusion, injection molding fiber spinning	Thermoplastic
Amorphous cross-linked	Not meltable, insoluble swellable	Shaping before vulcanization	Elastomer
Amorphous dense networks	Not meltable, insoluble hardly swellable	Shaping before hardening	Duromer
Semicrystalline cross-linked	Not meltable insoluble	Shaping before cross-linking	Cross-linked polymer

swelling agents, they may swell without dissolving. Processing of those materials is performed in a state at which chemical cross-links are not yet present. By the cross-linking process, which is called *vulcanization* in the case of rubbers, the shape is stabilized. This may be done also during injection molding. This process is then called *reaction injection molding* (RIM).

A very high cross-link density can be achieved by certain chemical reactions of oligomers. In such cases the formation of macromolecular chains and cross-links run in parallel and dense networks are formed. These materials are called *duromers*. Shaping occurs during the hardening process, after which duromers no longer can be molten, are insoluble, and resistant against a wide spectrum of ingredients. This property and low viscosity and good wettability of the oligomers before curing are the basic features for their importance as the polymeric component of modern fiber-reinforced composites.

Cross-linked semicrystalline polymers like polyethylenes have their largest applications in the field of cable jacketing and pipe manufacturing. The cross-linking is initiated after extrusion either by thermal treatment or by electron-beam irradiation.

In Table 4.2 some amorphous thermoplastics are listed, which have achieved larger practical importance. The first column of this table gives the official chemical designation as defined by various international standards committees [2], the second column the acronym, which is used in the literature. The third column gives the molar mass of the structural unit in g/mol and the fourth shows the structural unit.

PVC is an important mass product with a wide field of applications. Common PVC contains a large portion (about 65 %) of syndiotactic chain parts. Its properties are typical of an amorphous thermoplastic. However, sometimes a tendency to crystallization in the melt can be observed. The melting temperatures are below the usual processing temperatures and the crystallization rates are so low that crystallinity does not affect the transparency of items manufactured from PVC.

Table 4.2 Some amorphous thermoplastics [1]

Chemical name	Acronym	M_g, g/mol	Structural unit
Poly(vinyl chloride)	PVC	62.5	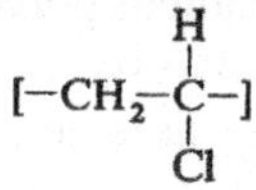
Poly(styrene)	PS	104.2	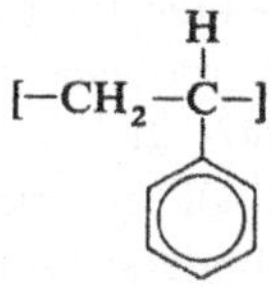
Poly(methyl methacrylate)	PMMA	100.1	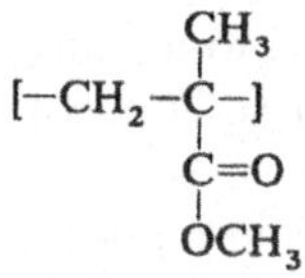
Poly(methyl acrylate)	PMA	86.1	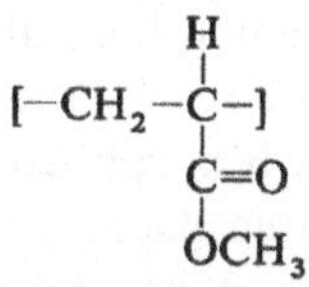
Poly(vinyl acetate)	PVAC	86.1	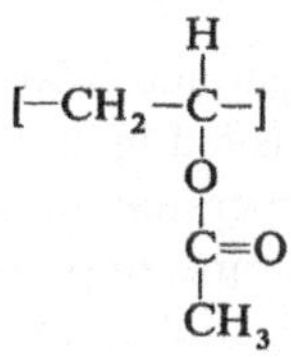
Poly(carbonate)	PC	254.3	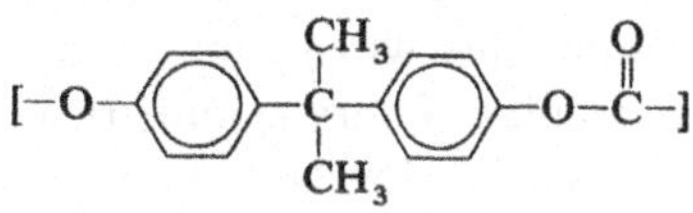
Poly(ether sulphone)	PES	232.1	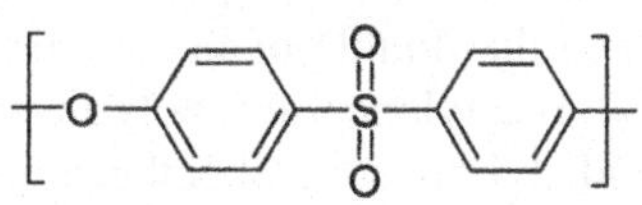

Nearly half of the PVC is used unmodified. Due to its high stiffness it is often called "*hard-PVC*" (PVC-U). PVC can be plasticized by adding phthalates or special esters of fatty acids. This "*plasticized PVC*" (PVC-P) has found a lot of applications as the glass temperature and consequently the viscosity can be varied in a rather wide range. Moreover, the modulus goes down but the impact strength can remarkably be increased. A disadvantage of these versatile materials is the gradual diffusion of the plasticizer out of a manufactured item and following from that the loss of flexibility and toughness.

PS produced by radical polymerization is an atactic product being amorphous. By stereospecific catalysts it is possible to produce isotactic or syndiotactic polystyrenes that are able to partly crystallize and show melting temperatures distinctly above 200 °C. However, the inferior mechanical properties of these materials have prevented their commercial success up to now. Common PS is a translucent brilliant product, especially appropriate for the manufacturing of injection molded parts. It is, however, very brittle and may break already when falling on a hard floor or being clamped too tight. Its low impact strength is overcome by copolymers on the base of styrene and butadiene (high impact polystyrene (HIPS)) or materials using particles of a polystyrene butadiene copolymer attached to a matrix of polyacrylonitrile by grafting ABS. The loss of transparency of these copolymers can be compensated for by reducing the butadiene domains to sizes smaller than the wavelengths of the visible light and suppressing its scattering by this modification. Such materials offer good impact strength combined with transparency.

A translucent amorphous thermoplastic with good mechanical properties is PMMA which, when polymerized under normal conditions, contains 60 % syndiotactic moieties. It is applied as a *polymeric glass*.

The excellent transparency of polycarbonate PC is the base for optical applications. Moreover, it has relatively high temperature resistance and excellent impact strength.

PES is transparent and belongs to the class of high temperature resistant polymers. It is distinguished by a glass transition temperature of 220 °C and also by a continuous use temperature of about 180 °C due to its excellent chemical stability.

Table 4.3 lists some important *semicrystalline thermoplastics*. Essential conditions for crystallization are the symmetry of the structural units and the strength of the interactive forces between different molecules. Structural units with large symmetries will fit into a crystal lattice if their side groups are not too bulky. Examples of very symmetrical molecules are the first four materials in Table 4.3.

Usually, for homopolymers, the *structural unit* is defined as the smallest repeating unit of the polymer chain. Exceptions to this rule are PE and PTFE, for which the double of the smallest repeating unit is defined as the structural unit.

Polyethylenes are polymers of the greatest economic importance. They are divided into several subgroups formally classified according to their density. HDPE stands for high-density polyethylene with densities between 0.935 and 0.960 g/cm^3 at ambient temperatures. The density allows some conclusions with respect to the structure of the polymer. HDPE reaches crystallinities up to 70 % due to the prevalent linear molecules that are able to form ordered segments. LDPE, the acronym for low density polyethylene, covers the density range adjacent to the lower values of the HDPE and further down to around 0.918 g/cm^3. The smaller densities point to a lower crystallinity due to the existence of *long-chain branches*. In case of *short-chain branches* that can reach up to six carbon atoms in commercial products one still speaks of linear materials, but as the side groups

Table 4.3 Some semicrystalline thermoplastics [1]

Chemical name	Acronym	M_g, g/mol	Structural unit
Poly(ethylene)	PE	28.1	$[-CH_2-CH_2-]$
Poly(tetrafluoroethylene)	PTFE	100.0	$[-CF_2-CF_2-]$
Poly(oxymethylene)	POM	30.0	$[-CH_2-O-]$
Poly(vinylidene chloride)	PVDC	96.9	$[-CH_2-CCl_2-]$
Poly(acrylonitrile)	PAN	53.1	$[-CH_2-CH(C{\equiv}N)-]$
Poly(propylene)	PP	42.1	$[-CH_2-CH(CH_3)-]$
Poly(vinyl alcohol)	PVAL	44.1	$[-CH_2-CH(OH)-]$
Poly(ethylene terephthalate)	PET	192.9	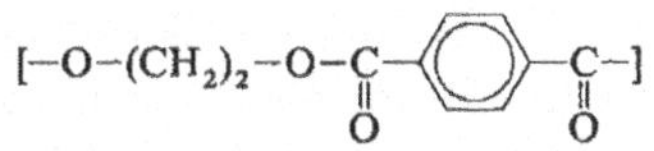
Poly(amide)	PA		
PA-6.6		226.3	$[-C(=O)-(CH_2)_4-C(=O)-N(H)-(CH_2)_6-N(H)-]$
PA-6.10		282.4	$[-C(=O)-(CH_2)_8-C(=O)-N(H)-(CH_2)_6-N(H)-]$
PA-6		113.2	$[-C(=O)-(CH_2)_5-N(H)-]$
Poly(phenylene sulfide)	PPS	108.1	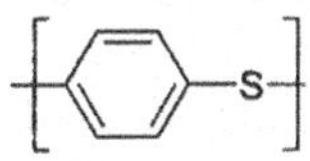
Poly(ether ether ketone)	PEEK	298.0	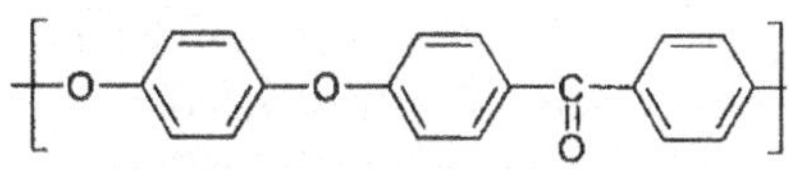

hinder a close packing of molecules the densities are smaller than those of HDPE. The densities are determined by the concentration of the comonomers built in and can reach values still lower than for the LDPE. According to their properties these products are called linear low density polyethylenes (LLDPE). HDPE and LLDPE are polymerized under relatively low pressures and moderate temperatures using catalysts, whereas LDPE is synthesized without any catalyst under pressures up to 3,000 bars and temperatures up to 300 °C. The main field of application for the various polyethylenes is the packaging sector, but tubes of different kinds and cable insulations are also their domains.

Most of the polypropylenes on the markets are linear products. Their moduli are larger than those of the polyethylenes and due to their smaller crystallinity and low crystallization rate the transparency can be rather good. Similar to polyethylenes of higher densities they are synthesized at moderate pressures and temperatures. Copolymers of propylene and ethylene are distinguished by a better impact strength than PP homopolymers and a higher modulus than PE. These two properties can be adjusted within certain limits by the ratio of the two components. Though PP has better temperature resistance than PE (it may be used up to temperatures of 120–130 °C), it is, however, less resistant against *UV*-light, as the hydrogen atoms bound to the tertiary C-atoms are the weak points for *UV-disintegration*. Big markets for such products are the automotive and appliance industries.

PTFE is a highly crystalline polymer for special purposes. It is chemically and especially thermally resistant, has a high melting temperature, low dielectric losses, high electric resistance, low friction coefficient, and good abrasion resistance. Its main applications are in the chemical, electrical, and electronics industries.

POM is a highly crystalline polymer. It has a special hardness and good form stability. It is used in technical products and replaces metals there.

A complete symmetry of the molecules is not unconditionally necessary for crystallization. If the molecule contains polar groups with strong interaction, even less symmetrical molecules will tend to crystallize. Examples are PA, PET, and PAN. The strongly polar groups are in these cases C=O, N–H, and C≡N.

PA and PET have a wide range of applications. Special grades are used for spinning of *synthetic fibers*. PET is the preferred material for soft-drink bottles. Although semicrystalline, this material is suitable for applications requiring some degree of transparency as the rate of crystallization is low enough to keep it in the amorphous state during cooling after processing. The modulus is sufficiently high to reach a satisfactory stiffness of items at wall thicknesses making the use of PET very economical. Furthermore, the excellent toughness prevents breakage even under large mechanical stresses.

PA is a versatile material as the chemical structure of its basic unit can be altered resulting in a change of properties. In Table 4.3 three modifications are presented. The numbers denote the C-atoms of the diamine and the dicarboxylacid, respectively. The low viscosities of polyamides are favorable for processing by *injection molding*; their good mechanical properties open up the widespread use in demanding technical applications. To enhance their moduli at higher temperatures, many grades of PA are filled with up to 50 wt. % short glass fibers. These fillers reduce the water uptake of the unmodified grades by several percent. Water improves the impact strength but is of disadvantage for the dimensional stability of manufactured items.

Polyphenylenesulfide (PPS) has a rather high modulus, particularly in combination with various fillers that are necessary to improve the inherent notch sensitivity. Due to the low viscosity of the basic resin even the compounds are suitable for injection molding. Continuous use temperatures up to 230 °C can be reached with special fillers. This property is the basis for its classification as a *high temperature resistant polymer*. Similar to other materials of this class, it possesses high resistance to inflammability and low smoke gas density. These properties have opened up demanding applications of PPS in the electrical, electronics, and appliance industries.

Polyaryletherketones are the top products among thermoplastic polymers. The largest market share has polyetheretherketone (PEEK) which is listed in Table 4.3. In spite of its high melting temperature of 330 °C, this material can be processed by common extruders or injection molding machines. Outstanding are the continuous use temperature of about 250 °C, high resistance against chemicals and mechanical abrasion, and low inflammability and smoke gas density. PEEK is used for items applied in harsh environments. Due to its excellent mechanical properties and stability against bodily fluids, PEEK has found its entrance into medical fields like bone replacement and orthopedic surgery.

Table 4.4 lists some polymers used for the preparation of *elastomers* (rubbers).

PIB may be prepared as an uncross-linked polymer with rather sharp molar mass distributions and widely different molar masses. Therefore, it has often been

Table 4.4 Some elastomers (rubbers) [1]

Chemical name	Acronym	M_g, g/mol	Structural unit
Poly(isobutene)	PIB	56.0	$[-CH_2-C(CH_3)_2-]$
Poly(dimethyl siloxane)	PDMS	74.1	$[-O-Si(CH_3)_2-]$
Poly(*cis*-isoprene) (natural rubber)	NR	68.0	$[-CH_2-C(CH_3)=CH-CH_2-]$
Poly(butadiene)	PB	54.0	$[-CH_2-CH=CH-CH_2-]$
Styrene-Butadiene	SBR	–	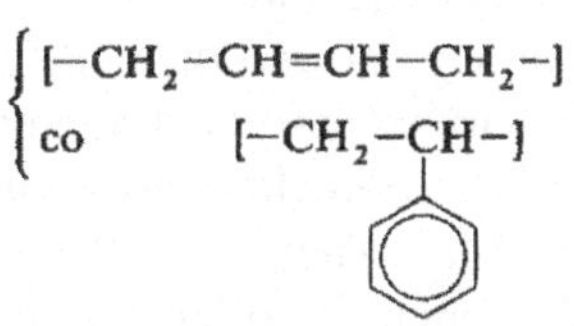
Poly(chloroprene)	CR	88.0	$[-CH_2-CH=CCl-CH_2-]$
Polyurethane rubber	PUR	–	$[-C(=O)-NH-R-NH-C(=O)-O-R'-O-]$

used for investigations into the influence of the molar mass on physical properties. In the uncross-linked state PIB is applied as adhesive and sealant. After copolymerizing isobutylene with small amounts (1–3 %) of isoprene, the product contains some double bonds and may be vulcanized with sulfur. In this way butyl rubber (BR) is obtained which is less sensitive against oxidation than natural rubber. Because of its small gas permeability butyl rubber is used for the production of gas hoses.

The rubbers with the largest economical significance are natural rubber (NR) and styrene-butadiene rubber (SBR). They are mainly used for the production of tires. Natural rubber is produced from poly(cis-isoprene), which is harvested as the *latex* of the hevea tree. It is cross-linked with sulfur (vulcanized). Some of the double bonds of the polyisoprene are replaced by sulfur bridges with length between 2 and 8 atoms. Lightly vulcanized NR contains about one sulfur bridge per 300 structural units of the polyisoprene molecule. NR is an amorphous, cross-linked polymer with *large elongations at rupture* and *large strengths* at break. Under elongation NR crystallizes. This process is mainly responsible for the large strength at break of this material.

The most important application of butadiene is its copolymer with 20 % styrene. This copolymer cross-linked with sulfur is the basis of *styrene-butadiene rubber* (SBR), which does not crystallize under stretching and is, therefore, much weaker than NR. For this reason, it is filled with carbon black for most applications. Nevertheless, more than half of all rubbers produced today are SBR, as it is cheaper.

From poly(chloroprene) (CR), a rubber is produced with large resistance against oils and fats. It is used for special purposes, e.g., sealings.

Polyurethanes (PUR) are versatile materials for many industrial applications. If cross-linked with trimethylolpropane and toluene diisocyanate (TDI) they form networks of adjustable mesh widths, which are decisive for mechanical properties of a manufactured item. Foaming and cross-linking in parallel lead to lightweight materials with a wide spectrum of applications.

All polymers listed in Table 4.4 emerge during the polymerization process as uncross-linked polymers and are vulcanized afterwards. In this way, mostly networks are formed with wide meshes leading to materials that are soft and strong at higher temperatures.

Among plastics, there are also materials with a molecular structure consisting of very narrow meshes. Examples are the *polycondensation products phenol formaldehyde resin* (PF), *urea formaldehyde resin* (UF), and *melamine formaldehyde resin* (MF). They are called *thermohardening polymers, thermosets,* or *duromers.*

Duromers are often applied in combination with fillers of different composition and used in the electrical industry. Epoxy resins (EP) and unsaturated polyester resins (UP) are of great technical importance as matrix materials for glass-fiber filled and carbon-fiber filled plastics. These materials have gained much importance because of their *high strength to weight ratio* and their *high stiffness to*

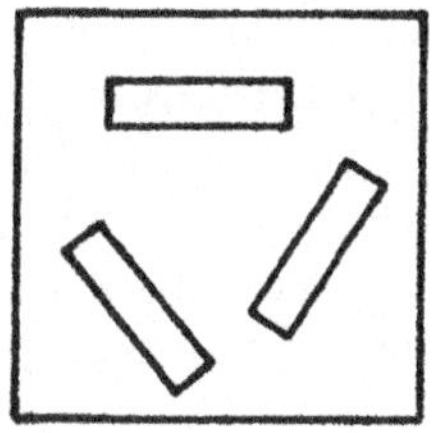

Fig. 4.1 Illustrating the concept of isotropy

weight ratio. They are on the way to replace more and more metallic constructional materials in the airplane and automobile industries. More detailed information on thermo hardening polymers can be found in [3].

By changing their states, many physical properties of polymers as electrical conductivity, dielectric behavior, heat conductivity, density, specific heat, and so on are altered. The strongest changes, however, occur in the mechanical deformation behavior. In order to describe these changes, some basic concepts are needed, which are introduced shortly in the following. A more exact definition with regard to the frequency dependence of moduli and loss tangent is given in Chap. 5.

We will consider the mechanical properties of *isotropic materials*. A material is called *isotropic* when its properties do not depend on the direction in which the specimen investigated was cut from the plate or block of the material.

An illustration is given in Fig. 4.1, which represents a plate of a polymer material from which three samples oriented in different directions were cut (or machined). If those specimens show identical properties the material is called isotropic with respect to this particular property.

During the following discussion, we use three characteristic quantities to describe the mechanical properties of materials, the *shear modulus*, the *compressive* or *bulk modulus,* and the *loss tangent*. The shear modulus is a measure of the resistance of a material against a change in shape at constant volume, the bulk modulus is a measure of the resistance of a material against a change in volume at constant shape, and the loss tangent is a measure for the dissipation of mechanical energy into heat during a deformation cycle.

For the definition of the shear modulus we consider Fig. 4.2. A small cube, oriented in parallel to a right-handed Cartesian coordinate system, is deformed into a rhomboid by shearing forces. The tangential forces act in x-direction on the two surfaces that are perpendicular to the y-direction. Their magnitude, divided by the

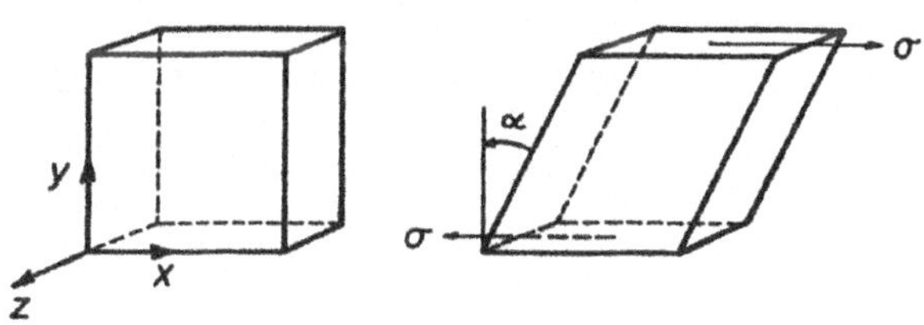

Fig. 4.2 Definition of simple shear and shear stress

surface area on which they act, is called the *tangential stress* or *shear stress* and is designated as σ.

The cube is deformed into a rhomboid with the same height and width. The edges, which were originally parallel to the y-axis, form an angle α with the y-axis after deformation. This angle is used as a measure for the magnitude of deformation and its tangent

$$\gamma = \tan\alpha \tag{4.1}$$

is called the *shear deformation* or *shear strain*. The type of deformation illustrated in Fig. 4.2 is designated as *simple shear*. The quotient σ/γ defines the *shear modulus*.

$$G = \sigma/\gamma \tag{4.2}$$

The definition of the *loss tangent* relates to an experiment with a harmonic simple shear strain:

$$\gamma(t) = \gamma_0 \sin\omega t \tag{4.3}$$

with γ_0 being the shear amplitude, $\omega = 2\pi v$ the *angular frequency*, v the *frequency* of the shear oscillation, and t the time. The shear stress is, under certain conditions, also a harmonic function of time and precedes the shear strain by a *phase angle* δ

$$\sigma(t) = \sigma_0 \sin(\omega t + \delta). \tag{4.4}$$

The tangent of δ is called the *loss tangent* and describes the ratio of the mechanical energy dissipated during one period of the oscillation to the maximum mechanical energy stored during one period.

For the definition of the *bulk modulus* we refer to Fig. 4.3. A cube with volume V_0 is deformed under a hydrostatic pressure p, acting on all six surfaces of the cube, into a smaller cube with volume $V_0 - \Delta V$.

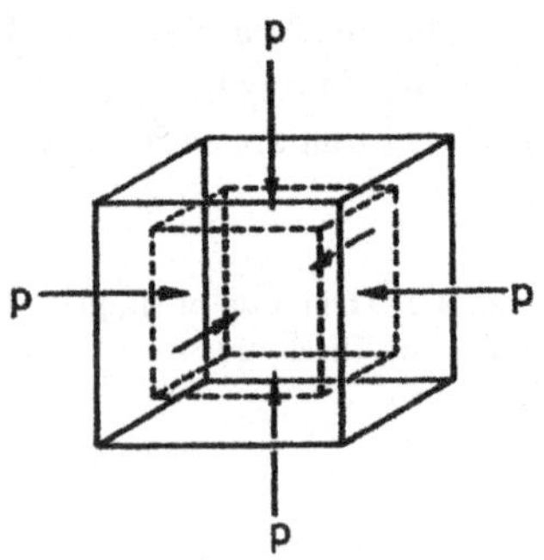

Fig. 4.3 Defintion of isotropic compression and bulk modulus

The *bulk modulus* or *compression modulus* is defined as

$$K = p/(\Delta V/V_0) \tag{4.5}$$

The type of deformation illustrated in Fig. 4.3 is designated as *isotropic compression*.

For *elastic* materials, G and K are independent of the duration of the acting stresses and of the frequency in case of an oscillatory experiment. The loss tangent is zero. Examples of almost elastic materials are glass, ceramics, and metals below their limits of plasticity. However:

> Polymers are *viscoelastic* materials, i.e., their moduli G, K, and their loss tangent tan δ depend on the frequency of the oscillation in case of an oscillatory experiment, or on the duration of the stresses applied.

For the following discussion of the relations between the states of polymers and the temperature dependence of their mechanical properties we therefore consider an oscillatory experiment with the fixed frequency of 1 Hz and discuss the temperature dependence of the quantities $G_{1\text{Hz}}(T)$, $K_{1\text{Hz}}(T)$, and tan $\delta_{1\text{Hz}}(T)$.

4.2 Molecular Structure of Amorphous Polymers

We look at the structure of amorphous uncross-linked polymers. In solution and in the melt the single macromolecule has the shape of a coil. We assume that this structure persists during the process of cooling and solidification. The apparent density of the macromolecule in the coil is small (compare Eq. 3.44 of Sect. 3.2) viz., of the order of 0.01 g/cm^3, while the density of the polymer in the solid state is about 1 g/cm^3 To reach this value, we have to assume that different coils penetrate each other. Consequently, entanglements have to occur between them. Chemical bonds between the different molecules do not exist. This consideration leads to the picture of Fig. 4.4 for the *molecular structure* of an *amorphous uncross-linked polymer*.

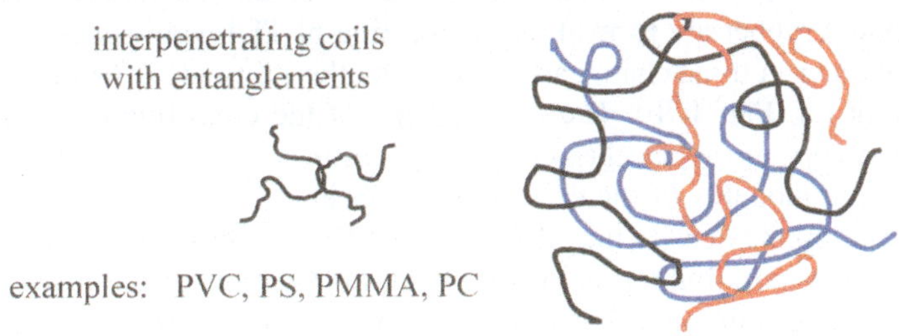

Fig. 4.4 Molecular structure of an amorphous uncross-linked thermoplastic

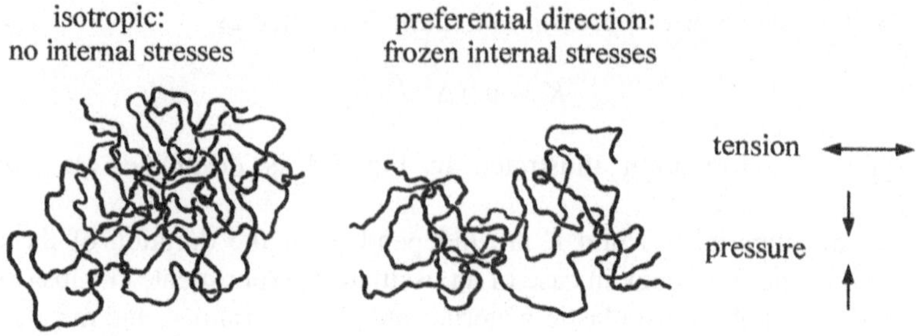

Fig. 4.5 Illustration of the distribution of preferential directions for the molecular segments and the occurrence of internal stresses

With respect to the *orientation* of the different molecular segments, the following remark should be made. If the polymer is isotropic, all directions of the molecular segments will occur with an equal probability. If however, preferential directions for the segments exist, the material is called *oriented* or *non-isotropic*; *stresses* are *frozen*, and specifically

tensile stresses	in the direction of the orientation and
compressive stresses	in the direction perpendicular to the orientation.

Whether a polymer contains frozen stresses depends on the conditions during the processing phase. The *degree of orientation* increases in the succession casting, pressing, extrusion, injection molding (Fig. 4.5).

If the polymer is heated above its glass transition temperature (compare Sect. 4.3), it will shrink in the direction of frozen internal stresses and extend in the perpendicular direction.

The *molecular structure* of an *amorphous cross-linked polymer* can be assumed to be similar to the structure of an amorphous thermoplastic with the difference that the various coils are connected by valence bonds between different chains, which are called *cross-links*. The distribution of the directions of the segments will be isotropic if the process of cross-linking has been performed in the isotropic non-oriented state.

The topography indicated in Fig. 4.6 corresponds to natural rubber, lightly cross-linked with sulfur. In this case, the short sulfur bridges connect the two halves of neighboring polymer molecules. This may be designated as a 4-functional cross-linking point, as indicated in Fig. 4.7a. Contrarily, the cross-links of poly(urethane) rubber are formed by the reaction of TDI with trimethylol propane (TMP) and connect three chain parts with each other, forming a 3-functional cross-linking point (cf. Fig. 4.7b). The *functionality* of the cross-linking points plays a part in the theory of rubber elasticity and determines together with the number of cross-links the modulus of the cross-linked rubber.

The cross-linking density may differ widely. For a lightly cross-linked rubber, it amounts to one cross-linking point for about 1,000 atoms of the main chain, and for a well hardened duromer it can achieve up to one cross-linking point for about 20 atoms of the main chain. Besides the cross-links, entanglements will exist. But

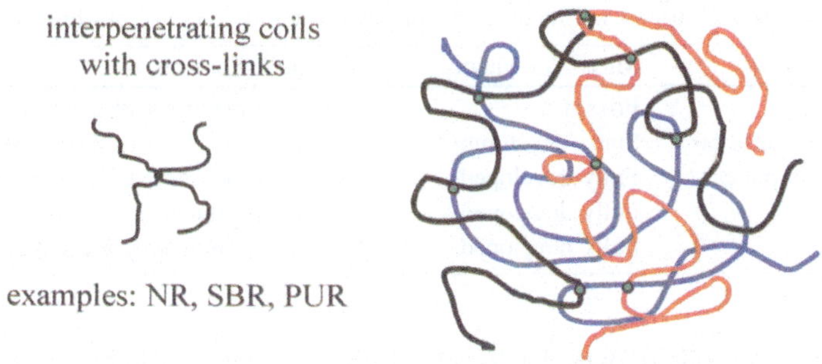

Fig. 4.6 Molecular structure of an amorphous lightly cross-linked elastomer

Fig. 4.7 Connections between different polymer molecules: **a** 4-functional cross-link (sulfur bridges in NR), **b** 3-functional cross-link (PUR cross-linked with TMP and TDI), **c** entanglement (uncross-linked polymer)

they play a significant role for the mechanical behavior only, if the number of cross-links is low.

Entanglements may be classified as 4-functional connection points (cf. Fig. 4.7c). They are not fixed at a position, but can slip along the entangled chains. In the rubbery plateau of the thermoplastics, entanglements play a role similar to that of cross-links for elastomers.

4.3 States of Order of Uncross-Linked Amorphous Polymers

Amorphous uncross-linked polymers show *three states of order* and *two transition regions* connecting them. They are listed in Table 4.5 together with a molecular interpretation.

At temperatures below the *glass transition temperature* T_g, the polymer is found in the *glassy state*. The micro-Brownian motion is frozen and no change in the shape of the molecular chains between adjacent entanglements occurs. The shape of the parts between entanglements is fixed. At very low temperatures the thermal motion only effects a vibration of the molecules around the equilibrium positions of valence bonds and angles. Additionally, vibrations around the

Table 4.5 States, transitions and molecular motions for amorphous uncross-linked polymers

State or transition	Micro-Brownian motion	Entanglements
Glassy state	Frozen	Have the effect of fixed network points
Glass-rubber transition	Starts developing	Have the effect of fixed network points
Rubber elastic plateau	Fully developed	Have the effect of fixed network points
Flow region	Fully developed	Are continuously formed and loosened
Melt	Fully developed	Are continuously formed and loosened

equilibrium positions of van der Waals bonds between neighboring parts of different macromolecules occur. No conformational changes or rotations of side groups occur.

Under the action of external forces, distances between and angles around equilibrium positions are changed and twisted elastically. At low temperatures this results into a *high value of the shear modulus*, a *high value for the strength* at *break* σ_b, a *small value* for the *elongation at break* ε_b, and a *low value* for the *loss tangent*

$$G \approx 3{-}4 \cdot 10^9\,\text{Pa}, \quad \tan\delta \approx 0.001{-}0.01$$
$$\varepsilon_b \approx 0.1{-}1\% \quad \sigma_b \approx 50 - 100\,\text{N/mm}^2(\text{MPa})$$

The material is hard and brittle.

At somewhat higher temperatures, but still in the glassy state, motions of small parts of the molecules may start and contribute to deformation. These consist of either motions of small parts of the main chain or of changes in conformation or rotations of side groups of the principle chain. These regions in the glassy state are called *secondary transition regions* as the moduli are not changed remarkably there. The higher the temperature, at which these motions occur, the larger are the parts of the molecules involved in them. The contribution of these motions to the deformation manifests itself in a *dispersion step* of the *shear modulus* decreasing to quantities between three quarters and one half of its original value and in a *maximum* of the *loss tangent* between 0.01 and 0.1. The occurrence, the temperature position, and height of these *secondary transitions* are material specific. As the material becomes tougher, after passing the secondary transitions, these effects have some technological significance. Polymers with pronounced secondary dispersions possess an improved impact resistance in comparison to those without secondary dispersions, which are mostly brittle. We consider this aspect in detail in Sect. 6.4.

In Fig. 4.8 we have sketched *thermal motions* and *deformation mechanism* in the *glassy state*.

Around the *glass transition temperature* T_g we find a temperature region of 30–60 °K, the so-called *glass-rubber transition*, in which the *micro-Brownian* motion starts to develop. This region is also called *softening region* or *freezing*

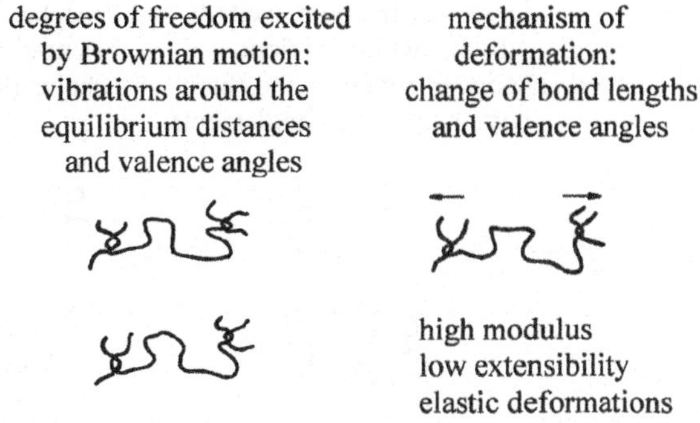

Fig. 4.8 Degrees of freedom and mechanism of deformation in the glassy state

region, depending on the direction in which it is passed. For the development of micro-Brownian motion, two conditions have to be fulfilled

(1) The internal rotation around the C–C bonds of the principle chain should be possible.
(2) There should be enough free volume next to the principle chain to enable the segments to rotate.

In the glassy state, at least one of both conditions is not fulfilled. Recent investigations support the opinion that in the glassy state the lack in free volume hinders the rotation of the single segments. Their rotation around the internal hindering potentials of the C–C bonds would be possible, if there were enough space.

As seen from Fig. 4.11 (on page 94), the changes in mechanical properties by passing the glass-rubber transition are enormous. The shear modulus decreases to a value of about 1/1,000 to 1/10,000 of that in the glassy state, the loss tangent runs through a broad maximum with values between 1 and 7. The *bulk modulus*, on the other hand, decreases only relatively weakly, showing a dispersion step of a factor between 2 and 3. The elongation at break passes through a maximum, and the strength at break decreases to about one-tenth of its value in the glassy state. In the softening region the relaxation phenomena are remarkable. The moduli depend strongly on the duration of the applied stresses or on the frequency in case of an oscillatory measurement, and on the temperature. Therefore, the materials are not appropriate for use in the glass transition region.

After the softening region, the *rubber-elastic plateau* follows. There, the micro-Brownian motion is fully developed. The parts of the chains between adjacent still fixed entanglements change their shape continuously under the thermally induced motion. The entanglements, however, are not yet loosened. Only the chain parts between entanglements are able to move (cf. Fig. 4.9).

Under the influence of forces, the parts of the chains between adjacent entanglements are oriented. As the oriented state of these chains is less probable than their isotropic state, the reduction in entropy yields repulsive forces which are proportional to the absolute temperature T. From the theory of rubber elasticity based on chain statistics the shear modulus in the rubber-elastic plateau follows as:

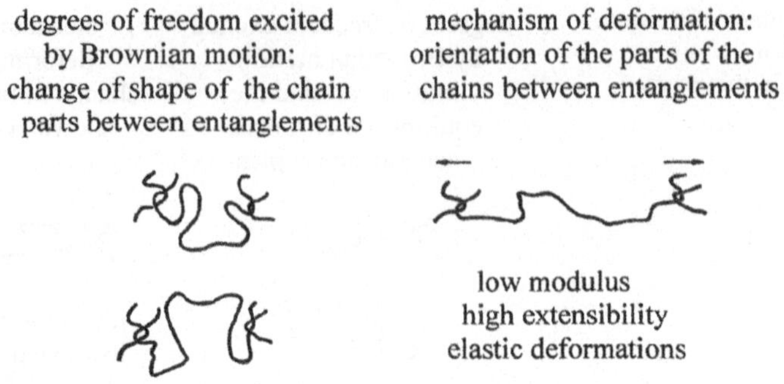

Fig. 4.9 Degrees of freedom and mechanism of deformation in the rubbery plateau

$$G \approx n_e \cdot kT = N_e \cdot \rho kT = \rho RT/2M_e \tag{4.6}$$

with n_e being the *number of entanglements per unit volume*, ρ the density of the polymer, N_e the *number of entanglements per unit mass* of the polymer, and M_e the *average molar mass* of the parts of the *chain between adjacent entanglements*. For the second part of Eq. (4.6) we used the fact that for entanglements (4-functional movable cross-links) the number of chain parts is just the double of the number of entanglements.

The value of the shear modulus in the *rubbery plateau* is of the order of $G \approx 10^5$ Pa. In contrast to Eq. (4.6) it decreases with increasing temperature. This experimental result is explained by a gradual decrease of the number of effective entanglements with increasing temperature. For cross-linked elastomers, however, the number of cross-links does not change with temperature and the resulting shear modulus increases proportional to the absolute temperature as expected from formula (4.6). The loss tangent runs through a minimum at the beginning of the rubber-elastic plateau with values that depend on molar mass and molar mass distribution. The extensibility at rupture is high.

At the *flow temperature* T_f the thermally induced motion becomes so strong that many entanglements disappear and the polymer changes to the *molten state*. There, entanglements are periodically destroyed and formed again by the thermal motion. That means, each entanglement possesses a finite lifetime, which decreases with increasing temperature. Under the influence of stresses, disentangled molecules may change their spatial position and shift with their center of gravity into the direction of the external forces. Such behavior will lead to irreversible flow together with the setup of elastic stresses from molecules that have been moved, but not yet slipped out of their entanglements completely. This is indicated in Fig. 4.10.

Under external stresses, polymer molecules will be uncoiled and oriented. The orientation is the source of elastic stresses in the melt. The flow of polymer melts will therefore always be connected with the occurrence of elastic stresses. The

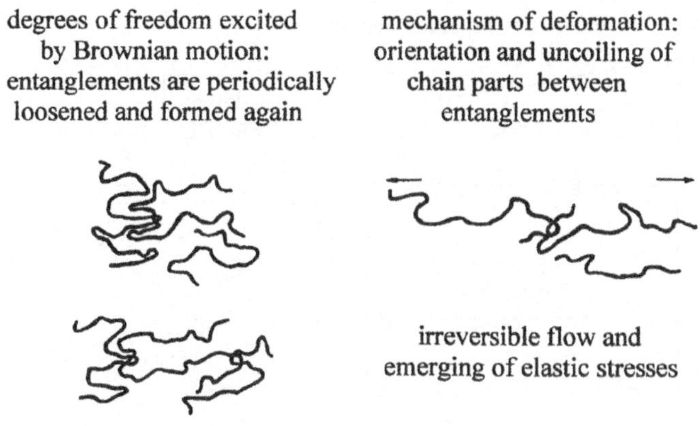

Fig. 4.10 Degrees of freedom and mechanism of deformation in the molten state

flow of polymer melts is *viscoelastic*, part of the deformation energy is dissipated as heat and part of it is stored by the orientation of the chain segments.

In the molten state, the shear modulus decreases steeply with increasing temperature. The proper quantity to describe the rheological behavior of a polymer melt is the viscosity, which is of the order of $\eta \approx 10^4$ Pa s, and also decreases with increasing temperature. The loss tangent goes up with temperature in the molten state and reaches values between 1 and 20.

A survey of the shear modulus, the bulk modulus and the loss tangent for an uncross-linked amorphous polymer, is given in Fig. 4.11. The moduli and loss tangent are plotted logarithmically versus a linear temperature scale.

The bulk modulus does not exhibit such steep transition steps as the shear modulus. Indeed, also the bulk modulus shows small transitions in the regions, where the transitions in shear occur, but they remain much smaller, and the compression modulus stays within the limits

$$10^9\ \text{Pa} < K < 10^{10}\ \text{Pa} \tag{4.7}$$

in the entire temperature region covered in Fig. 4.11. It follows that for all temperatures above the glass transition temperature $G \ll K$, i.e., the thermoplastic behaves there like an *incompressible* material. If shear stresses and compressive stresses are of the same order of magnitude, the changes in volume will be about 1000 times smaller than the changes in shape and may be neglected.

Two other quantities are important for the description of the deformation behavior of isotropic materials, the *modulus of elasticity,* also called *Young's modulus E,* and *Poisson's ratio* μ. They are defined by simple extension in one direction under the action of a tensile stress. E is defined as the ratio of the tensile stress to the longitudinal strain, μ as the ratio of the lateral contraction of the sample to its longitudinal deformation. If two of the four quantities E, G, K, μ are known, the other two may be calculated.

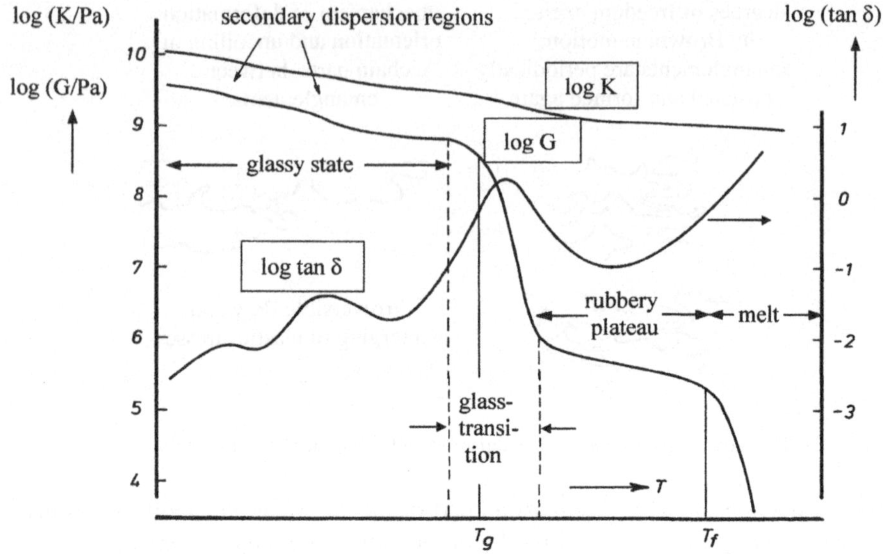

Fig. 4.11 Logarithmic plot of the shear modulus G, the bulk modulus K, and the loss tangent tan δ, at a constant frequency of 1 Hz versus the temperature T for an uncross-linked amorphous polymer (schematically). The positions of the secondary dispersion regions, the glass transition temperature T_g and the flow temperature T_f are indicated

If G and K are known as functions of temperature, the temperature dependence of the other two is easily given. The course of E with T is similar to that of G with T. In the glassy state $E < 3G$. At the glass transition temperature E approaches 3 times G and keeps this value in the rubbery plateau and in the melt. Poisson's ratio is about 1/3 at low temperatures, grows with increasing temperature, and abruptly attains 0.5 at $T = T_g$.

4.4 Influence of Molar Mass and Cross-Linking Density

In the glassy state ($T < T_g$), cross-linking density and molar mass only weakly influence G, K, E, and tan δ of amorphous polymers. These parameters do not affect the glass transition temperature except in cases of low molar masses or of high cross-linking densities.

For $T > T_g$, however, the influence of cross-linking density and molar mass on the temperature dependence of the shear modulus is significant, as demonstrated in Fig. 4.12. The effect of the molar mass for uncross-linked polymers becomes obvious from the three curves designated by (1), (2), and (3) corresponding to a decreasing order of the molar mass. The larger the molar mass, the more

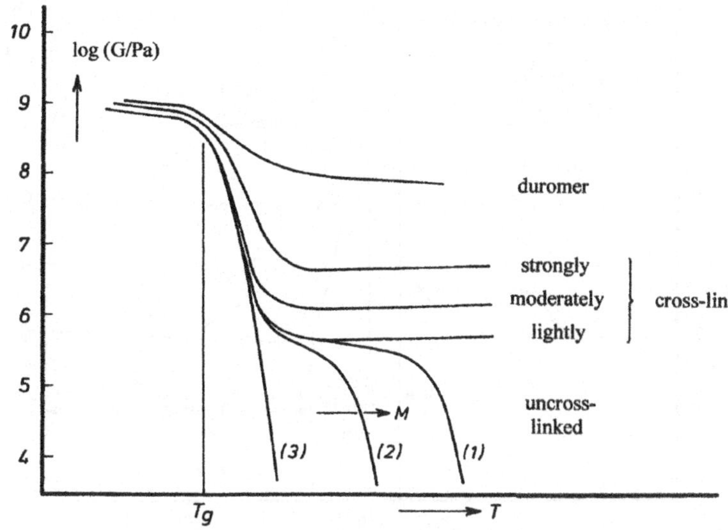

Fig. 4.12 Logarithm of the shear modulus as a function of temperature for cross-linked and uncross-linked amorphous polymers and for duromers

entanglements a single molecule contains. On average, the *number of entanglements per molecule* N_0 will be proportional to its molar mass M

$$N_0 = M/M_e \tag{4.8}$$

M_e the average *molar mass between two adjacent entanglements* does not depend, in a first approximation, on the molar mass of the molecules. The larger the number N_0, the more difficult it will be for a complete molecule to get rid of all the restrictions by the entanglements, and the higher becomes the value of the flow temperature T_f. The extension of the rubber-elastic plateau will therefore increase with the number of entanglements per molecule N_0. For the special case $M \leq M_e$ of curve (3), a rubber-elastic plateau will not exist at all and the modulus goes down directly from the glassy level to very low values of the molten material. Curve (1) designates a material with a high molar mass and curve (2) one with a moderate molar mass. M_e is called *entanglement molar mass*, and has been found for different polymers between $M_e \approx 1$ and 20 kg/mol.

For *cross-linked polymers* (elastomers), cross-links connect the molecules with each other, besides entanglements. As the cross-links are not loosened even at very high temperatures, the rubber-elastic state persists up to the temperature at which a chemical decomposition of the molecular structure starts to occur. Consequently, neither a flow temperature nor a molten state is observed for elastomers. Therefore, one speaks of a *rubber-elastic state* and not of a *rubber-elastic plateau* as in the case of thermoplastics.

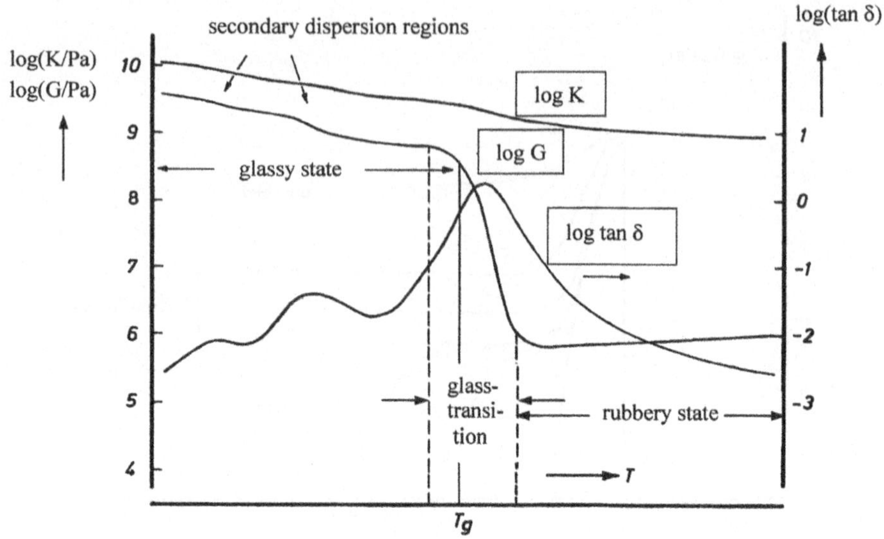

Fig. 4.13 Logarithm of the shear modulus *G*, the bulk modulus *K* and the loss tangent tan *δ* at a constant frequency of 1 Hz, versus the temperature *T* for a lightly cross-linked amorphous polymer (elastomer), schematically. The positions of the secondary dispersion regions and the glass transition temperature T_g are indicated

For the rubber-elastic state the shear modulus becomes

$$G \cong n_c \cdot kT \tag{4.9}$$

with n_c being the *number of cross-linking points per unit volume*. Consequently, *G* depends strongly on the cross-linking density in the rubber-elastic state and increases proportional to the absolute temperature *T*.

The case of very high degrees of cross-linking is represented by the *duromers*, which show a small dispersion step at their glass temperature only and high modulus in their "rubbery state."

Figure 4.13 displays schematically the shear modulus, bulk modulus, and loss tangent as a function of the temperature for a lightly cross-linked amorphous polymer. Below the glass transition, the temperature dependence of these quantities is principally the same as in Fig. 4.11. Above the glass transition in the rubber-elastic state, the shear modulus increases proportional to the absolute temperature. The loss tangent passes through a high maximum at the glass-rubber transition and then, decreases with increasing temperature to very low values (down to 0.01).

The difference between an amorphous thermoplastic and an elastomer is presented in Fig. 4.14. Instead of the shear modulus as a function of temperature Young's modulus *E* is plotted. For a thermoplastic, the temperature region of usage is in its glassy state, for an elastomer in its rubber-elastic state. Consequently, for thermoplastics, the glass transition temperature forms the upper limit

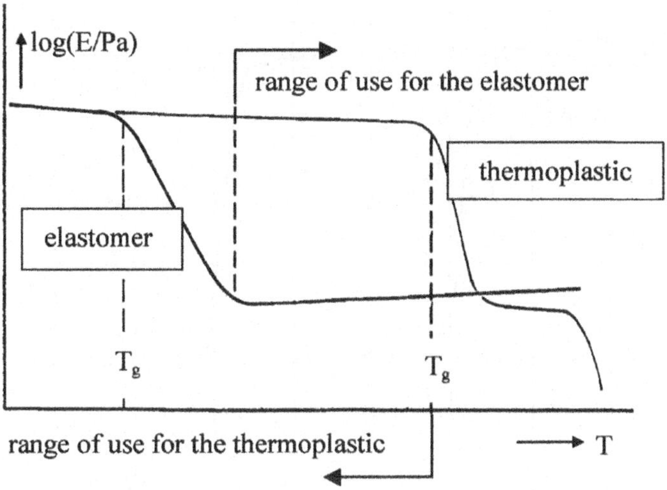

Fig. 4.14 Logarithm of Young's modulus as a function of temperature *T*, for an amorphous thermoplastic and an elastomer (schematically). T_g represents the upper temperature limit for the use of an amorphous thermoplastic and the end of the glass-rubber transition represents the lower temperature limit for the use of an elastomer

of the usage temperature region, for elastomers, the end of the glass transition region is the lower temperature limit for its usage. Each thermoplastic reaches its rubbery plateau region at high temperatures at which it loses the stiffness necessary for applications. Each rubber attains its glassy state at low temperatures at which it cannot be used as a rubber any longer.

Figures 4.15 and 4.16 show measurements of various materials at a fixed frequency of 0.64 Hz. Figure 4.15 displays the shear moduli of two elastomers and three amorphous thermoplastics as a function of temperature and Fig. 4.16 the loss tangent of the same materials. All three thermoplastics exhibit a secondary dispersion region, PS a weak one at low temperatures, PVC at −60 °C, and PMMA at +30 °C. The latter two are responsible for the good impact resistance of PVC and PMMA. The glass transition temperatures may be recognized by the sharp transition steps in the moduli and by the high maxima of the loss tangent. Their temperature positions increase in the order of NR, PUR, PVC, PS, and PMMA.

As already mentioned, the behavior in the rubber-elastic plateau and in the melt does not only depend on the chemical nature of the polymer, but also on its molar mass and molar mass distribution. The measurements in the rubbery plateau of PMMA shown in Figs. 4.15 and 4.16 were performed by Masuda et al. [4] on a sample with $M_w = 350$ kg/mol and $M_w/M_n = 1.5$. This material shows a significant and broad rubber-elastic plateau. The PS is the commercial product Styron 678 with $M_w = 240$ kg/mol and $M_w/M_n = 2.5$. Its rubber-elastic plateau is much less pronounced than that of the PMMA.

Concerning the loss tangent shown in Fig. 4.16, it should be mentioned that the different maxima of an amorphous polymer are generally *designated* with *Greek*

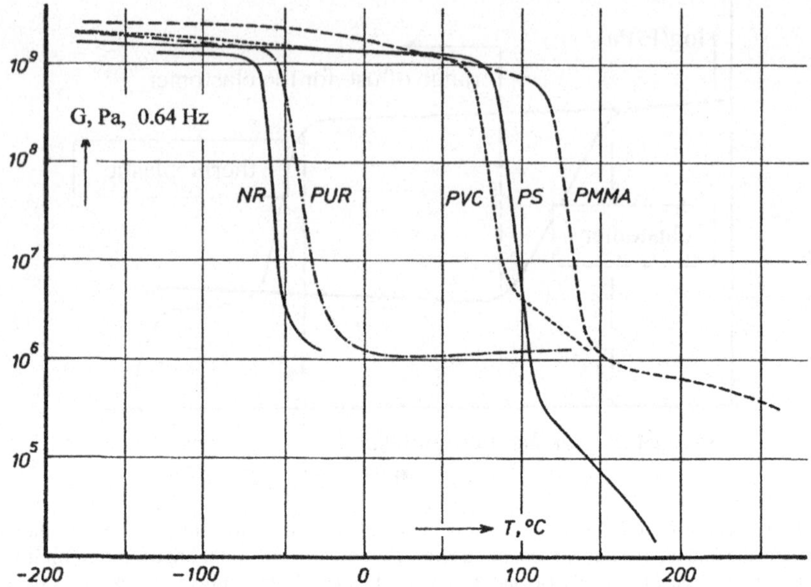

Fig. 4.15 Shear modulus G at 0.64 Hz as a function of temperature T for two elastomers and three amorphous thermoplastics [1]

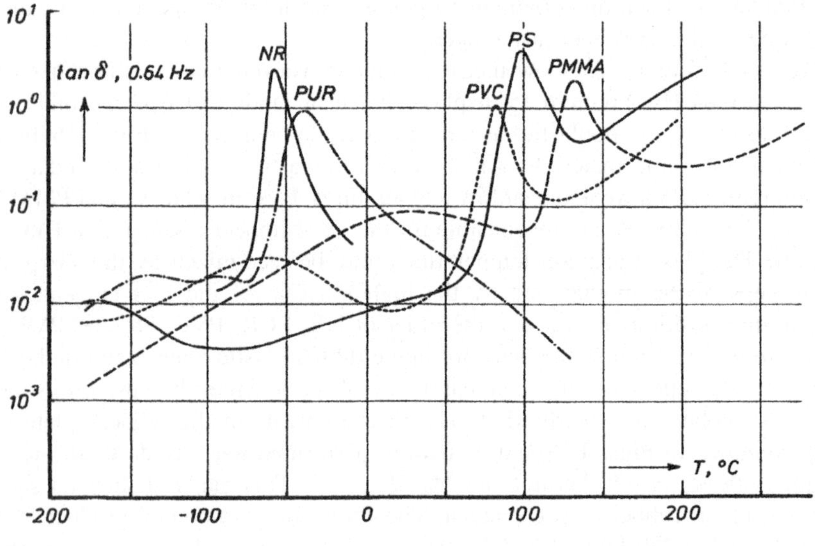

Fig. 4.16 Loss tangent tan δ at 0.64 Hz as a function of temperature T for two elastomers and three amorphous thermoplastics [1]

letters. Starting with the maximum at the glass transition temperature which is designated as the *α-maximum*, the further maxima are numbered as β-, γ-, δ-*maximum* in the order of decreasing temperature. The secondary maxima shown in Fig. 4.16 are known as the β-maximum of PMMA, the β-maximum of PVC, etc. This procedure has been generally adopted. It is of great advantage for describing and identifying relaxation processes in the literature, which are not yet interpreted molecularly. By this practice it is possible to speak of and discuss a molecular process, which still lacks molecular interpretation. Also, the corresponding molecular process is often named by the same letter. Thus, one speaks about the *α-process* of an *amorphous polymer*, meaning the molecular causes for its *glass-rubber transition*.

4.5 Semicrystalline Polymers

4.5.1 Features of Crystallinity

Crystallinity is distinguished by a special order of atoms or molecules that can be transformed into a disordered state by heat. This process is called melting. Metals are prototypes of crystalline materials. In the field of polymers, crystalline species often show a distinct glass transition in addition to the melting point, i.e., they contain amorphous parts and are called, therefore, *semicrystalline*. An interesting but not easy task is to get an insight into the structure of the ordered molecular regions. Even an evident quantity like the *degree of crystallinity* is not easy to determine with high accuracy as amorphous and crystalline regions can penetrate each other.

The simplest assumption is, the so-called *two phase model*, i.e., the mass m of the polymer is additively composed of the mass of the amorphous portion m_a and the mass of the crystalline portion m_c.

$$m = m_a + m_c \tag{4.10}$$

For the volumes, a corresponding relation

$$V = V_a + V_c \tag{4.11}$$

is assumed to hold. From this model two different degrees of crystallinity may be defined. One is based on the *mass* of the crystalline portion related to the total mass, i.e.,

$$x_m = m_c/m \tag{4.12}$$

and the other on the *volume* of the crystalline phase related to the total volume

$$x_v = V_c/V \tag{4.13}$$

Introducing the specific volume v of the bulk material and v_c and v_a as the specific volumes of the crystalline and amorphous phases, respectively, Eq. (4.11) reads using Eq. (4.10)

$$mv = m_a v_a + m_c v_c = (m - m_c)v_a + m_c v_c \tag{4.14}$$

or

$$x_m = m_c/m = (v_a - v)/(v_a - v_c) \tag{4.15}$$

A corresponding relation exists between the volume-based degree of crystallinity x_v and the densities ρ, ρ_c, ρ_a of the bulk, the crystalline and the amorphous phases, namely

$$x_v = x_m \frac{v_c}{v} = (\rho - \rho_a)/(\rho_c - \rho_a) \tag{4.16}$$

The two measures of crystallinity differ. x_m/x_v becomes the higher, the larger the ratio between the density ρ_c of the crystalline phase and the density of the sample.

A helpful tool for getting an insight into the degree of crystallinity is the classical small-angle X-ray scattering technique (SAXS). A typical schematic diagram of a semicrystalline polymer is shown in Fig. 4.17. The peaks indicate a molecular order, the non-structured region below them is due to the scatter of the amorphous regions. The so-called amorphous halo is determined from a totally amorphous sample that may be obtained by a fast quenching from the melt.

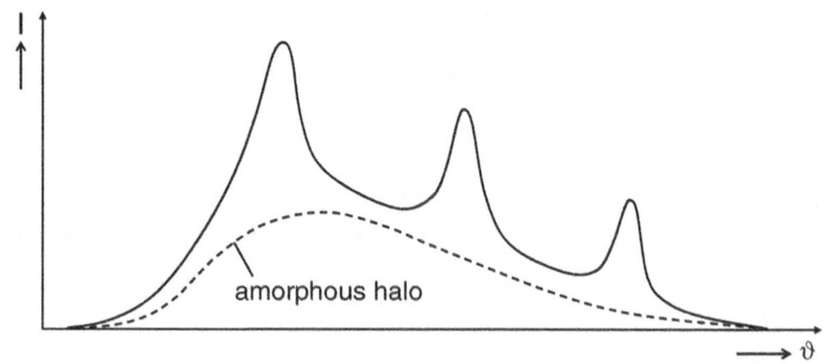

Fig. 4.17 Schematics of a small-angle X-ray scattering diagram of a semicrystalline polymer. I is the intensity and ϑ the scattering angle

Subtracting the area below the curve of the amorphous halo from that of the entire curve and relating this difference to the total area gives a value for the crystallinity. Due to the fact that the crystallinity follows from the areas under the measured curves and no calibration is necessary, SAXS is a direct method to determine the crystallinity of a material, which is designated as the *X-ray crystallinity* x_r.

In contrast to SAXS, the easy and often applied method to obtain the crystallinity from differential scanning calorimetry (DSC) is an indirect one. The amount of crystallinity is proportional to the melting enthalpy obtained from a DSC diagram as shown in Fig. 4.22. To draw any quantitative conclusions the enthalpy has to be related to the amount of crystallinity. For such a calibration SAXS could be used.

As Fig. 4.18 shows for polyethylenes, there exists a linear relationship between the crystallinity x_r measured by X-ray scattering and the specific volume v, which can be described by the equation

$$x_r = (v_a - v)/(v_a - v_c) \tag{4.17}$$

whose right-hand side is identical to that of Eq. (4.15).

This finding has two consequences:

1. The X-ray crystallinity x_r can be identified with the degree of crystallinity based on the mass portions at least in the case of the polyethylenes investigated.
2. The simple measurement of a specific volume allows quantitative conclusions with respect to the crystallinity of a sample, which decisively determines many mechanical properties. That is the reason for the important role of specific volume or density in specifications of semicrystalline polymers.

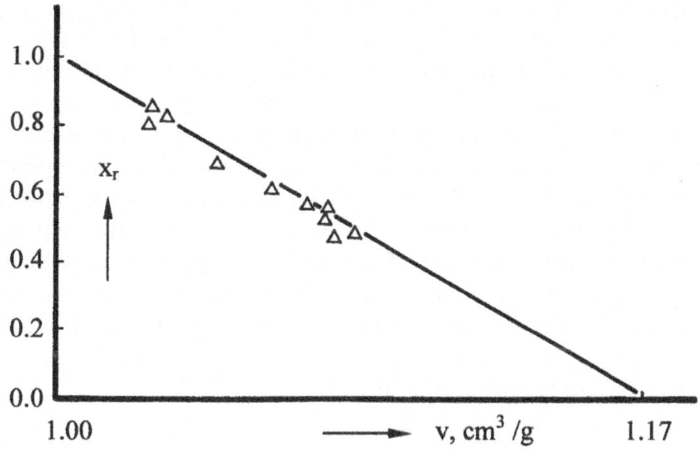

Fig. 4.18 Relationship between the crystallinity x_r determined by SAXS and the specific volume v for various polyethylenes [5]

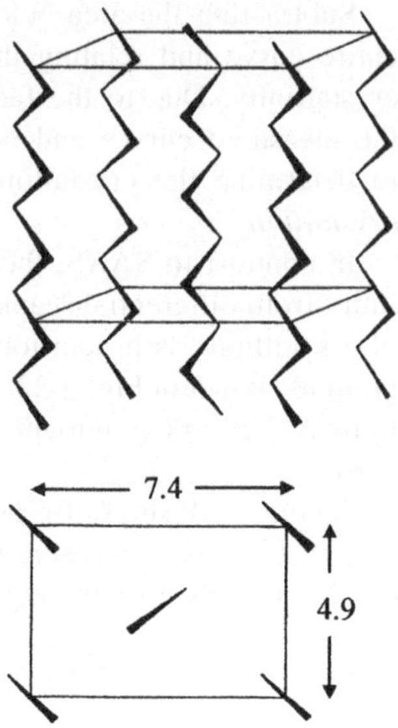

Fig. 4.19 Unit cell of ordered polyethylene chains (the numbers give the lengths in Angström) [6]

In detail, it follows from Fig. 4.18 that $v_c = 1\ \mathrm{cm^3/g}$ and $v_a = 1.17\ \mathrm{cm^3/g}$. Particularly, the extrapolated specific volume of the crystalline phase is a quantity that can be used for checking the compatibility of structural parameters for polyethylene. Already in 1939, the parameters of the unit cell of crystalline polyethylene presented in Fig. 4.19 were determined by X-ray scattering [6]. Five chains that lie parallel to each other belong to one unit cell. Four chains form the vertices of a rectangle, the fifth lies in its center and is shifted by one carbon bond. From the distances between the chains the volume of the unit cell can easily be determined and from the molar masses of the chain segments its mass follows. The resulting specific volume of 0.99 $\mathrm{cm^3/g}$ is in excellent agreement with the extrapolated value of 1.00 $\mathrm{cm^3/g}$ in Fig. 4.18 for the crystalline phase.

As the mobility of polymer segments is different in the amorphous and crystalline regions of a material, methods that are able to measure this property can be used, in principle, to get insight into its crystallinity. *Raman-spectroscopy*, e.g., [7] and *nuclear magnetic resonance*, e.g., [8] have been used for this purpose.

As the methods applied for the determination of the degree of crystallinity are based on different physical properties and as the morphology may not be uniform across a sample and between different specimens, it is evident that the crystallinity belongs to the quantities that can be determined with a limited accuracy, only. Nevertheless, even approximate values are frequently discussed as they allow some conclusions with respect to the structure of a polymeric material and the assessment of certain application properties.

Besides the degree of crystallinity, details on the structure of the crystalline regions are of interest. Investigations of this kind are rather complex, mainly because of two reasons. First, the expected dimensions of the crystallites are so small that transmission electron microscopy (TEM) has to be applied and,

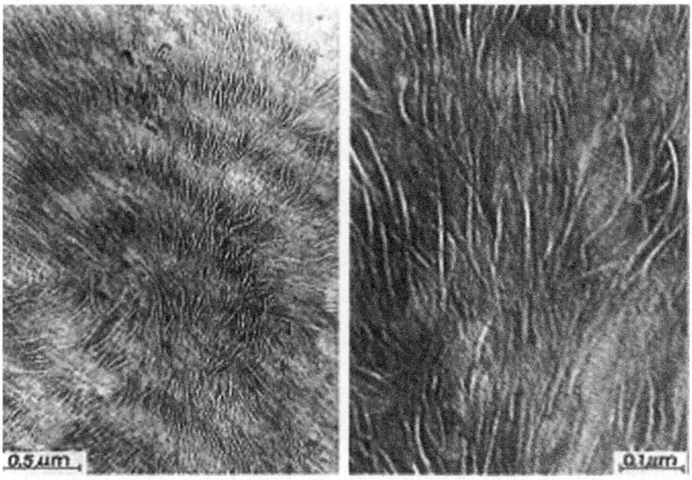

Fig. 4.20 TEM pictures of selectively stained ultrathin sections of an LDPE [9] (By permission of Carl Hanser Verlag, München)

therefore ultrathin cuts have to be available. Second, as the amorphous and crystalline regions consist of the same kind of atoms their interactions with the electrons of the beam are indistinguishable and differences between various regimes of order are not obvious. Staining with osmiumtetroxide, for example, is a well-known technique to generate contrasts within polymeric materials but it needs a lot of experience to make sure that the microscopic images of treated organic materials reflect the reality. In the case of polyethylenes it is assumed that the osmiumtetroxide attacks the amorphous regions, only, but leaves the crystalline ones unaffected. To what degree this assumption is fulfilled and how the regions have to be assessed which do consist of imperfect crystals are still questions of actual relevance.

Figure 4.20 shows TEM micrographs of selectively stained ultrathin sections of an LDPE in two different magnifications that were taken from [9]. The bright areas are interpreted as the crystalline regions not affected by the staining agent. They are called lamella. Their thickness is around 10 nm and lengths up to 1,000 nm can be found. As the left picture demonstrates, the lamellas are arranged in some order. Starting from a center they expand like a fan. This supramolecular feature, which can be observed for many semicrystalline polymers, is called *spherulites*. They are not discussed in detail here. More information on spherulites and their role for the deformation of semicrystalline polymers can be found in [9].

The question of how a lamella is built up does not seem to be unambiguously answered up to now. Two models are still under discussion. The so-called bundle-, fibril-, or fringe-model is based on the parallel arrangement of segments of different macromolecules (cf. left part of Fig. 4.21). The other model postulates sharp folds of the macromolecules leading to parallel parts of the same molecule as sketched in the right part of Fig. 4.21. So-called tie molecules may be able to

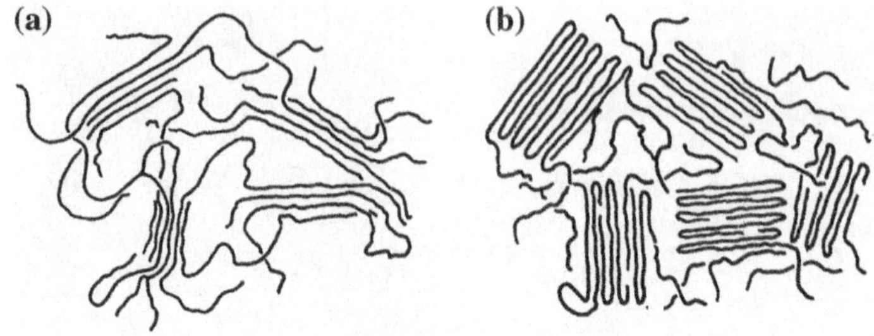

Fig. 4.21 Schematic pictures of the lamellar structures. **a** Bundle-, fibril- or fringe-model, **b** Chain folding

connect ordered regions. None of these two models is able, however, to describe all the features found for semicrystalline polymers. The majority of recent investigations assigns the folded chain to polymer molecules crystallized from the quiescent melt. Polyethylenes and polypropylenes are typical examples. Fibril structures are found for more rigid or highly extended macromolecules.

4.5.2 States of Order

The fact that semicrystalline polymers consist of an amorphous and a crystalline phase leads to a greater number of states than in the case of amorphous materials. In calorimetric measurements a melting peak is found in addition to the step in heat flow connected with the glass transition.

In Fig. 4.22 an example of such a measurement is presented for a polyethyleneterephthalate (PET) in the temperature range from −150 to 300 °C. At around 80 °C a distinct step of the heat flow is found that can be related to the glass-transition. Its value is listed in Table 4.7. The endothermic peak starting above 200 °C indicates the melting process of the crystallites. The peak temperature is denominated as the melting temperature T_m. Its value obtained under the experimental conditions is given in Table 4.7. Between T_g and T_m a flat exothermic maximum around 170 °C is found which is due to a recrystallization process taking place (designated as T_{rc}). From its occurrence it can be concluded that the crystallization rate of the PET investigated is smaller than the chosen cooling rate of 10 K/min of the crystallization process following the first heating run.

The range between the two characteristic temperatures T_g and T_m is of particular importance for properties of semicrystalline polymers. Above T_g the molecules of the amorphous region attain high mobility, but due to the crystallites that are still effective up to temperatures around the melting point T_m the material is not able to flow and preserves a certain level of mechanical properties. The existence of this state of order has the effect that the stiffness goes down but the toughness

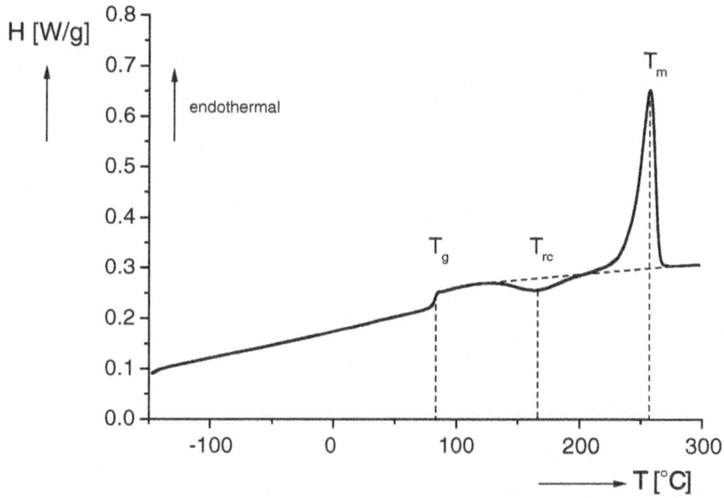

Fig. 4.22 Heat flow *H* as a function of temperature *T* for a PET measured by DSC at a heating rate of 10 K/min (Second heating)

Table 4.6 States of order in semicrystalline polymers

T	Polymer	Amorphous phase	Crystallites
$T < T_g$	Rigid state	Glassy state	Non-molten
$T_g < T < T_m$	Leatherlike state	Rubber-elastic state	Non-molten
$T \cong T_m$	Melting region	Rubber-elastic state	Melting
$T > T_m$	Melt	Melt	Molten

increases. Such a behavior is found more or less pronounced for all semicrystalline polymer materials. In comparison to amorphous products one has to distinguish and to discuss this additional state of order which is commonly called the *leather-like state* as it combines certain stiffness with some kind of toughness typical of the natural product leather. These relations are summarized in Table 4.6.

For $T < T_g$ the amorphous phase of the polymeric material is in its glassy state, i.e., it possesses the highest stiffness. For $T_g < T < T_m$ the rubber-elastic state of the molecules of the amorphous part contributes to the toughness of the material as long as the crystallites are intact. Above T_m the crystallites are molten.

For many polymers the picture of molecular processes and states based on calorimetric measurements has to be sharpened as follows from the dynamic-mechanical experiments (DMA) schematically presented in Fig. 4.23.

Characteristic is the steep decrease in the modulus at the melting temperature T_m, which is correlated with a distinct increase in tan δ corresponding to growing energy dissipation. The sharp kink schematically indicates that at the temperature at which the melting has come to an end the molecules still may be in their rubber-

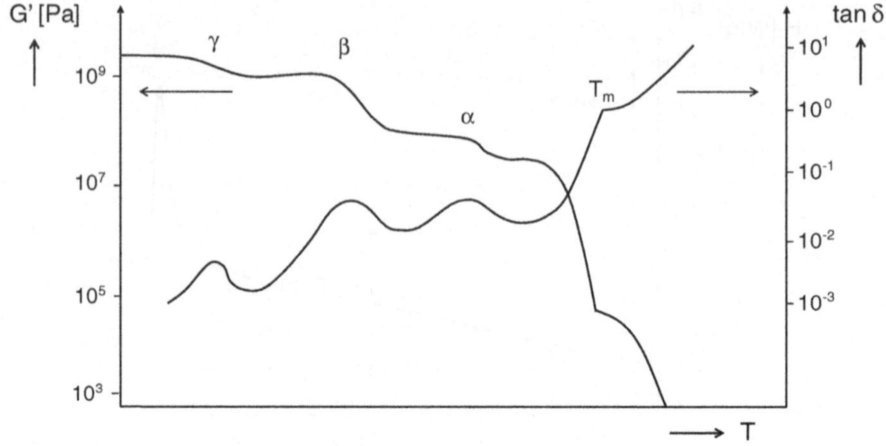

Fig. 4.23 Schematic diagram of the storage modulus G' and the loss tangent tan δ as a function of temperature T for a semicrystalline polymer

elastic state so that a further temperature increase is necessary to reach the state of flow. Such a situation is met for materials of high molar mass, only. In practice, ultrahigh molecular weight polyethylene (UHMWPE) is one of the rare examples. In reality such a transition is difficult to measure, however, due to the immense change in the modulus around the melting temperature.

For $T < T_m$ several dispersion steps denominated α, β, and γ are marked similar to the classification chosen for amorphous polymers in Sect. 4.4.

As measurements in Fig. 4.24 on a PP, LDPE, and HDPE demonstrate, these dispersion steps of the modulus can be difficult to distinguish and often superimpose each other in the case of real products.

From the representation of the corresponding *loss tangents* in Fig. 4.25 it is obvious that the maxima of tan δ are more distinct than the steps of the modulus and that is the reason why they are discussed in the following. The α-maximum that has been observed for a great number of semicrystalline polymers to be close to the melting point has been related to changes within the crystalline structure the exact origin of which is still a matter of discussion, however. Regarding the β- and γ-maximum there is a dispute in the literature which one should be related to the glass transition. In spite of this formal uncertainty it can be taken for granted that the low temperature maxima give an important hint to the impact strength of a semicrystalline polymer in the same way as discussed for amorphous materials (cf. Sect. 6.6). A maximum of tan δ indicates an energy dissipating process due to an increased mobility of molecule segments, which improves the impact strength.

From Fig. 4.25 it is obvious that the HDPE and the LDPE also exhibit a distinct γ-maximum around -100 °C. For PP, a sharp maximum at low temperatures is not found, but the flat maximum at around -60 °C can be interpreted as an indication of broad molecular transition processes that are not as pronounced, however, as in the case of the PE. This finding may be related to the distinctly higher low-

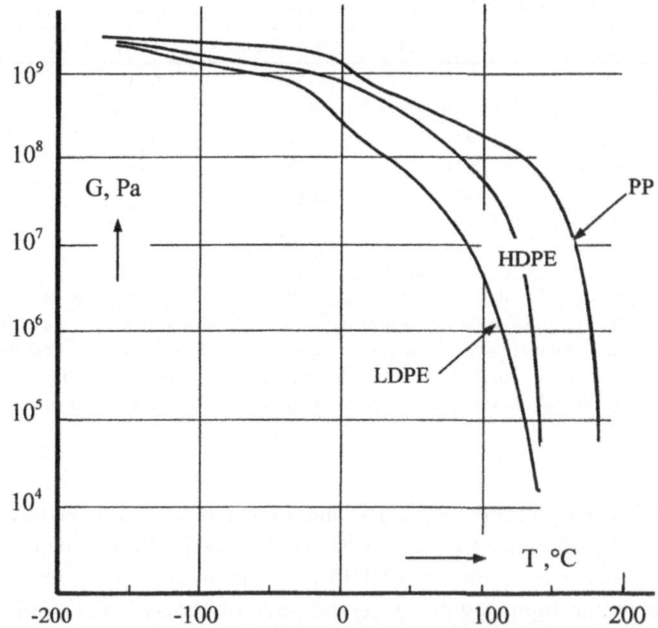

Fig. 4.24 Shear modulus G as a function of temperature T for a polypropylene (PP), a high density polyethylene (HDPE), and a low density polyethylene (LDPE) measured at a frequency of 1 Hz. The measurements represented by the full lines are taken from [10]

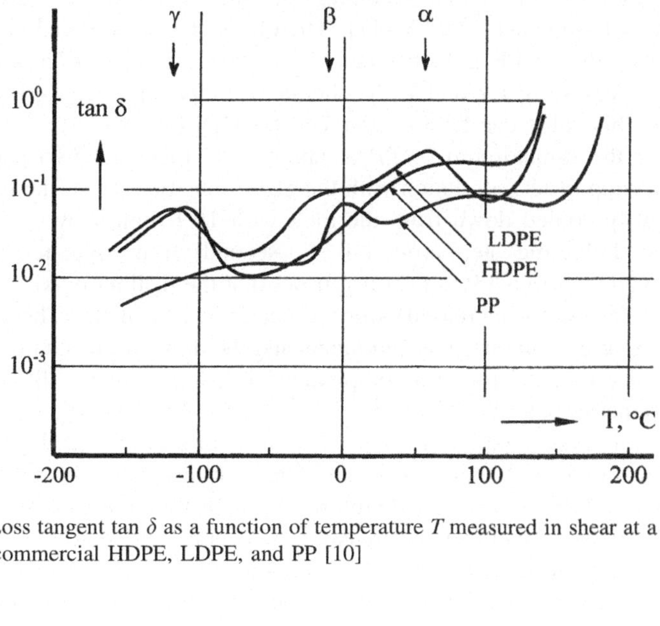

Fig. 4.25 Loss tangent tan δ as a function of temperature T measured in shear at a frequency of 1 Hz for a commercial HDPE, LDPE, and PP [10]

Table 4.7 Transition temperatures of various semicrystalline polymers

Polymer	T_m °C DSC	T_g °C DSC	T_α °C DMA	T_β °C DMA	T_γ °C DMA
HDPE	126–136	n.m	~90	n.m.	−110
LDPE	105–118	n.m.	70	~−10	−100
i-PP	160	n.m.	100	0	~−50
POM	160–170	n.m.	130	n.m.	−70
PA 6 (dry)	225	55	n.a.	50	−80
PET	250–260	84	n.a.	90	−80
PEEK	340	143	n.a.	150	−80

T_m Melting temperature, T_g Glass transition temperature, T_α, T_β, T_γ Transition temperatures from DMA according to Fig. 4.23. *n.m.* not found with the method applied, *n.a.* not achieved within the temperature range investigated. T_m and T_g were obtained from the second heating run of *DSC* measurements at a heating rate of 10 K/min after a preceding cooling from the molten state at a rate of 10 K/min. The DMA was performed at frequencies of around 1 Hz and at equilibrium temperatures

temperature impact strength of HDPE and LDPE in comparison to PP. The PP shows a distinct β-maximum around 0 °C. That of the LDPE can be assumed to lie at a slightly lower temperature. For HDPE, a β-maximum is not seen. This finding may be due to the high degree of crystallinity of the HDPE resulting in small contributions of the amorphous phase only. Calorimetric measurements on PE do not show any step in the heat flow from which conclusions with respect to a glass transition could be drawn. Clearly visible is the melting peak only, which is listed in Table 4.7. But copolymers of ethylene with about 10 % of propylene, hexane, or octane exhibit an acceptable agreement between temperatures of the glass transition found by differential scanning calorimetry (DSC) and those of the β-maximum determined by DMA [11]. Therefore, it can be concluded that the β-maximum of the LDPE hidden under the plateau of tan δ at around −10 °C indicates the glass transition and as the chemical components of LDPE and HDPE are the same this value can be also assumed for HDPE

Regarding the β-maximum of PP at about 0 °C there are two experimental findings that support its correlation with the glass transition. First, the heat flow of a PP previously cooled down from the melt with 10 K/min shows a minute but significant endothermic step around 0 °C, second, from dynamic-mechanical measurements on various PP in [12] it follows that the distinct maximum of tan δ close to 0 °C markedly increased with the decrease in density. These examples demonstrate that calorimetric measurements and the dynamic-mechanical analysis have to be used together to get a comprehensive insight into the different transitions of polymeric materials.

Melting and crystallization are thermodynamic processes with high relevance for the application of polymeric materials. The melting process of semicrystalline polymers is relatively sharp and, therefore, the melting temperature is a distinct quantity for the assessment of the transition between solid-state and rheological properties. The melting temperature T_m significantly depends on the chemical structure of a material, but experimental parameters like heating rate influence its determination within certain limits. In Table 4.7 values for T_m of some engineering polymers are given. Data from various sources differ, which may be due to the experimental methods used and slight variations in the chemical composition of the materials.

In the case of HDPE, for example, a minute comonomer content may be the reason for deviations, for LDPE differences in long-chain branching may change T_m. Nevertheless, the clear differences between the melting behavior of some products allow an assessment of some of their properties. In addition, T_g is given in Table 4.7, too. For PA 6, PET, and PEEK clear distinctions are possible according to a pronounced step in heat flow. The differences found in the literature can either be due to experimental conditions not comparable or to traces of impurities within a sample. It is well known for polyamides, for example, that a few percent of water can lower T_g by several degrees.

For the two polyethylenes and in some way for the polypropylene DSC measurements are not sensitive enough to monitor the heat input connected with the glass transition of the amorphous phase. DMA is more suitable in some cases to detect these processes but as discussed above the assignment of the different peaks to the glass transition may not be evident and needs supplementary considerations. From the above discussion it is probable that the ß-transition corresponds to the glass transition.

The difficulties arising from a correct interpretation of data from the DMA exemplarily become evident from the results of POM, PA 6, and PET presented in Fig. 4.26 in comparison to HDPE. The values of T_β of PET and PA 6 are in reasonable agreement with T_g if one assumes that the α-maximum was not

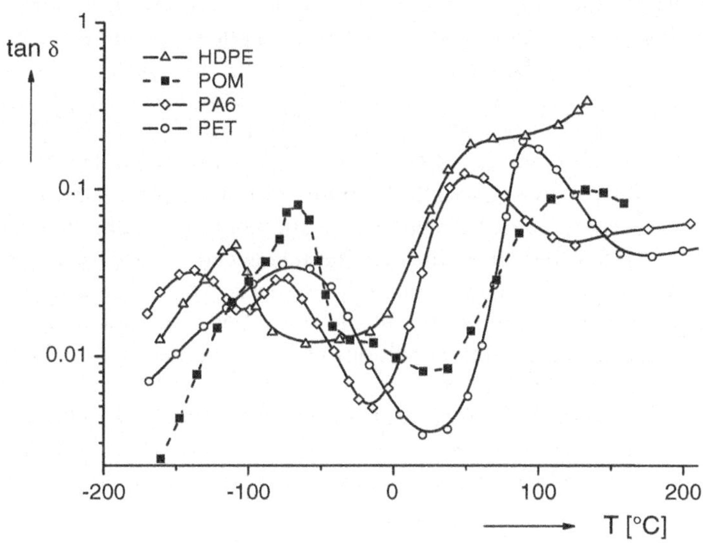

Fig. 4.26 Loss tangent tan δ as a function of the temperature T for four different polymer materials measured at a frequency around 1 Hz and a small heating rate [10]

achieved at the highest temperatures applied. The same holds for the β-transition of PEEK [13] (data not presented in Fig. 4.26). These assignments are not so obvious in the case of POM. One reason is that its glass transition cannot clearly be determined thermodynamically, the other that the difference between the melting point and the highest maximum temperature of 130 °C is close to that between T_m and T_α of the HDPE. Postulating similar relations for the POM investigated, only an α- and a γ-transition may be defined. The β-process can perhaps be assumed to be hidden under the plateau in the region around −10 °C, but such an interpretation is somewhat speculative, of course.

Much clearer is the determination of the γ-maxima that come to lie around −80 °C for the non-olefinic polymers (cf. Table 4.7). This low transition temperature is decisive for the good impact strength of these polymeric materials. In the case of the polyamide a second low-temperature maximum is found at about −140 °C the origin of which is not understood till now. Another still open question is the specific molecular motions underlying the γ-maximum.

4.5.3 Crystallization

Crystallization takes place in a wide temperature region between T_m and T_g. Around T_m first nuclei are formed but their lifetime is short as the difference between the free enthalpies of the melt and the crystal are small. Around T_g the crystal nuclei are thermally more stable, indeed, but the molecular motion is too small for an effective formation of ordered regions.

The crystallization rates of semicrystalline polymers are thermodynamic properties interesting from fundamental and application points of view. The quantitative agreement between various sources of the literature is still much worse than for T_m or T_g. Significant differences allow some interesting conclusions, however. The extremely high crystallization rate of HDPE makes it impossible to get amorphous samples even by quenching in liquid nitrogen. In the case of the distinctly lower crystallization rate of polypropylene a wide variety of crystallinity can be obtained depending on the cooling conditions chosen. Furthermore, the size of the crystallites can be influenced. At high cooling rates the crystallites are generally smaller than at lower ones. As the size and, particularly, the amount of crystals significantly affect optical and mechanical properties of a polymeric material the cooling conditions can become decisive for properties of items processed from the molten state.

From the low crystallization rate of the PET as demonstrated in Fig. 4.22 by the occurrence of *recrystallizaton*, it becomes clear that this material is suitable for the manufacture of optical transparent bottles widely used in the beverage industry.

4.6 The Specific Volume of Polymers

4.6.1 The Specific Volume of Amorphous Polymers

Polymers possess densities between 0.8 and 2.5 g/cm^3 at room temperature, values of about 1 g/cm^3 are most frequent ranging them as light construction materials. Their thermal expansion behavior differs from that of classical construction materials. Let v be the *specific volume* and $\rho = 1/v$ the *density*, then

$$\alpha = \frac{1}{v}\left(\frac{\partial v}{\partial T}\right)_p \tag{4.18}$$

is called the (volume) *expansion coefficient* at constant pressure. Normally, α is measured at an atmospheric pressure of 1 bar.

For many classical construction materials, the expansion coefficient depends only weakly on the temperature. In contrast, polymers show a rather high volume expansion, which, moreover, changes significantly at the glass-transition temperature T_g. Below T_g the expansion coefficient amounts to about one-third of the value beyond T_g. Table 4.8 lists the volume expansion coefficients of some materials.

From this table it may be seen that the expansion coefficient of steel is only about 1/5 of that of polymers in the glassy state. Consequently, we have to expect the occurrence of *strong thermal stresses* in all technical applications where steel and plastics are attached to each other.

The specific volume of amorphous polymers as a function of temperature depends on the conditions of the experiment (heating or cooling). On cooling a sample with a constant cooling rate β from the rubber-elastic plateau or state down to the glassy state, a contraction behavior is found as sketched in Fig. 4.27. The specific volume decreases linearly with the temperature, i.e., the expansion coefficient[1] α_r in the rubber-elastic plateau or state is constant.

The expansion coefficient in the glassy state α_g is significantly lower. At the temperature T_g the slope of the v–T curve changes rather abruptly. The intersection

Table 4.8 The volume expansion coefficient of some materials in 10^{-5} K^{-1}

Quartz glass	0.16	Ice	11	Amorphous	Polymers
Glass	2.5	H_2O, 20 °C	21	Glassy state $T < T_g$	$\alpha = \alpha_g \cong$ 10–20
Steel	3.5	Various fluids	20–60	Rubbery state $T > T_g$	$\alpha = \alpha_r \cong$ 50–60

[1] According to the measuring technique, the designation "contraction coefficient" would be more appropriate. For the matter of simplicity, the notation expansion coefficient is used here, too.

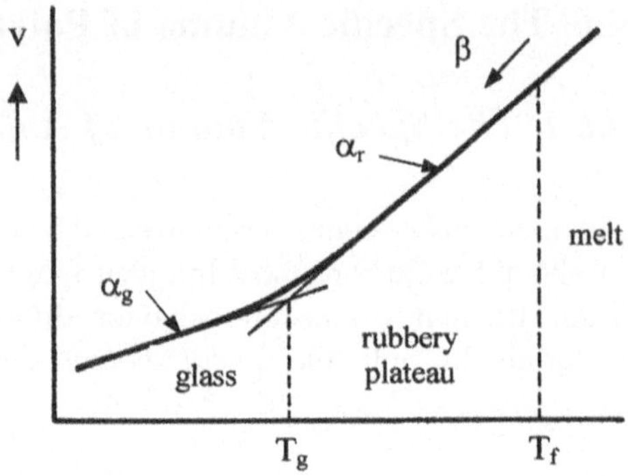

Fig. 4.27 Specific volume of amorphous polymers versus the temperature during cooling and the definition of the glass-transition temperature

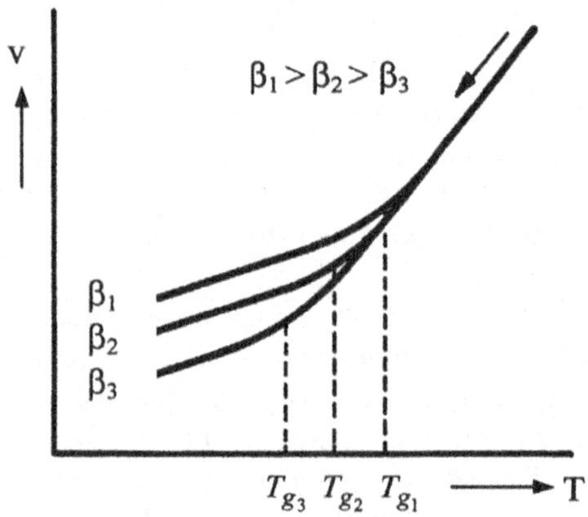

Fig. 4.28 Specific volume as function of temperature at various rates of cooling

of the two extrapolated straight lines defines the *volumetric glass-transition temperature* T_g. Such measurement was the method by which the glass-transition temperature was introduced first. If the experiment is repeated with larger cooling rates, neither α_r nor α_g change significantly, but the point of intersection shifts linearly with the logarithm of the cooling rate to higher temperatures and the level of the specific volume in the glassy state that is reached, becomes higher (cf. Fig. 4.28).

The following relation between the volumetric glass-transition temperature and the cooling rate is found for amorphous polymers

$$T_g = T_{g_0} + a \cdot \log(\beta/\beta_0) \qquad (4.19)$$

with T_{g_0} and T_0 being the glass-transition temperatures at the cooling rates β and β_0, respectively. Coefficient a is about 3 K (cf. Table 4.9).

Table 4.9 Some dilatometric data of amorphous polymers [14]

Polymer	T_g °C, $\beta = 1$ K/min	α_g, 10^{-4} K^{-1}	α_r, 10^{-4} K^{-1}	a, K
PS	95	2.30	5.73	3.3
PMMA	110	2.57	6.06	3.2
PVC	69	2.06	5.54	3.2
PC	141	2.15	6.04	2.5

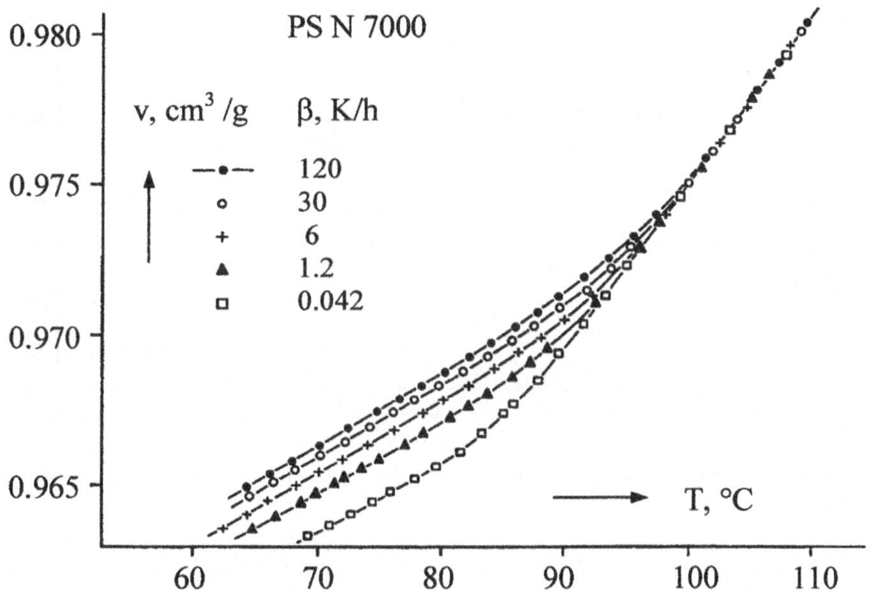

Fig. 4.29 Specific volume versus temperature for a PS at various rates of cooling after Greiner [15]

As an example, the specific volume of PS is shown in Fig. 4.29 as a function of the temperature in the vicinity of the glass-transition temperature over a wide range of cooling rates [15]. The measurements were performed by means of Hg-dilatometers that were immersed in a bath filled with silicone oil in a thermostat with transparent walls. The temperature of the thermostatic fluid was programmed to decrease with constant rate or regulated to constant temperature. The level of the mercury column was photographed.

Figure 4.30 shows the glass-transition temperatures as functions of the cooling rates for four different technical polymers. Note that for each polymer a different temperature scale had to be used, which is indicated by arrows in the figure, which proves the validity of Eq. (4.19) for the four polymers over a wide range of cooling rates.

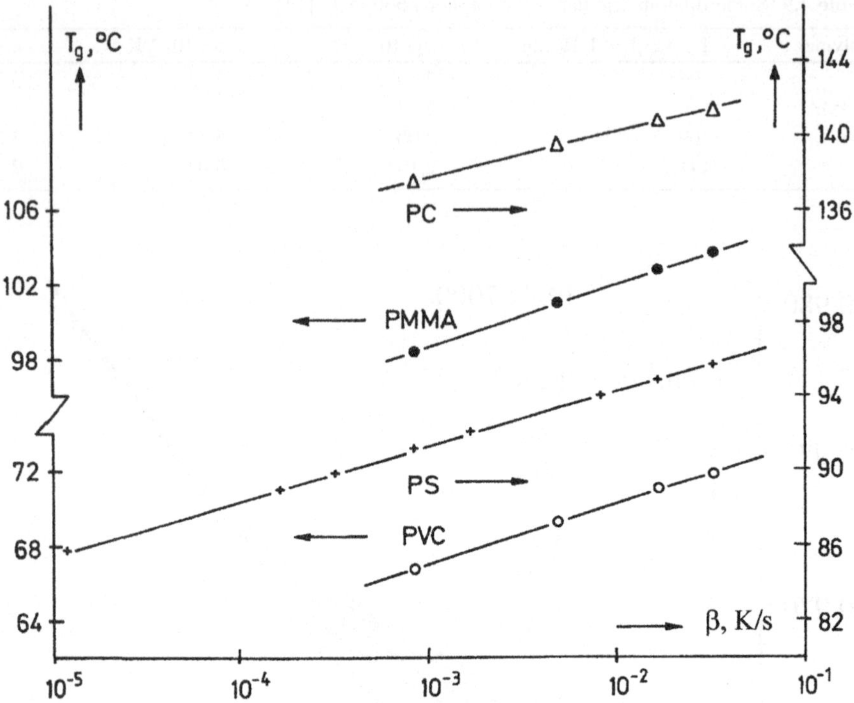

Fig. 4.30 Volumetric glass transition temperatures for four amorphous polymers versus the rate of cooling after Greiner [15]

In Fig. 4.31 the thermal expansion behavior is shown of four polymers at a cooling rate of 2 K/min. The measurements on polystyrene extend from the rubber-elastic plateau down to the temperature of liquid He (4 °K). The measurements indicated by filled circles have been performed by Hartwig using a helium cryostat and a laser dilatometer [16]. Measurements indicated by crosses were performed using a dilatometer especially constructed in which the length of a specimen was determined by transducers. The thermostatic medium was nitrogen gas of regulated temperatures [17]. Measurements indicated by open circles were performed by means of mercury dilatometers using the technique described above [17].

The change of the slope at the glass-transition temperature is clearly seen for all four polymers. The thermal expansion coefficient decreases somewhat in the glassy state of the polymers. For polystyrene, the thermal expansion coefficient shows a further obvious change at about −210 °C. Approximating the absolute zero, the thermal expansion vanishes completely.

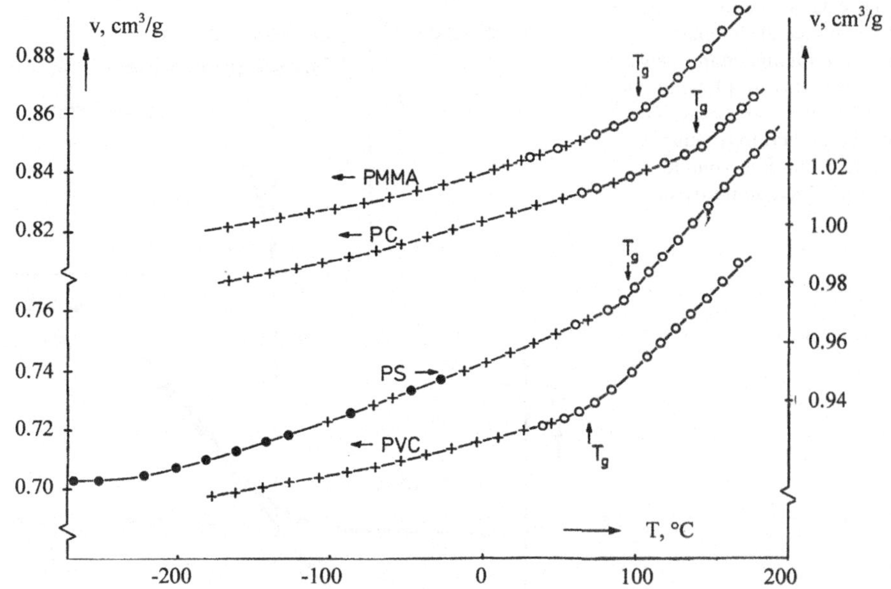

Fig. 4.31 Specific volume versus temperature at a rate of cooling of $\beta = 2$ K/min for the four amorphous polymers PMMA, PC, PS, and PVC after Greiner [16]

4.6.2 The Free Volume Theory

The expansion behavior of amorphous polymers may be explained by the *theory of free volume*. The specific volume of an amorphous polymer or a liquid may be assumed to be composed of three terms

$$v = v_0 + v_{\text{vib}} + v_f. \tag{4.20}$$

The "*Eigen*" *volume* v_0 is the volume taken by the molecules themselves and is, therefore, independent of the temperature. The *vibrational volume* v_{vib} arises from the thermal vibrations around the equilibrium positions of the van der Waals bonds between neighboring molecules. As the potentials of van der Waals bonds are asymmetric (cf. Fig. 2.12), the average distance between neighboring molecules will become larger with increasing amplitudes of the thermal vibrations, i.e., with increasing temperature. Consequently, the vibrational volume will increase with the temperature.

The thermal motion can sometimes move parts of different molecules so far that holes or vacancies occur, which form the *free volume* v_f. Each additional vacancy increases the internal energy of the polymer by the surface energy of the vacancies formed, but also contributes essentially to its entropy. The increase in free volume with temperature is stronger than that of the vibrational volume. In the thermodynamic equilibrium the course of the free volume with temperature might be as

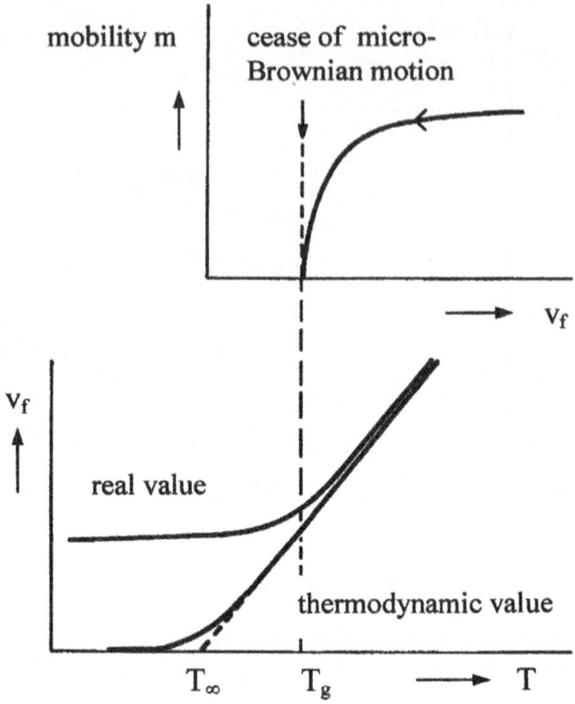

Fig. 4.32 Freezing of the free volume at the glass transition temperature. *Upper part* Mobility as a function of the free volume. *Lower part* Real and thermodynamic value of the free volume versus temperature, during cooling

indicated in the lower part of Fig. 4.32. Above T_g the thermodynamic value of v_f is assumed to increase linearly with the temperature.

The mobility and thus the micro-Brownian motion strongly depend on the presence of sufficient free volume. If the free volume falls short of a certain critical value, the mobility decreases suddenly and steeply and the micro-Brownian motion ceases, as is indicated in the upper part of Fig. 4.32. This figure, like the following illustrative example taken from all days life, is due to LCE Struik [18]: During the rush hours, getting on or getting off a bus or an underground car may become impossible due to overcrowding and the absence of free space.

When an amorphous polymer is cooled down from the rubber-elastic state or the rubber-elastic plateau to the glassy state, the free volume will decrease as long as it is able to adopt its thermodynamic value by means of the diffusion of surplus vacancies out of the material. If the free volume falls short of its critical value, the micro-Brownian motion ceases and diffusion of vacancies can no longer take place. The free volume will be frozen within the material, which had just been present at passing the critical temperature, the glass-transition temperature T_g. The temperature at which the freezing process occurs depends on the rate of cooling. The specific volume composed as the sum of v_0, v_{vib}, and v_f, will show a change in slope at the glass-transition temperature. The course of the real value of the free volume is indicated in the lower part of Fig. 4.32.

The sudden decrease in the *mobility* m with decreasing free volume may be described by the *Doolittle-equation* [19] which was successfully used to explain the temperature dependence of the viscosity of liquids of low molar mass.

$$\ln \eta_0 = \ln a_0 + \frac{B(v - v_f)}{v_f} = \ln a_0 + B\left(\frac{1}{f} - 1\right) \tag{4.21}$$

In this formula η_0 is the viscosity of the liquid with the fractional free volume $f = v_f / v$, a_0 the viscosity of a liquid with the—hypothetical—free volume fraction unity and B is a constant of the order unity. If m is defined as the logarithm of the ratio a_0/η_0, it will show a course similar to that in Fig. 4.32. It should be mentioned, however, that the shape of the curve m versus f strongly depends on the chosen value for B.

The real value of the fractional free volume as a function of temperature (c.f. lower part of Fig. 4.32) may be approximated by the equation

$$\begin{aligned} f &\cong f_0 \quad \text{for} \quad T < T_g \\ f &\cong f_0 + \alpha_f (T - T_g) \quad \text{for} \quad T > T_g \end{aligned} \tag{4.22}$$

with α_f being the expansion coefficient of the free volume and f_0 the unknown value of the free volume fraction frozen at T_g. The temperature T_∞ shown in Fig. 4.32 is the temperature at the point of intersection of the extended straight line of the equilibrium value of the free volume fraction with the temperature axis at the level, where the free volume fraction would vanish. Under the assumption that the difference in the volume expansion in the rubbery and glassy state is completely due to the increase in free volume, and that α_r and α_g remain constant within the considered temperature interval, we obtain

$$\alpha_f = \alpha_r - \alpha_g \tag{4.23}$$

However, as the value of f_0 is not known, the relevance of Eqs. (4.21)–(4.23) remains limited.

4.6.3 Volume Relaxation and Physical Aging

As polymers in the glassy state normally are not in their thermodynamic equilibrium, their specific volume will be too high. After quenching, they tend to assume their equilibrium-specific volume, without being able to reach it. This process is called *volume-relaxation*. It was studied intensively first by Kovacs [20]. An example for volume relaxation after quenching to various constant

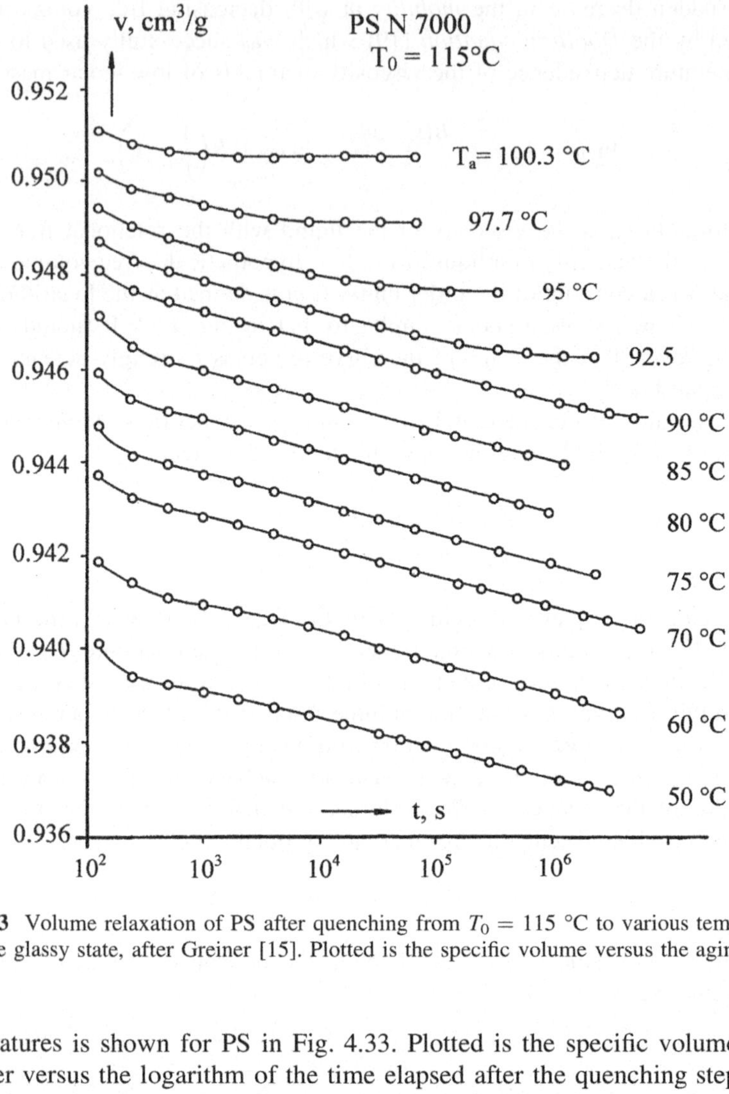

Fig. 4.33 Volume relaxation of PS after quenching from $T_0 = 115$ °C to various temperatures T_a in the glassy state, after Greiner [15]. Plotted is the specific volume versus the aging time t

temperatures is shown for PS in Fig. 4.33. Plotted is the specific volume of the polymer versus the logarithm of the time elapsed after the quenching step.

If the temperature at which the quenching was performed is only a few degrees below T_g, an equilibrium value of the specific volume is reached after some time. But already 25 °C below the glass-transition temperature, the equilibrium is not reached within reasonable times. The specific volume continues to decrease approximately proportional to the logarithm of the aging time.

Volume relaxation is accompanied by the change in various physical properties of the polymer, a process called *physical aging* and intensively studied by Struik [18]. The specific and the free volume of a polymer depend on its age (*Age* = the time elapsed between the last quenching into the glassy state and the time of

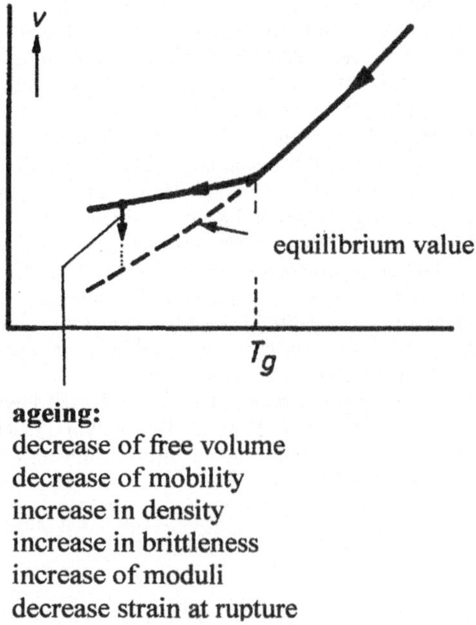

Fig. 4.34 Volume relaxation and physical aging

observation). Some important physical and technological properties of the polymer change during aging, as indicated in Fig. 4.34.

Especially the viscoelastic behavior is influenced by physical aging. The time-position of the glass-transition is no longer a function of the temperature only, but depends on the age too. In particular, shortly after the quenching the proceeding aging during a creep measurement changes the shape of the creep compliance, with the effect that the shifting law is violated and the compliance curves not only shift, but also get flatter.

Struik [21] has presented a theory to describe this effect by introducing the dependence of the retardation times on aging. Though this theory is clear and straightforward, its presentation would be outside the scope of this book. We recommend the reader the monograph of Struik [21] for this purpose.

References

1. Schwarzl FR (1990) Polymermechanik. Springer, Berlin
2. Elias HG (1999) Abbreviations for thermoplastics, thermosets, fibers, elastomers and additives. In: Brundrup J, Immergut EH, Grülke EA (ed) Polymer Handbook vol 2 VIII/1. Wiley Interscience, New York
3. Elias HG (1984) Macromolecules 1. Plenum Press, New York
4. Masuda T, Kitagawa K, Onogi S (1970) Polymer J 1:418
5. Hendus H, Schnell G (1961) Kunststoffe 51:69

6. Bunn CW (1939) Trans Farad Soc 35:482
7. Mutter W, Stille W, Strobl G (1993) J Polym Sci Polym Phys Ed 31:99
8. Van Krevelen DW (1997) Properties of polymers. Elsevier, Amsterdam
9. Michler GH (1992) Kunststoff-Mikromechanik. Carl Hanser Verlag, München Wien
10. Heijboer J (1969) Brit Polym J 1:3
11. Mäder D, Heinemann J, Walter P, Mülhaupt R (2000) Macromolecules 33:1254
12. Flocke HA (1962) Kolloidzeitschrift 180:118
13. Gensler R (1995) Diploma Thesis, University Erlangen-Nürnberg
14. Greiner R, Schwarzl FR (1984) Rheol Acta 23:378
15. Greiner R (1978) Zum Ausdehnungs- und Volumenrelaxationsverhaltens eines amorphen Polymers. Doctoral thesis, University Erlangen-Nürnberg
16. Greiner R (1987) Private communication
17. Hartwig G (1989) KFZ Karlsruhe private communication
18. Struik LCE (1977) Physical aging in amorphous polymers and other materials. Doctoral thesis, Technical University Delft
19. Doolittle AK, Doolittle DB (1957) J Appl Phys 28:901
20. Kovacs AJ (1964) Fortschr. Hochpolym. Forschung 3:394
21. Struik LCE (1990) Internal stresses, dimensional instabilities and molecular orientations in plastics. Wiley, New York

Chapter 5
Linear Viscoelastic Deformation Behavior in Simple Shear

5.1 Theoretical Description of the Deformation Behavior of Polymers

A general phenomenological theory of the mechanical deformation behavior of polymers does not exist up to now, as the various states of polymers are very different and demanding with regard to their description. Such a theory would ask for the proper description of large deformations in combination with the influence of the history of stress and temperature, without any simplifications as superposition principle, linearity, elasticity, or even isotropy. As this task cannot be accomplished by a mathematical scheme sufficiently simple to be useful in practice, the setup of a general theory is not pursued further. Instead one looks at the various states of polymers, and tries to setup theoretical descriptions, dealing with the essential features, which polymers show in the corresponding states. In each case simplifications are made, in order to come to manageable relations.

The most important ones are

(I) The material is *linear*; i.e., in the most simple case $\sigma = G \cdot \gamma$ with G being independent of deformation. For viscoelastic materials, there exists a simple and very useful generalization, the *superposition principle* of *Boltzmann* which will be treated in extension in this chapter. This principle forms the basis of the *technical mechanics of polymeric materials*, in their linear range.

(II) The material is *elastic*; this means that the stress is a unique function of the strain at the same time, independent of the time history of the strain: $\sigma = f(\gamma)$, where f denotes an arbitrary function of the strain. In this case, there is no energy dissipation.

(III) The deformations are *infinitesimal*, i.e., $\gamma \ll 1$ and therefore $\gamma^2 \cong 0$.

H. Münstedt and F. R. Schwarzl, *Deformation and Flow of Polymeric Materials*,
DOI: 10.1007/978-3-642-55409-4_5,

Some of the theories, based on the postulates mentioned above, are

(1) The *Hookean theory of elasticity*, which forms the basis of the technical mechanics of classical construction materials: (I), (II), (III) are assumed to be valid. This theory is useful for metals below their limits of plasticity, for glass, ceramics, concrete, etc.
(2) The *theory of large elastic deformations*: (II) is valid, (I) and (III) are not needed. This theory is useful for the description of the deformation behavior of cross-linked rubbers.
(3) The *theory of the linear viscoelastic deformation behavior at small deformations*: In this theory postulate (I) is assumed to be valid in the form of the superposition principle and the postulate (III) is used. The assumption (II) is not needed. This theory forms the basis of the technical mechanics of plastics in the glassy state, and partially also in the rubbery state. In the case of polymer melts, however, this theory can offer only a rough picture of the mechanical behavior, as one of the most important assumptions, the infinitesimal deformations are not met by most applications.
(4) For the description of polymer melts one starts with the description of the kinematics of large deformations, and then tries to set up rheological relations based on more or less simple molecular concepts.

5.2 Creep, Creep Recovery, and Stress Relaxation

The theory of the linear visco-elastic behavior describes the mechanical behavior of polymers in all states under the restriction of small deformations and low stresses. Therefore it is valid within certain limits of the strain, ε_l and the stress σ_l, which are called *limits of linearity*. They will be discussed in more detail in Sect. 5.3. The basis for the theory of linear viscoelasticity is the superposition principle of Boltzmann, which will be introduced in Sect. 5.3.

The theory finally offers a correlation between a three dimensional state of stress and strain for isotropic and anisotropic materials. We will start with the consideration of one type of deformation, namely simple shear. Later in Chap. 7, we will generalize the theory to three dimensional states of stress.

The definition of simple shear follows from the Fig. 4.2. A cube, oriented parallel to a right-handed Cartesian coordinate system, is deformed into a rhomboid by shearing forces. The tangential forces act in x-direction on the two surfaces which are perpendicular to the y-direction. Their magnitude, divided by the surface on which they act, is called the *tangential stress* or *shear stress* and is designated as σ_{yx}. The cube is deformed into a rhomboid, which has the same height and width as the cube. The edges which were originally parallel to the y-axis, form an angle α with the y-axis after the deformation. This angle is used as a measure for the magnitude of the deformation and its tangent is defined as shear strain

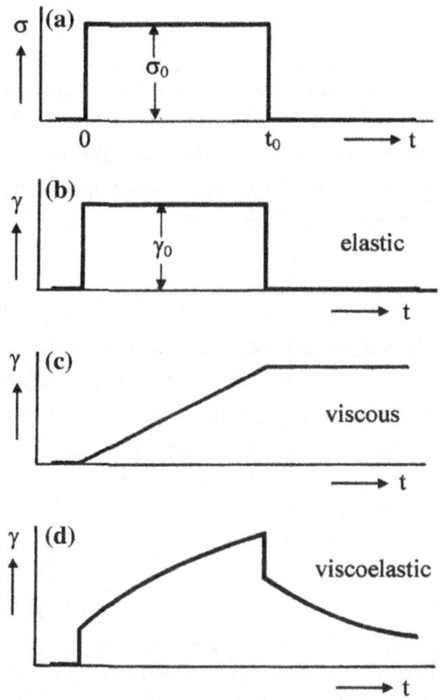

Fig. 5.1 Creep and recovery of an elastic, a viscous, and a viscoelastic material

$$\gamma_{yx} = \tan\alpha \tag{4.1}$$

For a discussion of the relations between σ_{yx} and γ_{yx} in simple shear the indices y and x are omitted in this chapter. Furthermore, we will neglect the influence of the inert mass of the sample on the movements of the specimen.

For an illustration of the concepts *elastic*, *viscous*, and *visco-elastic* we consider the *creep* and *creep recovery experiment*, as defined by the stress history

$$\begin{array}{ll} \sigma = 0 & \text{for } t \leq 0 \\ \sigma = \sigma_0 = \text{const} & \text{for } 0 < t < t_0 \\ \sigma = 0 & \text{for } t \geq t_0 \end{array} \tag{5.1}$$

and sketched in part a of the Fig. 5.1. Under the constant stress σ_0 the shear deformation as a function of time is measured in a *creep experiment*. After unloading of the sample at the time $t = t_{0,}$ the recovery of the deformation is observed in the *recovery experiment*.

An e*lastic* deformation behavior exists, if the deformation is a unique function of the stress at the same time t, i.e., if

$$\gamma = F(\sigma) \tag{5.2}$$

Consequently, γ has to be constant as long as σ remains constant and γ has to become zero if $\sigma = 0$. An elastic material *does not creep* and shows *complete* and *instantaneous recovery* in the recovery experiment. (Compare Fig. 5.1b). The special case, in which γ is proportional to σ, defines the *linear elastic* or *Hookean deformation behavior*:

$$\gamma = \sigma / G_0 = J_0 \cdot \sigma \tag{5.3}$$

G_0 is called the *elastic shear modulus* and $J_0 = 1/G_0$ the *elastic shear compliance*. Classical construction materials as steel, glass, ceramics or concrete behave linear elastic under small stresses and may be described by the *classical theory* of *elasticity* in good approximation.

Viscous flow means that the rate of deformation is a unique function of the stress, i.e.,

$$\frac{d\gamma}{dt} \equiv \dot{\gamma} = F(\sigma) \tag{5.4}$$

From $\sigma =$ constant, it follows $\dot{\gamma} =$ constant or $\gamma = at + b$. Consequently, a viscous material shows a *linear increase* of the deformation in *creep* with time and *no recovery* at all during the recovery period. A viscous deformation is completely *irreversible* (see Fig. 5.1c). The special case

$$\frac{d\gamma}{dt} \equiv \dot{\gamma} = \sigma / \eta_0 = \varphi_0 \cdot \sigma \tag{5.5}$$

defines the *linear viscous* or *Newtonian flow behavior*. The constant η_0 is called *Newtonian viscosity* and its reciprocal φ_0 the *Newtonian fluidity*. Gases and low molecular liquids behave linear viscous and may be described by *classical fluid mechanics*.

Polymeric substances may not even approximately be described as elastic or viscous, but show a *visco-elastic deformation behavior* (Fig. 5.1d). Under a constant stress, one part of the deformation occurs instantaneously, another part of the deformation develops under the action of the stress (creep). In the recovery experiment one part of the deformation recovers instantaneously at the time $t = t_0$, a further part with some retardation during the recovery period, and a third part does not recover at all, but remains permanent.

In the special case of a *linear viscoelastic behavior*, $\gamma(t)$ in Fig. 5.1d is proportional to the magnitude of the constant stress σ_0 during the creep period. The deformation in the creep experiment may then be simply written as

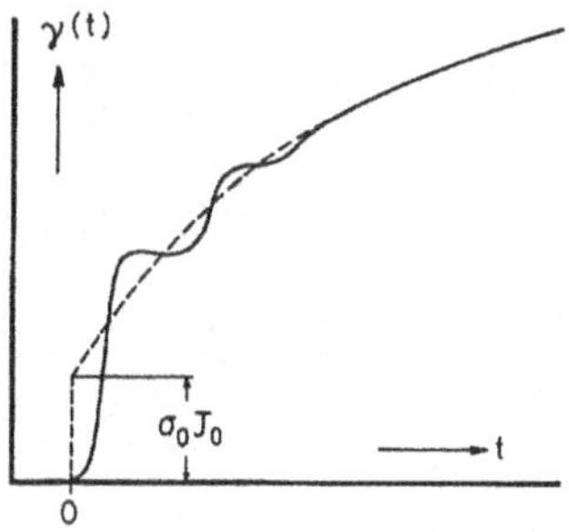

Fig. 5.2 Schematic example of a creep experiment (*full line*) and its approximate description by Eq. (5.6) (*dashed line*)

$$\gamma(\sigma_0, t) = \sigma_0 \cdot J(t) \quad \text{for} \quad t > 0 \tag{5.6}$$

with $J(t)$ being a time function independent of the magnitude of the stress. Creep curves under different constant stresses coincide if dividing them by the values of the corresponding stresses. The function $J(t)$ describes the creep behavior of a linear viscoelastic material completely. It is called the *creep function* or the *time-dependent creep compliance.*

The creep function is a positive, *monotonously increasing* function of the time, defined for positive arguments only.

$$J(t) > 0 \quad \text{and} \quad \frac{dJ(t)}{dt} \equiv \dot{J}(t) \geq 0 \quad \text{for all} \quad 0 \leq t < \infty \tag{5.7}$$

The limit of this function for $t \to +0$ is designated as J_0 and called the *instantaneous creep compliance.*

$$J_0 = \lim_{t \to +0} J(t) > 0 \tag{5.8}$$

To avoid mathematical difficulties we assume J_0 to be always larger than zero. That means we exclude materials, which behave infinitely stiff directly after loading. This is no restriction, as no materials are known, which are infinitely stiff. Even diamond shows a shear compliance of $J_0 > 10^{-13}$ Pa^{-1}. It should be mentioned, however, that Eq. (5.8) does not offer a realistic possibility to determine J_0 experimentally. So far, we have neglected the influence of the inert mass of the sample on its deformation behavior. This is not correct directly after loading, as the specimen has to be accelerated first and then will oscillate around the creep curve, until the initial vibrations have been damped out. Only then, the description by means of Eq. (5.6) is admitted. To determine J_0 one has to extrapolate, which in most cases will not be possible (compare Fig. 5.2).

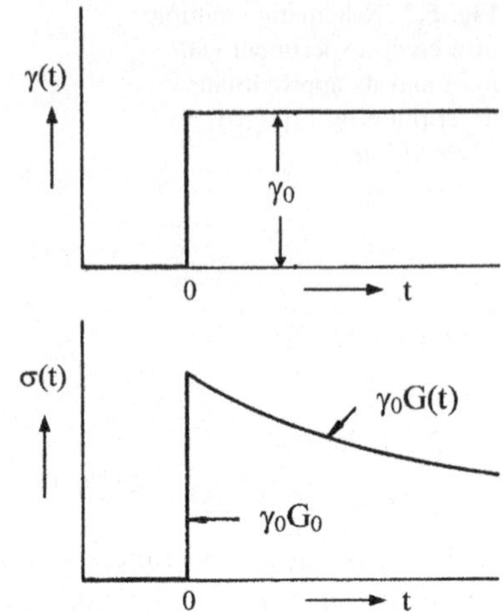

Fig. 5.3 The stress relaxation experiment, schematically

The counterpart of the creep experiment is the *stress relaxation experiment*, shown in Fig. 5.3: Upon the sample, which had been free of stress before $t = 0$, a sudden deformation is enforced at the time $t = 0$ and kept constant further.

$$\gamma(t) = 0 \quad \text{for } t<0 \quad \text{and } \gamma(t) = \gamma_0 \quad \text{for } t\geq 0 \tag{5.9}$$

The stress necessary to keep up this deformation is measured. For a linear viscoelastic material the stress is the product of the constant strain and a time-dependent function

$$\sigma(\gamma_0, t) = \gamma_0 \cdot G(t) \quad \text{for t}\geq 0 \tag{5.10}$$

called the *relaxation modulus G*(*t*). The relaxation modulus is a positive, *monotonously decreasing* function of the time,

$$G(t) > 0 \quad \text{and} \quad \frac{dG(t)}{dt} \equiv \dot{G}(t) \leq 0 \quad \text{for all} \quad 0\leq t<\infty \tag{5.11}$$

Its limit for small positive times

$$G_0 = \lim_{t\to +0} G(t) > 0 \tag{5.12}$$

is designated as the *instantaneous relaxation modulus*. Similar to J_0, also G_0 is not directly measurable. It will be shown later, that

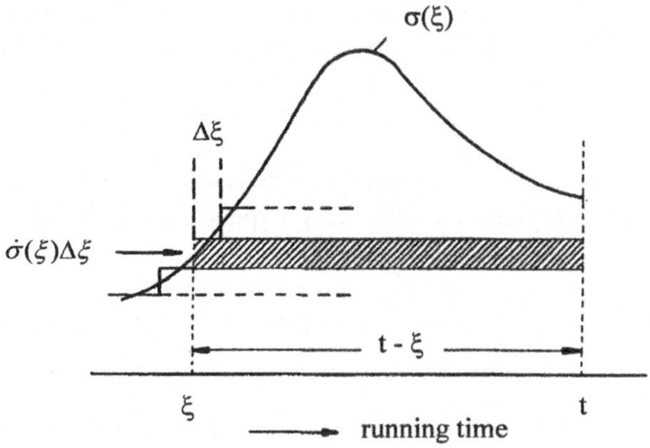

Fig. 5.4 Decomposition of the stress history into horizontal sectors

$$J_0 \cdot G_0 = 1 \tag{5.13}$$

Therefore, from the condition that J_0 remains finite, also follows that G_0 remains finite.

5.3 The Principle of Superposition

The basis of the theory of the linear viscoelastic behavior is the *superposition principle of Boltzmann*, formulated already in 1874 [1]. Its statement is about the following: Let $\sigma_1(t)$ be an arbitrary, time-dependent stress, causing a strain history $\gamma_1(t)$ on the sample, and let $\sigma_2(t)$ be another, time-dependent stress, causing the strain history $\gamma_2(t)$ on the sample, then the sum of both stress histories $\sigma_1(t) + \sigma_2(t)$ will just effect the sum of both strain histories $\gamma_1(t) + \gamma_2(t)$.

Cast in mathematical shape, this principle allows the calculation of the strain at the present time t, if the stress history from the past up to the time t is known. The stress history in the time interval $(-\infty, t]$ is given by

$$\sigma(\xi) \quad \text{for} \quad -\infty < \xi \leq t$$

Stress and strain are assumed to vanish both at the time $\xi = -\infty$. Further we assume $\sigma(\xi)$ to be continuous and piecewise differentiable in the complete interval. For deriving the mathematical form of the superposition principle, we approximate the stress history by a stepwise curve with step widths $\Delta\xi$, as indicated in Fig. 5.4.

The area under the stress-time curve may be built up by narrow horizontal sectors. The hatched sector starting at the time ξ has the step height $\dot{\sigma}(\xi)\Delta\xi$,[1] its

[1] $\dot{\sigma}$ indicates here and in the following the derivative of the function σ with respect to its argument.

effect on the deformation will be that of a creep experiment starting at the time ξ under the stress $\dot{\sigma}(\xi)\Delta\xi$. At the time t, its contribution to the deformation will be

$$\Delta\gamma = \dot{\sigma}(\xi) \cdot J(t-\xi)\Delta\xi$$

Adding all contributions between time $-\infty$ and t and proceeding to the limit $\Delta\xi \to 0$ yields the following expression for the deformation at time t

$$\gamma(t) = \int_{-\infty}^{t} J(t-\xi)\dot{\sigma}(\xi)d\xi \tag{5.14}$$

By partial integration of (5.14) one obtains the following form for the superposition principle

$$\boxed{\gamma(t) = J_0\sigma(t) + \int_{-\infty}^{t} \dot{J}(t-\xi)\sigma(\xi)d\xi} \tag{5.15}$$

with $\dot{J}(t-\xi)$ representing an abbreviation for the expression $(dJ(y)/dy)_{y=t-\xi}$.

In the form of Eq. (5.15), the principle of superposition may also be applied to a *stress history*, which is only *stepwise differentiable* and contains a *limited number* of *finite discontinuities*.

The *dual* form of the superposition principle, in which the deformation is considered as the "stimulus" and the stress as the "response", allows in a similar way the calculation of the stress at time t, if the strain history between $-\infty$ and t is known. Let $\gamma(\xi)$ be the deformation history for $-\infty < \xi \leq t$. Then, the stress at the time t is given by the integral

$$\sigma(t) = \int_{-\infty}^{t} G(t-\xi)\dot{\gamma}(\xi)d\xi \tag{5.16}$$

Partial integration again yields a form for the superposition principle

$$\boxed{\sigma(t) = G_0\gamma(t) + \int_{-\infty}^{t} \dot{G}(t-\xi)\gamma(\xi)d\xi} \tag{5.17}$$

in which the prescribed strain history may be piecewise differentiable and may contain a limited number of finite discontinuities. As before, $\dot{G}(t-\xi)$ represents an abbreviation for the expression $(dG(y)/dy)_{y=t-\xi}$.

The Eqs. (5.15) and (5.17) contain all essential features of the theory of linear viscoelastic behavior in shear. The two functions $J(t)$ and $G(t)$ are called

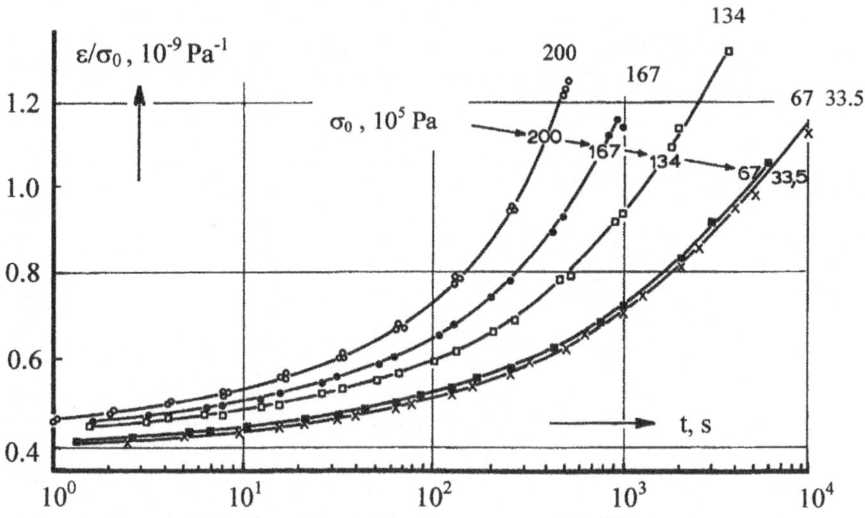

Fig. 5.5 Non-linear creep behavior of PVC at 60 °C under different tensile stresses

characteristic functions, as they describe the viscoelastic behavior in shear completely. It is seen that for the calculation of the strain at time *t*, the complete history of the stress up to the time *t* is involved. The form of the function *J*(*t*) determines how long the memory has an effect or, after which time lag the history has been forgotten.

The superposition principle in the form of Eqs. (5.15) and (5.17) represents the fundamentals of the technical mechanics of polymeric materials in the linear range of deformation. Therefore, the important question regarding the limits of validity of the superposition principle arises. Generally, it is assumed, that the superposition principle offers a correct description of the mechanical behavior of polymeric materials, as long as the applied stresses are not too high, and as long as the occurring deformations are not too large, that means as long as $|\sigma| < \sigma_l$ and $|\gamma| < \gamma_l$. The limits of the range of validity of the superposition principle σ_l and γ_l are called *linearity limits*. Their values can only be determined approximately and will depend, amongst others on the accuracy of the measuring techniques. The following estimates are given under the assumption of an experimental error of about 3 %. This is the accuracy, which at present may be reached for the determination of moduli or compliances. The value of the limits of linearity will also depend on the type of the experiment, which is considered. The following data refer to uniaxial extension. The limits of linearity in shear can be expected to be of the same order of magnitude.

The change from the linear to the non-linear creep behavior in tension is shown in Fig. 5.5 for a PVC at 60 °C. This figure shows the creep compliances in extension as a function of the logarithm of the creep time, under the various constant tensile stresses indicated. In the case of a linear creep behavior, all

Table 5.1 Linearity limits σ_l and ε_l under uniaxial stress in creep and stress relaxation

Amorphous polymers	T_g °C	T °C	σ_l in MPa	ε_l in %	σ_b in MPa	Lit.
NR cross-linked		35	4.7	150		[3]
	−72	70	5.6	150		[3]
		100	1.5	100		[4, 5]
PIB uncross-linked	−70	−70	0.8	10		[6]
		−45	0.4	50		[6]
		30	0.27	100		[7]
PC	149	23	25	1.2		[2]
		85	22.7	1.2		[2]
		100	17.7	1.1		[2]
		120	12.8	0.85		[2]
		130	10	0.80		[2]
		167	0.2	25		[8]
PVC	80	20	15	0.5	54	[9]
PMMA	105	20	17	0.6	74	[9]
Semicrystalline polymers						
LDPE		20	2.0	1.0	11	[9]
PP		20	4.9	0.4	32	[9]
PA 6.6		20	28	0.9	59	[9]
POM		20	17	0.6	67	[9]

compliances should coincide at one common curve. As can be seen that is only the case up to tensile stresses of $67 \cdot 10^5$ Pa. Under higher stresses, the compliance curves shift systematically into the direction of shorter times. We conclude that the linearity limit for this material is about $5 \cdot 10^6$ Pa. This result is in agreement with the direction of the deviation from linearity in the case of dispersion regions: The transitions are shifted to shorter times by higher stresses.

A systematic survey on the limits of linearity has been reported by Yannas [2], who has performed creep and relaxation measurements on NR, PIB, and PC. In the glassy state, the limits of linearity were found for the stresses between 10 and 20 N/mm^2, for the strains at about 1 %.

Table 5.1 lists some limits of linearity from the literature. The table also contains in some cases, the tensile stresses σ_b for rupture after 100 s creep time. A comparison between σ_b and σ_l shows, that only a small part of the deformation of the polymer fulfils the conditions for the description by means of the theory of linear viscoelasticity. For PVC, for instance, the stress limit of linearity only amounts to one fourth of the stress to rupture.

5.4 Relaxation and Retardation Spectra

Considering the time dependences of the creep compliance and the relaxation modulus, two cases may be distinguished:

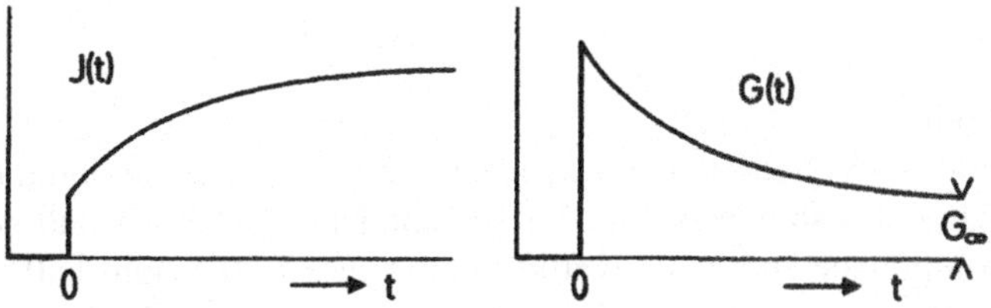

Fig. 5.6 Creep compliance and stress relaxation modulus, case A viscoelasticity without flow

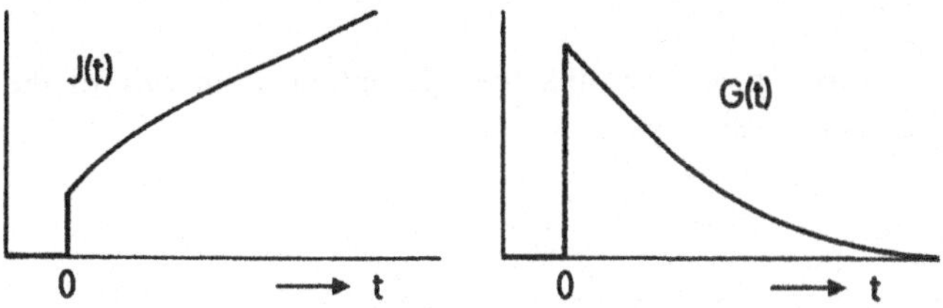

Fig. 5.7 Creep compliance and stress relaxation modulus, case B viscoelasticity with flow

(A) *viscoelastic behavior without flow* and
(B) *viscoelastic behavior with flow.*

The creep compliance $J(t)$ is a positive function, which increases monotonously in the entire positive time interval. Its slope is maximal at the time $t = 0$ and decreases monotonously with increasing time. After long measuring times, $J(t)$ either attains a straight line with positive slope (case B Fig. 5.7) or reaches a horizontal saturation value (case A Fig. 5.6). In both cases we define the slope of the creep compliance after infinite long-creep times as *fluidity* φ_0 and its reciprocal as *viscosity* η_0.

$$\lim_{t\to\infty} \dot{J}(t) = \varphi_0 = 1/\eta_0 \tag{5.18}$$

Difficulties may arise with the determination of the limiting slope in Eq. (5.18), as very often it is not possible to measure long enough to decide with which one of the cases A or B one has to deal. In those cases, apparent values for φ_0 and η_0 are found, which depend on the duration of the creep measurement.

The relaxation modulus is a positive, monotonously decreasing function of time. Its slope has the strongest negative value at $t = 0$ and increases monotonously until, after infinite times, the slope reaches the value zero. After long times, $G(t)$ relaxes either to zero (case B) or to a positive limit (case A). In both cases we define

$$\lim_{t\to\infty} G(t) = G_\infty \tag{5.19}$$

and call G_∞ the *equilibrium value of the modulus.*

Fluidity and equilibrium value of the modulus are related to each other, as for viscoelastic behavior without flow $\varphi_0 = 0$ and $G_\infty > 0$ and for viscoelastic behavior with flow $\varphi_0 > 0$ and $G_\infty = 0$ or

$$\varphi_0 \cdot G_\infty = G_\infty/\eta_0 = 0 \tag{5.20}$$

is valid in both cases.

The functions $J(t)$ and $G(t)$ and their derivatives play an important role in the following. They are shown in Fig. 5.8 schematically. The left half of the figure shows the compliance and its derivatives in the case B, the right-half the modulus and its derivatives in the case A. The first derivative of the creep compliance is a positive, monotonously decreasing function of time, the second derivative will be negative everywhere and tends to zero with increasing time, the third derivative is again positive, etc.

It is assumed as a postulate, that the alternating monotony of the derivatives remains valid, also for all higher orders, i.e.,

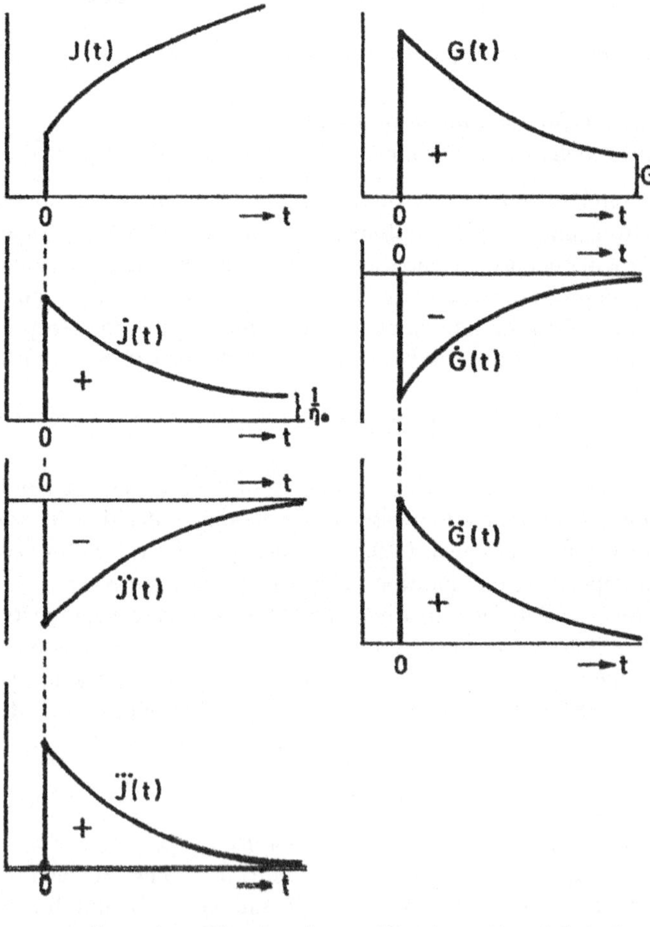

Fig. 5.8 Creep compliance (case B), relaxation modulus (case A) and their time derivatives

$$J(t) > 0,\ \dot{J}(t) > 0,\ \ddot{J}(t) < 0,\ \dddot{J}(t) > 0, \quad \text{or}$$
$$(-1)^{n-1} J^{(n)}(t) > 0 \ \text{for} \ 0 < t < \infty \ \text{and for} \ n = 1, 2, 3, \ldots \tag{5.21}$$

whereby $J^{(n)}(t)$ denotes an abbreviation for the nth derivative of the function $J(t)$. A positive function whose derivatives change its sign with its order, is called *totally monotonous*. The postulate (5.21) expresses the demand that $\dot{J}(t)$ should be a totally monotonous function.

The first time derivative of the relaxation modulus is a negative, monotonously increasing function of time and tends to zero for long times. The second derivative is positive, monotonously decreasing. Again, we postulate that, in this case, $G(t)$ itself should be a *totally monotonous function* of the time

$$(-1)^n G^{(n)}(t) > 0 \ \text{for} \ 0 < t < \infty \ \text{and for} \ n = 0, 1, 2, 3, \ldots \tag{5.22}$$

Because of the superposition principle, the two postulates (5.21) and (5.22) are not independent of each other, from the validity of one of them follows the validity of the other and vice versa. Of course, the validity of a postulate of the form (5.21) can never be proven by a finite number of experiments. All what may be stated is that the consequences of these postulates have been checked in a very large number of experiments with polymers, and so far never have been found to be violated. Therefore, we assume their validity.

From the postulates of the total monotony, far reaching consequences may be drawn concerning the mathematical form of creep compliance and relaxation modulus. The simplest example for a totally monotonous function is the exponential function

$$G(t) = a e^{-t/\tau}$$

with positive values for the constants a and τ. A finite sum of exponential functions with positive constants a_i and τ_i also constitutes a totally monotonous function of t.

$$\boxed{G(t) = G_\infty + \sum_{i=1}^{n} a_i e^{-t/\tau_i}} \tag{5.23}$$

A further possibility to create a totally monotonous function is to replace the finite sum in (5.23) by an integral

$$\boxed{G(t) = G_\infty + \int_0^\infty g(\tau) e^{-t/\tau} d\tau} \tag{5.24}$$

The integration variable τ has the dimension of a time and is called the *relaxation time* and $g(\tau)$ is a nonnegative function of the relaxation time and is called the *relaxation spectrum*.

Vice versa, it has been shown by Bernstein [10], that each totally monotonous function may either be written in a unique manner in the form (5.23) or (5.24) or as a sum of both, with a nonnegative relaxation spectrum. In the case of the representation in the form of Eq. (5.23) one speaks of a line spectrum or discrete spectrum, and $\tau_1, \tau_2, \tau_3, \ldots$ are called *relaxation times*, the nonnegative coefficients a_1, a_2, a_3, *relaxation strengths* and the entity of the n couples (a_i, τ_i) and G_∞ is the *line spectrum* or the *discrete relaxation spectrum*. The function $g(\tau)$ is designated as *continuous relaxation spectrum.*

The difference between a discrete spectrum and a continuous spectrum is a matter of mathematical formality. Though, a mathematical proof can be given, that each completely known totally monotonous function may be represented uniquely by either (5.23) or (5.24), this is not true for experimentally obtained moduli. These functions are always known only for a *limited time interval*, and only *with finite accuracy*. For those functions a discrete spectral representation is always even accurate as a continuous representation. Moreover, by the finite accuracy of the measurement, the uniqueness of the representation by the spectrum is lost. Each experimentally measured relaxation function may be represented by an infinite number of different spectra, which all describe the experiment within a given small, but finite, experimental error. This is meant, when one speaks about the determination of spectra from experiments as an *ill-defined mathematical problem.*

For the mathematical description of the creep compliance, we start from the postulate, that $\dot{J}(t)$ should be a totally monotonous function and, therefore, can be presented as

$$\dot{J}(t) = \int_0^\infty s(\tau) e^{-t/\tau} d\tau + \frac{1}{\eta_0}$$

with a nonnegative function $s(\tau)$. Integration of this equation with respect to t and designating the function $\tau s(\tau)$ by $f(\tau)$ leads to the formula

$$\boxed{J(t) = J_0 + \int_0^\infty f(\tau)[1 - e^{-t/\tau}] d\tau + t/\eta_0} \tag{5.25}$$

The integration constant J_0 turns out to be the instantaneous compliance of Eq. (5.8). The integration variable τ is called here the *retardation time* and the nonnegative function $f(\tau)$ the *retardation spectrum*. In the case of a line spectrum we obtain instead of (5.25)

$$\boxed{J(t) = J_0 + \sum_{i=1}^{m} b_i [1 - e^{-t/\bar{\tau}_i}] + t/\eta_0} \tag{5.26}$$

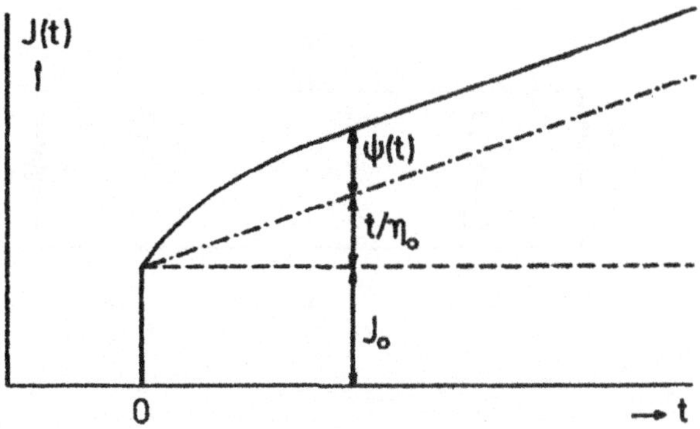

Fig. 5.9 Decomposition of the creep compliance into its three components

$\bar{\tau}_1, \bar{\tau}_2, \bar{\tau}_3, \ldots$ are called *retardation times*, the nonnegative coefficients b_1, b_2, b_3, the *retardation strengths* and the entity of the m couples (b_i, $\bar{\tau}_i$) and J_0 and η_0 is the *line spectrum* or the *discrete retardation spectrum.*

Equations (5.25) and (5.26) suggest a decomposition of the creep compliance into three components, which behave differently in a creep and a recovery experiment.

$$J(t) = J_0 + \psi(t) + t/\eta_0 \tag{5.27}$$

with

$$\psi(t) = \int_0^\infty f(\tau)[1 - e^{-t/\tau}]d\tau \tag{5.28}$$

or

$$\psi(t) = \sum_{i=1}^{m} b_i[1 - e^{-t/\bar{\tau}_i}] \tag{5.29}$$

J_0 is the *instantaneous creep compliance*, which is time independent and mechanically reversible, $\psi(t)$ the *retarded viscoelastic compliance*, which is time dependent and mechanically reversible, and t/η_0 is the *flow term*, which is time dependent and mechanically irreversible.

If the creep compliance has been measured completely, i.e., if J_0 and η_0 both have been determined, this decomposition may be performed as explained in Fig. 5.9. First, the constant value of J_0 is subtracted from the compliance. Then a straight line with the slope $1/\eta_0$ is drawn through the point (0, J_0). The remainder $J(t) - J_0 - t/\eta_0$ is the retarded viscoelastic compliance $\psi(t)$. In practice, often not sufficient experimental information is available to perform this decomposition of the creep compliance reliably.

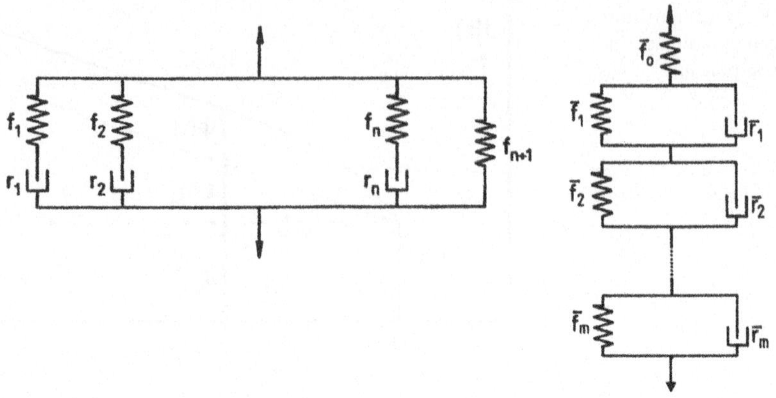

Fig. 5.10 *Left side* The generalized Maxwell model, showing viscoelastic behavior without flow and *right side*: Its mechanically equivalent Kelvin model with $m = n$

The viscoelastic deformation behavior is often described by *mechanical circuit diagrams*, consisting of *linear springs* and *linear dashpots*, which are linked together, the so-called *mechanical viscoelastic models*. The elongation of the model represents the resulting strain, the force on the upper and lower end the resulting stress. A spring and a dashpot arranged in series is called a *Maxwell model*, their parallel link a *Kelvin model*.

Figure 5.10 shows, on its left-hand side, the so-called *generalized Maxwell model*, which consists of n Maxwell-models and one single spring linked in parallel.

It is easily seen that the stress relaxation modulus of the generalized Maxwell model is given by Eq. (5.23), if we assume the following relations between the constants of the model and the discrete relaxation spectrum

$$a_i = f_i; \quad \tau_i = r_i/f_i \text{ for } i = 1, 2, 3, \ldots n \text{ and } G_\infty = f_{n+1} \tag{5.30}$$

This model shows viscoelastic behavior without flow (case A).

The right-hand side of Fig. 5.10 shows a model composed of n Maxwell elements and one spring linked in series. This model may show the same viscoelastic behavior as the model on the left-hand side, on condition that a number of relations is valid between the constants of the generalized Maxwell model (f_i, r_i) and those of the model on the right-hand side ($\bar{f}_i, \bar{r}_i$). In this case we call the two models *mechanically equivalent*. For a detailed description of the models and of the mentioned relations, we refer to the contributions in [11–13].

Figure 5.11 shows the so-called *generalized Kelvin model*, which consists of m Kelvin elements, one spring and one dashpot, all linked in series. The creep compliance of the generalized Kelvin model is seen to be identical with Eq. (5.26), if we assume the following relations between its constants and the discrete retardation spectrum

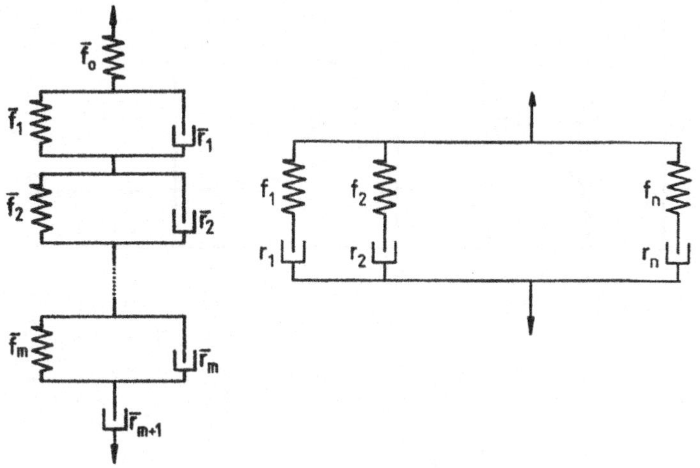

Fig. 5.11 *Left side* The generalized Kelvin model, showing viscoelastic behavior with flow and *right side*: its mechanical equivalent Maxwell model with $n = m + 1$

$$J_0 = 1/\bar{f}_0; b_i = 1/\bar{f}_i;\ \ \bar{\tau}_i = \bar{r}_i/\bar{f}_i \text{ for } i = 1, 2, \ldots m \text{ and } \eta_0 = \bar{r}_{m+1}; \tag{5.31}$$

This model describes viscoelastic behavior with flow (case B).

The generalized Kelvin model shown on the left-hand side of Fig. 5.11 behaves under stress exactly as a generalized Maxwell model without the last spring, as shown on the right-hand side of the Fig. 5.11. Again, a number of relations have to be valid between the constants of both models, and $n = m + 1$.

5.5 The Creep Recovery Experiment

Figure 5.12 shows the stress and strain during a creep recovery experiment. Between $t = 0$ and $t = t_0$, the sample is loaded with a constant stress σ_0 and its strain is measured. At $t = t_0$ the load is suddenly removed and the *residual strain* of the sample during the *recovery period*, $\gamma(t_0, t')$ is measured as a function of the *recovery time* t'.

The *residual strain* depends on both, the *creep time* t_0 and the *recovery time* t'. Its magnitude is found by inserting the definition of the creep recovery experiment (5.1) into the superposition principle (5.15) as

$$\gamma(t_0, t') = \sigma_0 \cdot [J(t) - J(t - t_0)] = \sigma_0 \cdot [J(t_0 + t') - J(t')] \tag{5.32}$$

The *retraction* $\gamma_r(t_0, t')$ of the specimen during the recovery period is found by subtracting the residual strain $\gamma(t_0, t')$ from the strain which had been reached just before unloading $\sigma_0 J(t_0)$. The retraction (*the recoverable strain*) during the recovery phase is found as

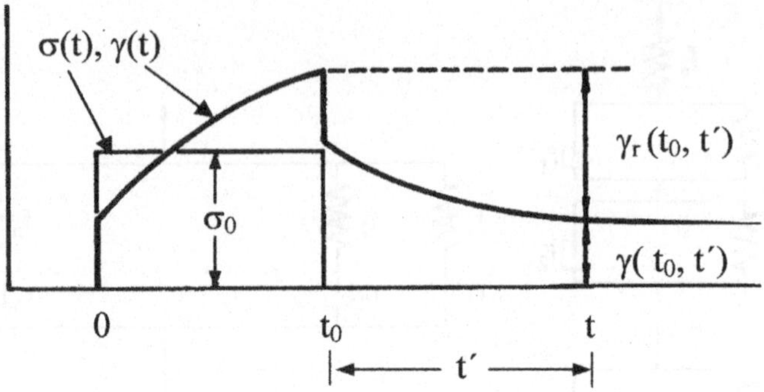

Fig. 5.12 The creep recovery experiment, the retraction $\gamma_r\ (t_0,t')$ and the residual strain $\gamma(t_0,t')$

$$\gamma_r(t_0, t') = \sigma_0 \cdot [J(t_0) + J(t') - J(t_0 + t')] \tag{5.33}$$

Introducing the decomposition (5.27) into these equations yields

$$\gamma(t_0, t') = \sigma_0 \cdot [\psi(t_0 + t') - \psi(t')] + \sigma_0 t_0/\eta_0 \tag{5.32a}$$

$$\gamma_r(t_0, t') = \sigma_0 \cdot [J_0 + \psi(t_0) + \psi(t') - \psi(t_0 + t')] \tag{5.33a}$$

Expressing ψ by the retardation spectrum (5.28) leads to the following formulae for the retraction and for the residual strain divided by the creep stress

$$\gamma(t_0, t')/\sigma_0 = \int_0^\infty f(\tau)e^{-t'/\tau}[1 - e^{-t_0/\tau}]d\tau + t_0/\eta_0 \tag{5.34}$$

$$\gamma_r(t_0, t')/\sigma_0 = J_0 + \int_0^\infty f(\tau)[1 - e^{-t_0/\tau}][1 - e^{-t'/\tau}]d\tau \tag{5.35}$$

$\gamma_r(t_0, t')/\sigma_0$ is a monotonously increasing function of both times, the creep time t_0, and the recovery time t'. It will therefore reach its maximum value after infinitely long creep times and infinitely long recovery times. This maximum value is called the *stationary recoverable compliance* J_e^0 and has attracted considerable attention of experimentalists, recently, as its value offers information about the deformation and orientation of the macromolecules during stationary flow.[2]

[2] The stationary recoverable compliance is designated here and throughout Chaps. 5 and 6 by either J_e or by J_e^0. Later, in considering non-linear creep behavior, the designation J_e will be used only for the non-linear stationary recoverable compliance.

$$J_e^0 = \lim_{\substack{t_0 \to \infty \\ t' \to \infty}} \gamma_r(t_0, t')/\sigma_0 = J_0 + \int_0^\infty f(\tau) d\tau \tag{5.36}$$

In the case of a discrete retardation spectrum, the stationary recoverable compliance becomes instead of (5.36)

$$J_e^0 = J_0 + \sum_{i=1}^{m} b_i \tag{5.37}$$

Especially the question, whether t_O and t' have been chosen long enough to guarantee sufficient accuracy in the approximation of the limit in Eq. (5.36) or (5.37) has been a matter of interest. This however, is a very delicate question, as it depends on the detailed structure of the retardation spectrum in the long-time region. For, the error in the determination of J_e will be equal to

$$J_e^0 - \gamma_r(t_0, t')/\sigma_0 = \int_0^\infty f(\tau)[e^{-t_0/\tau} + e^{-t'/\tau} - e^{-(t_0+t')/\tau}] d\tau \tag{5.38}$$

which may be written in the case of a discrete retardation spectrum as

$$J_e^0 - \gamma_r(t_0, t')/\sigma_0 = [\psi(\infty) - \psi(t_0)] + [\psi(\infty) - \psi(t')] - [\psi(\infty) - \psi(t_0 + t')] \tag{5.38a}$$

In replacing the negative exponential function by a unit step function, which drops from unity to zero at the time $t = \bar{\tau}_i$, the differences of the retarded viscoelastic compliance may be approximated as follows:

$$\psi(\infty) - \psi(t') = \sum_{i=1}^{m} b_i e^{-t'/\bar{\tau}_i} \cong \sum_{i=r+1}^{m} b_i \tag{5.39}$$

if $\bar{\tau}_r < t' < \bar{\tau}_{r+1}$. Introducing this approximation into Eq. (5.38a) yields the following approximation for the error in the determination of the stationary recoverable compliance under the assumption $t' < t_0$

$$J_e^0 - \gamma_r(t_0, t')/\sigma_0 \cong b_{r+1} \cdots + b_s + 2b_{s+1} + \cdots + 2b_u + b_{u+1} + \cdots + b_m \tag{5.40}$$

where the indices are defined by the equations $\bar{\tau}_r < t' < \bar{\tau}_{r+1}$, $\bar{\tau}_s < t_0 < \bar{\tau}_{s+1}$ and $\bar{\tau}_u < t_0 + t' < \bar{\tau}_{u+1}$. Every retardation strength, whose retardation time exceeds either t_0 or t', will contribute once or twice to an error in the determination of the limits in Eq. (5.36) and (5.37). The calculated limits will *always* be *too small*.

If the creep time t_0 has been chosen long enough, the first bracket in (5.35) will reduce to unity, and the retraction per unit stress will tend to the limit

$$\lim_{t_0 \to \infty} \left[\gamma_r(t_0,t')/\sigma_0\right] = J_R(t') = J_e^0 + \int_0^\infty f(\tau)[1 - e^{-t'/\tau}]d\tau = J(t') - t'/\eta_0 \quad (5.41)$$

$J_R(t')$ is called the *recoverable compliance*, which is equal to the creep compliance, without the flow term. The problem in reaching the limit in Eq. (5.41) is similar to that of the limit in (5.36).

After long-creep times and long-recovery times, the residual strain (5.34) will have decreased to its *lower limit*

$$\lim_{t_0 \to \infty, t' \to \infty} \gamma(t_0,t') = \sigma_0 t_0/\eta_0 \quad (5.42)$$

which equals the viscous contribution of the strain during the creep period. This equation is sometimes used to determine the viscosity by means of a recovery experiment. Again, difficulties in connection with the limiting process in (5.42) will arise.

Finally, a further way for the determination of the viscosity is to plot the creep compliance, divided by the creep time, versus the logarithm of the creep time. After creep times long enough a lower limit for the quotient will be reached, which is equal to the inverse viscosity

$$\lim_{t \to \infty} [J(t)/t] = 1/\eta_0 = \varphi_0 \quad (5.43)$$

It is highly recommended, in this limiting process as in the other mentioned cases, to use logarithmic time scales for the abscissa, to elucidate whether the limiting value was really attained.

5.6 The Creep Compliance of Amorphous Polymers

Summarizing the considerations on the decomposition of creep compliance and relaxation modulus, we state: The quantities J_0, G_0, $\varphi_0 = 1/\eta_0$, G_∞, J_e (or J_e^0) and the function $\psi(t)$ are defined by mathematical extrapolation processes, and, therefore, apart from exceptional cases not directly accessible by the experiment. Quantities, which may be always determined, are the creep compliance and the relaxation modulus as a function of the time, but within a limited time interval, only,

$$J(t), \quad G(t) \qquad \text{for} \qquad t_{min} < t < t_{max}$$

with t_{min} being the time needed between loading of the specimen and getting the first reliable measuring point, while t_{max} depends on the patience of the experimentalist, the stability of the measuring system and of the material. In favorable cases, for creep or stress relaxation $t_{min} \approx 0.1$ s and $t_{max} \approx 3 \times 10^7$ s (1 year).

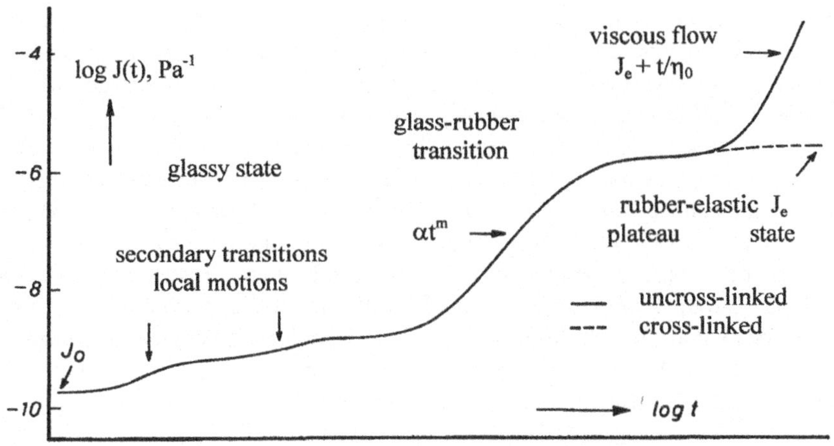

Fig. 5.13 Schematic dependence of the creep compliance on time in a double-logarithmic plot for amorphous, uncross-linked polymers (*full line*) and for amorphous cross-linked polymers (*dashed line*)

Though more than 8 decades of the time scale will be accessible as *experimental window*, this is this still insufficient to cover all relaxation processes which may occur in an amorphous polymer at one single temperature.

The creep compliance of amorphous polymers changes in such a wide range and within such a wide time region, that a comprehensive survey can be presented only in a double logarithmic plot. A schematic picture is given in Fig. 5.13. A scale for the logarithmic time axis is not given for reasons which will become clear soon.

In the double-logarithmic plot, the creep compliance shows a *number* of *dispersion steps*. Each step represents a relaxation phenomenon, which just develops in the corresponding time region and starts there to contribute to the deformation process. From left to right, in the Fig. 5.13, the size of the parts of the molecules, which participate in these processes, increase.

At very short creep times, the compliance of the order of $J_0 \cong 3\text{–}5 \times 10^{-10}$ Pa^{-1} is very low. On loading, only van der Waals' bonds, valence bonds and valence angles are distorted. The material behaves stiff, elastic, and brittle. As van der Waals' bonds are much weaker than covalent bonds, the van der Waals' bonds will chiefly determine the value of the short-time compliance.

At somewhat longer creep times, dispersion regions are found, which lead to a moderate increase in the compliance (the compliances increase there to the 1.5–2 fold of their value before the process). These dispersions are related to the motion of small parts of the macromolecules: Either rotations or changes in the conformation of side groups or of small parts of the principle chains are involved. These dispersions are called *secondary dispersions* in the glassy state, indicating their moderate influence on the change of mechanical properties in comparison to the much stronger influence of the glass-transition. Secondary transitions are *material specific*; various polymers show different secondary transitions, according to their chemical structure. The material is still in the glassy state, though, it may loose some of its brittleness after having passed through one of the secondary transitions.

After the glassy state, a very strong dispersion region is observed, which is common to *all amorphous polymers*, though the details of the dispersion may change with the molecular structure. This dispersion region is called the *glass-rubber transition* or the *softening region*, as it leads from the glassy state to the much softer rubbery state or rubbery plateau with a shear compliance which is about a factor 1000–10000 times higher than the compliance in the glassy state. The glass-rubber transition extends over 6–8 decades in the logarithmic time scale, the slope of $J(t)$ versus t in the double-logarithmic plot is approximately constant and reaches values, which differ from polymer to polymer, but generally are found between 0.50 and 0.90. An approximate description of the compliance- time relation in the glass-rubber transition is therefore

$$\log J \cong m \cdot \log t + C \quad \text{or} \quad J(t) \cong \alpha \cdot t^m \tag{5.44}$$

The molecular process responsible for the glass-rubber transition is the uncoiling and change in shape of the polymer chains between adjacent entanglements or cross-links, leading to much higher extensibilities of the chains and therefore to much higher values of the compliance (cf. Fig. 4.9).

After the glass-rubber transition follows the *rubber-elastic plateau* (for uncross-linked polymers) or the *rubber-elastic state* (for cross-linked polymers). There, the values for the compliance increase much less with creep time than in the glass-rubber transition and are of the order of

$$J \cong 1/(\text{nkT}) \cong 10^{-6}\,\text{to}\;\, 10^{-5}\text{Pa}^{-1} \tag{5.45}$$

with n being explained below.

For amorphous cross-linked polymers (the dashed line in Fig. 5.13), the compliance remains approximately constant as a function of time and approaches the limiting value $J(\infty) = J_e$, because those polymers do not show any viscous flow. The molecular process governing the deformation in the rubber-elastic state is the change in shape of the macromolecules under the influence of mechanical forces. The magnitude of the deformation is limited by the entropic forces of the temperature kicks, which drive the molecules back into their isotropic shape. These forces may be calculated by means of the theory of rubber elasticity and are found to be proportional to the absolute temperature and to the number of cross-links per unit volume, $n = n_c$ (cf. Eq. (5.45)).

Also uncross-linked polymers with sufficiently high-molar masses show rubber-elastic behavior in a certain time region, which we call the *rubber-elastic plateau* in this case. The reason for it is the presence of entanglements, which—for some time—have a similar effect as cross-links. In this case Eq. (5.45) is applicable too, with $n = n_e$ being the number of effective entanglements per unit volume.

At longer times, however, the macromolecules will slip out of the entanglements and will move into the direction of the shear forces, giving rise to viscous

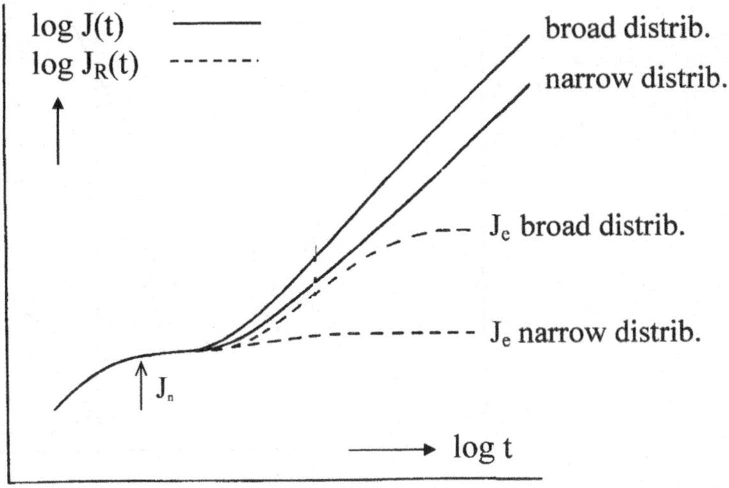

Fig. 5.14 Creep compliances (*full lines*) and recoverable creep compliances (*dashed lines*) as a function of the time, for uncross-linked polymers with a narrow and a broad molar mass distribution

flow behavior. At the end of the flow region, the shear behavior of the melt may be approximated by

$$J(t) \cong J_e^0 + t/\eta_0 \tag{5.46}$$

and the double logarithmic slope of J versus t finally approaches the value unity.

The picture of Fig. 5.13 is a simplification, as between the rubber-elastic plateau and the flow region at least one additional dispersion region occurs, which has been called the *entanglement transition*, as it is supposed to be connected with some change of the structure of the entanglement network. If only the creep compliance is observed, this transition is often masked by the flow term which has already reached a considerable magnitude there. It may be clearly observed, however, if the recoverable creep compliance, $J_R(t)$, which does not contain the flow term, is measured instead of the creep compliance.

In Fig. 5.14, the quantities $J(t)$ and $J_R(t)$ are sketched in a double logarithmic representation from the beginning of the rubber-elastic plateau to the flow region. The curves describe the behavior of two uncross-linked polymers with about the same weight average molar mass, one with a narrow molar mass distribution and the other with a broad molar mass distribution. The step height of the transition depends on the molar mass distribution. While the compliance starts at the beginning of the plateau with a value of about J_n, it reaches at the end of the plateau a value J_e. The ratio J_e/J_n may amount to a factor 2–3 for narrowly distributed polymers, it may increase however, for broadly distributed polymers up to values of 10–100. In the creep behavior this difference is reflected by the width of the entanglement transition, which is much more pronounced for the broadly distributed polymers.

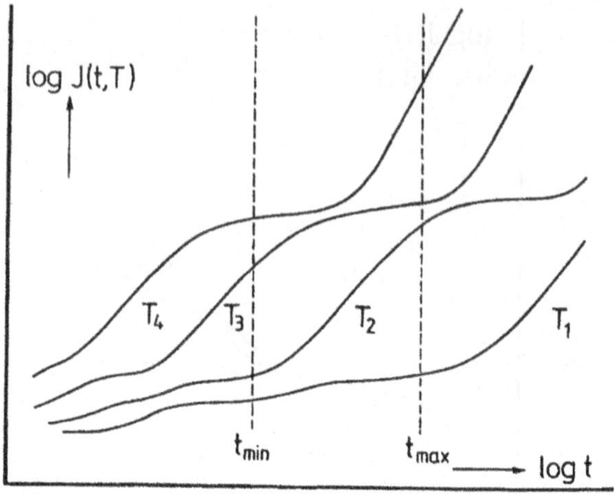

Fig. 5.15 Creep compliances as a function of creep time for an uncross-linked amorphous polymer at different temperatures ($T_1 < T_2 < T_3 < T_4$)

Figure 5.13 describes the compliance as a function of time at one fixed temperature. The influence of the temperature consists in a shift of the position of the various dispersion regions on the logarithmic time scale without changing much the heights of the dispersion steps and their shape. This empirical finding is called the *time-temperature superposition principle* which will be discussed in more detail in Chap. 6. An increase in temperature shifts all relaxation phenomena to shorter times on the logarithmic timescale. The magnitude of this shift is about equal for the softening region and for the flow transition. For secondary transitions this shift is smaller. Figure 5.15 shows the influence of the temperature on the time dependence of the compliances for an amorphous uncross-linked polymer. Because of the different amounts in the shift for the glass-rubber transition and the secondary transitions, the dispersion steps come closer together with higher temperatures. In the same figure, we have exemplarily indicated the width of the experimental window [$t_{\min}$, $t_{\max}$], which covers only a small part of the interesting time scale. It is therefore impossible, to measure the curves which were presented in Fig. 5.13, at one single temperature. On the contrary, different parts of the curves have been determined at different temperatures and these parts have been composed to one single curve, using the time-temperature superposition principle.

In the following, the creep compliance of a technical polystyrene with a broad molar mass distribution, which has been presented in Fig. 2.6, is discussed. Figure 5.16 shows its creep compliance at different temperatures as a function of the creep time in a double-logarithmic plot in all the various states of this polymer from extremely low temperatures up to the melt and the limit of thermal decomposition. The picture shows the compliance over a logarithmic time axis of 7 decades, while the value of the measured compliance itself changes by about 9 decades.

In order to obtain reliable results within such a broad experimental window and within all states of consistency, an extreme experimental effort was necessary. Measurements in shear on materials which are able to keep their shape under the

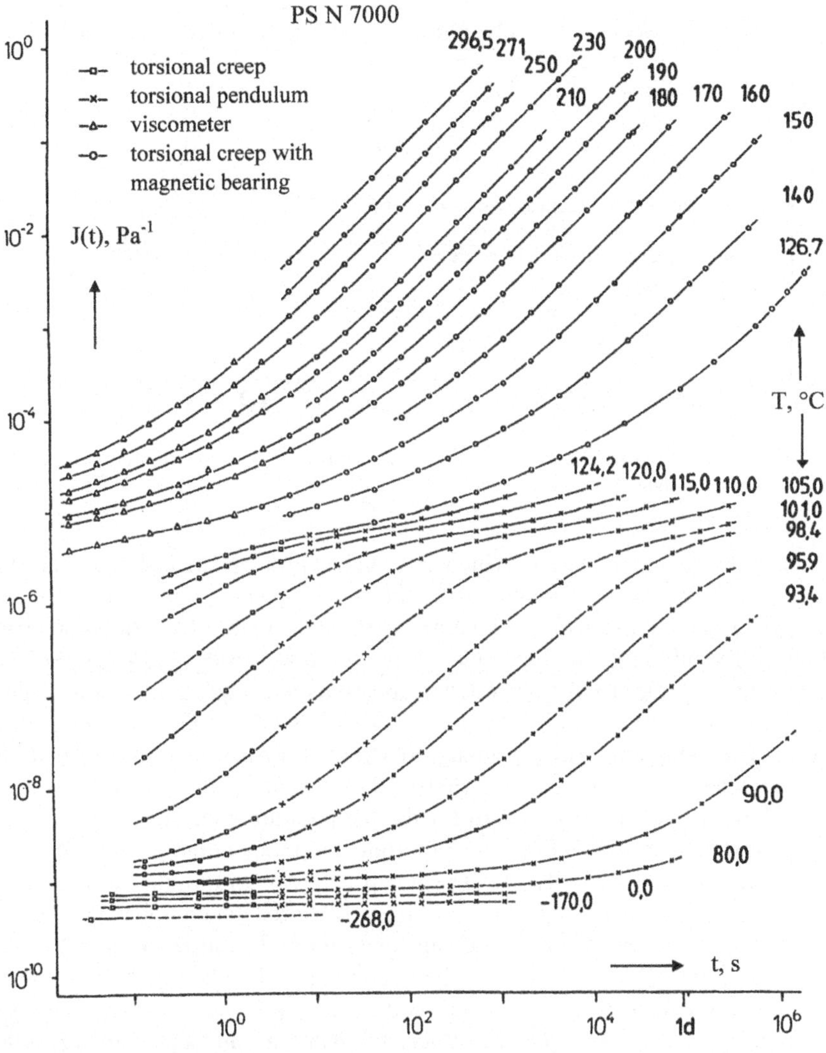

Fig. 5.16 Shear creep compliance versus creep time for PS N 7000 at various temperatures in a double-logarithmic representation

forces of gravity (the measurement of compliances with a value below 10^{-6} Pa^{-1}), can best be performed accurately by the torsion of cylindrical rods with a length to diameter ratio of about 10. For this purpose a specially developed torsional creep apparatus was used [14]. Furthermore, for measurements within the glass-rubber transition, a very accurate and stable thermostating system is necessary. This was constructed based on the recirculation of heated nitrogen gas [15]. Its long-term stability and reproducibility is about 0.2 °K with respect to the absolute value of the temperature.

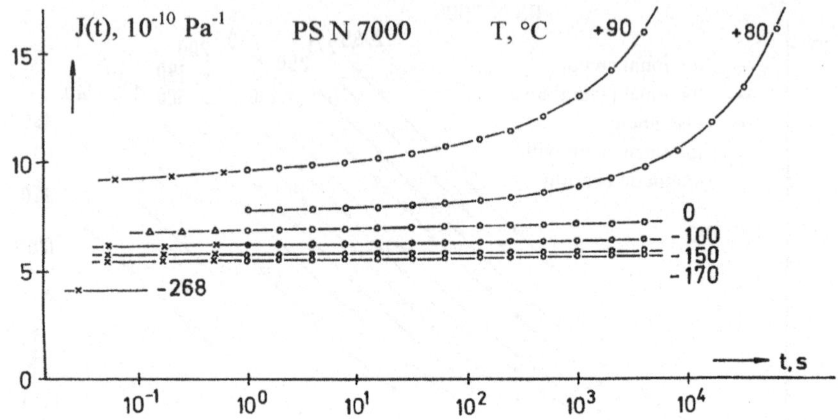

Fig. 5.17 Compliance on a linear scale versus the creep time on a logarithmic scale for PS N 7000 in the glassy state

Measurements on materials which are unable to keep their shape against the forces of gravity (measurements within the rubber-elastic plateau or in the melt) were performed by means of a dynamic viscometer with a Couette geometry by Pfandl [16] and with an apparatus originally developed by Plazek [17] and reproduced by Link [18, 19]. The latter apparatus was equipped with a magnetic bearing in order to minimize the friction and the disturbing moments of the bearing. This apparatus was especially constructed for the direct determination of the recoverable compliance in the rubber-elastic plateau. It uses the torsion of a disk and an electric heating system for the temperature control.

The dependence of the compliance on time and temperature is very different in the various states of the polymer. The creep behavior in the glassy state is once more shown in Fig. 5.17, but now on a linear scale for the compliance. The weak dependence of the compliance on creep time and temperature in the glassy state becomes obvious. From −170 °C to the beginning of the glass-transition (+80 °C) the creep compliance does not depend on the creep time and only weakly depends on temperature. The PS may be described there as an approximately elastic material with a small temperature dependence. This is due to the fact that PS does not show secondary transitions in this temperature range. A value for the compliance at −268 °C, has been indicated [20].

The compliance in the glass-rubber transition is well documented in Fig. 5.16. The complete dispersion step extends over 7 decades of the time axis, the step height amounts to 3.6 decades of the compliance. The double-logarithmic slope of the compliance versus time reaches a maximum value of 0.89. The time-temperature shift is the strongest here. A change in temperature of 3 K shifts the creep compliance curves by a factor of ten on the time axis. In this region, the compliance of the material strongly depends on temperature and creep time, i.e., the viscoelastic character of the material is especially pronounced there.

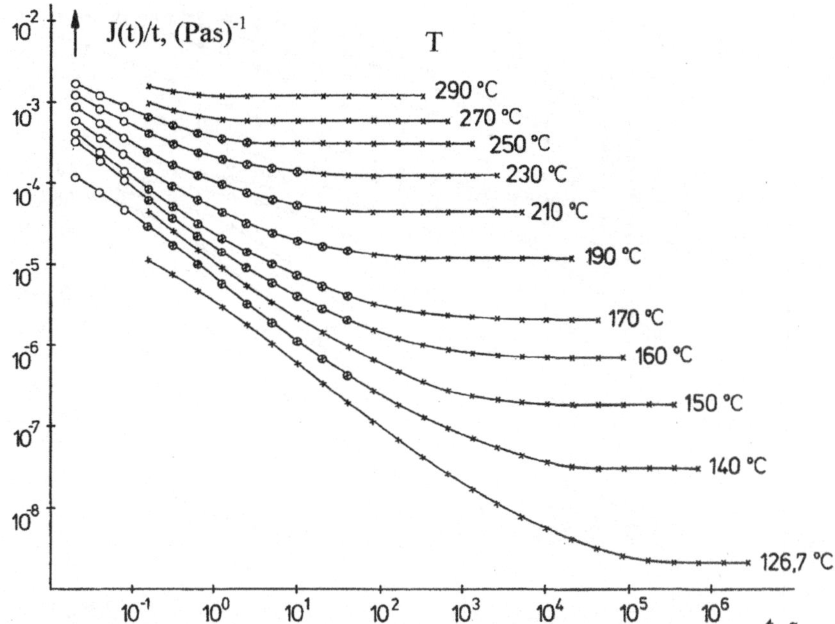

Fig. 5.18 $J(t)/t$ versus t in double logarithmic representation for PS N 7000 in the rubber-elastic plateau and in the melt, after Pfandl [16]

The creep behavior in the rubber-elastic plateau and in the melt becomes also obvious from Fig. 5.16. The rubber-elastic plateau is not very pronounced for the polystyrene measured, however, $J(t)$ in the plateau region is not constant, but slightly increases with time. As already mentioned, this behavior is related to the broad molar mass distribution. For approximately monodisperse polymers, a much more distinct rubber-elastic plateau is found.

At high temperatures and at long creep times, the polymer is in the so-called *molten state*. There the compliance is represented in a double-logarithmic scale by a straight line with the slope one. The viscosity may be obtained by making use of the extrapolation based on Eq. (5.43). In Fig. 5.18 the quotient $J(t)/t$ is shown in a double-logarithmic representation versus the creep time for PS N 7000 in the rubber-elastic plateau and in the melt. $J(t)/t$ is a monotonously decreasing function of the creep time, which finally reaches a constant horizontal plateau indicating that the *stationary state of flow* has been reached. The viscosity may be calculated in the temperature range between 126 and 290 °C using Eq. (5.43). Figure 5.18 is a very convincing illustration of the time scale needed to get reliable viscosity data from the application of Eq. (5.43).

The recoverable compliance $J_R(t)$ as a function of the recovery time t has been measured using the creep apparatus with magnetic bearing and is shown for the PS N 7000 in a double-logarithmic representation in Fig. 5.19 [18]. A direct measurement was possible only for temperatures above 140 °C.

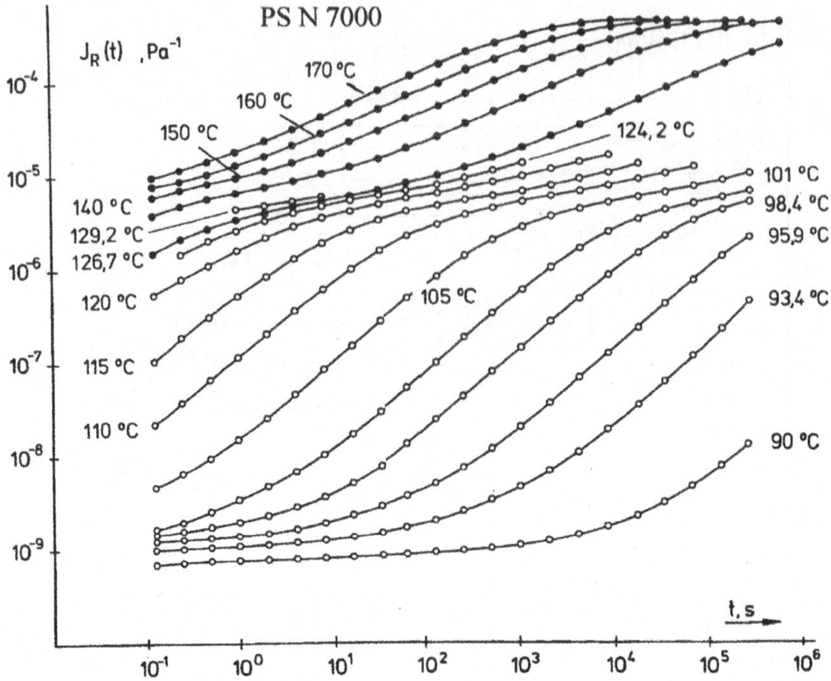

Fig. 5.19 Recoverable compliance versus the recovery time for PS at various temperatures in a double-logarithmic representation, after Link [18]

For lower temperatures the so-called *Leaderman technique* [21] has been used. The stationary state of flow has been realized by performing a creep experiment at the temperature of 170 °C, than the specimen was cooled to the desired temperature and the retraction of the unloaded specimen was measured. The basis of this technique is the assumption that a stationary state which had been reached at higher temperatures, remains undisturbed through cooling, as long as the specimen stays under constant stress. If this hypothesis is true it is possible to reach a stationary state of flow within a reasonable time and to determine afterward the viscosity from the shear rate and the recoverable compliance at lower temperatures.

At temperatures below 126 °C just the creep compliance has been measured as a function of the creep time, as in this temperature region the contribution of the flow term to the creep compliance is negligible and therefore $J_R(t)$ equals $J(t)$. Consequently, in Fig. 5.19 the recovery time has been used as abscissa for the representation of J_R and designated by t, while for the measurements below 126 °C the creep compliance is plotted as a function of the creep time.

Figure 5.19 shows the existence of a further distinct relaxation process between the glass-rubber transition and the flow transition for the recoverable compliance $J_R(t)$, the *entanglement transition*. The time-temperature shifts of the three transitions shown in Figs. 5.16 and 5.19 will be discussed in detail in Chap. 6.

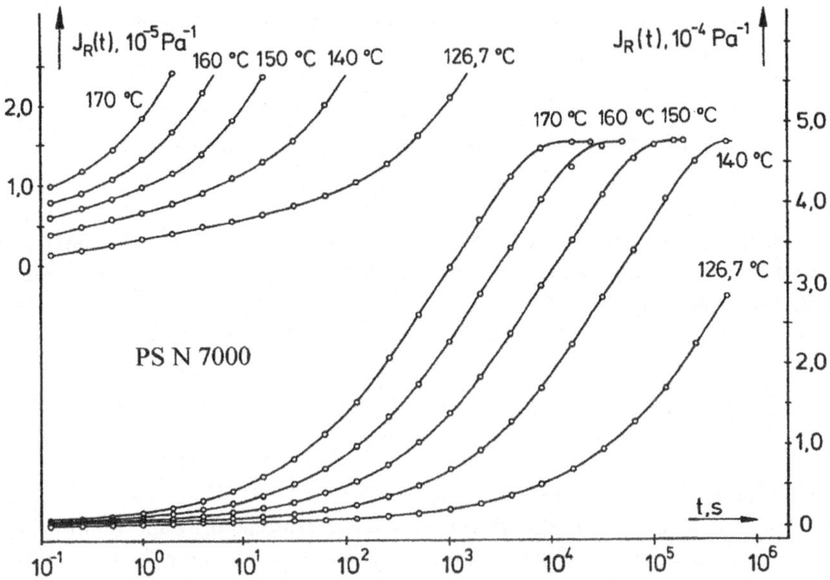

Fig. 5.20 Entanglement transition for PS in semi-logarithmic representation after Link [18]. The curves on the *left* represent a tenfold magnification of the curves on the *right*

This transition becomes still more pronounced if it is represented in a semi-logarithmic plot, as in Fig. 5.20. From this figure, the very large step height of this transition becomes evident, as well as the temperature-independent level of the limiting value $J_R(\infty) = J_e$. The enlarged section of this figure shows the beginning of the entanglement transition and the end of the glass transition.

5.7 Relations Between Creep and Stress Relaxation

As the two equations of the superposition principle (5.15) and (5.17) which express a relation between the stress history and the strain history, have to be consistent with each other, the two characteristic functions $J(t)$ and $G(t)$, cannot be independent. They are related by an integral equation which is derived as follows. Equation (5.17) is used to calculate the stress which belongs to a creep experiment under the constant stress σ_0 after the deformation history $\gamma(t) = 0$ for $t < 0$ and $\gamma(t) = \sigma_0 \cdot J(t)$ for $t > 0$. Inserting $\sigma(t) = \sigma_0$ and $\gamma(t) = \sigma_0 \cdot J(t)$ in (5.17) and division by the constant stress σ_0 lead to

$$1 = G_0 J(t) + \int_0^t \dot{G}(t - \xi) J(\xi) d\xi \qquad (5.47)$$

By partial integration and introduction of a new integration variable $t - \xi = \zeta$ we obtain the equation

$$1 = J_0 G(t) + \int_0^t \dot{J}(t-\xi)G(\xi)d\xi = J_0 G(t) + \int_0^t \dot{J}(\xi)G(t-\xi)d\xi \qquad (5.47a)$$

in which the creep compliance and the relaxation modulus are interchanged with each other. Both equations may be integrated with respect to t and one obtains the symmetrical relation

$$\int_0^t G(t-\xi)J(\xi)d\xi = \int_0^t J(t-\xi)G(\xi)d\xi = t \quad for \quad 0 \leq t < \infty \qquad (5.48)$$

Differentiating of (5.48) with respect to t yields either (5.47) or (5.47a). The four Eqs. (5.47), (5.47a), and both sides of (5.48) are equivalent. Each relation represents an integral equation, from which the creep compliance may be derived, if the relaxation modulus is known and vice versa. If the creep compliance is known in an analytical form, the stress relaxation modulus may sometimes be calculated using one of these equations, but an analytic solution is known only in very simple cases. Two useful examples are given below.

For $t = 0$, Eq. (5.47) reads $1 = G_0 \cdot J_0$, which is the relation between the instantaneous values of creep compliance and modulus (cf. Eq. (5.13)). Adding the identity

$$-J(t)G(t) = -J_0 G(t) - \int_0^t G(t)\dot{J}(\xi)d\xi$$

to Eq. (5.47a), we find

$$1 - J(t)G(t) = \int_0^t \dot{J}(\xi)[G(t-\xi) - G(t)]d\xi \qquad (5.49)$$

As the creep compliance is a monotonously increasing function of time and the relaxation modulus is a monotonously decreasing function of the time, both factors of the integrand in Eq. (5.49) are positive, and therefore the integral in Eq. (5.49) cannot be negative. From this relation the inequality of *Zener* [22] follows for the product of creep compliance and relaxation modulus.

$$0 < J(t) \cdot G(t) \leq 1 \text{ for all } t \geq 0 \tag{5.50}$$

Deviations of this product from unity may occur only in one direction. It will be shown that such deviations occur in the dispersion and flow regions.

In the glass transition region, the creep compliance may be approximated by the expression $J_A(t)$ (5.44):

$$J(t) \cong J_A(t) = \alpha \cdot t^m \tag{5.44a}$$

with α being a positive constant and $0 < m < 1$ (cf. Fig. 5.13). It is shown in the following, that the modulus in the same interval can be described then by the approximation

$$G(t) \cong G_A(t) = \beta \cdot t^{-m} \tag{5.51}$$

For the proof we insert (5.44) and (5.51) into Eq. (5.48) and obtain after the substitution $\xi = t \cdot y$

$$t = \int_0^t J_A(t-\xi) \cdot G_A(\xi) d\xi = t\alpha\beta \int_0^1 y^{-m}(1-y)^m dy$$

The integral is the product of two Γ-functions and may be written as (see [23])

$$\int_0^1 y^{-m}(1-y)^m dy = \Gamma(1+m)\Gamma(1-m) = \frac{m\pi}{\sin(m\pi)}$$

The integral Eq. (5.48) is satisfied, if we set $\alpha\beta = \sin(m\pi)/(m\pi)$. Therefore in the *glass-transition region* the approximation follows

$$J(t) \cdot G(t) \cong \frac{\sin(m\pi)}{m\pi} \quad \text{for the glass transition} \tag{5.52}$$

For the slopes $m = 0.5, 0.7, 0.9$ which are found in practice within glass transitions, this product will be equal to 0.64, 0.37, 0.11.

As an approximation, this equation may also be used in cases, for which the double-logarithmic slope of the creep compliance stays only nearly constant within a dispersion region. This is often the case for *secondary dispersions*, for which the slope m is small compared to unity. We then find by the development of the sin-function

$$J(t) \cdot G(t) \cong 1 - 1.6\ \text{m}^2 \quad \text{for secondary dispersion regions} \tag{5.53}$$

From this equation it follows that in the glassy state the product $G(t) \cdot J(t)$ may deviate only slightly from unity, as for $m \cong 0.1$ we find $G \cdot J \cong 0.98$.

A further important application of Eq. (5.48) is the approximation of $J(t)$ in the flow region of uncross-linked polymers. There the following approximation for the creep compliance is found

$$J(t) \cong J_A(t) = J_e^0 + t/\eta_0 = J_e^0[1 + t/\tau_e] \quad \text{for the flow region} \qquad (5.46a)$$

In the right-hand side of this equation we used the definition for the *longest relaxation time* τ_e

$$\tau_e = J_e^0 \eta_0 \qquad (5.54)$$

In terms of the generalized Maxwell model of Fig. 5.11a, τ_e is equal to $\tau_n = r_n/f_n$, if the Maxwell elements are arranged in the order of increasing relaxation times. The corresponding stress relaxation modulus is found to be

$$G(t) \cong G_A(t) = \frac{1}{J_e^0} e^{-t/\tau_e} \quad \text{for the flow region} \qquad (5.55)$$

as may be seen by inserting (5.46) and (5.55) into (5.48). While the creep compliance as a function of time approaches a straight line with the slope unity in a double logarithmic plot in the flow region, the relaxation modulus decreases to zero with an ever growing negative slope which tends to minus infinity with increasing time. Therefore the product $G(t) \cdot J(t)$ will vanish very quickly in the flow region with increasing time.

A schematic representation of the relaxation modulus as a function of time in a double-logarithmic plot is given in Fig. 5.21. The relaxation modulus is approximately the mirror picture of the creep compliance, apart from the behavior in the

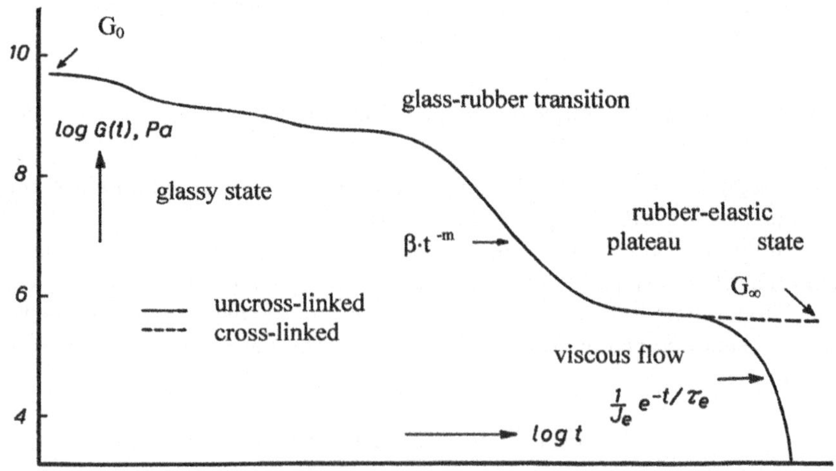

Fig. 5.21 Relaxation modulus versus time in a double-logarithmic representation for an amorphous uncross-linked polymer (*full line*) and for an amorphous cross-linked polymer (*dashed line*)

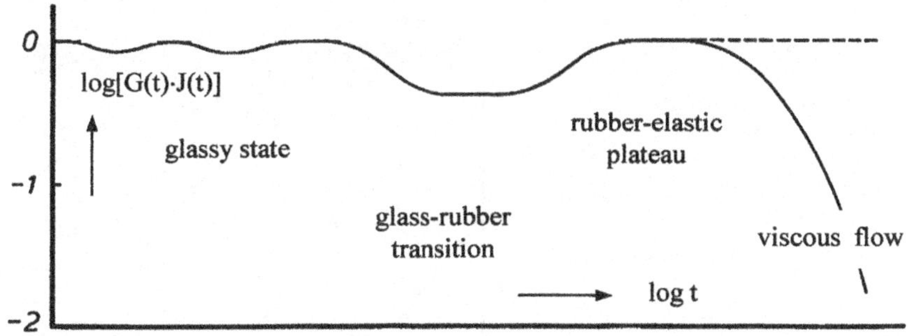

Fig. 5.22 The "Zener product" $J(t){\cdot}G(t)$ versus time, in a double-logarithmic representation for amorphous uncross-linked polymers (*full line*) and for amorphous cross-linked polymers (*dashed line*)

flow region (cf. Fig. 5.13). To each dispersion step of the compliance, there corresponds a dispersion step of the modulus. On the double-logarithmic scale, the heights and the positions of the corresponding dispersion steps are approximately equal. At very short times, the modulus attains constant values of the order $G_0 \cong 2$ to 3×10^9 Pa. At longer times, one can observe the dispersion steps of the secondary transitions. After the glassy state, the softening region follows, in which the modulus decreases in a double-logarithmic representation approximately with a constant slope of $-m$. After the glass-rubber transition, the rubber-elastic plateau or the rubber-elastic state is found. For cross-linked amorphous polymers, the modulus remains constant at the level of $G_\infty = 1/J_e$, which depends on the absolute temperature and the cross-linking density. For amorphous uncross-linked polymers, the flow region follows with the modulus steeply falling to zero.

According to the foregoing considerations, the *Zener product* in a double-logarithmic representation follows as given in Fig. 5.22. This product is always smaller or at most equal to unity. At very short times, $G(t){\cdot}J(t) \cong G_0{\cdot}J_0 = 1$. At the secondary transitions in the glassy state, small minima occur, which only deviate by some percent from unity (5.53). In the glass-rubber transition, the Zener product assumes values, which are significantly lower than unity (between 0.1 and 0.6). After this region, the product increases again and tends to unity. For cross-linked polymers it stays there, for uncross-linked polymers, however, the Zener product falls steeply to zero with increasing time.

Stress relaxation measurements in linear extension have been performed by Tobolsky and coworkers [24–26]. The relaxation modulus in extension behaves similar to the relaxation modulus in shear.

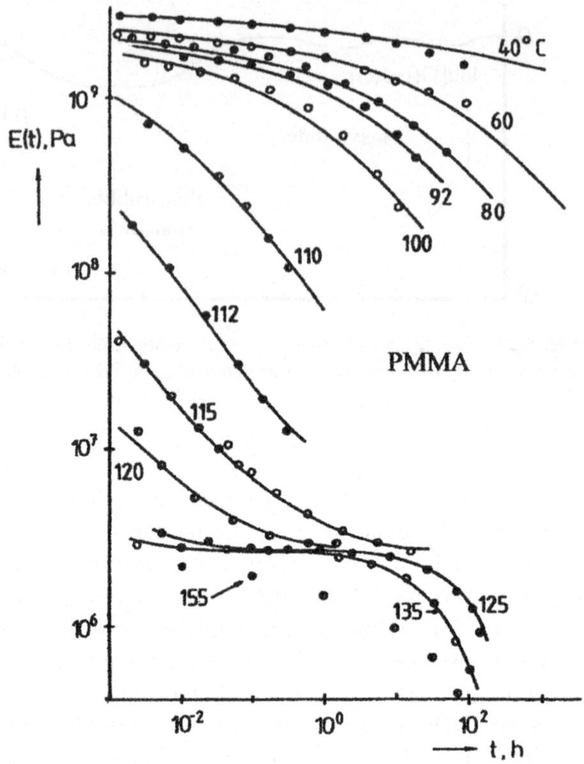

Fig. 5.23 Extensional modulus versus time in a double-logarithmic representation for PMMA at various temperatures, after Mc Loughlin and Tobolsky, reproduced from [24] by permission from Elsevier, UK

Figure 5.23 shows the modulus in extension as a function of the time, in a double-logarithmic representation, for PMMA. In this figure, all four states of order are present, the glassy state, the glass-rubber transition, the rubber-elastic plateau and the molten state. The extensional modulus for a PIB with a high-molar mass is presented in Fig. 5.24. Using the time-temperature shift, Tobolsky constructed from these data $E(t)$ over a wide range of time at $T = 25$ °C. In Fig. 5.25 these data are shown for three PIB with different molar masses. In the glassy region and in the glass-rubber transition, no influence of the molar mass on the moduli is visible. The width of the rubber-elastic plateau however, increases with increasing molar mass.

For amorphous polymers, $E(t)$ equals about 2.5–2.7 times the shear modulus $G(t)$ in the glassy state, while in the softening region, in the rubber-elastic plateau and in the molten state $E(t)$ is exactly three times $G(t)$.

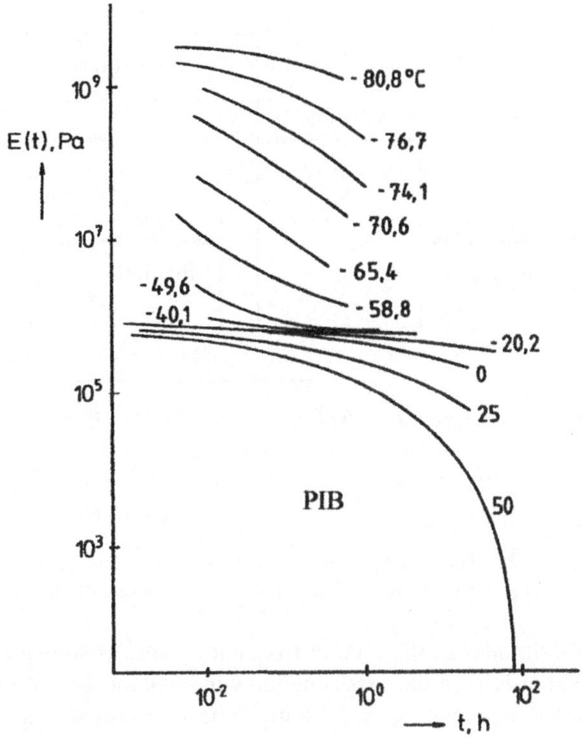

Fig. 5.24 Extensional modulus versus time, in a double-logarithmic representation for PIB with $M_v = 6.60 \times 10^3$ kg/mol at various temperatures after Tobolsky and Catsiff, reproduced from [25] by permission from John Wiley and Sons, New York

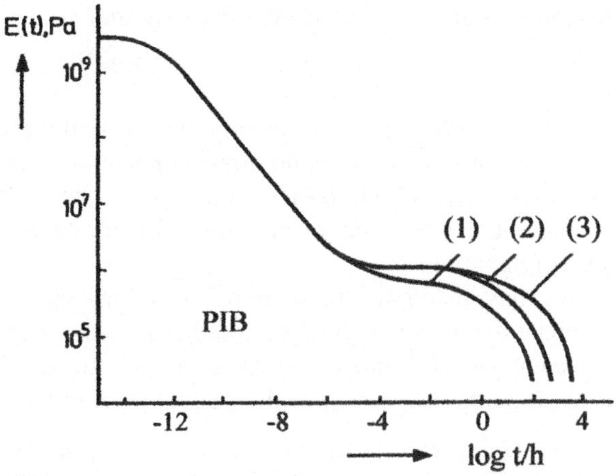

Fig. 5.25 Extensional modulus versus time in a double-logarithmic representation after a temperature shift to 25 °C for three PIB of different molar masses (*1*) $M_v = 1.36 \times 10^3$ kg/mol, (*2*) $M_v = 2.80 \times 10^3$ kg/mol, (*3*) $M_v = 6.60 \times 10^3$ kg/mol, after Tobolsky and Mc Loughlin, reproduced from [26] by permission from John Wiley and Sons, New York

5.8 Oscillatory Experiments

It is possible to enlarge the experimental window considerably, if besides creep and stress relaxation, also oscillatory experiments are used for the characterization of the viscoelastic behavior. The measuring techniques known today extend over

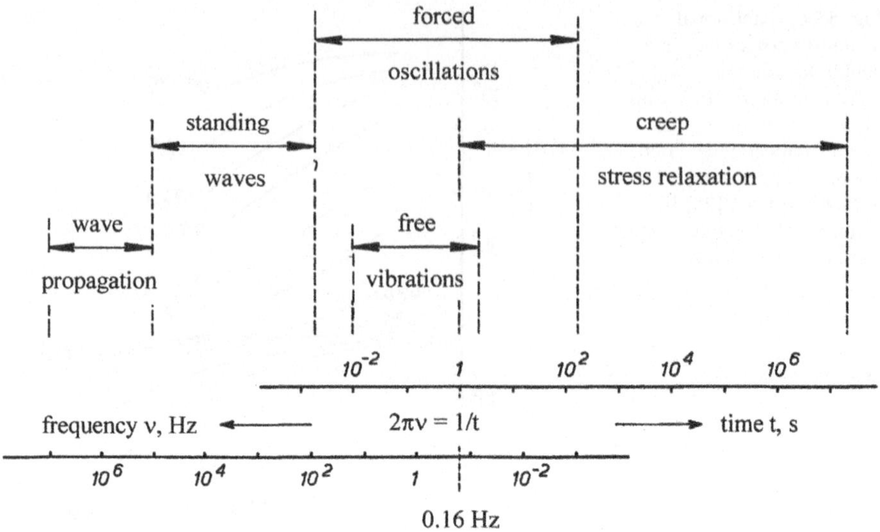

Fig. 5.26 The position and the experimental window of various mechanical measuring techniques for the determination of the viscoelastic behavior. (Ref 1 from Chap. 4)

14 decades of the time or frequency scale, as shown in Fig. 5.26. In this figure, the extensions of the experimental windows for the different measuring techniques are presented over a logarithmic time or frequency axis, respectively. The relation between the time (in seconds) in a creep or stress relaxation experiment and the frequency ν in s^{-1} in an oscillatory experiment has been set by the equation

$$\omega = 2\pi\upsilon = 1/t \tag{5.56}$$

ω is the *angular frequency*, the number of oscillations in 2π s. The relation (5.56) is an arbitrary one and has no direct connection with the duration of one period of the oscillation, which is $1/\nu$. It is shown later that this relation yields good agreements between the results of oscillatory experiments and those of creep and stress relaxation.

We may distinguish between four different types of vibration experiments, if we compare the wavelength of the oscillation with the dimensions of the specimen. As wavelength and frequency of an oscillation are reciprocal quantities, this division is at the same time one after the frequency of the measuring technique considered.

Forced Oscillations Below the Resonance Frequency

If the wave length of the oscillation is large against all dimensions of the specimen, the latter does not form an oscillating system. However, the behavior of the material under harmonic oscillations may be investigated by exciting the specimen to vibrations with a frequency below the resonance frequency. The harmonic strain and stress occurring are compared with each other with respect to their amplitude and phase. In this experiment, the frequencies have to be chosen low enough that the inertia moments of the specimen or of moving parts of the instrument do not play any role.

Free Damped Vibrations with Additional Mass or Inertia Moment
For this type of measurements an additional mass or an additional moment of inertia is added to the specimen. The system is deflected out of its equilibrium position and then released. It returns by means of damped free vibrations into its initial position. The eigenfrequency and the logarithmic decrement of the free damped vibration are measured. For an evaluation of this experiment, the inertia forces of the additional mass have to be considered.

Resonance of Standing Waves
If the wave length of the elastic waves is of the order of the largest dimension of the specimen, the system will form an oscillation system without any need for additional masses or moments of inertia. The specimen can be excited to torsional, bending or longitudinal vibrations by a wave generator with variable frequency, and the amplitude of the occurring vibration is determined. If the exciting frequency attains the resonance frequency, the latter is determined together with the half width of the resonance curve.

Wave Propagation
If the wave length of the excited vibration is small compared to all dimensions of the specimen, elastic waves propagate through the sample. The velocity of propagation and the attenuation of the wave amplitude are measured. For evaluating this experiment, one needs the density of the material, just as in the case of evaluating the resonance of standing waves.

All those experiments yield the magnitude of a modulus or a compliance, respectively, and an internal damping expressed by the tangent of a phase angle, which will be introduced in detail below.

Evaluation of the Forced Oscillation Experiment
The experiment of a forced oscillation may be started as follows. The specimen free of any stress until the time $t = 0$, is excited by a harmonic stress of an amplitude σ_0 and an angular frequency ω at the time zero. The prescribed stress history reads as

$$\begin{aligned} \sigma(t) &= 0 && \text{for} \quad t \leq 0 \\ \sigma(t) &= \sigma_0 \cos(\omega t) && \text{for} \quad t > 0 \end{aligned} \tag{5.57}$$

The deformation as a function of the time is obtained by inserting (5.57) into the superposition principle (5.15) and making use of the decomposition according to (5.27). One gets

$$\gamma(t) = \sigma_0[A'(\omega, t)\cos(\omega t) + A''(\omega, t)\sin(\omega t)] \quad \text{for} \quad \text{t} > 0 \tag{5.58}$$

with A' and A'' being abbreviations for the following integrals

$$A'(\omega, t) = J_0 + \int_0^t \dot{\psi}(\xi)\cos(\omega\xi)d\xi \tag{5.59}$$

$$A''(\omega, t) = \frac{1}{\omega\eta_0} + \int_0^t \dot{\psi}(\xi)\sin(\omega\xi)d\xi \tag{5.60}$$

The deformation (5.58) is not a harmonic function of the time, as long as the quantities A' and A'' still depend on the time t. If one waits long enough, the integrals tend to limits, which are independent of the time t and depend on the parameter ω, only.

$$J'(\omega) = \lim_{t\to\infty} A'(\omega, t) = J_0 + \int_0^\infty \dot{\psi}(\xi)\cos(\omega\xi)d\xi \tag{5.61}$$

$$J''(\omega) = \lim_{t\to\infty} A''(\omega, t) = \frac{1}{\omega\eta_0} + \int_0^\infty \dot{\psi}(\xi)\sin(\omega\xi)d\xi \tag{5.62}$$

The existence of these limits follows from the fact that $\dot{\psi}(\xi)$ is a monotonously decreasing function of ξ which becomes zero for $\xi \to \infty$ (cf. Fig. 5.8). If these limiting values have been reached with sufficient accuracy, one says that the *stationary state of the harmonic oscillation* has been attained. In this case, the shear strain reads as

$$\gamma(t) = \sigma_0[J'(\omega)\cos(\omega t) + J''(\omega)\sin(\omega t)] \tag{5.63}$$

or

$$\gamma(t) = \sigma_0 J_d(\omega) \cdot \cos(\omega t - \delta) \tag{5.64}$$

with

$$J_d(\omega) = \sqrt{[J'(\omega)]^2 + [J''(\omega)]^2} \tag{5.65}$$

and

$$\tan\delta(\omega) = J''(\omega)/J'(\omega)] \tag{5.66}$$

$$J'(\omega) = J_d(\omega)\cos\delta \tag{5.67}$$

$$J''(\omega) = J_d(\omega)\sin\delta \tag{5.68}$$

The quantities $J'(\omega)$ and $J''(\omega)$ are called *storage compliance* and *loss compliance*, $J_d(\omega)$ the *absolute (value of the) dynamic compliance* and *tan* $\delta(\omega)$ the *internal damping* or the *loss tangent.* All these quantities are functions of the angular frequency ω of the prescribed stress, but independent of its amplitude σ_0. $J_d(\omega)$ is also often designated as $|J^*(\omega)|$, the *magnitude of the complex compliance* J^*.

Figure 5.27 shows just one and a half period of stress and strain as functions of the time in the *stationary state* of a *harmonic oscillation.* The deformation may be

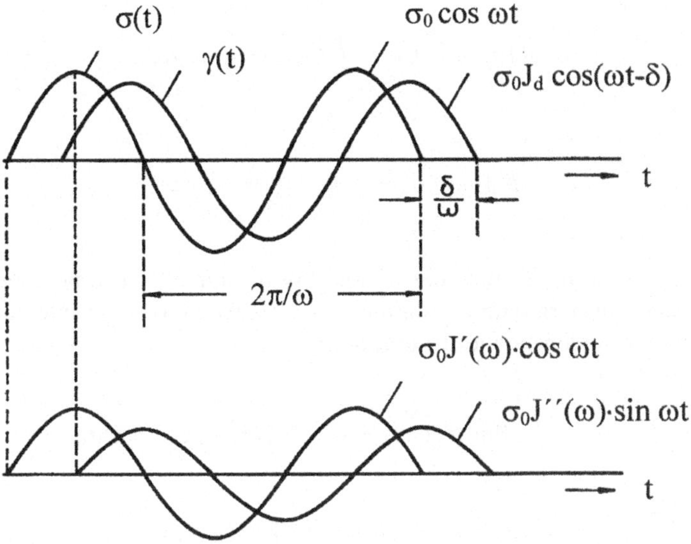

Fig. 5.27 Decomposition of the stationary harmonic deformation in an oscillatory experiment

described according to (5.64) as a harmonic oscillation with the same angular frequency as the stress, which lags behind it by a phase angle δ. $J_d(\omega)$ is the ratio of the amplitude of the deformation to the amplitude of the stress. Due to (5.63) the deformation may be decomposed into a sum of two harmonic oscillations of the same angular frequency, one of them being in phase with the stress and having the amplitude $\sigma_0 J'(\omega)$, the other lagging behind by 90° and having the amplitude $\sigma_0 J''(\omega)$.

A complex quantity can be defined whose real part is the storage compliance and whose negative imaginary part the loss compliance

$$J^*(\omega) = J'(\omega) - iJ''(\omega) = J_d(\omega)e^{-i\delta} \tag{5.69}$$

which is called the *complex compliance*.

An oscillatory experiment started under harmonic strain at the time $t = 0$ is defined by

$$\begin{aligned} \gamma(t) &= 0 && \text{for} \quad t \leq 0 \\ \gamma(t) &= \gamma_0 \sin(\omega t) && \text{for} \quad t > 0 \end{aligned} \tag{5.70}$$

The stress as a function of the time is obtained by inserting (5.70) into (5.17)

$$\sigma(t) = \gamma_0 [B'(\omega, t) \sin(\omega t) + B''(\omega, t) \cos(\omega t)] \quad \text{for} \quad t > 0 \tag{5.71}$$

B' and B'' are abbreviations for the following integrals

$$B'(\omega,t) = G_0 + \int_0^t \dot{G}(\xi)\cos(\omega\xi)d\xi \tag{5.72}$$

$$B''(\omega,t) = -\int_0^t \dot{G}(\xi)\sin(\omega\xi)d\xi \tag{5.73}$$

(5.71) is not a harmonic function as long as B' and B'' are time-dependent. A stationary state of harmonic oscillation can be reached as these integrals tend to limits which are independent of the time, i.e.,

$$G'(\omega) = \lim_{t\to\infty} B'(\omega,t) = G_0 + \int_0^\infty \dot{G}(\xi)\cos(\omega\xi)d\xi \tag{5.74}$$

$$G''(\omega) = \lim_{t\to\infty} B''(\omega,t) = -\int_0^\infty \dot{G}(\xi)\sin(\omega\xi)d\xi \tag{5.75}$$

The convergence of these integrals follows from the fact that $\dot{G}(\xi)$ is a monotonously increasing negative function of ξ which becomes zero for $\xi \to \infty$ (cf. Fig. 5.8). Due to the minus sign in the Eq. (5.75), G'' becomes a positive quantity. In the stationary state the harmonic stress can then be written as

$$\sigma(t) = \gamma_0[G'(\omega)\sin(\omega t) + G''(\omega)\cos(\omega t)] \tag{5.76}$$

or

$$\sigma(t) = \gamma_0 G_d(\omega)\cdot\sin(\omega t + \delta) \tag{5.77}$$

with

$$G_d(\omega) = \sqrt{[G'(\omega)]^2 + [G''(\omega)]^2} \tag{5.78}$$

$$\tan\delta(\omega) = G''(\omega)/G'(\omega)] \tag{5.79}$$

and

$$G'(\omega) = G_d(\omega)\cos\delta \tag{5.80}$$

$$G''(\omega) = G_d(\omega)\sin\delta \tag{5.81}$$

The quantities $G'(\omega)$ and $G''(\omega)$ are called *storage modulus* and *loss modulus*, the quantity $G_d(\omega)$ the *absolute dynamic modulus*. The latter is also designated as $|G^*(\omega)|$, the *magnitude of the complex modulus* G^*. The identity of the quantity defined in (5.79) with the loss tangent defined in (5.66) will be shown after Eq. (5.82).

The stress in the stationary state of a harmonic oscillatory strain may be described according to (5.77) as a harmonic function with the same angular frequency as the strain, which precedes the strain by a phase angle δ. $G_d(\omega)$ is the ratio of the amplitude of the stress to the amplitude of the strain. According to (5.76) the stress may be decomposed into a sum of two harmonic components with the same angular frequency, one of them being in phase with the stress and having an amplitude $\gamma_0 G'(\omega)$, the other preceding the stress by 90° and having an amplitude $\gamma_0 G''(\omega)$.

A *complex dynamic modulus* is defined by the equation

$$G^*(\omega) = G'(\omega) + iG''(\omega) = G_d(\omega)e^{+i\delta} \tag{5.82}$$

Relations between the compliances and moduli in forced oscillations can be derived from the following consideration. The stationary state of harmonic oscillations, which occurs after some time has to be the same, independent of the way in which this state was reached, viz., by prescribing a harmonic stress at the time $t = 0$ or by prescribing a harmonic strain at the time $t = 0$. In both cases, the resulting expressions for stress and strain should be the same, apart from a phase shift in the time scale. Therefore, the phase angle between stress and strain has to be the same, as well as the ratio of their amplitudes, i.e.,

$$\tan\delta = J''(\omega)/J'(\omega) = G''(\omega)/G'(\omega) \tag{5.83}$$

$$\sigma_0/\gamma_0 = G_d(\omega) = 1/J_d(\omega) \tag{5.84}$$

Comparing (5.69) with (5.82) shows that complex modulus and complex compliance are inverse quantities

$$J^*(\omega) \cdot G^*(\omega) = 1 \tag{5.85}$$

Separating the real and imaginary parts of this equation yields

$$G' = J'/J_d^2 \tag{5.86}$$

$$G'' = J''/J_d^2 \tag{5.87}$$

$$J' = G'/G_d^2 \tag{5.88}$$

$$J'' = G''/G_d^2 \tag{5.89}$$

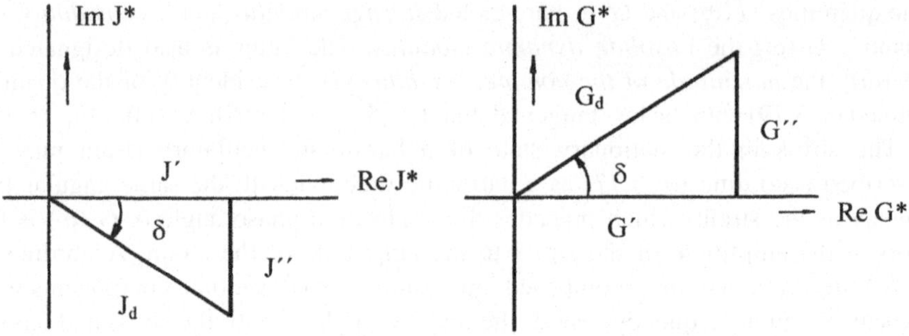

Fig. 5.28 Representation of the complex compliance (*left*) and the complex modulus (*right*) in the complex plane

Equations (5.65)–(5.68) and (5.78)–(5.81) can geometrically be interpreted. Consider the representation of the complex compliance and the complex modulus in the complex plane as shown in the left and right half of Fig. 5.28.

In the representation of the complex compliance, J' and J'' are the abscissa and the negative value of the ordinate of the point J^*, in the representation of the complex modulus, G' and G'' are the abscissa and the ordinate of the point G^*. The corresponding rectangular triangles include the angle δ between their adjacent leg and hypotenuse, while J_d and G_d form their hypotenuses.

The characteristic quantities as functions of time or angular frequency, respectively, are compared in Fig. 5.29 in a region in which a new molecular process develops and starts to contribute to the deformation. The shape of the curves $J(t)$, $J'(\omega)$, and $J_d\,(\omega)$ in the dispersion region is similar. Prior to the relaxation process, $J(t)$ shows a more or less horizontal plateau, in the dispersion region $J(t)$ increases to another plateau which is reached after a full development of the new process. $J_d\,(\omega)$ and $J'(\omega)$ show the same plateau values before and after the relaxation process. Within the dispersion region the validity of the inequality

$$J'(\omega) \leq J_R(t) \leq J(t) \tag{5.90}$$

can be shown from the corresponding inequality of their intensity functions (cf. (5.119) and (5.120)) and from Eq. (5.41). Regarding the order of $J_d(\omega)$ and $J(t)$ no general statement can be made. $J_d(\omega)$ running below $J(t)$ as indicated in this figure does not pretend to have any general validity. $J''(\omega)$ passes through a maximum in the middle of the dispersion step, *tan* $\delta(\omega)$ also passes through a maximum which is situated left to that of $J''(\omega)$.

The dispersion steps of the modulus functions appear as mirror pictures of the dispersion steps of the compliances. The dispersion steps start at high values of the moduli and end at low values. For the moduli, the inequality

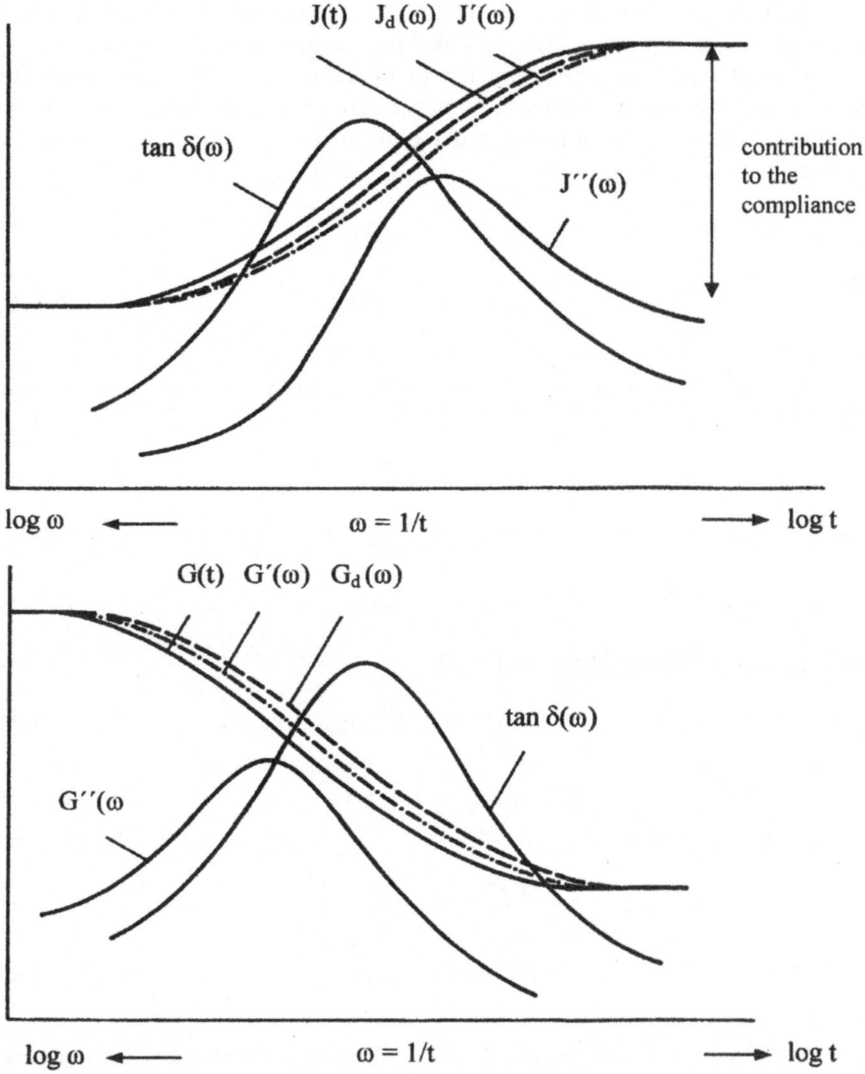

Fig. 5.29 Development of a molecular process and its contribution to the compliance functions (*above*) and to the modulus functions (*below*). The position on the time or frequency axis is the same for tan δ in both representations, the positions of the maxima in J″(ω) and G″(ω) are, respectively, *right* and *left* of the maximum of tan δ

$$G(t) \leq G'(\omega) \leq G_d(\omega) \tag{5.91}$$

is valid everywhere. The maximum of the loss modulus is located at lower times (higher frequencies) than the maximum of the loss tangent and the latter is situated at lower times than the maximum of the loss compliance. The height and the shape of the dispersion steps may strongly differ for various molecular processes.

A molecular process with characteristic functions which are completely different from those shown in Fig. 5.29, is the *flow transition*. There, the approximations for the creep compliance and for the relaxation modulus are using the Eq. (5.54).

$$J_A(t) \cong J_e^0 + t/\eta_0 \tag{5.46b}$$

$$G_A(t) \cong \frac{1}{J_e^0} e^{-t/\tau_e} \tag{5.55a}$$

As (5.46b) contains only a constant term and a flow term, we find from Eqs. (5.61) and (5.62)

$$J'_A(\omega) \cong J_e \tag{5.92}$$

and

$$J''_A(\omega) \cong 1/\omega\eta_0 \tag{5.93}$$

and from (5.65), (5.66), (5.86) and (5.87)

$$\tan\delta_A(\omega) \cong 1/(\omega\tau_e) \tag{5.94}$$

$$J_{dA}(\omega) \cong J_e^0\sqrt{1 + 1/(\omega\tau_e)^2} \tag{5.95}$$

$$G'_A(\omega) \cong \frac{1}{J_e^0}\frac{\omega^2\tau_e^2}{1+\omega^2\tau_e^2} \tag{5.96}$$

$$G''_A(\omega) \cong \frac{1}{J_e^0}\frac{\omega\tau_e}{1+\omega^2\tau_e^2} \tag{5.97}$$

In the flow region, the storage compliance tends with decreasing angular frequency to the stationary recoverable compliance, the loss compliance and the loss tangent grow proportional to the inverse angular frequency and the storage modulus and the loss modulus are in the limit $\omega\tau_e \ll 1$ proportional to the second, respectively, first power of the angular frequency.

Finally, the storage and loss components of the dynamic compliances are represented as integral transforms of the retardation spectrum $f(\tau)$. We differentiate (5.28) with respect to t and insert the result into (5.61) and (5.62). Changing the order of the differentiations after ξ and τ and making use of the relations (see [27])

$$\int_0^\infty \cos(\omega\xi)e^{-\xi/\tau}d\xi = \frac{\tau}{1+\omega^2\tau^2} \quad \text{and} \quad \int_0^\infty \sin(\omega\xi)e^{-\xi/\tau}d\xi = \frac{\omega\tau^2}{1+\omega^2\tau^2} \tag{5.98}$$

yields

$$J'(\omega) = J_0 + \int_0^\infty f(\tau)\frac{1}{1+\omega^2\tau^2}d\tau \tag{5.99}$$

$$J''(\omega) = \int_0^\infty f(\tau)\frac{\omega\tau}{1+\omega^2\tau^2}d\tau + \frac{1}{\omega\eta_0} \tag{5.100}$$

The corresponding equations for the storage and loss modulus in terms of the relaxation spectrum are found by differentiating (5.24) with respect to t and inserting the result into (5.74) and (5.75). One gets

$$G'(\omega) = G_0 - \int_0^\infty g(\tau)\frac{1}{1+\omega^2\tau^2}d\tau \tag{5.101}$$

$$G''(\omega) = \int_0^\infty g(\tau)\frac{\omega\tau}{1+\omega^2\tau^2}d\tau \tag{5.102}$$

Equation (5.101) may be written in a slightly different form. From Eq. (5.24) it follows for $t \to 0$

$$G_0 - G_\infty = \int_0^\infty g(\tau)d\tau \tag{5.103}$$

Inserting (5.103) into (5.101) yields

$$G'(\omega) = G_\infty + \int_0^\infty g(\tau)\frac{\omega^2\tau^2}{1+\omega^2\tau^2}d\tau \tag{5.104}$$

From (5.102) and (5.104), the limiting slopes of G'' and G' for $\omega \to 0$ follow as

$$\lim_{\omega\to 0}\left[G''(\omega)/\omega\right] = \int_0^\infty \tau g(\tau)d\tau = \tau_e/J_e^0 = \eta_0 \tag{5.105}$$

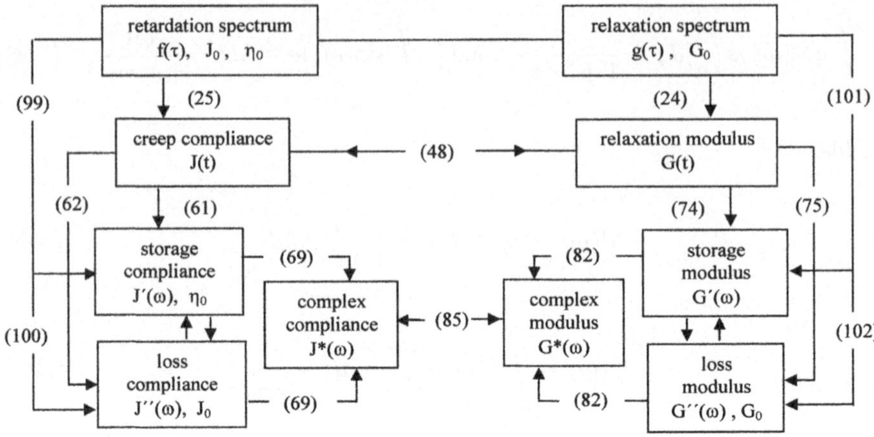

Fig. 5.30 The structure and the interconnections of the theory of the linear viscoelastic behavior

$$\lim_{\omega \to 0} [G'(\omega)/\omega^2] = \int_0^\infty \tau^2 g(\tau) d\tau = \tau_e^2 / J_e^0 = \eta_0 \tau_e = \eta_0^2 J_e^0 \tag{5.106}$$

The second parts of these equations follow from (5.96) and (5.97) considering (5.54).

If oscillatory measurements are performed on polymer melts, instead of the complex modulus, a *complex viscosity* η^* is often used for the description of the results, which is defined by

$$\eta^*(\omega) = \frac{G^*(\omega)}{i\omega} = \eta'(\omega) - i\eta''(\omega) \tag{5.107}$$

with $\eta'(\omega)$ being the *real part* and $\eta''(\omega)$ being the negative of the *imaginary part of the complex viscosity*. Comparison with Eq. (5.82) yields the positive quantities

$$\eta'(\omega) = G''(\omega)/\omega \tag{5.108}$$

and

$$\eta''(\omega) = G'(\omega)/\omega \tag{5.109}$$

In Fig. 5.30 an overview of the links between different characteristic functions and quantities of the theory of linear viscoelastic behavior is given. A *characteristic function* characterizes the viscoelastic behavior completely, so that knowing it makes the calculation of all other viscoelastic functions possible. For instance, the retardation spectrum alone does not constitute a characteristic function, as one needs the additional information of J_0 and η_0 to calculate *J*(*t*).

Thus the entity of $f(\tau)$, J_0 and η_0 together, forms the *characteristic function*. Relations between the characteristic functions are marked by connecting lines. The corresponding equation number is indicated on this line. If the number is missing, there exists a relation, which, however, has not been discussed in this context. For this case we refer the reader to [12, 13].

5.9 Approximate Relations Between Measurable Viscoelastic Functions

Most of the equations cited in Fig. 5.30 are integral transforms and may be used only, if an analytic expression for the function under the integral exists, which describes the experimental results with sufficient accuracy and is simple enough to allow the explicit performance of the integral transformation in a closed form. In practice, this is very seldom the case. A further complication of the problem are the *truncation errors*. For executing the transformations, the function under the integral has to be known theoretical within its complete region of definition, e.g., from $t = 0$ to $t = \infty$ in the case of Eq. (5.61) or (5.62), or from $\omega = 0$ to $\omega = \infty$ in the case of the calculation of $J(t)$ from $J'(\omega)$ and $J''(\omega)$. (The corresponding equations are not shown here.) As measurements can only be performed over a finite time or frequency scale, experimental information may sometimes be lacking for the complete evaluation of the integrals. This deficiency may lead to the so-called *truncation errors*, which can be large in some cases. Thus, it will often be necessary to evaluate the integral transforms by numerical methods and to estimate the possible influence of truncation errors on the result.

Therefore, the question arises whether it is possible to develop *numerical approximate equations* for the calculation of values of a measurable characteristic quantity, if a number of data points of the starting function are known. The most simplest of those equations are summarized in Table 5.2.

These are so-called two point approximations. Two points of the starting function are needed to calculate one point of the function which is to be determined. The arguments of the starting points have to differ by a factor 2 which corresponds to a constant logarithmic distance log 2. This system may be used in recursive form and allows to calculate $n - 1$ successive points of the desired function from n successive points of the starting function in a simple way. Starting, for instance, from n successive points of the creep compliance which differ on the time axis by a factor of 2, Eq. (5.110.1) may be used to calculate $n - 1$ successive points of the storage compliance, which have a constant logarithmic distance corresponding to a factor of *2* for the angular frequency, and Eq. (5.111.1) may be used to calculate $n - 1$ successive points of the loss compliance, which however are shifted by the factor 2 on the angular frequency axis in comparison to the calculated points of the storage compliance. From the $n - 2$ data for $J'(\omega)$ and $J''(\omega)$, which are known at the same angular frequencies, using the exact

Table 5.2 Some simple approximations for the conversion of viscoelastic functions

Approximate equation ($\omega = 1/t$)	Bounds for the relative error, %	Equation
$J'(\omega) \cong J(t) - 0.86\,[J(2t) - J(t)]$	$\pm 15 \tan\delta$	(5.110.1)
$J''(\omega) \cong 2.12\,[J(t) - J(t/2)]$	Uncertain	(5.111.1)
$J(t) \cong J'(\omega) + 0.57\,J''(\omega/2) - 0.20\,J''(\omega)$	$\pm 8 \tan\delta$; ± 7.6	(5.112.2)
$G'(\omega) \cong G(t) + 0.86\,[G(t) - G(2t)]$	$\pm 15 \tan\delta$	(5.113.1)
$G''(\omega) \cong 2.12\,[G(t/2) - G(t)]$	Uncertain	(5.114.1)
$G(t) \cong G'(\omega) - 0.57G''(\omega/2) + 0.20G''(\omega)$	$\pm 8 \tan\delta/(1-\tan\delta)$	(5.115.2)

Eqs. (5.86) and (5.87), $n - 2$ logarithmic equidistant points of storage and loss modulus may be calculated. Finally, using the latter set together with Eq. (5.115.2) it is possible to arrive at $n - 3$ logarithmically equidistant points for the relaxation modulus G(t).

The equations of Table 5.2 are sometimes used in a slightly different form. By a Taylor development it is possible to show that

$$J(2t) - J(t) \cong t \cdot \frac{dJ}{dt} = \frac{dJ}{d(\ln t)} = J \cdot \frac{d(\log J)}{d(\log t)}$$

Inserting this approximation into the conversion formulae (5.110.1) and (5.111.1) of Table 5.2, one gets

$$J'(\omega) \cong J(t)\left[1 - 0.86 \cdot \frac{d(\log J)}{d(\log t)}\right] \tag{5.116}$$

$$J''(\omega) \cong 1.06\frac{dJ}{d(\ln t)} = 0.46\frac{dJ}{d(\log t)} \tag{5.117}$$

If the double-logarithmic slope of the creep compliance is small against unity, we further have

$$\tan\delta(\omega) \cong \frac{J''(\omega)}{J(t)} \cong 1.06\frac{d(\log J)}{d(\log t)} \tag{5.118}$$

From these equations we conclude

(1) $J'(\omega)$ is always somewhat smaller than $J(t)$; the difference $J(t) - J'(\omega)$ increases proportional to the double logarithmic slope of the creep compliance as a function of time and is maximal in the dispersion regions.
(2) $J''(\omega)$ is proportional to the slope of $J(t)$ as a function of the logarithm of the creep time and is maximal in the dispersion regions.
(3) if $d(\log J)/d(\log t) \ll 1$, $\tan\delta(\omega)$ is proportional to the double logarithmic slope of $J(t)$ versus t and is maximal in the dispersion regions.

For the derivation and error bounds of these formulae we refer to the original literature [28, 30–33]. Here we want to mention only that the representation of the measurable characteristic functions by means of the spectra is the basis for all approximations. For instance, for the derivation of Eq. (5.110.1) we use the abbreviation $t/\tau = x$ and write (5.25), (5.99), and (5.100) as integral transforms of the retardation spectrum $f(\tau)$ and a known intensity function, which is a function of x only, as follows:

$$J'(1/t) = J_0 + \int_0^\infty f(\tau)\frac{x^2}{1+x^2}d\tau = J_0 + \int_0^\infty f(\tau)\chi'(x)d\tau \tag{5.119}$$

$$J(t) = J_0 + \int_0^\infty f(\tau)[1-e^{-x}]d\tau + \frac{t}{\eta_0} = J_0 + \int_0^\infty f(\tau)\varphi(x)d\tau + \frac{t}{\eta_0} \tag{5.120}$$

$$J(2t) - J(t) = \int_0^\infty f(\tau)e^{-x}[1-e^{-x}]d\tau + \frac{t}{\eta_0} = \int_0^\infty f(\tau)\psi(x)d\tau + \frac{t}{\eta_0} \tag{5.121}$$

$$J''(1/t) = \int_0^\infty f(\tau)\frac{x}{1+x^2}d\tau + \frac{t}{\eta_0} = \int_0^\infty f(\tau)\chi''(x)d\tau + \frac{t}{\eta_0} \tag{5.122}$$

Instead of an approximation for $J'(1/t)$ in terms of $J(t)$ and $J(2t)$, we derive an approximation for the difference $J(t) - J'(1/t)$ in terms of the difference $J(2t) - J(t)$. For this purpose, the intensity function $\varphi(x) - \chi'(x) = 1/(1+x^2) - e^{-x}$ which is positive everywhere and increases as x for small x, decreases as $1/x^2$ for large x and shows a maximum around $x \cong 0.46$, is approximated by the intensity function $\psi(x) = \varphi(2x) - \varphi(x) = e^{-x}(1-e^{-x})$ which is positive everywhere. It increases as x for small x, decreases as e^{-x} for large x and shows a maximum around $x \cong 0.69$ We arrive at the following approximation for the intensity function $\varphi(x) - \chi'(x) \cong 0.86 \cdot \psi(x)$ from which the approximation (5.110.1) follows by multiplying with $f(\tau)$ and integrating over τ. The error bound of the formula follows from an estimate of the error of this formula in terms of the intensity function of $J''(1/t)$, $\chi''(x) = x/(1+x^2)$ and the fact that $f(\tau)$ is a nonnegative function of the retardation time. The coefficient 0.86 is obtained by trial and error in trying to minimize the error bound. This procedure for the derivation of approximation formulae has been first introduced by Ninomiya and Ferry [29]. The derivation of formulae (5.110.1) and (5.111.1) is explained in full detail in [30].

Despite its simplicity, the approximate formula (5.110.1) may be very useful in cases of low damping. If *tan* $\delta < 0.1$ (a condition which is fulfilled in the complete glassy state and in the rubber-elastic state of cross-linked elastomers), formula (5.110.1) may be applied, as its error bound is ± 1.5 % at the maximum. Similar remarks apply to formulae (5.112.2), (5.113.1), and (5.115.2).

Table 5.3 Coefficients of the approximations for the calculation of the storage compliance from the creep compliance with Eq. (5.110)

a	b	c	d	e	f	h	Bounds for the relative error %	Equation
				0.855			+14.6 tan δ −14.6 tan δ	(5.110.1)
			0.445	–	0.376		+7.8 tan δ −7.7 tan δ	(5.110.2)
		−0.099	0.608	–	0.358		+7.5 tan δ; 5.2 −7.5 tan δ; −9.1	(5.110.3)
		−0.119	0.680	–	0.225	0.0429	+2.1 tan δ; 5.9 −2.1 tan δ; −2.0	(5.110.4)
	0.0108	−0.168	0.734	–	0.235		+8.8 tan δ; 1.5 −1.9 tan δ; −1.5	(5.110.5)
	0.0109	−0.169	0.739	–	0.214	0.0451	+2.3 tan δ;1.6 −2.3 tan δ; −1.6	(5.110.6)
−0.000715	0.0185	−0.197	0.778	–	0.181	0.0494	+3.1 tan δ; 0.8 −3.1 tan δ; −0.8	(5.110.7)

If $\tan\delta > 0.1$, the accuracy of formula (5.110.1) is no longer sufficient. One of the more complicated formulae of Table 5.3 has to be selected then. For the calculation of the storage compliance from the creep compliance, formulae of the following type may be used

$$\begin{aligned} J'(\omega) \cong J(t) &- a[J(32t) - J(16t)] - b[J(16t) - J(8t)] \\ &- c[J(8t) - J(4t)] - d[J(4t) - J(2t)] - e[J(2t) - J(t)] \\ &- f[J(t) - J(t/2)] - h[J(t/4) - J(t/8)] \end{aligned} \tag{5.110}$$

where $a, b, c, \ldots$ are constant coefficients which have been chosen to minimize the relative error of the corresponding approximation, and are listed in Table 5.3.

This table shows seven different conversion formulae, indicated by formula numbers which are double indexed. The first part of the number, viz., (5.110) refers to the type of the conversion problem, the increasing end digit of the number relates to the increasing accuracy of the formula. The column before last lists upper and lower bounds for the relative error of the formula. Equations (5.110.1) and (5.110.2) only have bounds for the relative error, which depend on the value of *tan* δ at the angular frequency $\omega = 1/t$, for which $J'(\omega)$ is to be calculated. The other conversion formulae have error bounds which are proportional to *tan* δ and those, which are independent of the value of *tan* δ. In these cases, of course, the more favorable one is to be used. Formulae with a higher number of each table are more accurate, but require more experimental information (extend over a *wider experimental window*).

As to be seen from Table 5.3, it is always possible to find a conversion formula with an error bound smaller than 1 %. It should be noticed that a possible error originating from truncation is already included in the given error bounds. The

Table 5.4 Coefficients of the approximations for the calculation of the loss compliance from the creep compliance with Eq. (5.111)

d	e	f	g	h	j	l	n	Bounds for the relative error, %	Equation
		2.12						8[1 + 1/tan δ]; 26 −8[1 + 1/tan δ]	(5.111.1)
−0.470	1.715	–	0.902					0.7[1 + 1/tan δ]; 2.3 −4.6/tan δ	(5.111.2)
−0.505	1.807	–	0.745	–	0.158			1.1[1 + 1/tan δ]; 3.5 −1.3/tan δ	(5.111.3)
−0.470	1.674	0.196	0.627	–	0.194			0.7[1 + 1/tan δ]; 1.3 −2.5[1 + 0.5/tan δ]	(5.111.4)
−0.470	1.674	0.197	0.621	0.011	0.172	0.0475		0.7[1 + 1/tan δ]; 2.3 −2.5[1 + 0.12/tan δ]	(5.111.5)
−0.470	1.674	0.198	0.620	0.012	0.172	0.0430	0.0122	0.7[1 + 1/tan δ]; 2.7 −2.5[1 + 0.03/tan δ]	(5.111.6)
−0.470	1.674	0.198	0.620	0.012	0.172	0.0433	0.0108	0.7[1 + 1/tan δ]; 2.7 −2.7; −2.7/tan δ	(5.111.7)

conclusion can be drawn, that the problem of the conversion from $J(t)$ to $J'(\omega)$ is solved in a simple way. At low damping, the problem is especially easy, at higher damping, at the utmost, an experimental window from $t/8$ to $32\,t$ is needed to apply a conversion formula which is better than 1 %.

There is, however, one regime for which the conversion will be very difficult. This is the flow region of polymers, in which the double-logarithmic slope of the creep compliance as a function of time approaches unity. Then the conversion (5.110.7) requires the addition and subtraction of terms of equal order in order to arrive finally at a result for $J'(\omega)$, which is one or two orders of magnitude smaller. In this case, the propagation of the experimental error will prohibit the conversion (cf. Fig. 5.34, on page 180).

Equation (5.110) and its error bounds remain valid, if the creep compliance $J(t)$ is substituted by the recoverable creep compliance $J_R(t)$. Then, the calculation may be performed without problems, presuming the recoverable compliance is known accurately enough from the experiment.

The conversion of $J(t)$ into $J''(\omega)$ is much more difficult. The approximate equations are given in formula (5.111).

$$
\begin{aligned}
J''(\omega) \cong\, & d[J(4t) - J(2t)] + e[J(2t) - J(t)] + f[J(t) - J(t/2)] \\
& + g[J(t/2) - J(t/4)] + h[J(t/4) - J(t/8)] + j[J(t/8) - J(t/16)] \\
& + l[J(t/32) - J(t/64)] + n[J(t/128) - J(t/256)] + \cdots
\end{aligned}
\tag{5.111}
$$

The constant coefficients $d, e, f, \ldots$ have been chosen to optimize the error bounds of the formulae. Their values, together with the error bounds are given in Table 5.4.

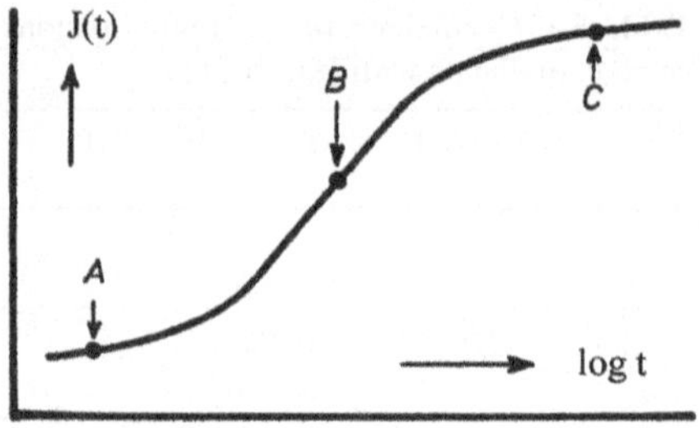

Fig. 5.31 The creep compliance in the vicinity of a dispersion region and the truncation problem in the calculation of $J''(\omega)$ at the points *A*, *B*, and *C*

The most simplest of these equations, viz.,

$$J''(\omega) \cong 2.12[J(t) - J(t/2)] \tag{5.111.1}$$

provides only a rough estimate of the loss compliance. Even at the favorable case of high damping ($\tan\delta = 1$), the error bounds for this approximation will be as high as ±16 %. In order to arrive at higher accuracies, more complicated formulae have to be used. If we ask for bounds of ±3 % for the relative error, we may use the approximation (5.111.2) in the tan δ-region $[1.5 < \tan\delta < \infty]$, the formula (5.111.3) in the region $[0.45 < \tan\delta < \infty]$, the formula (5.111.5) in the region $[0.2 < \tan\delta < \infty]$ and the formula (5.111.6) in the region $[0.075 < \tan\delta < \infty]$.

There exists a formula which is accurate for all values of tan δ, viz., (5.111.7). This formula is an infinite series. After the term with the coefficient 0.0108 an infinite number of further terms follow whose coefficients are smaller by a factor 4 and which are shifted to shorter times by a factor 4. In this equation $J''(\omega = 1/t)$ is controlled by the logarithmic slope of the creep compliance within the entire time interval left from the point of calculation, *t*. The influence of those terms decreases only weakly with their distance, namely inverse proportional to their distance from the point of calculation. In this formula, the short time truncation problem occurs explicitly in its structure. The relative error of this formula is limited between −2.7 % and +2.7 % for all values of tan δ. It is not necessary in all cases, to know the complete behavior of the creep compliance for times smaller than the point of calculation. Sometimes it will be sufficient to know upper limits for the logarithmic derivatives of the creep compliance of the short-time tail to limit the possible truncation error. In those cases, the short-time tail of the formula (5.111.7) may be omitted in the calculation.

How many terms of the formula have to be taken into account, depends on the special conditions for the conversion. This is illustrated by Fig. 5.31, in which $J(t)$ is shown in a dispersion region. For the calculation of $J''(\omega)$ at the beginning or in the middle of the dispersion region (points A or B) only a few terms of the conversion formula will be necessary for a reliable result. However, at the point C at the end of a dispersion region the calculation will be especially troublesome, as the principal term $J(t) - J(t/2)$ will be small there, while the terms of the short-time tail in the formula will contribute extraordinarily much to the result.

Table 5.5 Coefficients of the approximations for the calculation of the creep compliance from the storage compliance and the course of the loss compliance with Eq. (5.112)

b	c	d	e	f	g	h	Bounds for the relative error %	Equation
		0.446					22.3 tan δ; 7.9 −10.8 tan δ; −1.0	(5.112.1)
		0.566	−0.203				8.0 tan δ; 7.6 −7.9 tan δ; −7.0	(5.112.2)
		0.482	–	−0.0920			8.2 tan δ; 8.7 −8.2 tan δ; −8.2	(5.112.3)
	0.0872	0.319	–	−0.0532			7.5 tan δ; 3.9 −7.5 tan δ; −3.9	(5.112.4)
	0.103	0.278	–	–	–	−0.0166	3.5 tan δ; 3.6 −3.3 tan δ; −3.6	(5.112.5)
0.0198	0.0375	0.339	–	−0.0122	–	−0.0152	3.6 tan δ; 2.1 −3.6 tan δ; −2.1	(5.112.6)
0.0509	−0.116	0.635	−0.254	0.0383	0.0547	−0.0404	1.3 tan δ; 1.3 −1.3 tan δ; −1.3	(5.112.7)

If there is not enough information about the short-time behavior available, it is not possible to apply the complete formula (5.111.7). In those cases, one of the shorter formulae, the choice of which depends on the extension of the experimental window, has to be chosen. As those formulae were obtained by skipping the short-time tail, they will fail if the truncated part exceeds the error bound. This is the reason, why for all formulae but one of Table 5.3 the lower bound for the relative error tends to −100 % with vanishing tan δ.

We conclude that the calculation of $J''(\omega)$ from the creep compliance constitutes a problem, which is the more difficult, the lower the damping at the point of calculation is. The most simple of the formulae of Table 5.3 does not yield values for $J''(\omega)$ which are much too high, it may, however, yield values which are much too low.

The calculation of the creep compliance from the dynamic compliances constitutes an easier problem. In cases, for which the simplest formulae (5.112.1) and (5.112.2) in Table 5.2 are not sufficient, conversions based on the following equation may be used

$$\begin{aligned} J(t) \cong J'(\omega) + bJ''(\omega/8) + cJ''(\omega/4) + dJ''(\omega/2) \\ + eJ''(\omega) + fJ''(2\omega) + gJ''(4\omega) + hJ''(8\omega) \end{aligned} \tag{5.112}$$

with $b, c, d, \ldots$ being constant coefficients, which have been chosen to minimize the relative error of the corresponding approximation. They are listed in Table 5.5 [30]. It is seen that a formula with error bounds around 1 % may always be found.

Table 5.6 Coefficients of the approximations for the calculation of the storage modulus from the relaxation modulus with Eq. (5.113)

a	b	c	d	e	f	g	k	bounds for the relative error, %	Equation
		0.855						14.6 tan δ; 21 –14.6 tan δ	(5.113.1)
	0.444	–	0.378					7.8 tan δ; 17.7 –7.8 tan δ	(5.113.2)
–0.138	0.690	0.119	–	0.161				3.2 tan δ; 26 – 3.2 tan δ	(5.113.3)
–0.142	0.718	0.044	0.102	0.101	–	0.00855		1.0 tan δ; 6.4 – 1.0 tan δ	(5.113.4)
–0.142	0.717	0.046	0.099	0.103	0.001	0.00716	0.000451	1.0 tan δ; 6.9 – 1.0 tan δ; – 6 .9	(5.113.5)

The calculation of the creep compliance from the storage compliance and the course of the loss compliance is a simple problem, especially simple at low damping. No problems arise, neither with the width of the experimental window, nor with a propagation of the experimental error.

A simple and accurate equation should be mentioned here, which uses the values of $J'(\omega)$ and $J''(\omega)$ at two frequencies [31] (bounds for the relative error ± 3 %.)

$$J(t) \cong J'(\omega) + 0.485 J''(\omega/2) + 0.200[J'(\omega/2) - J'(\omega)] \tag{5.123}$$

This formula is especially appropriate for cases of high damping, at which other approximations with a narrow experimental window fail.

Formulae for the calculation of the storage modulus from the stress relaxation modulus have the following form [32]

$$\begin{aligned} G'(\omega) \cong\ & G(t) + a[G(4t) - G(8t)] + b[G(2t) - G(4t)] \\ & + c[G(t) - G(2t)] + d[G(t/2) - G(t)] + e[G(t/4) - G(t/2)] \\ & + f[G(t/8) - G(t/4)] + g[G(t/16) - G(t/8)] + k[G(t/64) - G(t/32)] + \ldots \end{aligned} \tag{5.113}$$

The coefficients together with the error bounds are listed in Table 5.6.

The formula (5.113.1) is *dual* to the formula (5.110.1). The other formulae of the Table 5.6 are not dual to the formulae of Table 5.3.[3] In deriving Eqs. (5.113.1 to 4), it was tried to minimize the error bounds in terms of tan δ. For those

[3] Two equations, approximate equations or inequalities are designated as dual, if they turn into each other by the following simultaneous substitutions $\sigma(t) \leftrightarrow \gamma(t), \gamma(t) \leftrightarrow \sigma(t), J(t) \leftrightarrow -G(t)$; $J'(\omega) \leftrightarrow -G'(\omega)$; $J''(\omega) \leftrightarrow +G''(\omega)$.

formulae, no absolute bounds for the relative error exist. The formula (5.113.5) is an infinite series. After the term with the coefficient k follow further terms, each being shifted by a factor 4 to smaller times and having a coefficient, which is smaller by a factor 16. This formula has apart from the bounds proportional to tan δ, the absolute bounds ± 6.9 %. It is a second example for a formula for which the short time truncation shows up explicitly.

The calculation of the storage modulus from the stress relaxation modulus constitutes a simple problem. Very troublesome, on the contrary, is the calculation of the loss modulus from the stress relaxation modulus. The only formula for which bounds for the relative error may be given is the one which is dual to Eq. (5.111.7) [33].

$$\begin{aligned} G''(\omega) \cong & -0.470 \cdot [G(2t) - G(4t)] + 1.674 \cdot [G(t) - G(2t)] \\ & + 0.198 \cdot [G(t/2) - G(t)] + 0.620 \cdot [G(t/4) - G(t/2)] \\ & + 0.012 \cdot [G(t/8) - G(t/4)] + 0.172 \cdot [G(t/16) - G(t/8)] \\ & + 0.0433 \cdot [G(t/64) - G(t/32)] + 0.0108 \cdot [G(t/256) - G(t/128)] + \cdots \end{aligned} \tag{5.114.7}$$

The bounds for the relative error of this formula, which again represents an infinite series, are ± 2.7 %. For the application of this formula, the same remarks about the truncation error apply which have been discussed in case of Eq. (5.111.7).

If the value of the storage modulus is known at one frequency together with the course of the loss modulus, it is possible to calculate the stress relaxation modulus by means of the following formulae [33]:

$$\begin{aligned} G(t) \cong & G'(\omega) - aG''(\omega/16) - bG''(\omega/8) - cG''(\omega/4) \\ & - dG''(\omega/2) - fG''(2\omega) - gG''(4\omega) - hG''(8\omega) \end{aligned} \tag{5.115}$$

The constants $a, b, c, \ldots$ are listed in Table 5.7. The column before last of this table contains upper and lower bounds for the *absolute* error, expressed in percents of the quantity $G''(\omega)$. It is also possible to present upper bounds for the ratio $G''(\omega)/G(t)$ as a function of tan δ. For details see [33].

A discussion of the error bounds of the formulae of Table 5.7 shows that the calculation of the relaxation modulus from dynamic measurements is an easy problem for low values of tan δ. For higher values of tan δ, however, the problem gets more and more difficult. This becomes especially obvious if one tries to apply these formulae to the flow region of polymers. The problem is additionally complicated there by the propagation of the experimental error: A large value of $G''(\omega)$ has to be subtracted from a large value of $G'(\omega)$ in order to arrive at a small value for $G(t)$.

Table 5.7 Coefficients of the approximations for the calculation of the relaxation modulus from one value of the storage modulus and the course of the loss modulus with Eq. (5.115)

a	b	c	d	e	f	g	h	Bounds for the absolute error %	Equation
			0.400					$20\ G''$ $-20\ G''$	(5.115.1)
			0.566	−0.203				$8\ G''$ $-8\ G''$	(5.115.2)
			0.478	–	−0.0783			$8\ G''$ $-8\ G''$	(5.115.3)
			0.528	−0.112	−0.0383			$8\ G'''$ $-8\ G''$	(5.115.4)
		0.0825	0.369	−0.169	0.167	−0.0828		$4\ G''$ $-4\ G''$	(5.115.5)
		0.123	0.179	0.186	−0.168	0.0788	−0.0317	$3\ G''$ $-3\ G''$	(5.115.6)
0.00378	0.0309	−0.068	0.564	−0.186	−0.0017	0.0677	−0.0428	$1\ G''$ $-1\ G''$	(5.115.7)

The stress relaxation modulus may also be calculated, if only the frequency dependence of the storage modulus is known. For details we refer to [33].

A summary of the problems occurring at the conversion of viscoelastic functions is given in Table 5.8. Examples for the numerical conversion of viscoelastic functions will be given in Figs. 6.6, 6.7 and 6.8 in the next chapter.

Table 5.8 Problems occurring at the conversion of viscoelastic functions

$J(t) \rightarrow J'(\omega)$	Simple; somewhat more complicated at higher damping; at very high damping great difficulties due to the propagation of the experimental error. Solution: replace $J(t)$ by $J_R(t)$
$J(t) \rightarrow J''(\omega)$	Difficult; much information about the short-time behavior is requested; solution: Estimation of the magnitude of the short time tails. No problems with the propagation of the experimental error
$J'(\omega), J''(\omega) \rightarrow J(t)$	Easy; little information is sufficient for all values of the damping. no problems with the propagation of the experimental error
$G(t) \rightarrow G'(\omega)$	Somewhat more difficult than $J(t) \rightarrow J'(\omega)$; no problems with the propagation of the experimental error
$G(t) \rightarrow G''(\omega)$	Very difficult; the short-time tail of $G(t)$ has to be known. No difficulties with the propagation of the experimental error
$G'(\omega), G''(\omega) \rightarrow G(t)$	Easy for low values of the damping; difficult or even impossible for high values of the damping; Great difficulties with the propagation of the experimental error for high values of the damping

5.10 The Viscoelastic Behavior of Amorphous Polymers in Shear

The characteristic viscoelastic functions of amorphous cross-linked polymers are shown in the Figs. 5.32 and 5.33. Figure 5.32 is a double-logarithmic representation of the creep compliance as a function of the creep time, and the storage compliance, the loss compliance, and the loss tangent, as functions of the angular frequency, at a fixed temperature. The relation between the creep time t and the angular frequency is given by Eq. (5.56). Creep compliance and storage compliance show a similar behavior. To each dispersion step of the creep compliance corresponds one of the storage compliance. Outside the dispersion regions, both compliances coincide, within the dispersion regions the storage compliance is lower than the creep compliance. The difference between the two is maximal in the glass-rubber transition region. The limit of $J(t)$ for $t \to 0$ is the same as the limit of $J'(\omega)$ for $\omega \to \infty$, viz., the instantaneous compliance J_0. The limit of $J(t)$ for $t \to \infty$ is finite and the same as the limit of $J'(\omega)$ for $\omega \to 0$, viz., the stationary recoverable compliance J_e. $J_d(\omega)$ has not been shown in the picture. It

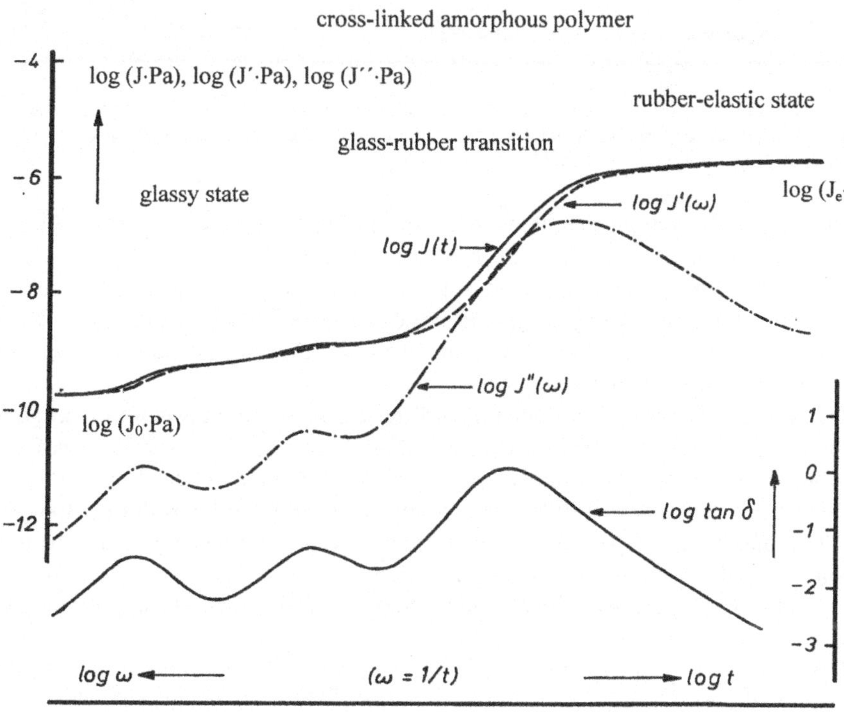

Fig. 5.32 Schematic double-logarithmic representation of the creep compliance, storage compliance, loss compliance and the loss tangent as functions of the time or angular frequency for cross-linked amorphous polymers at a fixed temperature

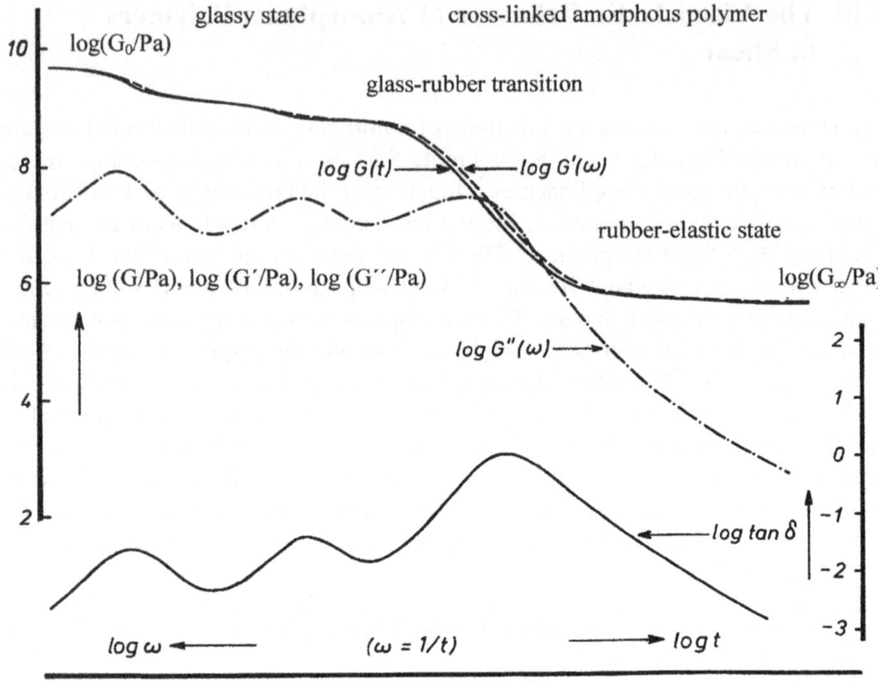

Fig. 5.33 Schematic double-logarithmic representation of the relaxation modulus, storage modulus, loss modulus, and the loss tangent as functions of time or angular frequency for cross-linked amorphous polymers at a fixed temperature

would even be closer to $J(t)$ than to $J'(\omega)$. To each dispersion step of $J(t)$ or $J'(\omega)$ corresponds a maximum of the loss compliance and the loss tangent. According to Eq. (5.118), tan δ is approximately proportional to the double-logarithmic slope of $J(t)$ resp. $J'(\omega)$. In the dispersion regions of the glassy state, $J(t)$ and $J'(\omega)$ differ only by a few percent, while the loss tangent shows there maxima between 0.05 and 0.1.

The half widths of those maxima may differ considerably between two decades on the angular frequency axis for a sharp relaxation process and six decades for a very broad one. In the glass transition, the value of $J(t)$ may exceed that of $J'(\omega)$ between 60 % and some 100 %, depending on the double-logarithmic slope of $J(t)$. The maximum of the loss tangent reaches values between 1 and 6 in this region. The half width of the damping maximum achieves three to five decades of the frequency axis. In the rubbery state, $J(t)$ and $J'(\omega)$ proceed approximately horizontally and *tan* δ decreases to values of the order of 0.001.

The moduli are shown in Fig. 5.33, again in a double-logarithmic plot at one fixed temperature. This picture is a mirror image of Fig. 5.32, apart from the run of the quantity log tan δ, which is the same in both figures. Again $G(t)$ and $G'(\omega)$ show a similar shape, but now according to Eq. (5.91) $G(t)$ is lower than $G'(\omega)$. In the secondary dispersion regions, $G(t)$ and $G'(\omega)$ differ by a few percents only. The

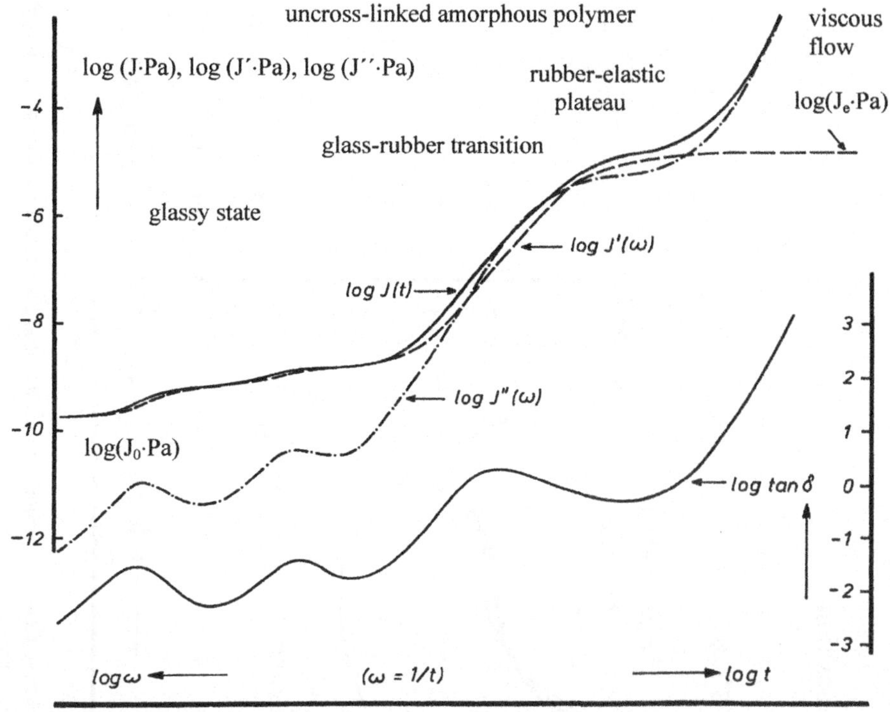

Fig. 5.34 Schematic double-logarithmic representation of the creep compliance, storage compliance, loss compliance, and the loss tangent as functions of the time or angular frequency for uncross-linked amorphous polymers at a fixed temperature

limit of $G(t)$ for $t \to 0$ is the same as the limit of $G'(\omega)$ for $\omega \to \infty$, viz., $G_0 = 1/J_0$. $G(t)$ shows a finite limit for $t \to \infty$, viz., $G_\infty = 1/J_e$.

The compliances of uncross-linked amorphous polymers are sketched in Fig. 5.34. The principal differences between Figs. 5.32 and 5.34 are the broader glass-rubber transition, the less extended rubbery plateau, and especially the flow region of the uncross-linked polymer. In the flow region, the creep compliance increases again, and finally becomes proportional to the creep time, according to Eq. (5.46). The storage compliance $J'(\omega)$ approaches a horizontal plateau with the limit J_e. The loss compliance at the beginning of the rubber-elastic plateau is lower than the storage compliance, but strongly increases in the flow region and approaches there asymptotically the creep compliance. In the limit of large t or small ω, the difference $J(t) - J''(\omega)$ becomes small compared to the values of $J(t)$ or $J''(\omega)$. The loss tangent tan δ shows a minimum in the rubber-elastic plateau. In the flow region the loss tangent increases again and finally tends to ∞ with $1/\omega$. In practice, values larger than 1000 for tan δ have not been reached, as chemical decomposition occurs before under the influence of the high temperatures due to the high-energy dissipation.

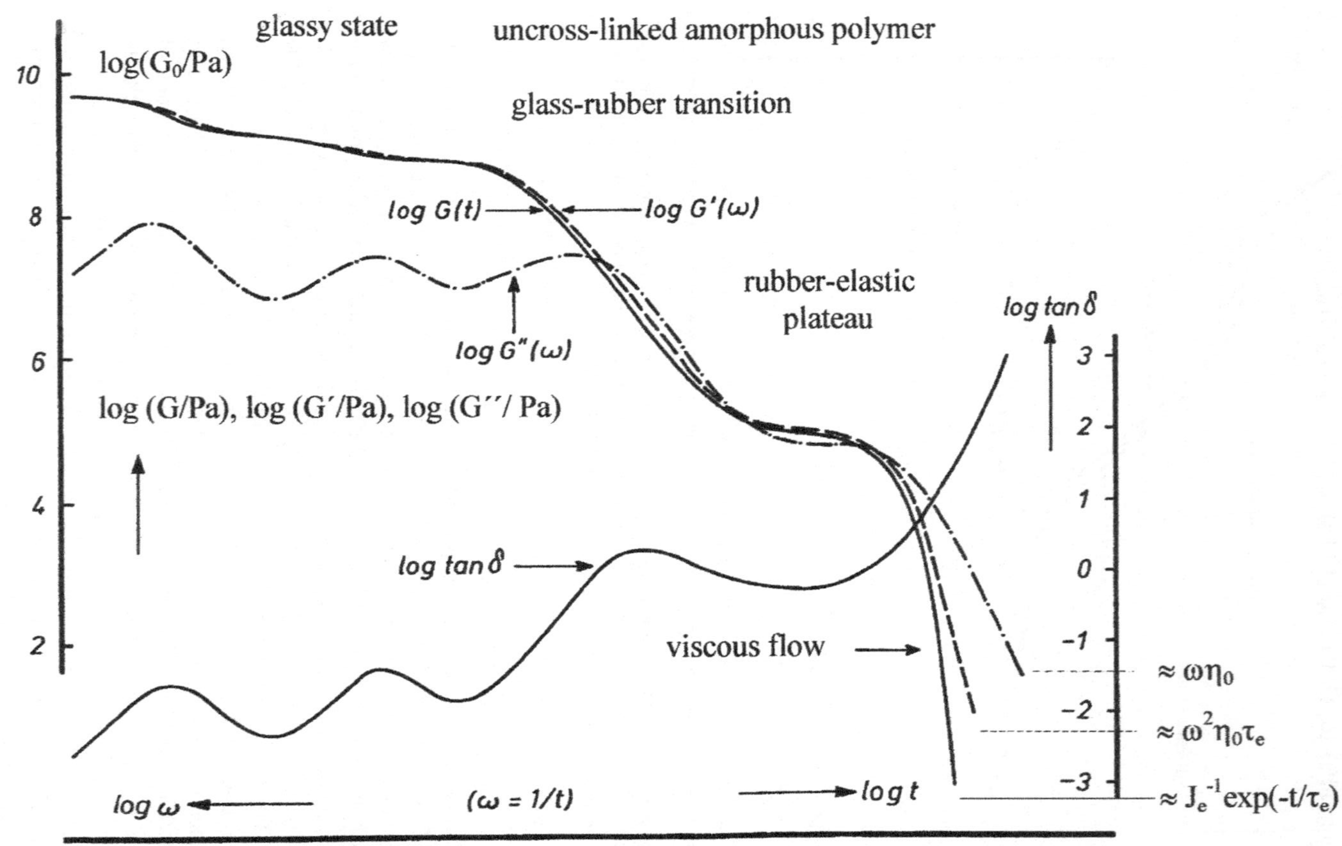

Fig. 5.35 The relaxation modulus, storage modulus, loss modulus and the damping factor as functions of the time or angular frequency for uncross-linked amorphous polymers at a fixed temperature, in a double-logarithmic schematic representation

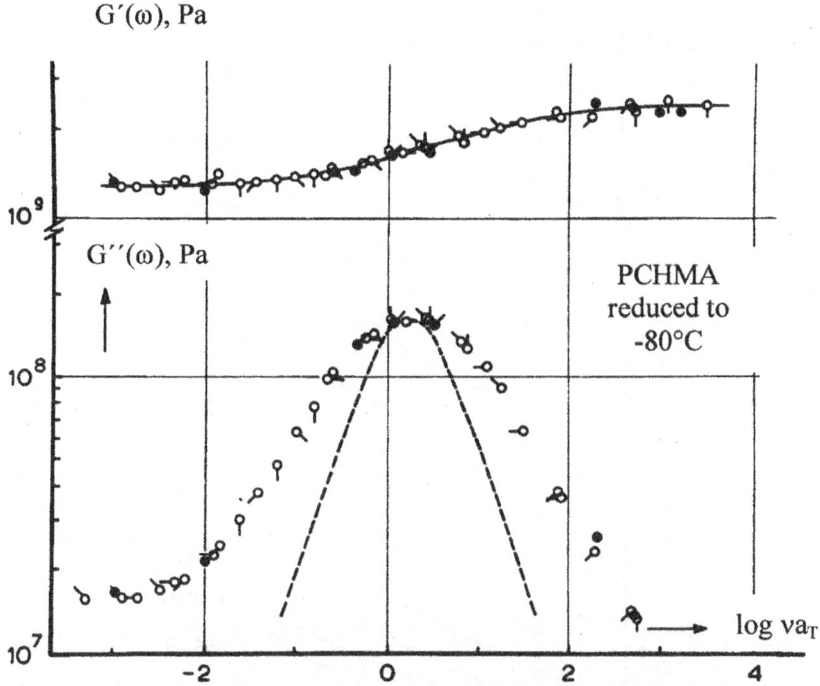

Fig. 5.36 Storage modulus and loss modulus of poly(cyclohexylmethacrylate) (PCHMA) as a function of the reduced frequency at the reference temperature $T_0 = -80$ °C after Heijboer [34]. The broken line indicates the loss modulus for a process with one single relaxation time

The moduli are shown in Fig. 5.35. In the whole range $G(t) \leq G'(\omega)$ and in some parts of the glass-transition even $G'(\omega) \leq G''(\omega)$ is valid. At the beginning of the rubber-elastic plateau, the loss modulus is lower than the storage modulus and the relaxation modulus. At the end of the rubber-elastic plateau, however, $G'(\omega)$ and $G(t)$ decrease so strongly with increasing time (decreasing frequency), that $G''(\omega)$ intersects the curves of $G'(\omega)$ and of $G(t)$. In the flow region, finally, the moduli scale according to $G(t) \leq G'(\omega) \leq G''(\omega)$. $G''(\omega)$ decreases to zero with $\omega \to 0$ as $\eta_0\omega$ (cf. Eq. (5.105), while $G'(\omega)$ decreases to zero with $\omega \to 0$ as $\eta_0^2 J_e \omega^2$ (cf. Eq. (5.106), and $G(t)$ decreases to zero as $J_e^{-1} \cdot \exp(-1/\omega\tau_e)$ (cf.

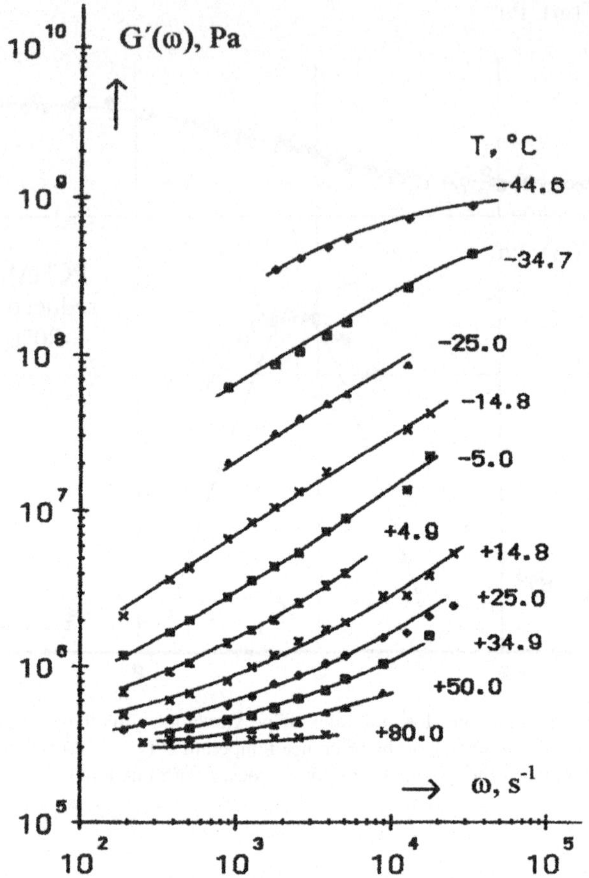

Fig. 5.37 Storage modulus versus angular frequency for a PIB in the glass-transition, reproduced from [35] by permission from AIP Publishing LLC

Eq. (5.55). With $\omega \to 0$ the ratio $G''(\omega)/\omega$ approaches the limit η_0, and the ratio $G'(\omega)/\omega^2$ approaches $\eta_0\tau_e = \eta_0^2 J_e$. Consequently, the ratio $G'(\omega)/[G''(\omega)]^2$ reaches the finite limit J_e.

An example for a relaxation process in the glassy state is shown in the Fig. 5.36. Storage and loss modulus of a poly(cyclohexylmethacrylate) (PCHMA) are plotted as a function of the reduced frequency in a double-logarithmic representation after Heijboer [34].

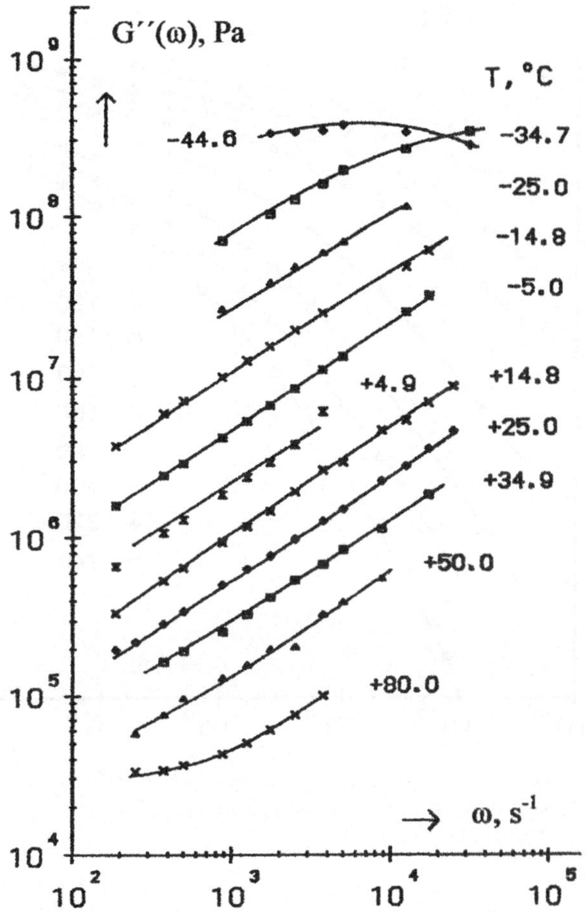

Fig. 5.38 Loss modulus versus angular frequency for a PIB in the glass-transition [35] by permission from AIP Publishing LLC

This dispersion process is caused by the change of the cyclohexyl ring from the one chair position into the other [34]. It is the sharpest relaxation phenomenon in polymers known today. Notwithstanding, it cannot be described by one single relaxation time, as demonstrated by the broken line in the figure, which indicates the loss modulus for a process with one single relaxation time. The half width of

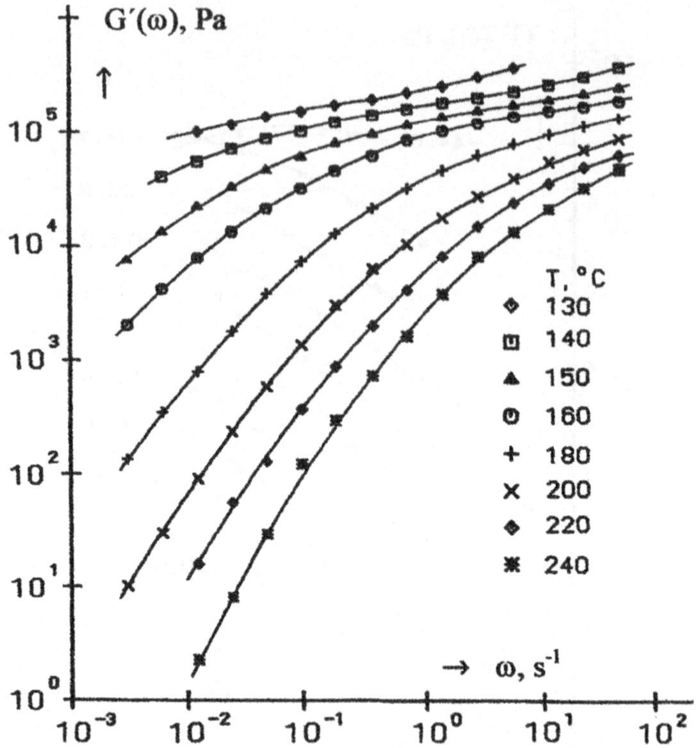

Fig. 5.39 Storage modulus versus angular frequency for PS N 7000 in the rubber-elastic plateau and in the flow region at different temperatures, after Lampel [36]

the G''-maximum of *PCHMA* extends to two decades on the frequency axis. The step height of the dispersion of $G'(\omega)$ corresponds to a factor of 2.

As an example for a dispersion step in the glass-rubber transition region, results of Fitzgerald, Grandine, and Ferry on poly(isobutene) are shown in Figs. 5.37 and 5.38 [35]. Storage and loss modulus are plotted versus the angular frequency in double-logarithmic representations. Comparing these curves with the presentations in Fig. 5.35, one should consider the other direction of the logarithmic frequency axis in the two latter figures. The maximum of the double logarithmic slope of $G'(\omega)$ and $G''(\omega)$ versus ω amounts to 0.64.

As an example for a measurement in the rubber-elastic plateau and in the flow region, in Figs. 5.39 and 5.40 the storage and loss modulus of PS N 7000 are

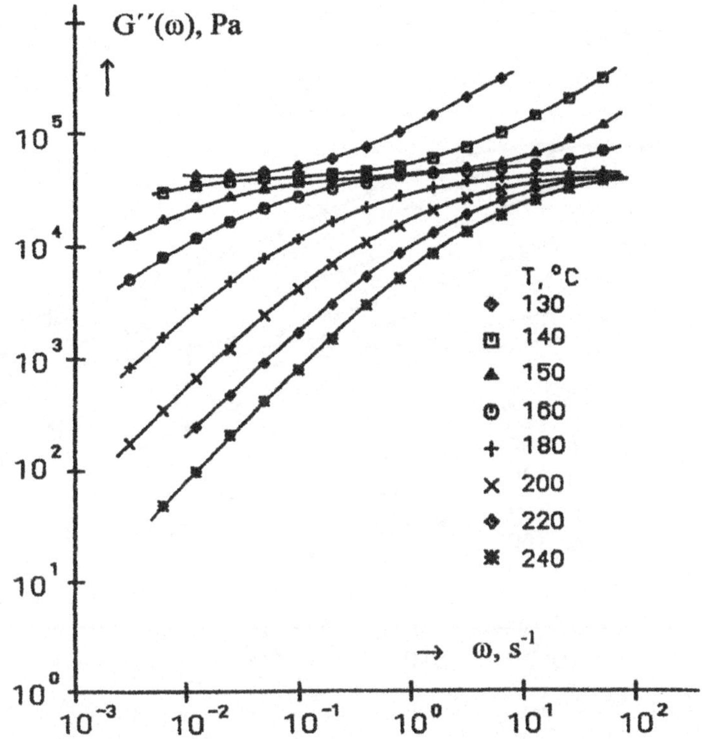

Fig. 5.40 Loss modulus versus angular frequency for PS N 7000 in the rubber-elastic plateau and in the flow region at different temperatures, after Lampel [36]

presented. In the limit of low frequencies, $G''(\omega)$ and $G'(\omega)$ are proportional to the first and second power of ω, respectively, i.e., on the double-logarithmic scale the curves approximate straight lines with the slope 1 and 2, respectively.

From these measurements, the values of $J'(\omega)$ were calculated according to Eq. (5.88) and are presented in Fig. 5.41. While for $G'(\omega)$ and $G''(\omega)$ the entanglement transition is masked due to the great influence of the flow term, it clearly shows up as a dispersion step of the storage compliance, which does not contain the flow term. The broad entanglement transition extends from the beginning of the rubber-elastic plateau nearly to the steady state value J_e at low frequencies. This is the same molecular process which had been observed for the recoverable compliance in Figs. 5.19 and 5.20.

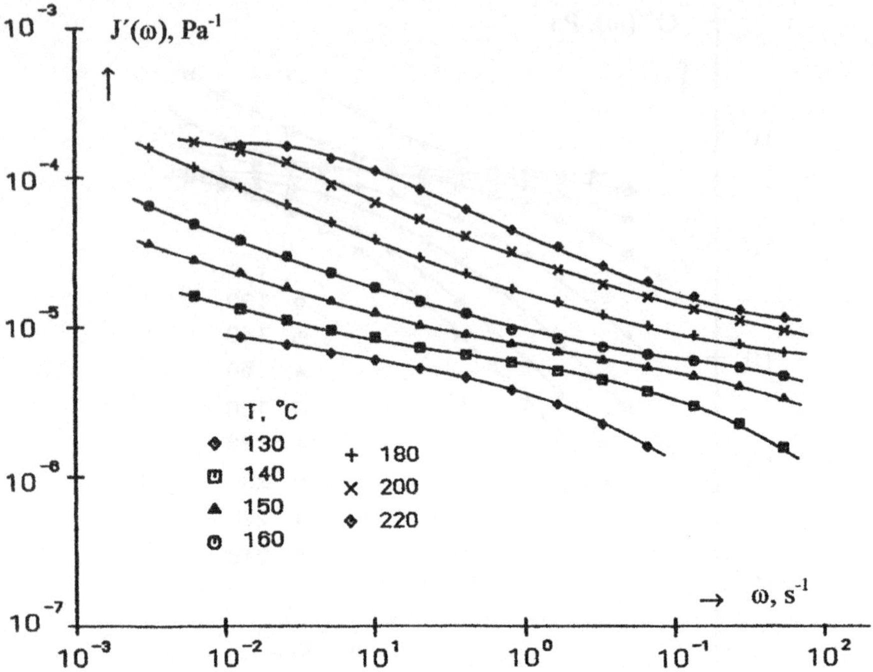

Fig. 5.41 Storage compliance calculated from the data shown in Figs. 5.39 and 5.40 versus angular frequency for PS N 7000 in the rubber-elastic plateau and in the flow region, after Lampel [36]

References

1. Boltzmann L (1874) Sitzber Kgl Akad Wiss Wien 2 Abt.70:275
2. Yannas IV, J Macromol Sci Phys (1970) B4: 603, (1971) B5: 487, (1972) B6: 91
3. Mooney M, Wolstenholme WE, Villars DS (1944) J Appl Phys 15:374
4. Tobolsky AV, Prettyman IB, Dillon JH (1944) J Appl Phys 15:380
5. Tobolsky AV, Andrews RD (1945) J Chem Phys 13:3
6. Brown GM, Tobolsky AV (1951) J Polymer Sci 6:165
7. Andrews RD, Hofman-Bang N, Tobolsky AV (1948) J Polymer Sci 3:669
8. Mercier JP, Aklonis JJ, Litt M, Tobolsky AV (1965) J Polymer Sci 9:447
9. Ogorkiewicz RM (ed) (1970) Engineering properties of thermoplastics. Wiley Interscience, London
10. Bernstein S (1929) Acta Math 52:1
11. Bland DR (1960) The theory of linear viscoelasticity. Pergamon Press, Oxford
12. Gross B (1968) Mathematical structure of the theories of viscoelasticity. Hermann & Cie, Paris
13. Tschoegl NW (1989) The phenomenological theory of linear viscoelastic behavior. Springer, New York
14. Kaschta J, Schwarzl FR, (1995) Proc. 1st T.D.M. conference Ljubljana p 98
15. Brather A, Link G, Luchscheider R (1980) Coll Polym Sci 258:1307

16. Pfandl W (1984) Doctoral thesis, University Erlangen Nürnberg
17. Plazek DJ (1968) J Polym. Sci A-2 6:621
18. Link G (1985) Doctoral thesis, University Erlangen Nürnberg
19. Link G, Schwarzl FR (1985) Rheol Acta 24:211
20. Hartwig G (1994) Polymer properties at room and cryogenic temperature. Plenum press, New York
21. Leaderman H, Smith RG (1954) J Polymer Sci. 14:47
22. Zener C (1948) Elasticity and anelasticity of metals. Univ Press, Chicago
23. Kuipers L, Timann R (1968) Handbuch der Mathematik, Walter de Gruiter, Berlin p 364–365
24. Mc Loughlin AR, Tobolsky AV (1952) J Colloid Sci 7:555
25. Tobolsky AV, Catsiff E (1956) J Polymer Sci 19:111
26. Tobolsky AV, Mc Loughlin JR (1952) J Polymer Sci 8:543
27. Rottmann K (1960) Mathematische Formelsammlung, Bibliographisches Institut AG, Mannheim, p 167
28. Schwarzl FR (1970) Pure and Appl Chem 23:219
29. Ninomiya K, Ferry JD (1949) J Colloid Sci 14:36
30. Schwarzl FR (1969) Rheol Acta 8:6
31. Schwarzl FR (1970) Rheol Acta 9:382
32. Schwarzl FR (1971) Rheol Acta 10:166
33. Schwarzl FR (1975) Rheol Acta 14:581
34. Heijboer J (1972) Doctoral thesis, University Leiden
35. Fitzgerald ER, Grandine LD Jr, Ferry JD (1953) J Applied Phys 24:650
36. Lampel H (1988) Diploma thesis University Erlangen Nürnberg

Chapter 6
Time-Temperature Shift of Mechanical Properties

6.1 The Significance of the Time-Temperature Shift for the Description of the Deformation Behavior of Polymers

It was already shown in the last chapter that the experimental time or frequency window at one single temperature is not broad enough to cover the complete dispersion step from the beginning to the end of a molecular process. Therefore, it would not have been possible to discuss the complete shape of most of the dispersion processes of amorphous polymers if not making use of the time-temperature shift phenomenon. Moreover, the overviews given in the Figs. 5.32–5.35 could never have been constructed without the application of this principle.

Most chemical, physicochemical and physical molecular processes are very sensitive to temperature changes. In many cases an increase in temperature enhances the rate of all molecular processes involved in a certain viscoelastic dispersion process by the same factor. That means, a change in temperature can be described by multiplying all relaxation times or retardation times of the process with the same constant factor. In a representation of the process over a logarithmic time scale or a logarithmic frequency scale, the characteristic functions of this process then shift without a change of their shape. In some cases, other minor alterations may take place by varying the temperature as a slight change in the step height or in the shape of the dispersion curve. But these effects are of subordinate importance and are often neglected—or overseen—when applying the time-temperature shift. If a pure shift of the characteristic functions in the logarithmic time or frequency-axis occurs without a change of their shape, the corresponding dispersion process is called a *thermorheologically simple* process [1]. Otherwise, it is called a *thermorheologically complex* process.

Furthermore, molecular processes as the various secondary dispersions in the glassy region, the glass-rubber transition or the flow transition are differently influenced by a temperature change and show, therefore, dissimilar shifts along the logarithmic time scale with a change of temperature, though each of them may shift without a change in shape of its "own" characteristic dispersion. That means the characteristic functions as given in Fig. 5.32, for example, never shift as a

H. Münstedt and F. R. Schwarzl, *Deformation and Flow of Polymeric Materials*,
DOI: 10.1007/978-3-642-55409-4_6,

whole with temperature, but changes according to the molecular process being dominant in the particular temperature range. Consequently, a characteristic function, governed by more than one single molecular process may not be expected to behave thermorheologically simple.

In general, the smaller the parts of the molecules are, which are involved in a certain relaxation process, the shorter is the time at which the dispersion region occurs and the smaller its shift with temperature. Therefore, the shape of characteristic functions as shown in Fig. 5.32 changes with temperature, but this change can be predicted, when the temperature shifts for the different processes are known.

First, the mathematical formalisms of the thermorheologically simple processes will be described and then the time-temperature shifts of the various molecular processes of amorphous and semicrystalline polymers.

6.2 The Time-Temperature Shift Principle

So far, only measurements at one fixed temperature have been discussed. If a measurement is repeated on the same material at higher temperatures, dispersion regions are shifted to shorter times. Some important examples have already been shown in Figs. 5.16, 5.19, 5.20, 5.23, 5.24, and 5.37–5.41. For a great number of materials, experimental evidence proved the validity of the *time-temperature shift principle* for which synonymously the expression *time-temperature superposition principle* is in use.

By an increase in temperature, the positions of dispersion regions of polymers shift to shorter times or higher frequencies, respectively. By this shift, the shape of the curves of $J(t)$ or $G(t)$ versus log t, respectively, $J'(\omega)$ and $J''(\omega)$ or $G'(\omega)$ and $G''(\omega)$ versus log ω is preserved. One has to deal with a simple parallel shift of these curves along the logarithmic time or frequency axis, only.

The mathematical description of this principle is illustrated by Fig. 6.1, which shows two creep compliance curves as a function of the logarithm of the creep time, one at the temperature T_0 and the other at the temperature T, whereby $T_0 > T$.

A parallel shifting is described by the equation

$$J(t, T) = J(x, T_0) \tag{6.1}$$

The distance log a_T between the corresponding points log t and log x does not depend on the creep time t, but on the temperatures chosen. T_0 is called the *reference temperature*, $a_T(T,T_0)$ the *time-temperature shift factor*, log $a_T(T,T_0)$ the *time-temperature shift function* and x the *reduced time*. From Fig. 6.1, one gets

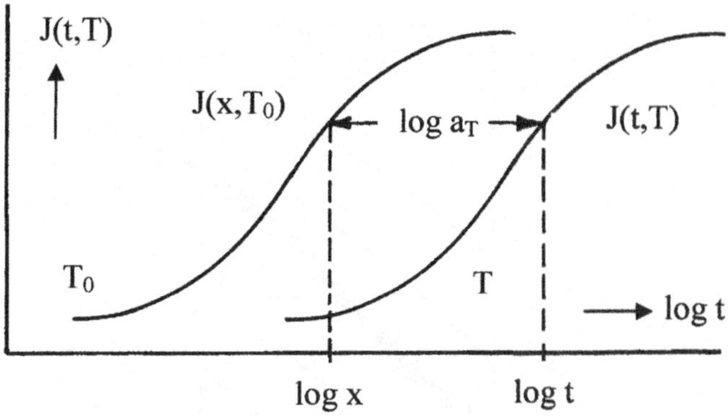

Fig. 6.1 Illustration of the time-temperature shift of the creep compliance for $T < T_0$

$$\log a_T(T, T_0) = \log t - \log x \tag{6.2}$$

or

$$x = t/a_T(T, T_0) \tag{6.3}$$

The direction of the shift along the logarithmic time axis is determined by the sign of the temperature difference $(T - T_0)$. For $T < T_0$ it follows $t > x$, $a_T > 1$, and $\log a_T > 0$. For an application of the time-temperature shift to oscillatory experiments, a *reduced angular frequency* y is defined by the equation

$$y = a_T(T, T_0) \cdot \omega \tag{6.4}$$

The mathematical description of the time-temperature shifting principle is then expressed by the set of equations:

$$\begin{aligned} &J(t,T) = J(x,T_0) = J_{T_0}(x) = J_{T_0}(t/a_T) \quad G(t,T) = G(x,T_0) = G_{T_0}(t/a_T) \\ &J'(\omega,T) = J'(y,T_0) = J'_{T_0}(a_T\omega) \qquad G'(\omega,T) = G'(y,T_0) = G'_{T_0}(a_T\omega) \\ &J''(\omega,T) = J''(y,T_0) = J''_{T_0}(a_T\omega) \qquad G''(\omega,T) = G''(y,T_0) = G''_{T_0}(a_T\omega) \\ &\tan\delta(\omega,T) = \tan\delta(y,T_0) = \tan\delta_{T_0}(a_T\omega) \end{aligned} \tag{6.5}$$

In these equations the creep compliance at the reference temperature is designated as $J_{T_0}(x)$, whereby x is the creep time at the temperature T_0. The function $J_{T_0}(x)$ is called the *master (curve of the) creep compliance*, $G_{T_0}(x)$ the *master (curve of the) relaxation modulus*, $J'_{T_0}(y)$ the *master (curve of the) storage compliance*, $G'_{T_0}(y)$ the *master (curve of the) storage modulus*, etc.

The existence of a time-temperature superposition makes an essential simplification of the description of the shear behavior of polymers possible. Instead of one function which depends on the two independent variables temperature and time or frequency, respectively, it is sufficient to handle two functions, each of

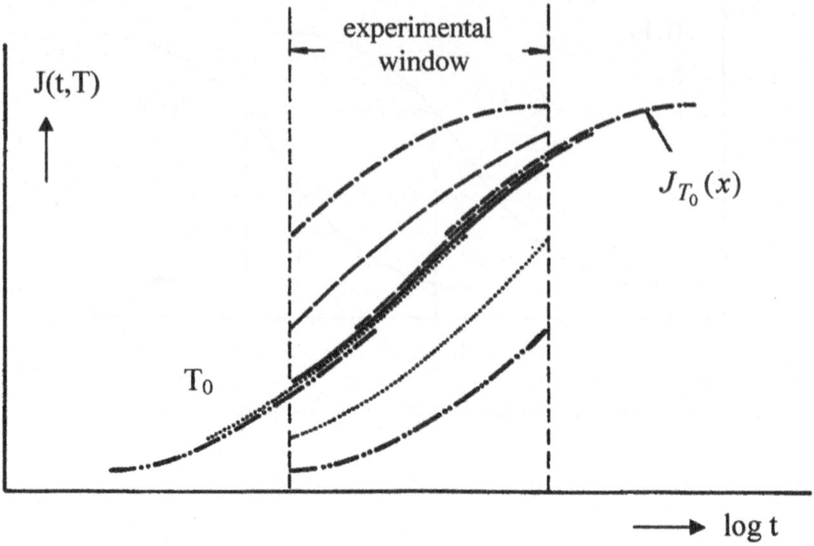

Fig. 6.2 Application of the time-temperature superposition principle to construct a master curve

them depending on one of the two variables, only. In the case of the creep compliance these are the master creep compliance $J_{T_0}(x)$ at the reference temperature T_0 and the temperature-dependent shift factor $a_T(T, T_0)$.

If the time-temperature superposition has been proven to be valid for one of the characteristic functions, it will be applicable to all the others. This is a consequence of the Boltzmann superposition principle presented by Eqs. (5.15) or (5.17).

The application of the time-temperature superposition principle to construct a so-called "master curve" is illustrated in Fig. 6.2 for the creep compliance as an example. Within the experimental window, creep compliances at a number of different temperatures were measured. One of those curves, preferably in the middle of the experimental window is selected as a master curve and the corresponding temperature as the reference temperature T_0. Then, the creep compliances measured at the other temperatures are shifted until they overlap smoothly with the reference curve. From the shift in time, $a_T(T, T_0)$ is obtained. The master curve constructed by the superposition of the creep compliances at various temperatures with respect to time extends the time window accessible.

The consequence of the validity of the time-temperature superposition is that knowing the shift factor and the time or frequency dependence of a material function at one temperature, the corresponding curves at other temperatures can be determined in a simple way.

If the shift function is known from an independent source, it is possible to plot the data measured at different temperatures versus the reduced time x in order to obtain the master curve. The time-temperature shift function for a certain process at the reference temperature T_0 can easily be changed to its equivalent at the reference temperature $\bar{T}_0$ by the equation:

$$\log a_T(T, \bar{T}_0) = \log a_T(T, T_0) - \log a_T(\bar{T}_0, T_0) \tag{6.6}$$

6.3 The Time-Temperature Shift of the Glass-Rubber Transition

The logarithm of the creep compliance as a function of the logarithm of the time in the glass-rubber transition has already been shown in the Figs. 5.16 and 5.19 for PS. Similar results have been obtained for PC, PMMA, PVC, NR, and for a cross-linked polyurethane rubber (PUR) [2].

As an example for a cross-linked polymer with a low glass transition temperature, the data for the PUR are shown in Fig. 6.3. Crosses indicate data points obtained by the technique of *free damped torsional vibrations*, circles those from a measuring device for *torsional creep*, which automatically registered compliance data at logarithmically equidistant creep times of a factor two. Such data are especially appropriate for evaluations by means of the conversion techniques described in Sect. 5.9. The agreement between the results of the two measuring techniques is convincing and yields an accurate picture of the strong dependence of the creep compliance on time and temperature in the glass transition region. Figure 6.3 shows the compliance within an experimental window of five decades on the logarithmic time scale.

It is obvious that the larger the *experimental window* of the original data, the more accurate are the results after the time-temperature superposition. Moreover, shifting the curves in the center of the transition results in a higher accuracy of the determination of the time-temperature shift function and of the master compliance than shifting the data in one of the flanks of the transition where the compliance runs flatter.

The strong influence of time and temperature on the value of the compliance in the glass transition is evident for all amorphous polymers investigated. This effect is particularly pronounced for polystyrene, for which an increase in temperature by 3 °C magnifies the compliance to the ninefold, or shifts it to shorter times by about one decade. Though the general course of the creep compliance with time is similar for all amorphous polymers in the glass transition region, significant differences exist with respect to a more quantitative description. The double-logarithmic slope $m = d\log J(t)/d\log t$ in the center of the glass transition and the step height H of the transition significantly differ as documented in Table 6.1 on page 194.

The measurements of Fig. 6.3 and those in the glass transition of other amorphous polymers permit a time-temperature superposition with a shift function of the following form

$$\log a_T(T, T_0) = -\frac{c_1(T - T_0)}{c_2 + T - T_0} \tag{6.7}$$

This relation has first been proposed by Williams, Landel, and Ferry [3] and is well known in the literature as the *WLF-equation*. It is a function of the temperature difference $T - T_0$, only. T_0 is an arbitrary reference temperature, and c_1 and c_2 are

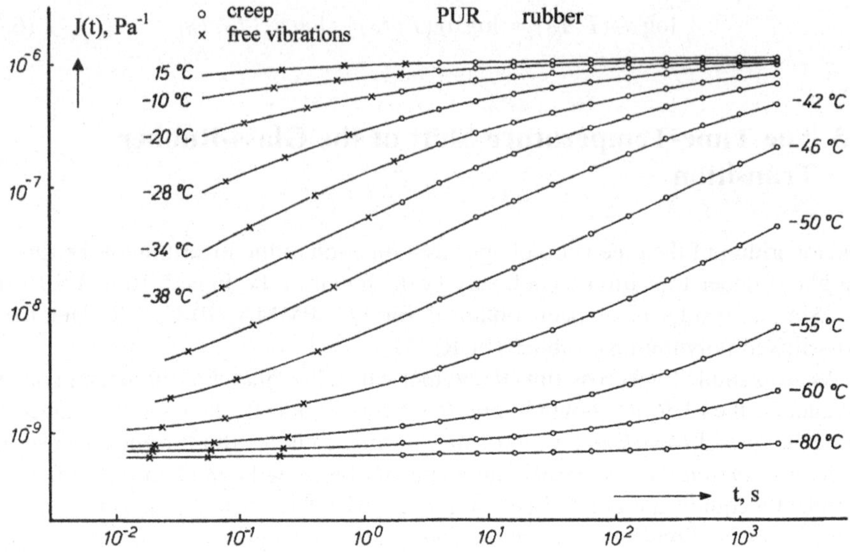

Fig. 6.3 Creep compliance in shear as a function of the creep time, in a double-logarithmic plot, for a cross-linked polyurethane rubber at various temperatures in the glass-rubber transition region reproduced from [2]

Table 6.1 Characteristic properties of the glass transition of some amorphous polymers

Polymer	H, decades	m	T_s °C	c_1^s	c_2^s K	$c_1 c_2$	T_∞, °C	T_g °C	Lit.
PS	3.6	0.88	105	8.9	36	320	64	91	[25]
PMMA	2.8	0.65	123.5	8.0	36	288	87.5	107	[2]
PVC	2.4	0.53	73.5	11.2	34.6	388	40	66	[2]
PC	2.7	0.70	146	12.7	44.5	566	101.5	141	[25]
PUR	3.1	0.48	−45	12.5	42.5	531	−87.5	−54	[2]
NR	3.0	0.72	−62	11.4	37.8	431	−99.8	–	[2]

two specific material constants. For a given material, the values of c_1 and c_2 depend on the reference temperature chosen. If a temperature $\bar{T}_0$ different from the reference temperature T_0 is selected, the shift function keeps the same form as (6.7), but with two different constants $\bar{c}_1$ and $\bar{c}_2$ which are related to c_1 and c_2 by

$$\bar{T}_0 - \bar{c}_2 = T_0 - c_2 \equiv T_\infty \tag{6.8}$$

and

$$\bar{c}_1 \cdot \bar{c}_2 = c_1 \cdot c_2 \tag{6.9}$$

These two combinations of the *WLF* parameters are independent of the choice of the reference temperature and are therefore called the *invariants* of the *WLF equation*.

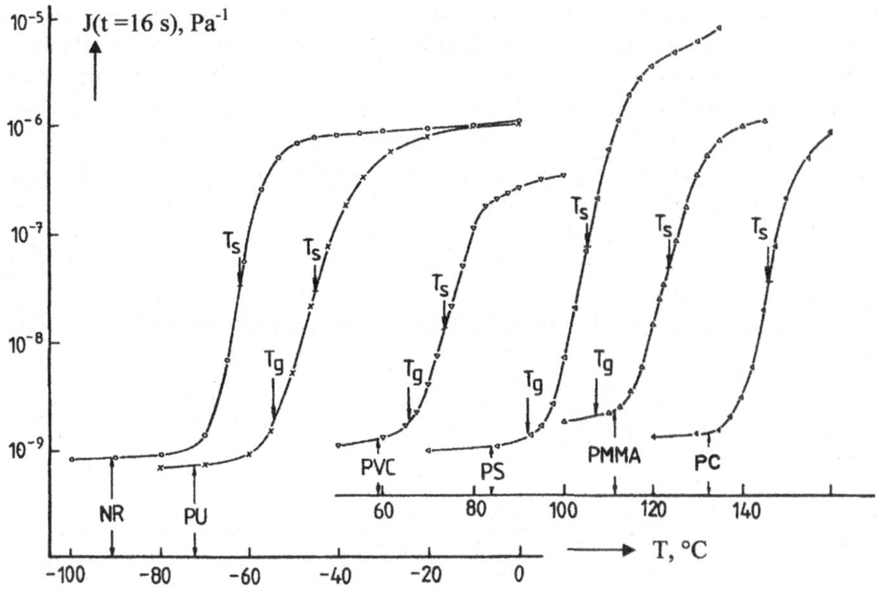

Fig. 6.4 Creep compliance after 16 s creep time versus the temperature for 6 amorphous polymers in the glass-rubber transition

Equation (6.7) may be written in an alternative form, which is obtained by adding and subtracting the term $c_1.c_2/(c_2 + T - T_0)$ to Eq. (6.7) resulting in

$$\log a_T(T, T_0) = -c_1 + \frac{c_1 c_2}{T - T_\infty}. \tag{6.10}$$

This equation has been used for the description of the temperature dependence of the viscosity of low molar mass liquids by Vogel [4], Fulcher [5], and Tammann and Hesse [6] and is therefore called the *VFTH-equation*. The temperature T_∞ is called the *Vogel temperature*. It is the temperature at which the shift factor tends to infinity.

For the characterization and comparison of the glass transition of various polymers, we choose a reference temperature in the following way. In Fig. 6.4, the logarithm of the creep compliance after 16 s creep time is plotted as a function of the temperature for four thermoplastics and two rubbers. On the logarithmic scale of the compliance, the glass transition occurs for all polymers as a pronounced step from the glassy to the rubbery level. The temperature of the middle of the logarithmic step height of the compliance is defined as the *softening temperature* T_s. The softening temperatures are indicated in Fig. 6.4 by arrows. They are chosen as the reference temperatures for the *WLF*-time-temperature shift of the corresponding polymer.

Table 6.1 lists the *logarithmic step height H* of *J* versus *T*, the *maximum double-logarithmic slope* $m = d\,\log J/d\,\log t$, the *softening temperature* T_s, the *WLF* constants c_1^s and c_2^s corresponding to the reference temperature T_s and the two *invariants* $c_1 \cdot c_2$ and T_∞. For some polymers, the dilatometric *glass transition temperature* T_g as measured at a cooling rate of 1 K/min has been listed, too.

It is not only possible to describe the temperature shift of the glass transition, but also its absolute position on the temperature scale. Let us designate the creep time necessary to reach the logarithmic middle of the glass transition as t_{glass} in seconds. Then, conform to our definition of the softening temperature, the *glass time* at the *softening temperature* equals 16 s

$$\log t_{\text{glass}}(T_s) = \log(16) = 1.20 \tag{6.11}$$

and the *glass time* at the arbitrary temperature *T*, in seconds, follows from the shift relation as

$$\log t_{\text{glass}}(T) = \log t_{\text{glass}}(T_s) + \log a_T(T, T_s) = c_0 + \frac{c_1^s c_2^s}{T - T_\infty} \tag{6.12}$$

with

$$c_0 = 1.20 - c_1^s \tag{6.13}$$

The application of this very useful rule of thumb only requires the knowledge of the *WLF* constants.

A similar rule of thumb holds for the temperature position of the frequency of the *tan* δ-maximum. The half of the logarithmic step height of the compliance $J(t,T)$ and the maximum of the loss tangent *tan* $\delta(\omega,T)$ are situated for the glass transition at the same position, if ω is set equal to $1/t$. This may be seen from Fig. 6.6 and also follows from Eq. 5.118. Therefore, the *frequency position of the maximum of the loss tangent* at the temperature *T* may be derived from Eq. (6.12) as

$$(\log \upsilon)_{\tan\delta,\,\max} = (\log \omega/2\pi)_{\tan\delta,\,\max} = -2.00 - \log a_T(T, T_s) \text{ with } \upsilon \text{ in } Hz \tag{6.14}$$

For a more detailed discussion of the glass transition the data of the glass-rubber transition of PS in Fig. 5.16 were shifted to a master curve at the reference temperature of $T_s = 105$ °C. The experimentally obtained shift factors marked by open triangles are shown in Fig. 6.5. The full line through these points represents the WLF-equation with the parameters indicated in the figure. The parameters were determined by plotting $-(T - T_s)/\log a(T,T_s)$ against $(T - T_s)$ and fitting the best straight line through these points. The WLF equation describes the experimentally obtained shift function with good accuracy.

The figure also demonstrates the significance of the two parameters of the WLF-equation, which constitutes a hyperbola. Its vertical asymptote is situated at the temperature T_∞ and has the distance c_2^s from the ordinate axis. Its horizontal asymptote has the ordinate $-c_1^s$. The invariant $c_1 \cdot c_2$ is a measure for its curvature.

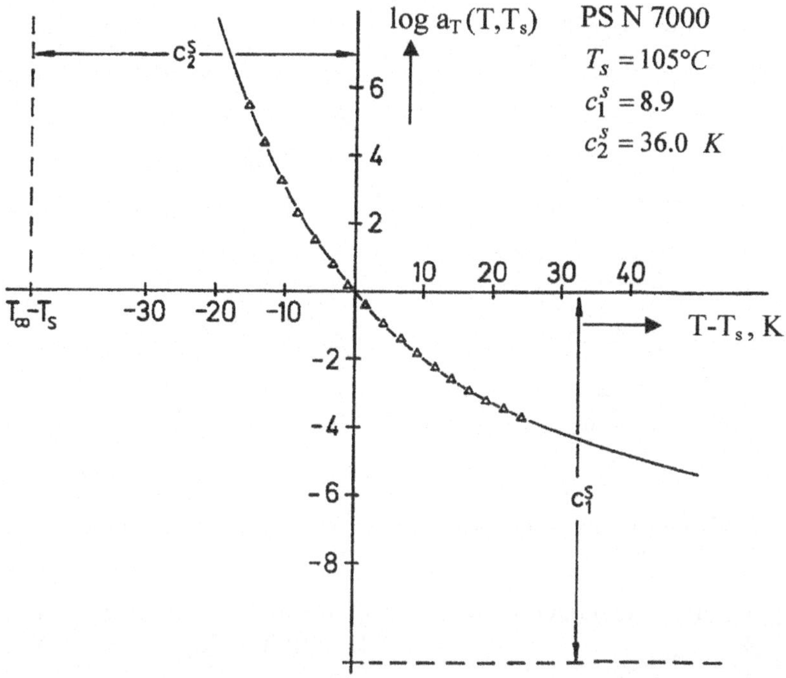

Fig. 6.5 The time-temperature shift function for the glass transition of PS and the significance of the *WLF* equation and its coefficients

The master curve of the creep compliance is shown versus the reduced creep time (t/a_T) in Fig. 6.6. The master curve has been smoothed and the values of the compliance at logarithmically equidistant time positions corresponding to a factor 2 have been determined by interpolation. They are shown as closed circles in Fig. 6.6. From the latter, the other characteristic functions were determined using the conversion equations of Sect. 5.9. In these calculations the formulae were used for which the experimental window of the starting data was just wide enough.

The various compliances are shown in Fig. 6.6, the moduli in Fig. 6.7 and the products $G'(a_T\omega)\cdot J'(a_T\omega)$ and $G(t/a_T)\cdot J(t/a_T)$ together with the *loss tangent* in Fig. 6.8. The creep compliance shows the well-known shape in the glass transition with a maximum double-logarithmic slope of 0.88. $J'(a_T\,\omega)$ deviates significantly from $J(t/a_T)$. In the glassy flank of the transition, $J'(a_T\,\omega)$ is considerably lower than $J(t/a_T)$, but then increases steeper in the transition region, before it almost catches up with the value of $J(t/a_T)$ at the end of the transition. It shows a maximum double-logarithmic slope of 1.28. At the beginning of the rubber-elastic plateau, $J'(a_T\,\omega)$ approaches $J(t/a_T)$, but runs closely below it.

The loss compliance comes very close to the creep compliance within the transition and then bends off to a flat maximum at the beginning of the rubber-elastic plateau. After a flat minimum at the end of the rubber-elastic plateau, it

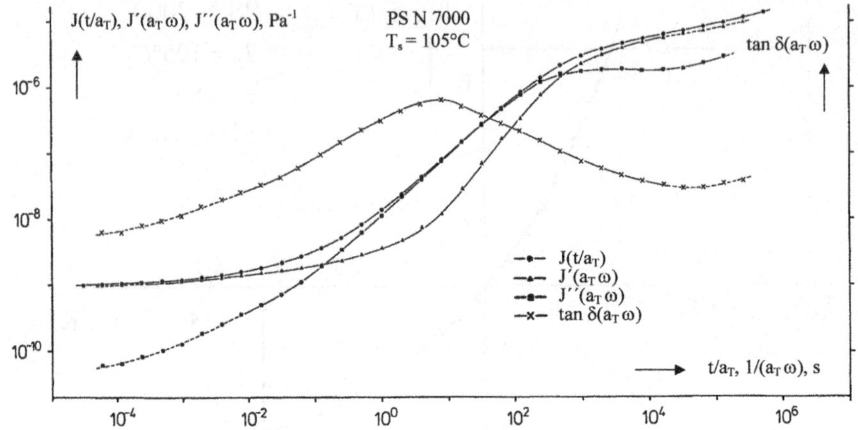

Fig. 6.6 Double-logarithmic plot of the smoothed experimental master curve of the creep compliance of PS in the glass transition versus the reduced creep time together with the calculated values of the storage compliance, the loss compliance, and the loss tangent versus the reduced angular frequency

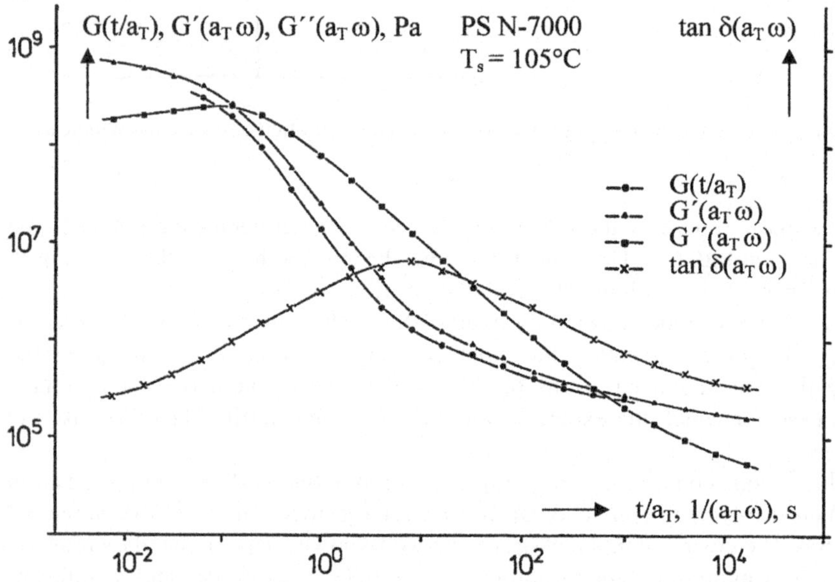

Fig. 6.7 Double-logarithmic plot of the calculated values of the relaxation modulus versus the reduced time and the storage modulus, the loss modulus, and the loss tangent versus the reduced angular frequency for PS in the glass-rubber transition

starts increasing again. The loss tangent shows a maximum of 6.2 at the position, at which the double-logarithmic slope of $J(t/a_T)$ attains its largest value.

The moduli are shown in Fig. 6.7. By the conversion, a part of the experimental window is lost; therefore, the experimental window of Fig. 6.7 (about six decades)

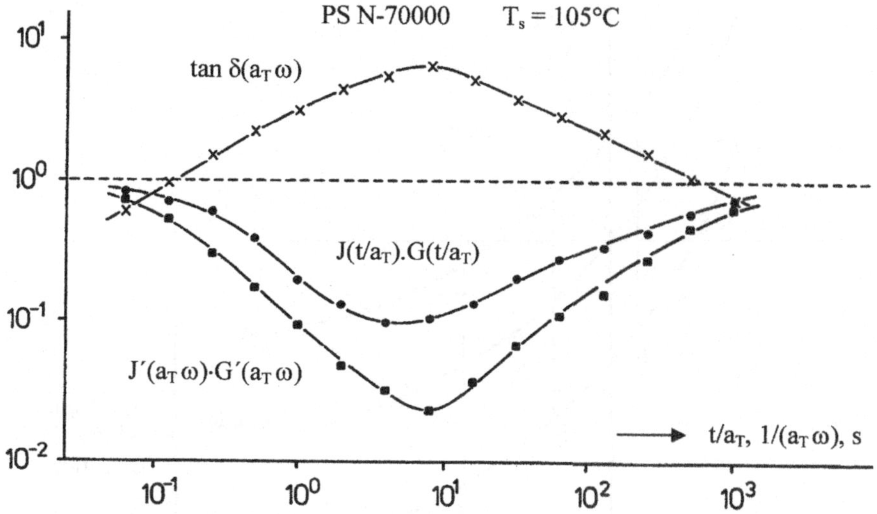

Fig. 6.8 Double-logarithmic plot of the calculated values of the Zener product $J(t/a_T)\cdot G(t/a_T)$ versus the reduced time, the product $J'(a_T\,\omega)\cdot G'(a_T\,\omega)$, and the calculated loss tangent versus the reduced angular frequency for PS in the glass-rubber transition

is much smaller than that of Fig. 6.6 (about 10 decades). The window for $G(t/a_T)$ even extends over four decades, only. The differences between $G(t/a_T)$ and $G'(a_T\,\omega)$ are much less than the differences between $J(t/a_T)$ and $J'(a_T\,\omega)$. The decrease of the moduli in the transition region is steeper than the increase of the compliances. The minimum values of the double-logarithmic slopes are -1.4 for $G(t/a_T)$, -1.3 for $G'(a_T\,\omega)$, and -0.9 for $G''(a_T\,\omega)$.

In Fig. 6.8 the Zener product $G(t/a_T)\cdot J(t/a_T)$ and the corresponding dynamic quantity $G'(a_T\,\omega)\cdot J'(a_T\,\omega)$ are compared with the loss tangent. The Zener product has a minimum value of 0.10 at a time somewhat shorter than the time position of the maximum of *tan* δ which amounts to 6.2. This result demonstrates, how much in error the "elastic" approximation $G\cdot J \approx 1$ would be, viz., by a full order of magnitude. Even more the product $G'(a_T\,\omega)\cdot J'(a_T\,\omega)$ deviates from unity with a minimum value of 0.025. The maximum of *tan* δ and the minimum of $G'\cdot J'$ have about the same position on the time and the reciprocal angular frequency scale, respectively.

So far, the characteristic functions in the glass transition in dependence on the logarithm of time or angular frequency were discussed. It was found that their shape is independent of the temperature at least within the accuracy of the shift procedures applied. Changing the temperature only shifts these curves on the logarithmic time or frequency scale.

Considering the *glass transition as a function of the temperature* at a constant creep time or angular frequency, does not leave the shape of the transition unchanged if the creep time or angular frequency are altered. This is demonstrated in Fig. 6.9.

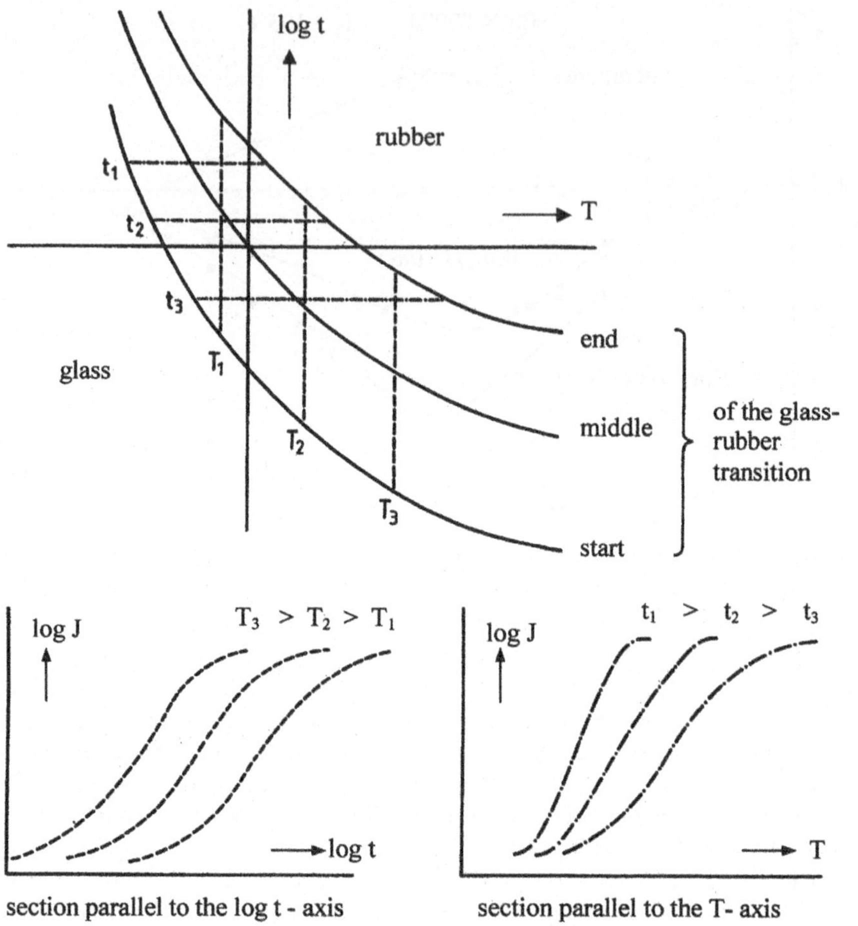

Fig. 6.9 The position of the glass transition in a time-temperature plane (*upper* part) and sections through a three-dimensional relief of the value of the logarithm of the compliance for three fixed values of the temperature (below *left*) and three fixed values of the creep time (below *right*). Notice the equal shape of the curves and following from that the validity of the time-temperature shifting law on the left, but the significant differences in the steepness of the curves on the right-hand side of the lower part of the figure

The figure shows, in its upper part, the position of the glass-rubber transition in a time-temperature plane. A coordinate system on this plane displays the temperature T as abscissa and the logarithm of the creep time t as ordinate. Above this plane, we imagine the built-up of a three-dimensional relief of the logarithm of the compliance within and around the glass transition region. The *glass transition* will show up in this *relief* as a steep increase from the low glassy level (left below) to the high rubber-elastic level (right above).

The position of the logarithmic middle of the glass transition is described by the *VFTH*-equation and shows the same shape as in Fig. 6.5. The temperature positions of the beginning and the end of the glass transition are given by curves which run in parallel to the curve for the middle of the glass transition and originate by a parallel shift of the VFTH-curve in the direction of the logarithmic time axis. These positions have been indicated in Fig. 6.9. The projections of three sections through the relief at constant temperatures $(T_1 < T_2 < T_3)$ on the basic plane are sketched, as well as the projections of three sections through the relief at constant creep times $(t_1 > t_2 > t_3)$. The lengths of the projections of the sections at constant temperatures are clearly equal, while the lengths of the projections of the sections at constant creep times increase significantly with decreasing creep time. This is a direct consequence of the positive curvature of the VFTH-equation.

Therefore, the sections through the relief at constant temperatures run parallel to each other as shown in the left lower part of Fig. 6.9, i.e., the time-temperature shifting law is fulfilled. On the contrary, $\log J(t)$ as a function of T at constant creep time runs more steeply through the transition with higher creep times (cf. Fig. 6.9 below right). This leads to the conclusion that the curve of the logarithm of the compliance as a function of temperature cannot simply be constructed by a shift along the temperature scale, but requires additional changes in shape.

In Sect. 4.6.2 we introduced the concept of the free volume fraction and the Doolittle-equation (4.21). It has been proposed by Williams et al. [3] that the time-temperature shift of the glass transition might be explained by the decrease in the real value of the free volume fraction above the glass transition temperature T_g (compare Fig. 4.32) together with the relation between the mobility m and the free volume fraction

$$\frac{m}{m_0} = e^{B/f} \tag{6.15}$$

where $f = v_f/v$ is the fractional free volume, $m_0 = e^B$ the maximum mobility for the hypothetical case that the total volume is free volume $(f = 1)$ and B is a parameter of the order of and nearby to unity.

Assuming that the mobility in the glass-rubber transition is determined by the free volume, i.e., by Eq. (4.21) and (6.15), respectively, one obtains for the time-temperature shift of the glass transition in the temperature region $T > T_g$

$$\log a_T(T, T_0) = 0.43B\left[\frac{1}{f} - \frac{1}{f_0}\right] \tag{6.16}$$

with f and f_0 being the free volume fractions at the temperatures T and T_0, respectively. Inserting Eq. (4.22) for the free volume fraction above T_g yields

$$\log a_T(T, T_0) = -\frac{0.43B}{f_0}\frac{(T - T_g)}{(f_0/\alpha_f) + (T - T_g)} \tag{6.17}$$

This is the *WLF*-equation (6.7) with the reference temperature T_g and the coefficients

Table 6.2 The constants of the Doolittle equation if applied to the glass transition of some amorphous polymers

	$c_1\,c_2$, K	α_f, 10^{-4} K^{-1}	B	$T_g - T_\infty$, K	f_0, 10^{-2}
PS	403	3.43	3.21	27	0.92
PMMA	288	3.49	2.33	19.5	0.68
PVC	388	3.48	3.13	26	0.90
PC	431	3.89	3.89	29	1.14

$$c_1 = 0.43B/f_0 \tag{6.18}$$

and

$$c_2 = f_0/\alpha_f \tag{6.19}$$

Its invariants are

$$c_1 c_2 = \frac{0.43B}{\alpha_f} \tag{6.20}$$

and

$$T_g - c_2 = T_g - \frac{f_0}{\alpha_f} = T_\infty \tag{6.21}$$

For the proof of the right-hand side of Eq. (6.21) we insert $T = T_\infty$ into Eq. (4.22). This yields $f = 0$ and therefore the following estimate for the *frozen free volume fraction*

$$f_0 = \alpha_f (T_g - T_\infty) \tag{6.22}$$

Using these equations andthe results of the measurements on the glass transition, it is possible to check the feasibility of the application of the free volume concept to the mobility at the glass transition.

From Table 4.9, we obtain the expansion coefficients and the expansion coefficient of the free volume fraction using (4.23), and from Table 6.1 we obtain the invariants. The results are summarized in Table 6.2.

These results are not very encouraging. The constant of the Doolittle equation B was shown due to calculations by Gibbs and Dimarzio [7] to be close to unity, while the experimental values are three times as large. The *frozen free volume fraction*, on the other hand is much lower, than expected, viz., 1 % instead of 2.5–3 vol% as estimated for instance by Ferry [8]. This leads to the conclusion that the free volume theory is either not the right explanation for the time-temperature shift of the glass transition, or at least there may be other molecular processes involved.

A different more promising proposal to explain the *WLF*-equation was given by Struik, who assumed an *energy barrier* described by the Arrhenius equation to be the cause of the time-temperature shift, but with a *temperature-dependent activation energy* [9]. With increasing temperature the mean distance between adjacent molecules increases, leading to a strong decrease of the energy barrier. Estimates by Struik of the temperature dependence of the corresponding activation energy led to the conclusion that the time-temperature shift function takes the shape of the *WLF*-equation at temperatures near the glass transition temperature, but the form of the Arrhenius equation at temperatures far above T_g.

6.4 The Time-Temperature Shift in the Flow Region of Amorphous Polymers

The time-temperature shift of the characteristic functions in the flow region differs from that in the glass-rubber transition region not only by the shift function itself but by the much wider temperature range it covers. Therefore, the contributions of the configurational entropy of the macromolecules to the moduli are expected to play a role. They are proportional to ρT according to the statistical theory of rubber elasticity with ρ being the density and T the absolute temperature. Consequently, within the scope of this theory, shifting laws are to be expected for $G(t,T)/\rho T$, $\rho T{\cdot}J(t,T)$ and other functions, correspondingly.

For example, instead of the equations

$$J(t,T) = J(x,T_0) \quad \text{and} \quad G(t,T) = G(x,T_0) \tag{6.1}$$

the relations

$$\rho T J(t,T) = \rho_0 T_0 J(x,T_0) \quad \text{and} \quad G(t,T)/\rho T = G(x,T_0)/\rho_0 T_0$$

are regarded for the shift procedures. Accordingly, the *reduced creep compliance* and the *reduced relaxation modulus* are introduced by means of the definitions

$$\begin{aligned} J_r(t,T) &= b(T,T_0)J(t,T_0) \\ G_r(t,T) &= G(t,T_0)/b(T,T_0) \end{aligned} \tag{6.23}$$

with

$$b(T,T_0) = \rho T/\rho_0 T_0 \tag{6.24}$$

The time-temperature shifting law in the flow region then reads

$$
\begin{aligned}
J_r(t,T) &= b(T,T_0)\cdot J(t,T) \equiv J_{T_0}(x) = J_{T_0}(t/a_T)\\
J'_r(\omega,T) &= b(T,T_0)\cdot J'(\omega,T) = J'_{T_0}(a_T\omega)\\
J''_r(\omega,T) &= b(T,T_0)\cdot J''(\omega,T) = J''_{T_0}(a_T\omega)\\
G_r(t,T) &= G(t,T)/b(T,T_0) = G_{T_0}(t/a_T)\\
G'_r(\omega,T) &= G'(\omega,T)/b(T,T_0) = G'_{T_0}(a_T\omega)\\
G''_r(\omega,T) &= G''(\omega,T)/b(T,T_0) = G''_{T_0}(a_T\omega)\\
\tan\delta(\omega,T) &= \tan\delta_{T_0}(y) = \tan\delta_{T_0}(a_T\omega)
\end{aligned}
\tag{6.25}
$$

In these definitions, the time-temperature shift factor a_T (T,T_0) for the flow region is to be inserted, which is different from the shift factor for the glass transition (cf. Fig. 6.13 on page 209).

The factor $b(T,T_0) = \rho T/\rho_0 T_0$, which distinguishes the shifting laws (6.5) and (6.25) from each other, is near to one in most cases. In the case of the PS whose flow properties may be determined within a temperature interval from T = 140 °C to T = 290 °C the factor $b(T,T_0)$ related to T_0 = 210 °C changes with increasing temperature from 0.89 to 1.11, i.e., by 22 %. In the same temperature interval, however, the shift factor $a_T(T)$ changes from 1 to 10^5. Thus, a horizontal shift of five decades is accompanied by a vertical shift of less than one-tenth of a decade.

For the glass transition, the factor $b(T,T_0)$ plays a still minor role. The shift of the glass transition may be investigated for PS between T = 90 and T = 120 °C. Within this window and with a reference temperature of 105 °C, the factor b changes from 0.97 to 1.03, i.e., by 6 %. The horizontal shift in this region amounts to more than seven decades.

As an example for the validity of the time-temperature shift of an amorphous polymer above the glass transition, the creep behavior of a technical polystyrene, which was studied in detail by Pfandl [10], Link [11], and Kaschta [12]), is discussed. Pfandl's data are presented in Fig. 6.10, which shows the reduced creep compliance as a function of the creep time in a double-logarithmic representation over a wide temperature range which covers the rubber-elastic plateau and the flow region.

The data in the flow region (compliances above 10^{-4} Pa^{-1}) of Fig. 6.10 can be brought to an excellent coincidence by a parallel shift along the logarithmic time axis and result in a master curve of the compliance covering four orders of magnitude of the reduced time as shown in Fig. 6.11. If, however, the shifting procedure is extended to the data within the rubber-elastic plateau (compliances between 10^{-5} and 10^{-4} Pa^{-1}), the superposition of the shifted curves becomes less good than in the flow region. Curves which could be brought to very good superposition in the flow region, fan out, when shifted down to the rubber-elastic plateau. The reason for this is that in the rubber-elastic plateau a further transition, the *entanglement transition*, occurs which follows a different time-temperature shifting law and disturbs the superposition there.

The superposition in the flow region was performed by the horizontal shifts at a compliance level of J_r (t) = 10^{-2} Pa^{-1}. This level is reached at the reference

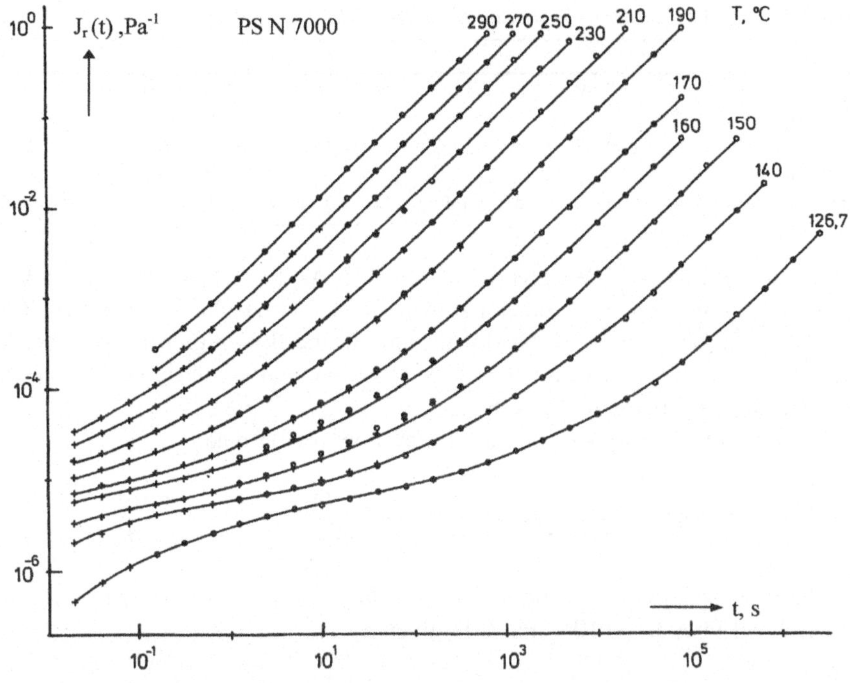

Fig. 6.10 Reduced creep compliance, $J_r\,(t,T)$, versus the creep time in a double-logarithmic plot, for a technical PS at various temperatures in the rubber-elastic plateau and in the flow region, after Pfandl [10]

Fig. 6.11 Reduced creep compliance J_r (x) versus the reduced time x in a double-logarithmic representation in the flow region of a technical PS, reference temperature $T_0 = 210$ °C, Pfandl [10]

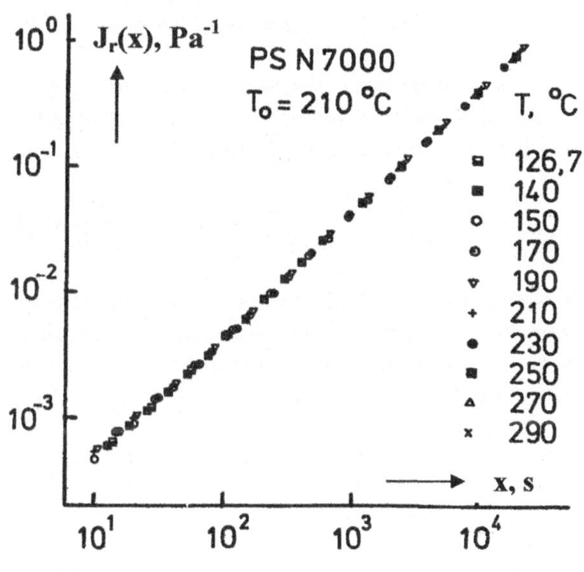

Table 6.3 Shift data for the three transitions of PSN 7000 [27, 29]

PSN 7000					
Glass transition	$c_0 = -7.68$	$c_1 = 8.9$	$c_2 = 36.0$	$c_1\, c_2 = 320$	$T_\infty^g = 69.0$
Entanglement transition	$e_0 = -4.45$	$e_1 = 3.96$	$e_2 = 153$	$e_1\, e_2 = 606$	$T_\infty^e = 57.0$
Flow transition	$d_0 = -1.86$	$d_1 = 4.3$	$d_2 = 164$	$d_1\, d_2 = 706$	$T_\infty^f = 45.7$

temperature of 210 °C after a creep time of 250 s. As the shifting law again a *WLF* equation is found, but with other parameters as for the glass-rubber transition. Therefore, the parameters of the shifting law in the flow region have to be distinguished from those valid for the glass transition, and are designated by d_1, d_2, and T_∞^f. They are listed in Table 6.3.

The time-temperature shifting law in the flow region reads

$$\log a_T(T, T_0) = -\frac{d_1(T - T_0)}{d_2 + T - T_0} = -\frac{d_1(T - T_0)}{T - T_\infty^f} = -d_1 + \frac{d_1 d_2}{T - T_\infty^f} \tag{6.26}$$

and the time position of the flow transition $\log t_{\text{flow}}(T)$ characterized by a compliance level of $J_r(t) = 10^{-2}\ \text{Pa}^{-1}$, is given by

$$\log t_{\text{flow}}(T) = \log t_{\text{flow}}(T_0) + \log a_T(T, T_0) = d_0 + \frac{d_1 d_2}{T - T_\infty^f} \tag{6.27}$$

with

$$d_0 = \log(250) - d_1 = 2.40 - d_1 \tag{6.28}$$

A relation between *the time-temperature shift for the flow transition* and the *temperature dependence of the viscosity* is found, when Eq. 5.46 is inserted into the first of Eqs. (6.25), resulting in

$$\eta_0(T)/\eta_0(T_0) = \frac{\rho T}{\rho_0 T_0} a_T(T, T_0) \tag{6.29}$$

It is, therefore, possible to derive the time-temperature shift in the flow region from the temperature dependence of the viscosity and vice versa. Similarly, the *reduced time* or the *reduced angular frequency* in the Eq. (6.25) may be replaced by the expressions

$$x = \frac{\rho T \eta_0(T_0)}{\rho_0 T_0 \eta_0(T)} t \tag{6.30}$$

$$y = \frac{\rho_0 T_0 \eta_0(T)}{\rho T \eta_0(T_0)} \omega \tag{6.31}$$

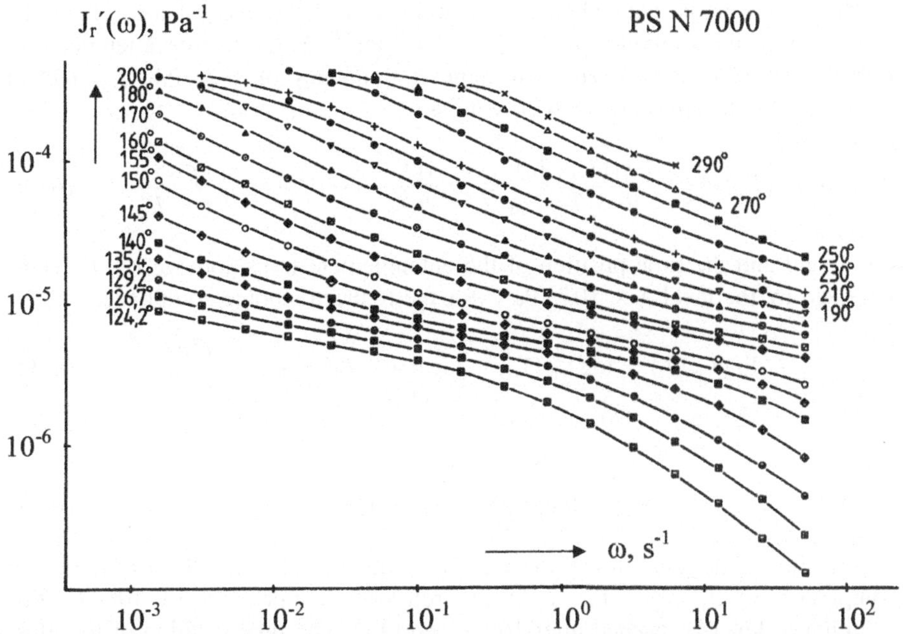

Fig. 6.12 Reduced storage compliance versus angular frequency calculated from oscillation experiments in double-logarithmic representation for a technical PS [10]

A time-temperature shift of the form (6.25) and the relation (6.29) has been introduced by Ferry [8].

In cases where the factor $b(T,T_0)$ is unknown but supposed to play a role, the correct shift function $a_T(T,T_0)$ may be determined, nevertheless, by shifting the curves for the loss tangent, as the shift for the loss tangent does not depend on the factor b. (Compare the last of Eq. (6.25)). This is always possible for the determination of the shift factor in the flow region of uncross-linked polymers whether amorphous or semicrystalline.

It was already mentioned that the molecular processes of the entanglement transition follow a time-temperature dependence different from that in the flow region. In a creep experiment the entanglement process is masked by the flow term (cf. Sect. 5.6). In order to determine the time-temperature shift of the entanglement transition, it has first to be separated from the flow term. This can be done in two ways. Instead of the creep compliance, the recoverable compliance can be measured (cf. Fig. 5.19), which very clearly shows the transition and its temperature dependence. Alternatively, from dynamic-mechanical experiments, the storage compliance can be calculated which is not influenced by the flow term and also exhibits the entanglement transition, as shown from Fig. 6.12. From this figure the frequency position of the entanglement transition may be determined as the level where the transition passes half of its logarithmic step in the temperature range between 140 and 270 °C.

The shifts for the entanglement transition were determined at a level for the reduced storage compliance of $J'_r = 5 \times 10^{-5}$ Pa^{-1}. At the reference temperature 210 °C, this level is reached at an angular frequency of $\omega \approx 0.8$ ($t = 1.25$ s) (cf. [10]). As shifting law the *WLF*-equation

$$\log a_T(T, T_0) = -\frac{e_1(T - T_0)}{e_2 + T - T_0} = -\frac{e_1(T - T_0)}{T - T^f_\infty} = -e_1 + \frac{e_1 e_2}{T - T^e_\infty} \tag{6.32}$$

was found and the time position of the entanglement transition $\log t_{ent}(T)$, characterized by the level 5×10^{-5} Pa^{-1} for the reduced storage compliance is

$$\log t_{ent}(T) = \log t_{ent}(T_0) + \log a_T(T, T_0) = e_0 + \frac{e_1 e_2}{T - T^e_\infty} \tag{6.33}$$

with

$$e_0 = \log(1.25) - e_1 = 0.097 - e_1 \tag{6.34}$$

The shift of the glass transition was determined in [11] at a level of the creep compliance of $J = 10^{-7}$ Pa^{-1}. At the reference temperature of 105 °C, this compliance level is reached after 16 s creep time. The time position of the glass transition is according to (6.12) and (6.1) given by

$$\log t_{glass}(T) = \log t_{glass}(T_s) + \log a_T(T, T_s) = c_0 + \frac{c_1^s c_2^s}{T - T^g_\infty} \tag{6.12}$$

with

$$c_0 = 1.20 - c_1^s \tag{6.13}$$

The three time-temperature shifts obtained for PS are compared in Fig. 6.13 which shows the *transition map* of the technical polystyrene PS N 7000. On a plane with the temperature as abscissa and the logarithm of the time positions of the transitions as ordinate, the glass-rubber transition, the entanglement transition, and the flow transition are plotted.

All three transitions may be described by a *WLF* equation, but with different values for the parameters, indicated as c_1, c_2, and $T^g_\infty = 69.0$ K for the glass transition, as e_1, e_2, and $T^e_\infty = 57$ K for the entanglement transition and as d_1, d_2, and $T^f_\infty = 45.7$ K for the flow transition. The values of the *WLF* constants of these transitions are listed in Table 6.3. The positions of the different transitions are given in Eqs. (6.12), (6.33), and (6.27). Though the three shifting laws seem to be similar, their shifting effect differs considerably. Within the usual experimental window (between 10 s and 10^5 s), a temperature change of 3 °C is sufficient to shift the reduced quantities over one decade in time for the glass transition. For the entanglement transition a change in temperature of 15 °C is necessary to achieve the same result, for the flow region, a temperature change of 30 °C.

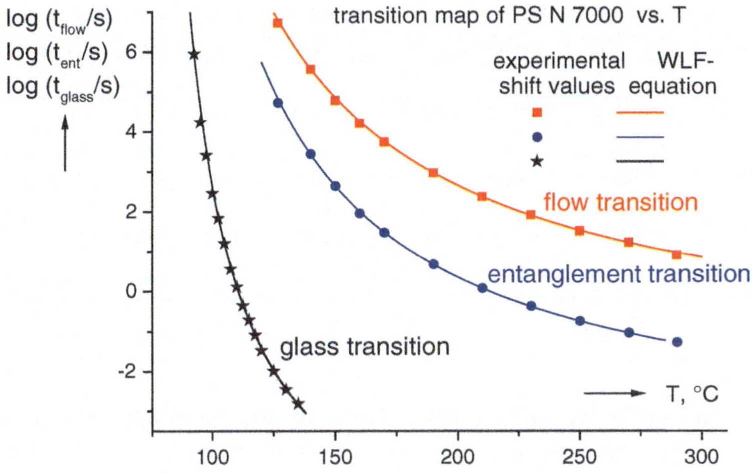

Fig. 6.13 Transition map with the temperature as abscissa and the logarithm of the time as ordinate, for PS N 7000 with the three transitions between the softening temperature and the flow region. The invariants of their WLF equations are given in Table 6.3

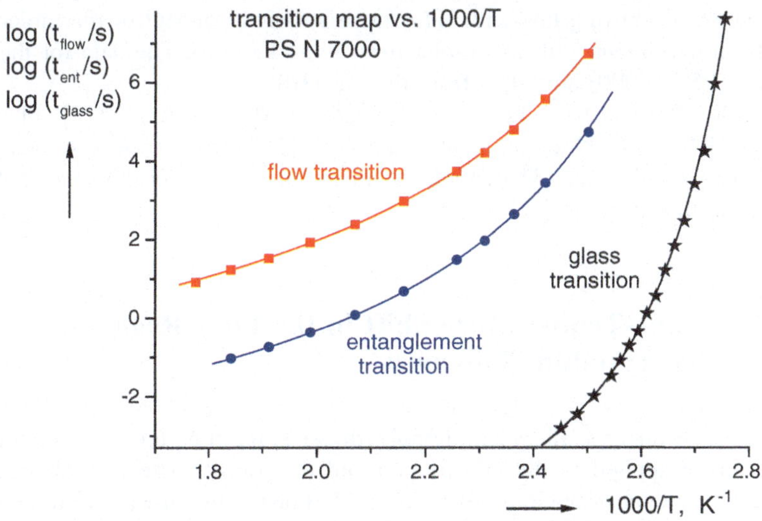

Fig. 6.14 Transition map with the inverse absolute temperature as abscissa and the logarithm of the time as ordinate, for PS N 7000

One further remark should be made here. For semicrystalline polymers, the shifting law for the flow transition is always found as an *Arrhenius* law. The question arises, whether the transitions shown above could be described by an Arrhenius law too. For this purpose, the transitions were plotted in a transition map with 1000/T as the abscissa in Fig. 6.14. In this plot, the positions of the transitions

Table 6.4 Shift data for the flow term of some amorphous polymers in the flow region

Polymer	M_n, kg/mol	M_w/M_n	T_0 °C	d_1	d_2 °K	d_1 d_1	Shifted quantity	T_∞^f K	Lit.
PSN 7000	170	2.2	210	4.3	164	706	$J_r(t)$	46	[10]
Anion. PS	21–616	1.1	190	5.6	142	800	G′ G″	48	[13]
Anion. PS	40–3000	1.05	180	5.7	134	760	G_r' G_r''	46.0	[14]
Anion. PS	34–750	1.11	180	5.2	124	644	G_r' G_r''	56	[15]
Anion. PS	95–1774	1.1–1.6	170	5.4	117	632	G′ G″	51	[16, 17]
Anion PMMA	20–203	>1.2	220	6.7	190	1270	G′ G″	30	[18]

should then appear as straight lines. None of the transitions plotted in this way occurs even approximately straight, but all show a very significant positive curvature over the entire experimental window in agreement with the description by the Eqs. (6.12), (6.33), and (6.27).

The *WLF* constants for the shift of the flow term of some amorphous polymers are listed in Table 6.4. Anionic PS have been investigated by Onogi et al. [13], Schausberger et al. [14], Wolf [15], and Hepperle [16]. All authors report the *WLF* equation as the shifting law with *WLF* invariants independent of the molar mass apart from specimens with low molar mass (<34 kg/mol). The data for the technical *PS* (PS N 7000) are taken from the ref. [10].

Anionic PMMAs have been investigated by Masuda et al. [17], blends of radicalic PMMAs by Onogi et al. [18] and Wolf [15]. Apart from specimen with a molar mass smaller than 31 kg/mol, the invariants were approximately independent of the molar mass.

6.5 The Time-Temperature Shift in the Flow Region of Semicrystalline Polymers

In the case of linear semicrystalline polyethylenes the time-temperature superposition can be applied in the flow region and the transition to the rubber-elastic regime as it becomes obvious from Fig. 6.15. For the two metallocene-catalyzed linear polyethylenes mHDPE 2 and mHDPE 3, an excellent overlap over a wide range of angular frequencies is found for the storage modulus G' and the loss modulus G'' as well, although they were not corrected by the vertical shift factor b_T (cf. Eq. 6.24). The experimentally obtained horizontal shift factors a_T (T,T_0) are semilogarithmically plotted as a function of the absolute reciprocal temperature 1000/T in Fig. 6.16 using a reference temperature of $T_0 = 150$ °C. All data points from the shift of G' and G'' for the two different samples come to lie on one straight line (full curve), i.e., they can numerically be described by the Arrhenius equation

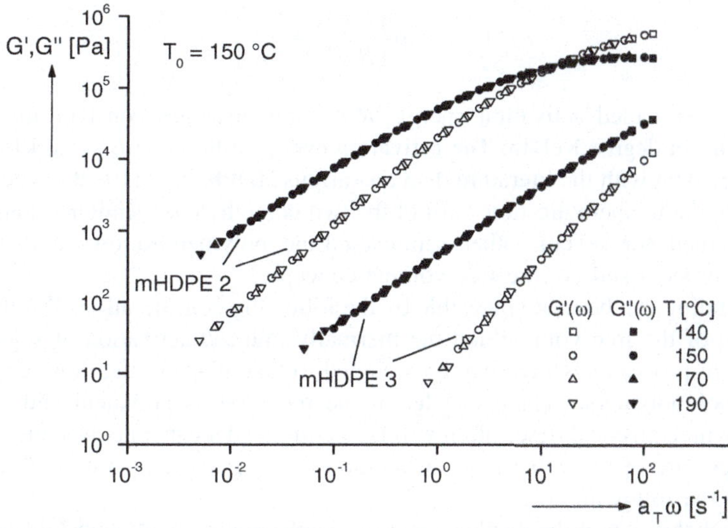

Fig. 6.15 Master curves of the storage modulus G' and the loss modulus G'' shifted by the factor $a_T(T)$ along the axis of the angular frequency ω. The measurements were performed on two linear polyethylenes at four different temperatures. The reference temperature is $T_0 = 150$ °C [28]

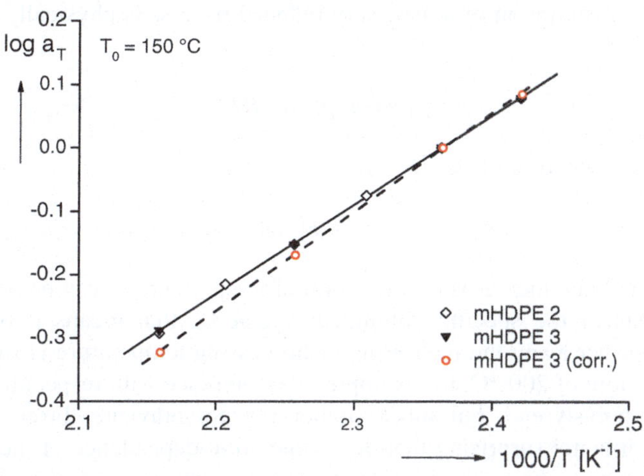

Fig. 6.16 Arrhenius plot of the shift factors a_T as a function of the reciprocal absolute temperature $1/T$ for the two linear polyethylenes of Fig. 6.15 with the reference temperature $T_0 = 150$ °C [28]. The broken line describes the temperature dependence of $log\ a_T$ if corrected for the vertical shift factor $b(T)$ according to Eq. (6.24). Its consequence for the activation energy is discussed later

$$a_T(T) = \exp\left\{\frac{E}{R}\left[\frac{1}{T} - \frac{1}{T_0}\right]\right\} \quad (6.35)$$

E is the so-called activation energy, R the universal gas constant and T the temperature in degree Kelvin. The activation energy follows as $E = 26$ kJ/mol in good agreement with the literature data on various HDPE, e.g., [19, 20]. The result means that the temperature-time shift of the two polyethylenes-similar results have been obtained for several other semicrystalline polymers-follows a thermally activated process and not the free volume concept.

This finding can be made plausible by the following consideration. The thermal expansion of the free volume and the thermally induced activation of segmental motions are processes which may be assumed to take place in the flow regime of all types of polymers. Which one has to be regarded as prevalent and, consequently, which physical description will be the more adequate one, is a question of the temperature of the experiment in relation to the glass temperature T_g as discussed in the following.

The probability p of the displacement of a molecule segment is postulated to be dependent on the probability p_1 to overcome the barrier potential for a motion and the probability p_2 to find sufficient free volume for the segment changing its position. As both conditions have to be fulfilled for a transposition, under the assumption of an independence of the two processes the total probability p follows as the product of the particular single probabilities, i.e., $p = p_1 p_2$.

For p_1, the assumption of a thermally induced process is physically reasonable, i.e.,

$$p_1 \approx \exp(-E/RT) \quad (6.36)$$

p_2 should be a function of the free volume v_f, i.e.

$$p_2 = f(v_f) \quad (6.37)$$

The free volume increases as a function of temperature as can be seen from the measurements of the specific volume in Fig. 4.31, that means it becomes the larger, the further away the measuring or processing temperature is from T_g. At a melt temperature of 200 °C, for example, the difference with respect to T_g is about 100 °C for polystyrene, but much higher for polyethylene. From this simple assessment, it is not surprising that the temperature dependence of the shift factor for the latter is given according to a thermally activated process as it can be assumed that enough free volume is available at the measuring temperatures and the probability p_2 is close to unity. In the case of the polystyrene, it is not unreasonable to postulate that at 100 °C from T_g the probability of the displacement of a molecule segment is still markedly dependent on the free volume available and, therefore, the *WLF* equation based on the free volume assumption provides a better numerical description of the temperature dependence of the shift factor than the Arrhenius equation.

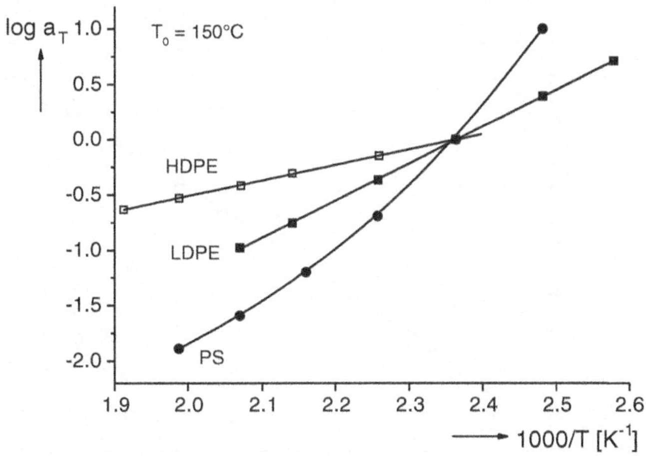

Fig. 6.17 Comparison of the shift factors *a*(*T*) for a HDPE, LDPE, and PS in an Arrhenius plot. The reference temperature was chosen as $T_0 = 150\ °C$ [28]

The various molecular processes underlying the temperature dependence of flow properties of amorphous and semicrystalline polymers lead to marked differences of the shift factors as shown in Fig. 6.17. Besides the well-known fact that the shape of a hyperbolic function as in the case of the *WLF* equation is different from that of an exponential one typical of an Arrhenius behavior, it becomes obvious that within the temperature range chosen the shift factors of the polystyrene change much more significantly than those of the polyethylenes indicating a stronger temperature dependence of the viscosity of the PS the knowledge of which can be helpful in understanding some aspects of processing.

The other interesting feature of Fig. 6.17 is that the shift factors of the LDPE follow an Arrhenius equation, too, but that its activation energy of $E = 66$ kJ/mol is distinctly higher than that of the HDPE. As the HDPE consists of linear molecules and the LDPE contains branches of a tree-like architecture this finding gives a hint to the potential of thermorheological properties for the characterization of polymers which will be discussed in detail in Chap. 14.

If one wants to use the activation energy for a reliable branching analysis, it has to be determined with high accuracy. One question is how the vertical shift factor *b*(*T*) introduced in Sect. 6.4 influences this quantity. As mentioned before, the effect of *b*(*T*) can be eliminated when $\tan\delta$ is shifted (cf. Eq. 6.25). As in the literature very often the moduli or the viscosities are used for a time-temperature superposition, however, it is of interest which changes of the activation energy may come into play in case of corrected and uncorrected data.

In Fig. 6.18, *b*(*T*) is presented as a function of the temperature for two classical LDPE, one linear low density polyethylene (mLLDPE), and one long-chain branched low density polyethylene (LCB-mLLDPE). The latter two were synthesized by metallocene catalysts. For all the four polyethylenes one gets in good approximation the same straight line increasing with rising temperature,

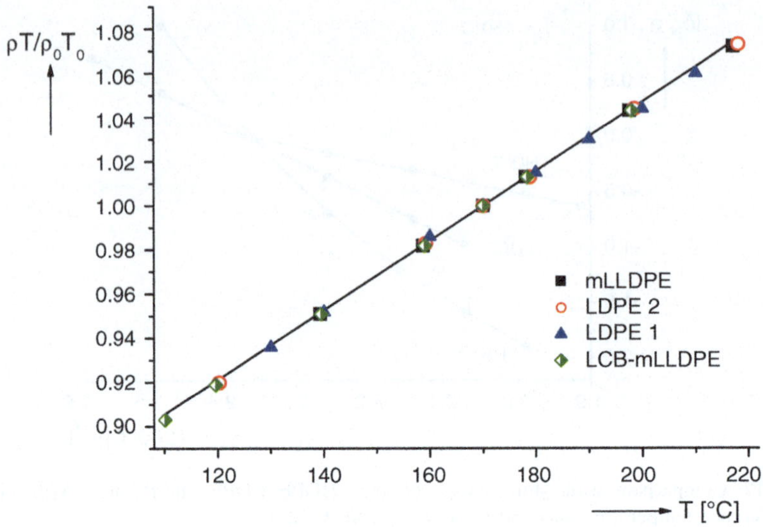

Fig. 6.18 $b(T,T_0) = \rho T/\rho_0 T_0$ (T in K) as a function of the temperature T in °C for four different polyethylenes. The reference temperature T_0 for the density correction coefficient was chosen as 443 K [21]

i.e., within the accuracy of the measurements the different molecular structures of the polyethylenes have no influence on the density changes of the melt. Over the temperature range of 80 °C which is normally used for the characterization of polyethylene melts the density corrections are around 5 % at the maximum with respect to a reference temperature T_0 chosen in the middle of this temperature interval. As these corrections lie within the accuracy limits of the measurements, the density correction is often neglected in the case of investigations on polyethylene melts.

In the following, it is demonstrated that activation energies calculated from the corrected data ($\bar{E}$) or uncorrected data (E), respectively, are somewhat different, indeed. Regarding the zero-shear viscosity, the shift factor taking $b(T)$ into account has to be determined according to Eq. (6.29), i.e.,

$$a_T(T,T_0) = \frac{\eta_0(T)}{\eta_0(T_0)} \frac{1}{b(T,T_0)} \tag{6.38}$$

In the case that a_T follows an Arrhenius equation as for the polyethylenes one gets

$$\ln(\eta_0(T)/\eta_0(T_0)) = \ln b(T,T_0) + \frac{\bar{E}}{R}\left[\frac{1}{T} - \frac{1}{T_0}\right] \tag{6.39}$$

Neglecting the vertical shift factor, i.e., $b(T,T_0) = 1$, an Arrhenius equation with a different activation energy E is found.

$$\ln(\eta_0(T)/\eta_0(T_0)) = \frac{E}{R}\left[\frac{1}{T} - \frac{1}{T_0}\right] \tag{6.40}$$

For the difference between E and $\bar{E}$, it follows

$$\frac{(E - \bar{E})}{R}\left[\frac{1}{T} - \frac{1}{T_0}\right] = \ln b(T, T_0) \tag{6.41}$$

and after differentiation with respect to $1/T$

$$\frac{(E - \bar{E})}{R} = \frac{d\ln b(T, T_0)}{d(1/T)} = -T^2\frac{d\ln b(T, T_0)}{dT} = -\frac{T^2}{b}\frac{db(T, T_0)}{dT} \tag{6.42}$$

Equation (6.42) has an interesting consequence. If b is introduced, the difference between the activation energies following from the shift of corrected and uncorrected data is temperature dependent. It cannot be said, however, whether E or $\bar{E}$ or both of them are dependent on the temperature as the temperature dependence is so little pronounced that it is not seen within the experimental accuracy of the measurements as the corrected and the uncorrected shift factors as functions of the reciprocal temperature in Fig. 6.16 follow a straight line in very good approximation.

From the straight line in Fig. 6.18, which describes the experimental results with good accuracy, the slope follows as

$$db(T, T_0)/dT = 1.6 \cdot 10^{-3}\mathrm{K}^{-1} \tag{6.43}$$

With this value one gets from Eq. (6.42) at the reference temperature $T_0 = 443$ K, i.e., $b(T,T_0) = 1$

$$\bar{E} - E = 2.6\,\mathrm{kJ/mol} \tag{6.44}$$

The activation energy $\bar{E}$ determined from viscosity data corrected by $b(T, T_0)$ is higher than the value obtained by using the uncorrected ones. The relative deviations obviously are the more pronounced the smaller the activation energies, i.e., they are expected to lie at around 10 % in the case of linear polyethylenes and decrease for the branched materials with higher values.

From the straight lines in Fig. 6.16, the activation energy following from the uncorrected and corrected shift factors can be calculated for the HDPE chosen. They follow as E = 26.0 kJ/mol and $\bar{E}$ = 28.2 kJ/mol. The corresponding data for the LDPE in Fig. 6.17 are E = 66.0 kJ/mol and $\bar{E}$ = 68.8 kJ/mol. The differences are in rather good agreement with Eq. (6.44).

These results point out that special care has to be taken if an analysis for slightly branched materials is undertaken from thermorheology. Furthermore, it becomes clear that a reliable comparability of activation energies from the literature requires a detailed description of the shifting method applied.

6.6 The Time-Temperature Shift of Secondary Transitions in the Glassy State

The relaxation processes occurring in the glassy state shift to shorter times with increasing temperature, too, but the shape of the curves *J(t)* versus *log t* is not preserved, as the dispersion steps, besides being shifted to shorter times, become sharper with increasing temperature. Notwithstanding a quantitative description of the time-temperature shifting is possible approximately on the basis of the last of Eq. (6.5). The shifting of the maximum of the loss tangent or of the midpoint of the dispersion steps can be described by a law different from (6.14), namely

$$(\log \upsilon)_{\tan \delta,\max} = \log \upsilon_0 - 0.43\frac{E}{RT} \quad \nu \,\text{in Hz} \tag{6.45}$$

In this equation, R is the gas constant, T the absolute temperature, and E the so-called *activation energy* characteristic for the molecular process under consideration, ν_0 is a frequency which according to Fig. 6.21 is about the same for all the secondary relaxation processes investigated, viz.,

$$\log(\upsilon_0/Hz) = 13.5 \pm 1.0 \tag{6.46}$$

Equation (6.45) is equivalent to the so-called *Arrhenius law* for the frequency position of the damping maximum

$$(\upsilon)_{\tan \delta,\max} = \upsilon_0 \cdot e^{-E/RT} \tag{6.47}$$

From Eq. (6.45), the time position t_{sec} of the logarithmic midpoint of the transition step in the compliance can be calculated as

$$(\log t)_{sec} = -0.80 - \log \upsilon_0 + 0.43\frac{E}{RT} \tag{6.48}$$

The time-temperature shifting function follows from (6.48) as

$$\log a_T(T, T_0) = \frac{0.43 \cdot E}{R} \cdot \left[\frac{1}{T} - \frac{1}{T_0}\right] \tag{6.49}$$

Equation (6.45) expresses the fact that the position of the *tan* δ-maxima of secondary relaxation processes may be described by straight lines in a *log* ν versus *1/T* diagram. This has convincingly been shown by Heijboer [24]. The most comprehensive data are those on polycyclohexylmethacrylate (PCHMA). This polymer exhibits a distinct secondary relaxation process (γ-process) in its glassy state which was extensively investigated and interpreted by Heijboer. Figure 6.19 presents the storage modulus and the loss tangent of this polymer as a function of

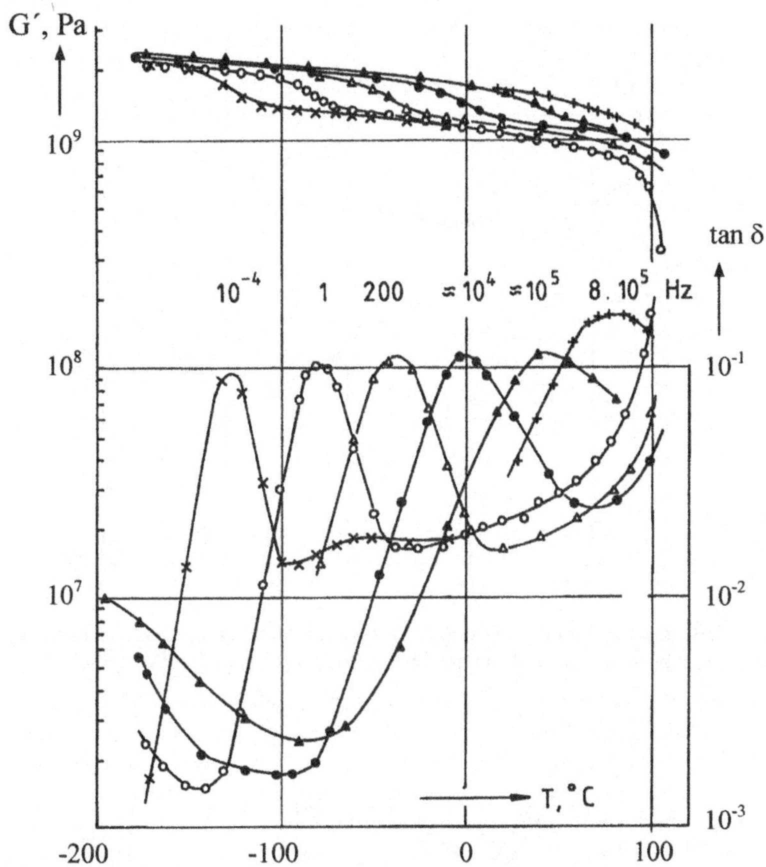

Fig. 6.19 Storage modulus and loss tangent for PCHMA at 6 frequencies as a function of the temperature, after Heijboer [22]

the temperature at various fixed frequencies from 10^{-4} Hz (calculated from creep experiments) to around 10^6 Hz (propagation of elastic waves). The frequency-temperature shift can clearly be observed for the dispersion steps in the storage modulus and the maxima of the loss tangent.

For the maxima of the loss tangent in Fig. 6.19, the inverse absolute temperature is plotted versus the logarithm of the frequency in Fig. 6.20. This figure proves the validity of the Arrhenius law for the temperature-frequency shift of the maximum of the loss tangent within an experimental window of 10 decades in frequency with high accuracy.

Secondary dispersion processes of other polymers in the glassy state have not been investigated within such a wide experimental window as PCHMA. Though, also for those polymers the shifting law (6.45) was established as to be seen from Fig. 6.21, in which the Arrhenius diagram of secondary dispersion processes of five other polymers is presented.

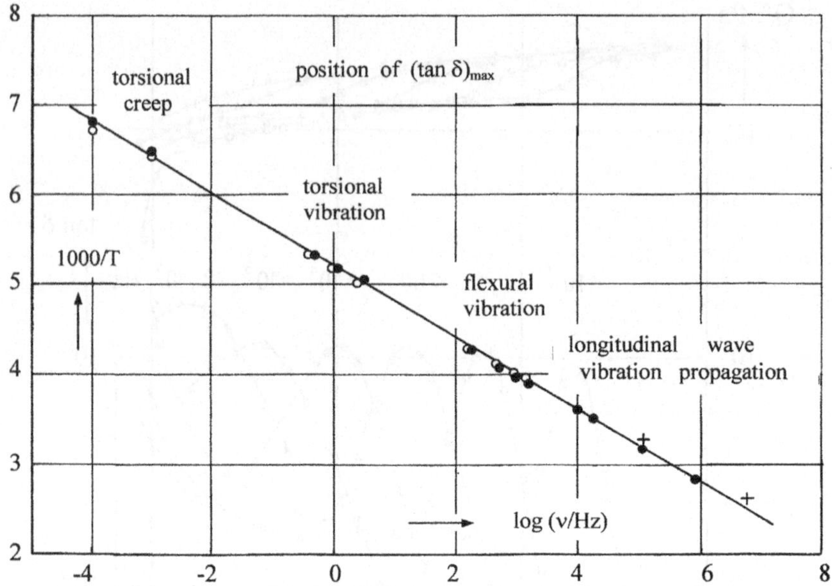

Fig. 6.20 Arrhenius diagram presented as the inverse absolute temperature versus the logarithm of the frequency of the position of the maxima of the γ-process of PCHMA, after Heijboer [22]

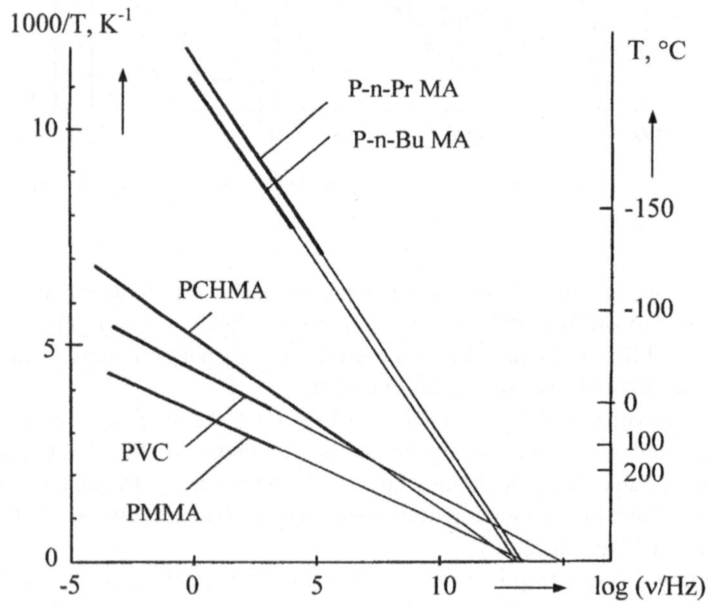

Fig. 6.21 Arrhenius diagram presented as the inverse absolute temperature versus the logarithm of the frequency for the position of the tan δ-maxima of various secondary relaxation processes in the glassy state of amorphous polymers, after Heijboer reproduced from [24] by permission from Taylor & Francis (UK)

Table 6.5 Some examples of secondary relaxation processes in the glassy state

Secondary relaxation process	T_{1Hz} °C	$(\tan\delta)_{max}$	E kJ/mol	E/T_{1Hz} T in °K	Lit.
β-max of PMMA	40	0.09	75	0.25	[24, 31]
β-max of PVC	−60	0.03	59	0.28	[24, 31]
β-max of PC	−110	0.03	50	0.31	[24, 31, 32]
γ-max of PCHMA	−81	0.1	47	0.24	[22]
γ-max of poly-n-butylmethacrylate	−180	0.06	23	0.25	[32]
γ-max of poly-n-propylmethacrylate	−188	0.06	22	0.26	[25, 32]
β-max of PS	−233	0.007	10	0.25	[23]

In Fig. 6.21 the bold lines indicate the regions, in which measurements were performed, the faint ones are extrapolations to the zero point of the ordinate. From this figure we conclude

(1) For all the secondary relaxation processes shown, Eq. (6.45) is valid within the experimental window investigated.
(2) For the determination of ν_0 an extrapolation of the measured data over many decades is necessary.
(3) Considering the uncertainty connected with an extrapolation over a wide scale, the value of v_0 given by Eq. (6.46) constitutes a reasonable estimate for all secondary relaxation processes.

If the general validity of Eqs. (6.45) and (6.46) is accepted, the position of the secondary relaxation processes in the time-temperature plane is fixed by the knowledge of one single constant of the process, namely the activation energy. According to Eq. (6.47) a simple relation follows between the activation energy E and the absolute temperature T_{1Hz} of the maximum of tan δ measured at 1 Hz.

$$E \cong 0.26 \cdot T_{1\mathrm{Hz}} \quad \text{in kJ/mol} \tag{6.50}$$

If the position of the tan δ-maximum at 1 Hz is known for a secondary relaxation process, its activation energy can easily be determined, and following from that the maximum temperatures at other frequencies. This important rule of thumb goes back to Heijboer, too [24]. From (6.50) it follows that the activation energies of secondary relaxation processes are the smaller the lower the temperature position of their maximum of tan δ.

In Table 6.5, some data on secondary relaxation processes are summarized. Listed are the temperature positions of the damping maxima at 1 Hz, the approximate values of the *tan* δ-maximum at 1 Hz, and the activation energies measured. From the column before last of this table it may be seen that the rule (6.50) is a reasonable estimate.

For three examples, the molecular interpretation of the motions underlying the secondary relaxation processes in the glassy state is illustrated in Fig. 6.22. The β-maximum of PMMA is caused by the rotation of the methyl-ester side group around the C–C bond which connects this side group with the main chain.

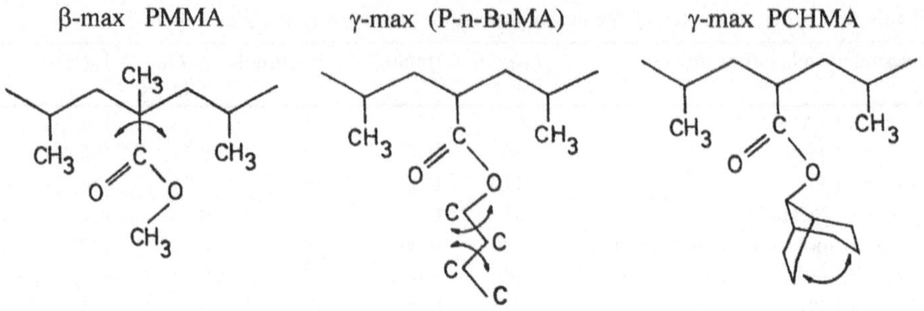

Fig. 6.22 Molecular interpretation of some secondary relaxation processes in the glassy state of amorphous polymers

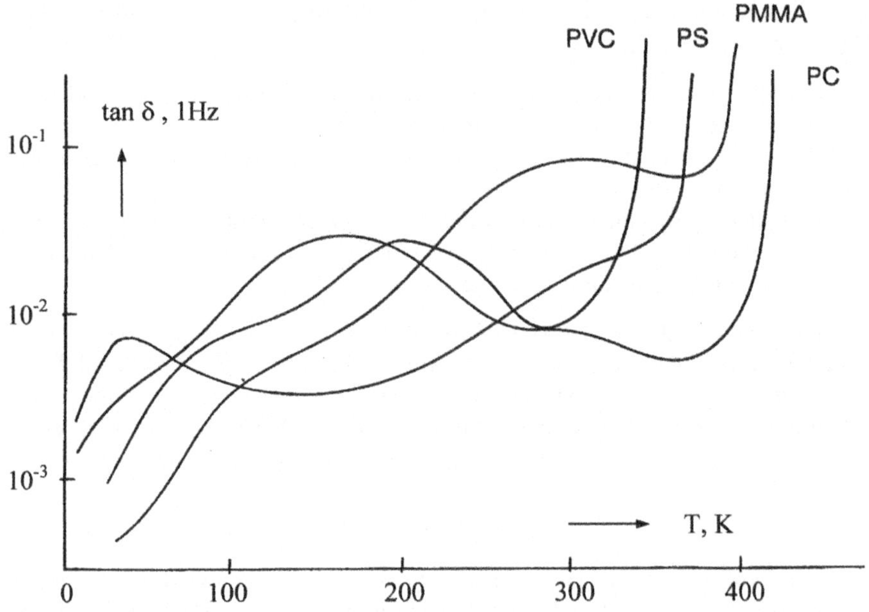

Fig. 6.23 Loss tangent as a function of the absolute temperature at a frequency of about 1 Hz for four polymers in their glassy state

P-n-BuMA shows besides the β-maximum (due to the rotation of the entire butylester side group) a γ-maximum at -180 °C. This is connected with the rotations around the *C–C* bonds of the alkyl moiety of the side group. The γ-maximum of PCHMA has been shown by Heijboer [22] to be caused by the change of the cyclohexyl side group from the one chair position to its opposite one. The β-maxima of PVC and PC have not been interpreted in detail so far, they have to be assigned to motions of short parts of the main chain, as those polymers do not have side groups.

The dependence of *tan δ* on the absolute temperature is shown in Fig. 6.23 for four amorphous polymers within the glassy region. For the curves at low temperatures (below 80 K), data measured by Hartwig [23] have been used, which

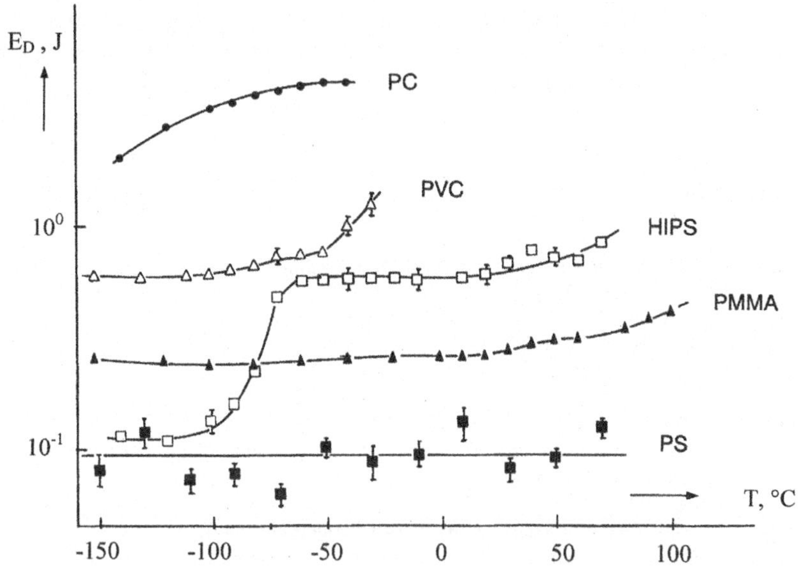

Fig. 6.24 Energy at break in a Charpy bending test versus the temperature for five amorphous polymers in their glassy state

were obtained at frequencies changing with temperature between 5 and 8 Hz. These data conform to data obtained at a fixed frequency of 0.64 Hz at higher temperatures. (In the region of very low temperatures, the temperature shift is of much less importance because of the small activation energies of the processes occurring in this region).

PS shows in its glassy state small *tan* δ-values, apart from a minor maximum at very low temperatures. This is ascribed to some motions of the phenyl ring. All the other three polymers exhibit a very distinct broad secondary maximum of *tan* δ in their glassy state.

These secondary damping maxima have an important influence on the impact resistance of the polymers. Polymers with secondary damping maxima are often (not always) impact resistant at ambient temperatures, while polymers without pronounced secondary damping maxima are brittle [26].

Figure 6.24 shows the energy at break on a logarithmic scale as a function of temperature of the four polymers presented in Fig. 6.23 together with an impact resistant PS. The data were obtained on non-notched specimens in a Charpy-bending test.

The impact resistance of the unmodified PS is very low in the entire glassy state, i.e., it is extremely brittle. To improve its impact resistance, it can be modified by copolymerization with a butadiene rubber. This high impact resistant PS (*HIPS*) shows a sharp transition of the energy at break around −75 °C corresponding to the maximum of *tan* δ of the butadiene rubber at −90 °C and 1 Hz [27].

The polymers PMMA, PVC, and PC show much higher impact resistances than the unmodified PS. But there is not such a clear correlation with the temperature position of the secondary damping maxima as in the case of HIPS.

References

1. Schwarzl FR, Staverman AJ (1952) J Appl Phys 23:283
2. Schwarzl FR, Zahradnik F (1980) Rheol Acta 19:137
3. Williams ML, Landel RF, Ferry JD (1955) J Am Chem Soc 77:3701
4. Vogel H (1921) Phys Z 22:645
5. Fulcher GS (1925) J Am Chem Soc 8(339):789
6. Tammann G, Hesse W (1926) Z Anorg Chem 156:245
7. Gibbs JH, DiMarzio J (1958) J Chem Phys 28:273
8. Ferry JD (1980) Viscoelastic properties of polymers, 3rd edn. Wiley, Chichester
9. Struik LCE (2007) Polymeerfysika, terugblik op 22 jaar Twente Afscheidcollege Universiteit Twente 12.10.2007
10. Pfandl W (1984) Doctoral Thesis, University Erlangen Nürnberg
11. Link G (1985) Doctoral Thesis, University Erlangen Nürnberg
12. Kaschta J (1991) Doctoral Thesis, University Erlangen-Nürnberg
13. Onogi S, Masuda T, Kitagawa K (1970) Macromol 3:109
14. Schausberger A, Schindlauer F, Janeschitz-Kriegl H (1985) Rheol Acta 24:220
15. Wolf M (1992) Doctoral Thesis, University Erlangen-Nürnberg
16. Hepperle J (2002) Doctoral Thesis, University Erlangen-Nürnberg, Shaker Aachen
17. Masuda T, Kitagawa K, Onogi S (1970) Polymer J 1:418
18. Onogi S, Masuda T, Toda N, Koya K (1970) Polymer J 1:542
19. Laun HM (1987) Prog Coll Polym Sci 75:111
20. Stadler FJ (2007) Doctoral Thesis, University Erlangen-Nürnberg, Sierke Göttingen
21. Kessner UM (2010) Doctoral Thesis, University Erlangen-Nürnberg, Sierke Göttingen
22. Heijboer J (1972) Doctoral Thesis, University Leiden
23. Hartwig G (1994) Polymer properties at room and cryogenic temperature. Plenum Press, New York
24. Heijboer J (1978) Secondary loss peaks in glassy amorphous polymers. In Meier DJ (eds) Molecular basis of transitions and relaxations. Gordon and Breach Science Publishers, London
25. Heijboer J (1968) J Polym Sci C 16:3755
26. Cigna G (1970) J Appl Polym Sci 14:1781
27. Pfandl W, Link G, Schwarzl FR (1984) Rheol Acta 23:277
28. Gabriel C (2001) Doctoral Thesis, University Erlangen-Nürnberg, Shaker Aachen
29. Link G, Schwarzl FR (1987) Rheol Acta 26:375
30. Ngai KL, Rendell RW (1988) Macromol 21:3396
31. Floudas G, Higgins JS, Meier G, Kremer F, Fischer EW (1993) Macromol 26:1676
32. Heijboer J, Pinari M (1982) Non-metallic materials and composites at low temperatures, 2nd ed. Evans D, Hartwig G (eds) Plenum Press, New York

Chapter 7
Linear Viscoelastic Deformation Under Three-Dimensional Stresses

7.1 The Stress Tensor and the Equations for the Balance of Forces

So far we only treated the viscoelastic behavior in simple shear. In order to generalize these results with respect to three-dimensional states of stress, we need the stress tensor and the deformation tensor under small strains, which are shortly explained now. For a more extensive treatment of this matter we refer the reader to textbooks on classical mechanics [1, 2].

The components of the stress tensor in a Cartesian right-handed coordinate system are defined referring to Fig. 7.1. Consider a small cube of material with edges parallel to the coordinate axis at the position (x, y, z). In the stressed state forces occur acting from one part of the sample to the other and being responsible for the changes in shape and volume. These forces act on the surfaces of the cube from the inside of the cube to its outside. Divided by the area of the surface on which they act these forces are called *stresses.*

Let $\underline{S}_1$ be the stress acting on the x-surface of the cube[1] in Fig. 7.1 and let $\sigma_{xx,}$ $\sigma_{xy,}$ σ_{xz} be its components in the x, y, z-direction. In a similar manner the stresses on the y and z-surface are defined.

	Components
Stress on the x-surface	σ_{xx} σ_{xy} σ_{xz}
Stress on the y-surface	σ_{yx} σ_{yy} σ_{yz}
Stress on the z-surface	σ_{zx} σ_{zy} σ_{zz}

The nine components σ_{xx}, σ_{xy} are called the Cartesian components of the stress tensor $\underline{\underline{S}}$. They are abbreviated by σ_{ik} whereby the indices run from 1 to 3 ($1 = x$, $2 = y$, $3 = z$). Stress components with equal indices are called tensile stresses and have a positive sign if they act from the inside to the outside of the surface. They are called pressures (compressive stresses) and are marked by a negative sign if

[1] A vector will be designated by an underlined letter, a tensor by a double underlined letter.

H. Münstedt and F. R. Schwarzl, *Deformation and Flow of Polymeric Materials*,
DOI: 10.1007/978-3-642-55409-4_7,

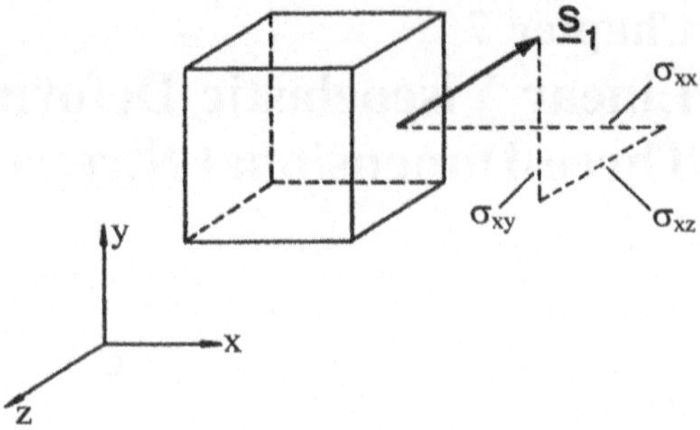

Fig. 7.1 Cartesian components of the stress tensor

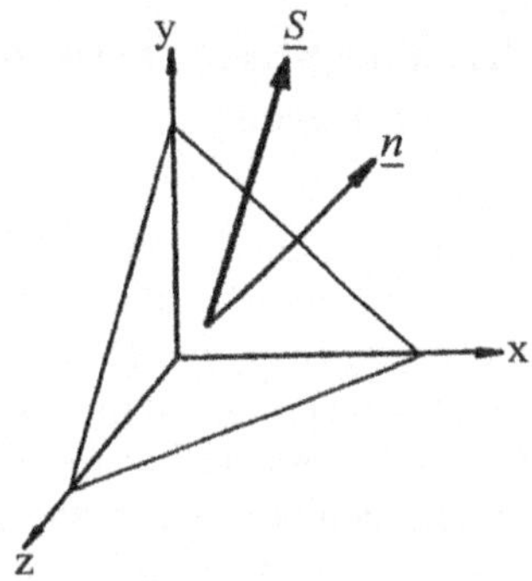

Fig. 7.2 Stress on a plane with an arbitrary orientation

they act the other way round. Stress components with unequal indices are called shear stresses. No sign is attributed to them.

If the components of the stress tensor are known at a fixed position, the stress on a plane with an arbitrary orientation at the same position can be derived (cf. Fig. 7.2).

Let the orientation of the plane be given by its normal vector $\underline{n}$ with the components n_i and let the stress on the plane be $\underline{S}$ with the components S_i then

$$\begin{aligned} S_1 &= n_1\sigma_{11} + n_2\sigma_{21} + n_3\sigma_{31} \\ S_2 &= n_1\sigma_{12} + n_2\sigma_{22} + n_3\sigma_{32} \\ S_3 &= n_1\sigma_{13} + n_2\sigma_{23} + n_3\sigma_{33} \end{aligned} \tag{7.1}$$

or, in tensor notation by using the convention that over indices occurring twice in a formula, summation from 1 to 3 is to be performed.

$$S_i = n_k\sigma_{ki} \tag{7.1'}$$

The index i is a so-called free index, to which each number between 1 and 3 may be attributed, while k is a summation index. Equation (7.1′) is an elegant shorthand notation of the system (7.1). Tensor notations are used from now on. The Cartesian coordinates are designated as $x = x_1$, $y = x_2$, $z = x_3$.

In general, the σ_{ik} are functions of the position $\sigma_{ik} = \sigma_{ik}\,(x_1, x_2, x_3)$; if the stresses on the sample are moreover time-dependent, the components of the stress tensor will depend on position and time $\sigma_{ik} = \sigma_{ik}\,(x_1, x_2, x_3; t)$.

The dependence of the stress tensor on position and time must not be chosen arbitrarily, but has to fulfill the conditions for the balance of forces of the Newtonian mechanics, which read

$$\boxed{\frac{\partial \sigma_{ik}}{\partial x_k} + F_i = \rho a_i} \tag{7.2}$$

These partial differential equations are called *equations of motion*. ρ is the density and a_i are the components of the acceleration of the material particles which just pass the position x_i at the time t. F_i are the components of the volume forces acting at the position x_i (for instance the forces of gravity).

In the frequently met case of static equilibrium without volume forces, Eq. (7.2) simplifies to

$$\frac{\partial \sigma_{ik}}{\partial x_k} = 0 \tag{7.3}$$

The conditions (7.3) are then often called *equilibrium conditions*. Equations (7.2) respectively (7.3) must always be fulfilled, independently of the consistency of the material considered. The dependence on the consistency first emerges by the relations between the stresses and the deformations (the so-called *rheological equations of state*).

Besides the conditions for the balance of forces also the condition for the balance of the angular momentum is to be met. It demands in its most simple form the symmetry of the stress tensor

$$\boxed{\sigma_{ki} = \sigma_{ik}} \tag{7.4}$$

In most cases the symmetry condition of the stress tensor is met. However, materials exist, for which the condition for the conservation of momentum takes a much more complicated form, the so-called Cosserat-media [3]. Fortunately, Cosserat-media do not play an essential role in rheology and need not to be discussed here.

In technical mechanics, it is often of advantage to introduce curved coordinate systems, as, e.g., a cylindrical or a polar coordinate system.

The cylindrical coordinate system (r, φ, z) is defined by the equations

$$x = r\cos\varphi \quad y = r\sin\varphi \quad z = z \tag{7.5}$$

In this case, the volume element sketched in Fig. 7.3 is formed by the r, φ, z-surfaces. The stress on the r-surface has the components σ_{rr}, $\sigma_{r\varphi}$, σ_{rz}, the stress on the φ-surface the components $\sigma_{\varphi r}$, $\sigma_{\varphi\varphi}$, $\sigma_{\varphi z}$, and the stress on the z-surface the

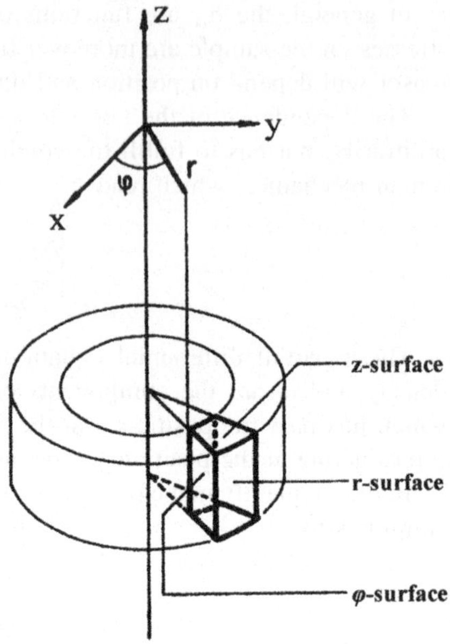

Fig. 7.3 Volume element in cylindrical coordinates

components σ_{zr}, $\sigma_{z\varphi}$, σ_{zz} . These nine numbers are the components of the stress tensor.

The equations for the balance of forces are in this case

$$\begin{aligned}\frac{\partial\sigma_{rr}}{\partial r}+\frac{1}{r}\frac{\partial\sigma_{r\varphi}}{\partial\varphi}+\frac{\partial\sigma_{rz}}{\partial z}+\frac{1}{r}[\sigma_{rr}-\sigma_{\varphi\varphi}]+F_r=\rho a_r\\ \frac{\partial\sigma_{\varphi r}}{\partial r}+\frac{1}{r}\frac{\partial\sigma_{\varphi\varphi}}{\partial\varphi}+\frac{\partial\sigma_{\varphi z}}{\partial z}+\frac{2}{r}\sigma_{\varphi r}+F_\varphi=\rho a_\varphi\\ \frac{\partial\sigma_{zr}}{\partial r}+\frac{1}{r}\frac{\partial\sigma_{z\varphi}}{\partial\varphi}+\frac{\partial\sigma_{zz}}{\partial z}+\frac{1}{r}\sigma_{zr}+F_z=\rho a_z\end{aligned}\tag{7.6}$$

From the condition for the balance of the angular momentum again the symmetry of the stress tensor follows, i.e., $\sigma_{r\varphi}=\sigma_{\varphi r}\quad\sigma_{\varphi z}=\sigma_{z\varphi}\quad\sigma_{zr}=\sigma_{rz}$.

For spherical coordinates one gets

$$\begin{aligned}x&=r\sin\vartheta\cos\varphi\\ y&=r\sin\vartheta\sin\varphi\\ z&=r\cos\vartheta\end{aligned}\tag{7.7}$$

The volume element is formed by the r,ϑ,φ-surfaces as sketched in Fig. 7.4.

The stress acting on the r-surface has the components σ_{rr}, $\sigma_{r\vartheta}$, σ_{rz} etc., and the conditions for the balance of forces are

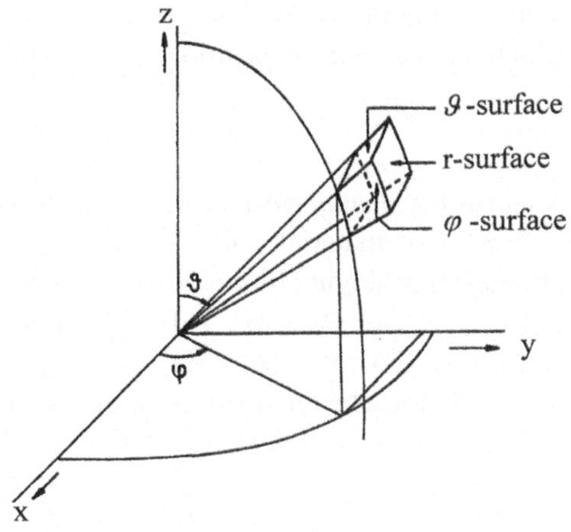

Fig. 7.4 Volume element in spherical coordinates

$$\begin{aligned}
\frac{\partial\sigma_{rr}}{\partial r}+\frac{1}{r}\frac{\partial\sigma_{r\vartheta}}{\partial\vartheta}+\frac{1}{r\sin\vartheta}\frac{\partial\sigma_{r\varphi}}{\partial\varphi}+\frac{1}{r}[2\sigma_{rr}-\sigma_{\vartheta\vartheta}-\sigma_{\varphi\varphi}+\sigma_{r\vartheta}\cot\vartheta]+F_r&=\rho a_r\\
\frac{\partial\sigma_{\vartheta r}}{\partial r}+\frac{1}{r}\frac{\partial\sigma_{\vartheta\vartheta}}{\partial\vartheta}+\frac{1}{r\sin\vartheta}\frac{\partial\sigma_{\vartheta\varphi}}{\partial\varphi}+\frac{1}{r}[(\sigma_{\vartheta\vartheta}-\sigma_{\varphi\varphi})\cot\vartheta+3\sigma_{\vartheta r}]+F_\vartheta&=\rho a_\vartheta\\
\frac{\partial\sigma_{\varphi r}}{\partial r}+\frac{1}{r}\frac{\partial\sigma_{\varphi\vartheta}}{\partial\vartheta}+\frac{1}{r\sin\vartheta}\frac{\partial\sigma_{\varphi\varphi}}{\partial\varphi}+\frac{1}{r}[3\sigma_{\varphi r}+2\sigma_{\varphi\vartheta}\cot\vartheta]+F_\varphi&=\rho a_\varphi
\end{aligned} \tag{7.8}$$

The condition for the balance of the angular momentum is again the symmetry of the stress tensor.

7.2 The Strain Tensor for Small Deformations

For the description of the deformation the strain tensor is used, which is defined here under the restriction of small deformations. The generalization to large deformations will be outlined in Sect. 8.2.

The position of all particles of the material considered is described by their coordinates in a space-fixed Cartesian coordinate system. We observe the motion of a particle, which was located at the position (x_1^0, x_2^0, x_3^0) at the time t_0, and is found due to the deformation process at the position (x_1, x_2, x_3) at the later time t. If the coordinates of all particles are given as a function of the time, the motion of all particles of the observed medium is documented by the equations

$$x_i = f_i(x_1^0, x_2^0, x_3^0; t) = x_i^0 + u_i(x_1^0, x_2^0, x_3^0; t) \quad \text{for} \quad i = 1, 2, 3 \tag{7.9}$$

with x_i being the components of the position vector $\underline{x}$ of the particle at the time t, which had been at the position $\underline{x}^0$ at the time t_0. The difference of the two vectors

$$\underline{x} - \underline{x}^0 = \underline{u} \tag{7.10}$$

is called the *displacement vector* $\underline{u}$ with the components (u_1, u_2, u_3).

Next, two material points P_1 and P_2 are considered which are situated in an infinitesimal neighborhood before the deformation. Their coordinates are $P_1 = (x_1^0, x_2^0, x_3^0)$ and $P_2 = (x_1^0 + \mathrm{d}x_1^0, x_2^0 + \mathrm{d}x_2^0, x_3^0 + \mathrm{d}x_3^0)$ at the time t_0, and $P_1' = (x_1, x_2, x_3)$ and $P_2' = (x_1 + \mathrm{d}x_1, x_2 + \mathrm{d}x_2, x_3 + \mathrm{d}x_3)$ at the time t after the deformation. Let $\mathrm{d}s_o$ be the distance of the two points P_1 and P_2 before the deformation and $\mathrm{d}s$ their distance after the deformation, then

$$(\mathrm{d}s_0)^2 = (\mathrm{d}x_1^0)^2 + (\mathrm{d}x_2^0)^2 + (\mathrm{d}x_3^0)^2 = \mathrm{d}x_i^0 \mathrm{d}x_i^0 \tag{7.11}$$

and

$$(\mathrm{d}s)^2 = (\mathrm{d}x_1)^2 + (\mathrm{d}x_2)^2 + (\mathrm{d}x_3)^2 = \mathrm{d}x_i \mathrm{d}x_i \tag{7.12}$$

Inserting the first term of the Taylor development of Eq. (7.9) with respect to the variables x_i^0 into Eq. (7.12), the following relation for $(\mathrm{d}s)^2$ is obtained

$$(\mathrm{d}s)^2 = C_{ik} \mathrm{d}x_i^0 \mathrm{d}x_k^0 \tag{7.13}$$

with

$$C_{ik} = \frac{\partial f_m}{\partial x_i^0} \frac{\partial f_m}{\partial x_k^0} \tag{7.14}$$

Equations (7.13) and (7.14) have been written in tensor notation. In (7.13) both indices i and k are summation indices. In (7.14) i and k are free indices and m is a summation index. The quantities C_{ik} form the nine components of a symmetric tensor, called the *Cauchy deformation tensor*. Expressing its components by the components of the displacement vector yields

$$C_{ik} = \delta_{ik} + \frac{\partial u_i}{\partial x_k^0} + \frac{\partial u_k}{\partial x_i^0} + \frac{\partial u_m}{\partial x_i^0} \frac{\partial u_m}{\partial x_k^0} \tag{7.15}$$

with δ_{ik} being the so-called *Kronecker-symbol* with the components $\delta_{ik} = 1$ if $i = k$ and $\delta_{ik} = 0$ if $i \neq k$.

The motion of a body is rigid, if it does not cause any change in distances of two points marked on it, i.e., $(\mathrm{d}s)^2 = (\mathrm{d}s_0)^2$ for all values of x_i^0 and t, viz., if $C_{ik} = \delta_{ik}$. Consequently, an appropriate measure of the deformation connected with the motion is the difference $(\mathrm{d}s)^2 - (\mathrm{d}s_0)^2$ which may be written by subtracting $(\mathrm{d}s_0)^2 = \delta_{ik}\, \mathrm{d}x_i^0\, \mathrm{d}x_k^0$ from (7.13) resulting in

$$(\mathrm{d}s)^2 - (\mathrm{d}s_0)^2 = 2e_{ik}\mathrm{d}x_i^0\mathrm{d}x_k^0 \tag{7.16}$$

with

$$e_{ik} = \frac{1}{2}\left[\frac{\partial u_i}{\partial x_k^0} + \frac{\partial u_k}{\partial x_i^0} + \frac{\partial u_m}{\partial x_i^0}\frac{\partial u_m}{\partial x_k^0}\right] \tag{7.17}$$

The special case of small deformations is characterized by the condition that all gradients of the displacement vector are small against unity, viz.,

$$\frac{\partial u_i}{\partial x_k^0} \ll 1 \quad \text{for all } i \text{ and } k \tag{7.18}$$

Therefore, the quadratic terms in (7.17) may be neglected and the *deformation tensor for small deformations* reads

$$\boxed{\varepsilon_{ik} = \frac{1}{2}\left[\frac{\partial u_i}{\partial x_k^0} + \frac{\partial u_k}{\partial x_i^0}\right]} \tag{7.19}$$

or, if written in components $x_0 = x_1^0$, $y_0 = x_2^0$, $z_0 = x_3^0$, and $u = u_1$, $v = u_2$, $w = u_3$.

$$\begin{aligned}
\varepsilon_{xx} &= \frac{\partial u}{\partial x_0} & \varepsilon_{xy} &= \frac{1}{2}\left[\frac{\partial u}{\partial y_0} + \frac{\partial v}{\partial x_0}\right] = \varepsilon_{yx} \\
\varepsilon_{yy} &= \frac{\partial v}{\partial y_0} & \varepsilon_{yz} &= \frac{1}{2}\left[\frac{\partial v}{\partial z_0} + \frac{\partial w}{\partial y_0}\right] = \varepsilon_{zy} \\
\varepsilon_{zz} &= \frac{\partial w}{\partial x_0} & \varepsilon_{zx} &= \frac{1}{2}\left[\frac{\partial w}{\partial x_0} + \frac{\partial u}{\partial z_0}\right] = \varepsilon_{xz}
\end{aligned} \tag{7.19'}$$

Tensor components with equal indices and positive sign are called tensile strains, those with equal indices and negative sign are called compressive strains, and those with different indices are called shear strains. Note that the shear γ defined in Eq. (4.1) is just the double of the shear component ε_{xy} as for simple shear

$$\varepsilon_{xy} = \varepsilon_{yx} = \frac{1}{2}\gamma \tag{7.20}$$

In the technical mechanics of polymers, the deformation tensor for small deformations (7.19) is mostly used.

Often the use of a curved coordinate system is of advantage. In the case of cylindrical coordinates, the components of the displacement vector are defined by

$$r = r_0 + u_r(r_0, \varphi_0, z_0; t) \quad \varphi = \varphi_0 + \frac{1}{r_0} u_\varphi(r_0, \varphi_0, z_0; t) \quad z = z_0 + u_z(r_0, \varphi_0, z_0; t) \tag{7.21}$$

with r_0, φ_0, z_0 being the cylinder coordinates of a point before the deformation and r, φ, z those after the deformation. The components of the deformation tensor (for small deformations) are then given by

$$\begin{aligned} \varepsilon_{rr} &= \frac{\partial u_r}{\partial r_0} & 2\varepsilon_{r\varphi} &= \frac{\partial u_\varphi}{\partial r_0} - \frac{u_\varphi}{r_0} + \frac{1}{r_0}\frac{\partial u_r}{\partial \varphi_0} = 2\varepsilon_{\varphi r} \\ \varepsilon_{\varphi\varphi} &= \frac{1}{r_0}\frac{\partial u_\varphi}{\partial \varphi_0} + \frac{u_r}{r_0} & 2\varepsilon_{rz} &= \frac{\partial u_r}{\partial z_0} + \frac{\partial u_z}{\partial r_0} = 2\varepsilon_{zr} \\ \varepsilon_{zz} &= \frac{\partial u_z}{\partial z_0} & 2\varepsilon_{\varphi z} &= \frac{\partial u_\varphi}{\partial z_0} + \frac{1}{r_0}\frac{\partial u_z}{\partial \phi_0} = 2\varepsilon_{z\varphi} \end{aligned} \tag{7.22}$$

For the case of polar coordinates, the components of the deformation tensor may be found in [1, p. 56].

7.3 The Rheological Equation of State for Isotropic Linear Elastic Materials (Hookean Theory of Elasticity)

The foundation of the technical mechanics of classical construction materials as metals below their yield stress, glass, ceramics, wood, and concrete is the *Hookean theory of elasticity*. In this theory, the components of the stress tensor are assumed to be linear functions of the components of the tensor of small deformations. If the material is assumed to be isotropic, this relation simplifies to (see for instance [4, p. 164])

$$\begin{aligned} \sigma_{xx} &= 2G\varepsilon_{xx} + [K - 2G/3] \cdot (\varepsilon_{xx} + \varepsilon_{yy} + \varepsilon_{zz}), \ldots \\ \sigma_{xy} &= 2G\varepsilon_{xy}, \ldots \end{aligned} \tag{7.23}$$

The system (7.23) consists of 6 equations, of which only two have been presented. The other four equations are found by a cyclic interchange of the indices ($x \to y, y \to z, z \to x$). Equations (7.23) contain two constants, the *shear modulus* G and the *bulk modulus* K, which for classical construction materials only weakly depend on temperature. Defining the inverse of these constants as $J = 1/G$, the *shear compliance* and $B = 1/K$ the *compressive compliance* (or the *compressibility*), the system of Eq. (7.23) may be inverted to express the components of the deformation tensor as linear functions of the components of the stress tensor

Table 7.1 The elastic constants E, K, G, μ as functions of two other elastic constants

	(G, E)	(G, K)	(E, K)	(G, μ)	(E, μ)	(K, μ)
Young's modulus E	–	$\frac{9G\cdot K}{3K+G}$	–	$2G\cdot(1+\mu)$	–	$3K\cdot(1-2\mu)$
Shear modulus G	–	–	$\frac{3E\cdot K}{9K-E}$	–	$\frac{E}{2(1+\mu)}$	$\frac{3K\cdot(1-2\mu)}{2(1+\mu)}$
Bulk modulus K	$\frac{E\cdot G}{9G-3E}$	–	–	$G\cdot\frac{2(1+\mu)}{3(1-2\mu)}$	$\frac{E}{3(1-2\mu)}$	–
Poisson's ratio μ	$\frac{E}{2G}-1$	$\frac{9K-2G}{2(3K+G)}$	$\frac{1}{2}\left[1-\frac{E}{3K}\right]$	–	–	–

$$
\begin{aligned}
\varepsilon_{xx} &= \frac{1}{2}J\sigma_{xx} + \frac{1}{3}[B/3 - J/2]\cdot(\sigma_{xx}+\sigma_{yy}+\sigma_{zz}),\ ..\\
\varepsilon_{xy} &= \frac{1}{2}J\sigma_{xy},\ ..
\end{aligned}
\tag{7.24}
$$

Two elastic constants are sufficient, viz., G and K, for the description of the behavior at small deformations of isotropic linear elastic materials. All the other moduli may be derived from these two. Further elastic constants which are often used are *Young's modulus E* and *Poisson's ratio* μ. Their relations with G and K are summarized in Table 7.1.

7.4 The Rheological Equation of State for Isotropic Linear Viscoelastic Materials at Small Deformations

A comparison of the second of Eqs. (7.23) with Eq. (5.17) leads to the generalization of (7.23) for isotropic linear viscoelastic materials at small deformations [5].

$$
\begin{aligned}
\sigma_{xx}(t) &= 2G_0\varepsilon_{xx}(t) + 2\int_{-\infty}^{t}\dot{G}(t-\xi)\varepsilon_{xx}(\xi)\mathrm{d}\xi + [K_0 - 2G_0/3]\cdot(\varepsilon_{xx}(t)+\varepsilon_{yy}(t)+\varepsilon_{zz}(t))\\
&\quad + \int_{-\infty}^{t}[\dot{K}(t-\xi) - (2/3)\dot{G}(t-\xi)]\cdot[\varepsilon_{xx}(\xi)+\varepsilon_{yy}(\xi)+\varepsilon_{zz}(\xi)]d\xi,\ ..\\
\sigma_{xy}(t) &= 2G_0\varepsilon_{xy}(t) + 2\int_{-\infty}^{t}\dot{G}(t-\xi)\varepsilon_{xy}(\xi)\mathrm{d}\xi,\ ..
\end{aligned}
\tag{7.25}
$$

Also this system consists of 6 equations, of which only two have been presented. The other four equations are found by a cyclic interchange of the indices. G and K are not constants now, but time-dependent characteristic viscoelastic functions, which characterize the linear viscoelastic behavior of the material completely. $G(t)$ is the *time-dependent shear modulus* already discussed in Chap. 5, $K(t)$ the *time-dependent bulk modulus*. G_0 and K_0 are the corresponding limits for $t \to 0$.

For an inversion of this system of equations, we define the *time-dependent shear compliance* $J(t)$ and the *time-dependent compression* (bulk) *compliance* $B(t)$, also called the *time-dependent compressibility*. Then the equations inverse to (7.25) are

$$\begin{aligned}
\varepsilon_{xx}(t) &= \frac{1}{2} J_0 \sigma_{xx}(t) + \frac{1}{2} \int_{-\infty}^{t} \dot{J}(t-\xi)\sigma_{xx}(\xi)\mathrm{d}\xi + [B_0/9 - J_0/6] \cdot (\sigma_{xx}(t) + \sigma_{yy}(t) + \sigma_{zz}(t)) \\
&\quad + \int_{-\infty}^{t} [(1/9)\dot{B}(t-\xi) - (1/6)\dot{J}(t-\xi)] \cdot [\sigma_{xx}(\xi) + \sigma_{yy}(\xi) + \sigma_{zz}(\xi)]\mathrm{d}\xi\,, \ldots \\
\varepsilon_{xy}(t) &= \frac{1}{2} J_0 \sigma_{xy}(t) + \frac{1}{2} \int_{-\infty}^{t} \dot{J}(t-\xi)\sigma_{xy}(\xi)\mathrm{d}\xi\,, \ldots
\end{aligned} \tag{7.26}$$

Equations (7.25) and (7.26) are *dual*, i.e., from one of the equations follows the other and vice versa. Of course, for this purpose the characteristic functions $B(t)$, $J(t)$, $K(t)$, and $G(t)$ have to be related to each other by the corresponding integral equations which are shown below. The conditions for the validity of the above rheological equations of state are:

(1) The validity of the superposition principle in its general form is fulfilled, i.e., not only two shear histories should superimpose undisturbed, but also a shear history and the history of the compressive stresses. So, for instance, the shear modulus should not be influenced by a superimposed hydrostatic pressure.
(2) The deformations should be infinitesimal, otherwise the use of the strain tensor (7.19) would not be justified.
(3) The polymer is isotropic. This condition might be violated, for example, if the material contains frozen stresses due to processing.

The field of application of Eqs. (7.25) and (7.26) is the technical mechanics of polymers in the hard or glassy state, partially also in the softening region. Their use in the rubber elastic state or in the flow region is problematic, as in those cases mostly the condition 2 will be not fulfilled. For polymers which are anisotropic as a result of processing, a generalization of the theory of linear viscoelasticity for anisotropic materials has been given. But this theory involves more than two independent viscoelastic characteristic functions, the number of which depends on the symmetry properties of the class of anisotropy in question. This theory is explained in detail in [6–8].

The Eqs. (7.25) or (7.26) may be written in tensor notation as

$$\sigma_{ij}(t) = 2G_0[\varepsilon_{ij}(t) - (1/3)\delta_{ij}\varepsilon_{kk}(t)] + K_0\delta_{ij}\varepsilon_{kk}(t)$$
$$+2\int_{-\infty}^{t} \dot{G}(t-\xi)[\varepsilon_{ij}(\xi) - (1/3)\delta_{ij}\varepsilon_{kk}(\xi)]d\xi + \delta_{ij}\int_{-\infty}^{t} \dot{K}(t-\xi)\varepsilon_{kk}(\xi)d\xi \quad (7.27)$$

and

$$\varepsilon_{ij}(t) = (1/2)J_0[\sigma_{ij}(t) - (1/3)\delta_{ij}\sigma_{kk}(t)] + (1/9)B_0\delta_{ij}\sigma_{kk}(t)$$
$$+(1/2)\int_{-\infty}^{t} \dot{J}(t-\xi)[\sigma_{ij}(\xi) - (1/3)\delta_{ij}\sigma_{kk}(\xi)]d\xi + (1/9)\delta_{ij}\int_{-\infty}^{t} \dot{B}(t-\xi)\sigma_{kk}(\xi)d\xi$$
$$(7.28)$$

Equations (7.25) and (7.26) remain valid for curvilinear coordinates. For cylindrical coordinates the indices x, y, z have to be replaced by r, φ, z, for polar coordinates by r, ϑ, φ.

7.5 Simple Shear

The case of simple shear has already been sketched in Fig. 4.2. Surfaces perpendicular to the y-direction are sheared in the x-direction. The Cartesian components of the displacement vector are

$$u_1 = \gamma \cdot x_2^0 \qquad u_2 = u_3 = 0 \quad (7.29)$$

with γ being the shear strain. (Compare Fig. 4.2). The shear γ may depend on the time t, but not on the position in space (x_1^0, x_2^0, x_3^0). From (7.19) the components of the deformation tensor follow as

$$\varepsilon_{xy} = \varepsilon_{yx} = \gamma/2 \qquad \varepsilon_{xx} = \varepsilon_{yy} = \varepsilon_{zz} = \varepsilon_{yz} = \varepsilon_{zx} = 0 \quad (7.30)$$

Inserting these deformations into Eq. (7.25) yields the components of the stress tensor

$$\sigma_{xx} = \sigma_{yy} = \sigma_{zz} = \sigma_{yz} = \sigma_{zx} = 0$$
$$\sigma_{xy}(t) = \sigma_{yx}(t) = G_0\gamma(t) + \int_{-\infty}^{t} \dot{G}(t-\xi)\gamma(\xi)d\xi \quad (7.31)$$

The components of the deformation tensor follow by inserting (7.31) into (7.26)

$$2\varepsilon_{xy}(t) = 2\varepsilon_{yx}(t) = \gamma(t) = J_0\sigma_{xy}(t) + \int_{-\infty}^{t} \dot{J}(t-\xi)\sigma_{xy}(\xi)d\xi \tag{7.32}$$

Equations (7.31) and (7.32) describe the special case of simple shear which was already discussed in Chap. 5. From Sect. 5.7 it is known that the relaxation modulus in shear $G(t)$ and the shear creep compliance $J(t)$ have to obey the integral equation (5.48) in order to admit the simultaneous validity of Eqs. (5.31) and (5.32).

The experiment of simple shear fulfills the conditions for the balance of forces, if no volume forces act and if inertia forces may be neglected. Under these conditions, Eq. (7.3) simply reads as

$$\frac{\partial \sigma_{yx}}{\partial y} = 0 \qquad \frac{\partial \sigma_{xy}}{\partial x} = 0$$

and the third equation is fulfilled identically. These equilibrium conditions are met because γ and therefore σ_{xy} do not depend on the position in space.

7.6 Isotropic Compression

Consider a cube with edges of the length a under *hydrostatic pressure* p, as sketched in Fig. 4.3. On each of the surfaces of the cube a normal force $-pa^2$ acts and the Cartesian components of the stress tensor are

$$\sigma_{xy} = \sigma_{yz} = \sigma_{zx} = 0 \qquad \sigma_{xx} = \sigma_{yy} = \sigma_{zz} = -p \tag{7.33}$$

Under the hydrostatic pressure, the cube is compressed in all directions and its edges are shortened to $(1-b)a$. If the origin of the Cartesian coordinate system is placed in the midpoint of the cube, the components of the displacement vector are

$$u_1 = -bx_1^0 \quad u_2 = -bx_2^0 \quad u_3 = -bx_3^0 \tag{7.34}$$

The cube is compressed from its original volume $V_0 = a^3$ to the volume ($b \ll 1$)

$$V = (a - ab)^3 = V_0(1-b)^3 \cong V_0(1-3b)$$

and therefore the *relative volume contraction* $\varDelta$ becomes

$$\varDelta = -\Delta V/V_0 = 3b \tag{7.35}$$

The components of the deformation tensor are found as

$$\varepsilon_{xx} = \varepsilon_{yy} = \varepsilon_{zz} = -b = -\Delta/3 \qquad \varepsilon_{xy} = \varepsilon_{yz} = \varepsilon_{zx} = 0 \tag{7.36}$$

The system of equations for the stress components obeys the equilibrium conditions, because the hydrostatic pressure was assumed to be independent of the spatial position. Inserting (7.33) and (7.36) into (7.25) and (7.26) results in

$$p(t) = K_0\Delta(t) + \int_{-\infty}^{t} \dot{K}(t-\xi)\Delta(\xi)d\xi \tag{7.37}$$

and

$$\Delta(t) = B_0(t)p + \int_{-\infty}^{t} \dot{B}(t-\xi)p(\xi)d\xi \tag{7.38}$$

These equations describe the principle of superposition for the case of isotropic compression and are analog to Eqs. (7.31) and (7.32). All definitions and conclusions which were drawn for the case of simple shear in Chap. 5 may be transferred to the case of isotropic compression. Instead of the shear modulus $G(t)$ the bulk modulus $K(t)$ is to be used and instead of the shear compliance $J(t)$ the compressibility $B(t)$. In particular, the relation (5.48) between the modulus and compliance is valid for $K(t)$ and $B(t)$ in the form

$$\int_0^t K(t-\xi)B(\xi)d\xi = \int_0^t B(t-\xi)K(\xi)d\xi = t \quad \text{for } 0 \le t < \infty \tag{7.39}$$

$K(t)$ may be measured in a stress relaxation experiment under isotropic compression, $B(t)$ in a creep experiment.

7.7 Uniaxial Tensile Stress

In the experiment under uniaxial tensile stress as indicated in Fig. 7.5, only the x-surfaces are assumed to be loaded by normal forces. Let K be the time-dependent tensile force on the sample, A_0 its cross-section in the undeformed state.

The stress components are

$$\sigma_{xx} = K/A_0 \qquad \sigma_{yy} = \sigma_{zz} = 0 \qquad \sigma_{xy} = \sigma_{yz} = \sigma_{zx} = 0 \tag{7.40}$$

This system of equations obeys the conditions for equilibrium as the stress tensor does not depend on the spatial position.

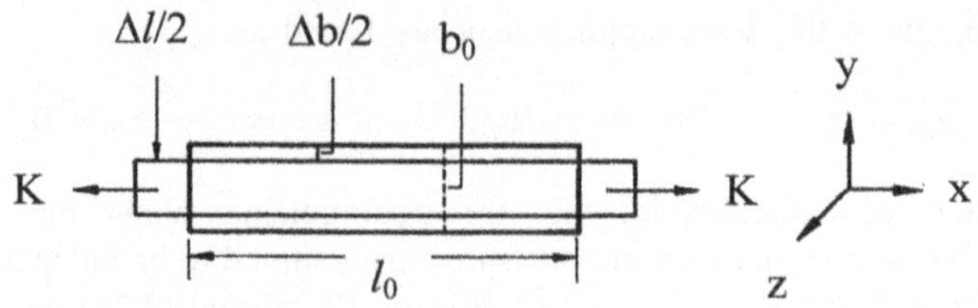

Fig. 7.5 Deformation under uniaxial tensile stress

The tensile stress causes the specimen to elongate in the x-direction and to contract in the y and z-direction. If $\Delta l/l_0$ is its relative longitudinal elongation and $\Delta b/b_0$ its relative lateral contraction, the components of the displacement vector are

$$u_1 = \frac{\Delta l}{l_0} x_1^0 \qquad u_2 = -\frac{\Delta b}{b_0} x_2^0 \qquad u_3 = -\frac{\Delta b}{b_0} b x_3^0 \tag{7.41}$$

The components of the deformation tensor read

$$\varepsilon_{xx} = \Delta l / l_0 \qquad \varepsilon_{yy} = \varepsilon_{zz} = -\Delta b / b_0 \qquad \varepsilon_{xy} = \varepsilon_{yz} = \varepsilon_{zx} = 0 \tag{7.42}$$

Inserting Eqs. (7.40) into (7.26) and defining a new viscoelastic function

$$F(t) = \frac{1}{9} B(t) + \frac{1}{3} J(t) \tag{7.43}$$

results in the following relation between longitudinal elongation and tensile stress

$$\varepsilon_{xx}(t) = F_0 \sigma_{xx}(t) + \int_{-\infty}^{t} \dot{F}(t - \xi) \sigma_{xx}(\xi) d\xi \tag{7.44}$$

This equation has the form of the superposition principle (cf. (5.15)) and $F(t)$ gets the meaning of a creep compliance under uniaxial tensile stress (the *tensile compliance*). Equation (7.43) shows that the tensile compliance is related to the shear compliance and the compressibility in a simple way. The equation which is dual to (7.44) is

$$\sigma_{xx}(t) = E_0 \varepsilon_{xx}(t) + \int_{-\infty}^{t} \dot{E}(t - \xi) \varepsilon_{xx}(\xi) d\xi \tag{7.45}$$

with $E(t)$ being the relaxation modulus under uniaxial tensile stress (the *tensile modulus*).

E_0 is its limit for $t \to 0$. Between $F(t)$ and $E(t)$ an equation analog to (5.48) is valid

$$\int_0^t E(t-\xi)F(\xi)d\xi = \int_0^t F(t-\xi)E(\xi)d\xi = t \quad \text{for} \quad 0 \le t < \infty \tag{7.46}$$

The relation between $E(t)$, $K(t)$, and $G(t)$, which may be derived from (7.25) and (7.26) is not as simple as (7.43), but a complicated integral equation [5].

For linear elastic materials, Poisson's ratio μ_0 is defined in a tensile experiment by the equation

$$\varepsilon_{yy} = -\mu_0 \varepsilon_{xx}$$

A generalization to viscoelastic materials is obtained by the introduction of a *time-dependent Poisson's ratio* $\mu(t)$. It is defined by the following equation which is valid between the lateral contraction and the longitudinal elongation under uniaxial tensile stress

$$-\varepsilon_{yy}(t) = -\varepsilon_{zz}(t) = \mu_0 \varepsilon_{xx}(t) + \int_{-\infty}^{t} \dot{\mu}(t-\xi)\varepsilon_{xx}(\xi)d\xi \tag{7.47}$$

The physical significance of the functions $E(t)$ and $\mu(t)$ becomes clear by considering a relaxation experiment under uniaxial tension which is defined by the equations $\varepsilon_{xx}(t) = 0$ for $t < 0$ and $\varepsilon_{xx}(t) = \varepsilon_0$ for $t > 0$ and by $\sigma_{yy} = \sigma_{zz} = 0$. From (7.44) and (7.47) follows

$$\sigma_{xx}(t) = \varepsilon_0 E(t) \qquad -\varepsilon_{yy}(t) = \varepsilon_0 \mu(t) \tag{7.48}$$

Poisson's ratio determines the ratio of lateral contraction to longitudinal extension in a stress relaxation experiment under uniaxial tensile stress. These relations are shown in Fig. 7.6.

No definite statements can be made about the time dependence of Poisson's ratio. From the scare experimental evidence available today, it is probable that $\mu(t)$ has the lower limit zero and the upper limit ½ and is a monotonously increasing function of the time.

$$0 < \mu(t) \le 1/2; \quad \dot{\mu}(t) \ge 0 \tag{7.49}$$

Further statements as for instance about a total monotony, which were possible for $G(t), K(t), E(t), \dot{J}(t), \dot{B}(t), \dot{F}(t)$, cannot be made for $\mu(t)$.

The following conclusion may be drawn: The deformation behavior of isotropic linear viscoelastic materials at small deformations is completely determined if two independent viscoelastic characteristic functions are known, as for example, the shear modulus $G(t)$ and the bulk modulus $K(t)$. All other characteristic functions may then be calculated, though not always in a simple manner.

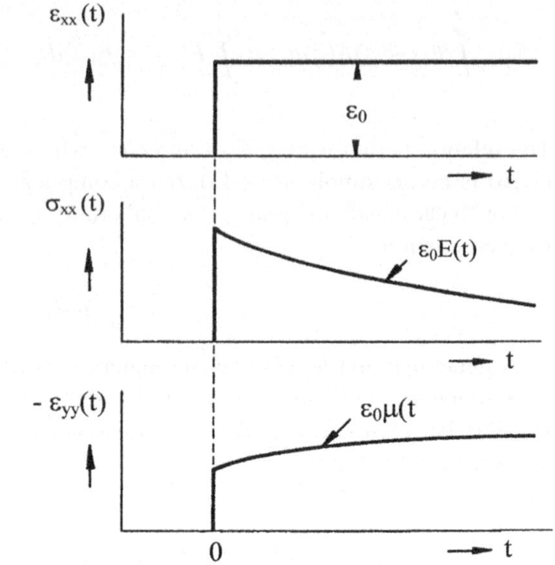

Fig. 7.6 Definition of $E(t)$ and $\mu(t)$

In Table 7.2, the different characteristic functions are listed. In column 2 the characteristic functions for simple shear are given, in column 3 the corresponding functions for isotropic compression and in column 4 those related to uniaxial tension. The relations between the quantities of the second column have been discussed in detail in Chap. 5 and in Fig. 5.30. The relations between the quantities of the third column and the fourth column may be found by using the corresponding equations from Chap. 5 under proper change of the symbols. The fifth column lists quantities in relation to $\mu(t)$ which are neither a modulus nor a compliance. Because of the assumption $\dot{\mu}(t) \geq 0$ the definitions for $\mu'(\omega)$, $\mu''(\omega)$, and $\mu^*(\omega)$ have been chosen in accordance with those of the compliances. Accordingly, it may be expected that $\mu''(\omega)$ is a positive function.

The relations between quantities of different columns of Table 7.2 are generally more complicated. The most simple equation is (7.43) between the creep compliances in tension, shear, and compression. Corresponding equations are valid between F', J', and B', between F'', J'', and B'', between F^*, J^*, and B^* and between the retardation spectra in tension, shear, and compression. Other relations are complicated integral equations and are discussed elsewhere [5].

There is a simple way to relate the dynamic quantities to each other. For the complex moduli G^*, K^*, E^*, and μ^* defined in Table 7.2, the equations of Table 7.1 remain valid, if all quantities are provided with a star. So for instance from

$$E^* = 2G^*(1 + \mu^*) \tag{7.50}$$

Table 7.2 Characteristic functions for isotropic linear viscoelastic materials

Experiment	Simple shear	Isotropic compression	Uniaxial tension	
Relaxation modulus	$G(t)$	$K(t)$	$E(t)$	–
Creep compliance	$J(t)$	$B(t)$	$F(t)$	$\mu(t)$
Storage modulus	$G'(\omega)$	$K'(\omega)$	$E'(\omega)$	–
Loss modulus	$G''(\omega)$	$K''(\omega)$	$E''(\omega)$	–
Loss tangent	$(\tan\delta)_s = G''/G'$	$(\tan\delta)_c = K''/K'$	$(\tan\delta)_e = E''/E'$	$(\tan\delta)_\mu = \mu''/\mu'$
Storage compliance	$J'(\omega)$	$B'(\omega)$	$F'(\omega)$	$\mu'(\omega)$
Loss compliance	$J''(\omega)$	$B''(\omega)$	$F''(\omega)$	$\mu''(\omega)$
Complex modulus	$G^* = G' + iG''$	$K^* = K' + iK''$	$E^* = E' + iE''$	–
Complex compliance	$J^* = J' - iJ''$	$B^* = B' - iB''$	$F^* = F' - iF''$	$\mu^* = \mu' - i\mu''$

two real equations for E', E'' follow, in which E' and E'' may be expressed in terms of G', G'', μ', and μ''.

7.8 The Viscoelastic Functions for Amorphous Uncross-Linked Polymers

Most experimental information on the viscoelastic behavior of polymers is available in shear or torsion, less in extension, and only a few results are known in isotropic compression. Generally $K(t)$, $K'(\omega)$, $K''(\omega)$, $(\tan\delta)_c$ and $G(t)$, $G'(\omega)$, $G''(\omega)$, $(\tan\delta)_s$ show dispersions in the same temperature and time or frequency regions, respectively. This fact is understandable, as the same molecular processes effect a change in shape and in volume. But the magnitude of these contributions may differ considerably. Dispersion steps in K are much less pronounced than those in G and the maxima of the loss tangent $(\tan\delta)_c$ are much lower than those of $(\tan\delta)_s$ (cf. Fig. 7.7). No flow occurs in isotropic compression, of course. Therefore, in contrast to E and G, $K(t)$ and $K'(\omega)$ do not show a steep decrease with temperature or time in the melt.

The viscoelastic functions are sketched in Fig. 7.7 for an uncross-linked amorphous polymer. The upper part of the figure shows the three storage-moduli G', E', and K' as functions of the temperature. In the glassy state the ratio E'/G' equals 2.6 to 2.7. Secondary dispersion regions show up in both, E' and G'. At the glass-transition temperature the ratio E'/G' increases abruptly to 3.0 and stays there at higher temperatures.

The ratio E'/K' is between 1.2 and 0.9 in the glassy state and decreases at each dispersion step, as the contributions to a change in shape are always (much) larger than those to a change in volume. In the glass-transition G' and E' show very pronounced dispersion steps (3 or more decades), while the dispersion in K' only

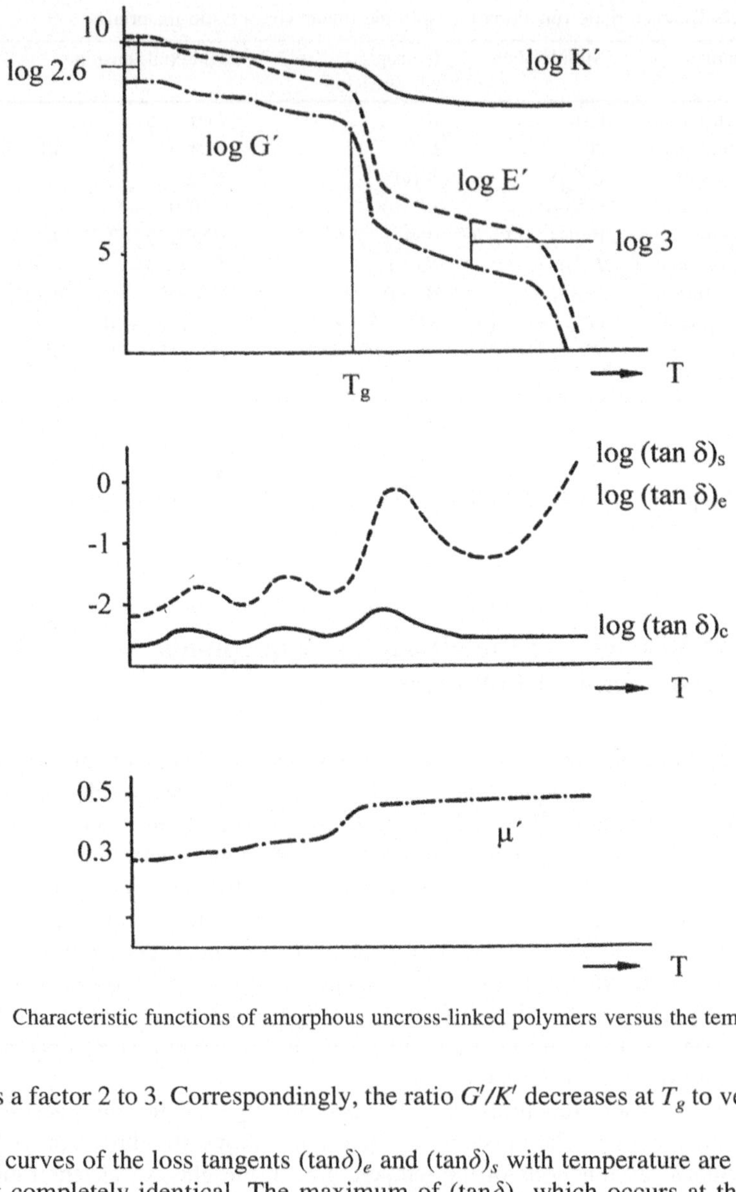

Fig. 7.7 Characteristic functions of amorphous uncross-linked polymers versus the temperature

reaches a factor 2 to 3. Correspondingly, the ratio G'/K' decreases at T_g to very low values.

The curves of the loss tangents $(\tan\delta)_e$ and $(\tan\delta)_s$ with temperature are similar but not completely identical. The maximum of $(\tan\delta)_c$ which occurs at the same temperature is much lower. In the melt, the loss tangents $(\tan\delta)_s$ and $(\tan\delta)_e$ increase steeply with temperature, while $(\tan\delta)_c$ remains low there. Poisson's ratio μ shows values between 0.30 and 0.35 in the glassy state. It increases moderately with temperature, but jumps to ½ at the glass transition.

Equation (7.43) is valid for all states of the polymer. If applied to the melt, at longer times the main contributions to the compliances are the flow terms, i.e.,

$$\frac{1}{\eta_e} = \frac{1}{9\eta_c} + \frac{1}{3\eta_s} \tag{7.51}$$

with η_e being the *extensional viscosity*, η_c the *volume viscosity* and η_s the *shear viscosity*. As no flow is observed in isotropic compression, $\eta_c \to \infty$ and Eq. (7.51) simplifies to

$$\eta_e = 3\eta_s \tag{7.52}$$

It should be emphasized that Eq. (7.52) is only valid for linear viscoelastic fluids at small deformations or under very low elongation rates. It is not generally applicable to polymer melts. In rheology, the ratio η_e/η_s is called the *Trouton ratio*.

Finally, in the temperature region above T_g many of the relations between the characteristic functions in shear and extension simplify considerably as $B \ll J$ and therefore

$$F(t) \cong J(t)/3 \qquad E(t) \cong 3G(t) \qquad \text{for } T \geq T_g \tag{7.53}$$

References

1. Love AEH (1944) A treatise on the mathematical theory of elasticity, 4th edn. Dover Publications, New York
2. Timoshenko SP, Goodier JN (1970) Theory of elasticity, 3rd edn. McGraw Hill, New York
3. Cosserat E, Cosserat F (1909) Théorie des Corps Déformables, p. 953. In Chwolson OD (ed) Traité de Physique, 2nd edn. trans. Daveaux E, Paris
4. Joos G (1977) Lehrbuch der theor. Physik. Akad. Verlagsges, Wiesbaden
5. Schwarzl F (1956) Kolloid Z 148:47
6. Biot MA (1954) J. Appl Phys 25:1385
7. Falk G, Meixner J (1956) Z f Naturf 11a:782
8. Buggisch H, Mazilu P, Weber H (1988) Rheol Acta 27:363

Chapter 8
Fundamentals of the Rheology of Large Deformations

8.1 Kinematics of Large Deformations

During processing, polymeric materials are mostly handled in the molten state ($T > T_g$ or $T > T_m$). Deformations and rates of deformation are generally large under these circumstances. Also rubber elastic materials can be subjected to large deformations before the occurrence of rupture. Therefore, the basics of large deformations have to precede the discussion of the deformation behavior of polymer melts and cross-linked rubbers.[1]

We consider the motion of material points of a body or a fluid in a space-fixed Cartesian coordinate system with the origin O. A certain point (or particle) of the body will be at the position $\underline{x}^0 = (x_1^0, x_2^0, x_3^0)$ [2] at the time t_0 in the undeformed state of the body. The same particle may be found at the position $\underline{x} = (x_1, x_2, x_3)$ at a later time t. If we specify the function $\underline{x} = \underline{f}(\underline{x}^0; t)$ for one fixed value of $\underline{x}^0$ and all $t \geq t_0$, the path of the particle $\underline{x}^0$ is known at all times. The particle is identified by its position vector at the beginning of the experiment. Therefore, we speak of the "particle $\underline{x}^0$ "and mean the particle which has been at the position $\underline{x}^0$ at the time t_0.

If we want to describe the paths of *all* particles of the body or fluid, we have to specify the three functions $x_i = f_i\left(x_1^0, x_2^0, x_3^0, t\right)(i = 1, 2, 3)$ designating the i-th coordinate of the position vector of the particle $\underline{x}^0$, for all values of the $\underline{x}^0$ and t. The three functions of the four independent variables x_1^0, x_2^0, x_3^0, t are called the *path-equations*

$$x_i = f_i(x_1^0, x_2^0, x_3^0; t) \tag{8.1}$$

[1] A very elegant introduction to the kinematics of large deformations in curvilinear coordinates is given in [1, 2].
[2] We will distinguish in the following vectors and tensors of the second rank from scalars by underlining them once or twice. The Cartesian components of a (positioning) vector $\underline{x}$ will be designated as x_i, of a tensor $\underline{\underline{f}}$ as f_{ik}.

H. Münstedt and F. R. Schwarzl, *Deformation and Flow of Polymeric Materials*,
DOI: 10.1007/978-3-642-55409-4_8,

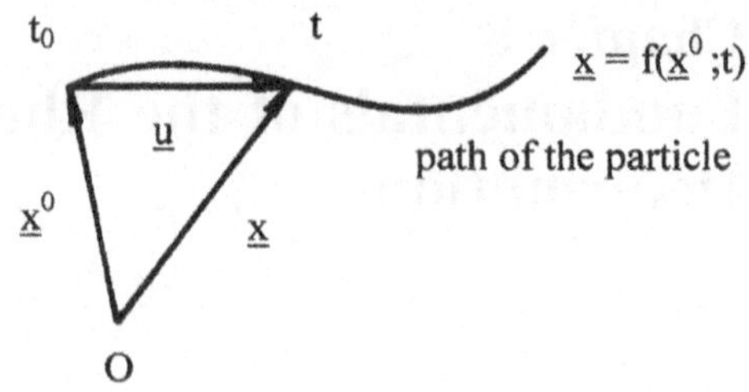

Fig. 8.1 Illustrating the path of the particle $\underline{x}^0$ with time. $\underline{u}$ is the displacement vector (cf. Eq. (7.10))

We assume these equations to be given. If there is no danger of a confusion, the function f_i is often designated by the same letter as the value x_i of this function by some authors. The path of the particle $\underline{x}^0$ is shown in Fig. 8.1.

The system of Eq. (8.1) has to be unique with respect to the position coordinates, otherwise a particle $\underline{x}^0$ would be at two different places at a later time. This would mean the occurrence of a rupture surface, a phenomenon which we want to exclude from our considerations.

Equation (8.1) may be solved for the variables x_i^0 at each fixed value of the time t, expressing the coordinates x_i^0 as functions of x_1, x_2, x_3, and t by means of the inverse equations

$$x_i^0 = g_i(x_1, x_2, x_3; t) \tag{8.2}$$

The functions g_i designate the position of the material point at the past time t_0 which is just at the position $\underline{x}$ at the current time t. Also this system of equations has to be unique regarding the position coordinates, otherwise a particle of the deformed state would have been at two different places at the beginning. Such a situation were only possible if two sides of a rupture surface would have been glued together during the deformation process.

The coordinates $\underline{x}^0 = \left(x_1^0, x_2^0, x_3^0\right)$ are called *Lagrange-coordinates* or *body-fixed coordinates*, the coordinates $\underline{x} = (x_1, x_2, x_3)$ *Euler-coordinates* or *space-fixed coordinates*. If the Euler-coordinates are kept constant, a fixed position in space is observed, if the Lagrange-coordinates are kept constant, a fixed particle is watched with time. The Lagrange-coordinates may be imagined as a coordinate system drawn on the deforming body and changing its metrics with the deformation.

An example from the daily life might be the highway and the cars driving on it. The milestones along the highway represent the Euler-coordinates, the number plates of the cars the Lagrange-coordinates.

The velocity of a particle $\underline{x}^0$, which is located at the position $\underline{x}$ at the time t is given by

$$\underline{\varphi} = \left(\frac{\partial \underline{f}}{\partial t}\right)_{\underline{x}^0}$$

and is obtained by differentiating the Eq. (8.1) after the time t while keeping the Lagrange-coordinates $\left(x_1^0, x_2^0, x_3^0\right)$ constant. Its components are

$$\varphi_i(x_1^0, x_2^0, x_3^0, t) = \frac{\partial f_i(x_1^0, x_2^0, x_3^0, t)}{\partial t} \tag{8.3}$$

Equation (8.3) is called the *velocity field* in the *representation* of *Lagrange*.

The acceleration of a particle $\underline{x}^0$ at the position $\underline{x}$ at the time t is given by

$$\underline{\alpha} = \left(\frac{\partial \underline{\varphi}}{\partial t}\right)_{\underline{x}^0} = \left(\frac{\partial^2 \underline{f}}{\partial t^2}\right)_{\underline{x}^0}$$

and is obtained by differentiating the Eq. (8.1) twice after the time t while keeping the values of $\left(x_1^0, x_2^0, x_3^0\right)$ constant. Its components are

$$\alpha_i(x_1^0, x_2^0, x_3^0, t) = \frac{\partial^2 f_i(x_1^0, x_2^0, x_3^0, t)}{\partial t^2} \tag{8.4}$$

Equation (8.4) is called the *acceleration field* in the *representation* of *Lagrange*. Equations (8.3) and (8.4) give the velocity and acceleration of one fixed particle in dependence on time.

It is also possible to ask for the velocity and acceleration of the particles which just pass the position x at the time t. We then deal with velocity and acceleration of different particles. To obtain these quantities the inverse Eqs. (8.2) are inserted into the arguments of (8.3) and (8.4). After rearranging the new independent variables x_1, x_2, x_3, and t two new functions are obtained which are designated by v_i and a_i:

$$v_i(x_1, x_2, x_3; t) \equiv \varphi_i[g_1(x_1, x_2, x_3; t), g_2(x_1, x_2, x_3; t), g_3(x_1, x_2, x_3; t); t] \tag{8.5}$$

$$a_i(x_1, x_2, x_3; t) \equiv \alpha_i[g_1(x_1, x_2, x_3; t), g_2(x_1, x_2, x_3; t), g_3(x_1, x_2, x_3; t); t] \tag{8.6}$$

$\underline{v} = \underline{v}(\underline{x}; t)$ and $\underline{a} = \underline{a}(\underline{x}; t)$ are called the *velocity field* and the *acceleration field* in the *representation* of *Euler*.

The velocity field in the Euler-representation is shown in Fig. 8.2. A fixed position $\underline{x}$ of the space is observed. During three successive times t_1, t_2, t_3 three different material particles pass through this point. The original positions of these particles at the time t_0 had been $\underline{g}(\underline{x}; t_1), \underline{g}(\underline{x}; t_2), \underline{g}(\underline{x}; t_3)$, their velocity vectors $\underline{v}(\underline{x}; t_1), \underline{v}(\underline{x}; t_2), \underline{v}(\underline{x}; t_3)$ are the tangents of the paths of these particles at the point $\underline{x}$.

Returning to the example of the highway: The velocity field in Lagrange representation are the velocities of the tachometers of the cars. The velocity field in Euler-representation is measured by the police at the border of the highway.

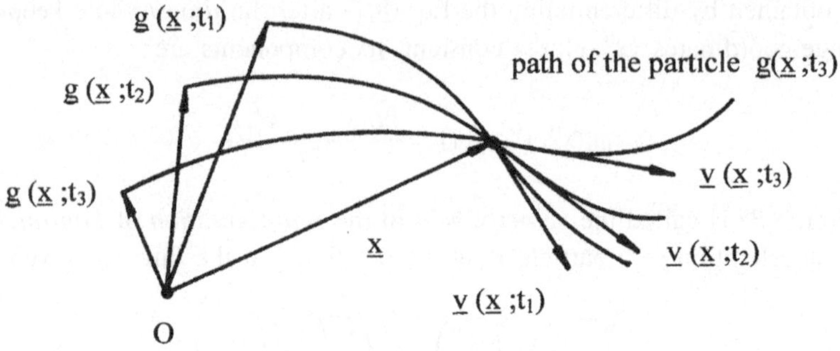

Fig. 8.2 Illustration of the velocity field in Euler-representation

If the velocity field is given in Euler-representation, the acceleration field in Euler-representation may be obtained in steps: First the path equations are inserted into the Euler-coordinates to yield the velocity field in Lagrange-coordinates, then differentiation after t while keeping the Lagrange-coordinates constant yields the acceleration field in Lagrange representation and finally by inserting the inverse path equations the acceleration field in Euler-representation is obtained. The question arises whether the acceleration field in Euler-coordinates may be directly derived from the velocity field in Euler-coordinates. This may be achieved by inserting the Lagrange-coordinates as independent variables into (8.5)

$$\varphi_i(x_1^0, x_2^0, x_3^0; t) = \upsilon_i[f_1(x_1^0, x_2^0, x_3^0; t), f_2(x_1^0, x_2^0, x_3^0; t), f_3(x_1^0, x_2^0, x_3^0; t); t]$$

Differentiating this equation with respect to t while keeping the values of (x_1^0, x_2^0, x_3^0) constant and using the chain rule of differentiation yields

$$a_i = \left(\frac{\partial \varphi_i}{\partial t}\right)_{\underline{x}^0} = \frac{\partial \upsilon_i}{\partial x_1}\frac{\partial f_1(x_1^0, x_2^0, x_3^0; t)}{\partial t} + \cdots + \left(\frac{\partial \upsilon_i}{\partial t}\right)_{\underline{x}}$$

Because of (8.3) this equation becomes

$$a_i(x_1, x_2, x_3; t) = \frac{\partial \upsilon_i(x_1, x_2, x_3; t)}{\partial t} + \sum_{k=1}^{3} v_k(x_1, x_2, x_3; t)\frac{\partial v_i(x_1, x_2, x_3; t)}{\partial x_k}$$

and using the summation convention we arrive at

$$a_i(\underline{x}; t) = \frac{\partial \upsilon_i(\underline{x}; t)}{\partial t} + v_k(\underline{x}; t)\frac{\partial v_i(\underline{x}; t)}{\partial x_k} \tag{8.7}$$

This equation expresses the acceleration components in Cartesian Euler-coordinates by the velocity components in Cartesian Euler-coordinates. The corresponding equations in cylinder coordinates and in polar coordinates may be found in [2].

If changes in temperature are considered, a *temperature field* in *Lagrange representation* $\Theta = \Theta(x_1^0, x_2^0, x_3^0; t)$, viz., the temperature of a particle $\underline{x}^0$ at the time t, and a temperature field in *Euler-representation*, $T = T(x_1, x_2, x_3; t)$, viz., the temperature of the particle which just passes the position $\underline{x}$ at the time t, should be distinguished, whereby.

$$T(x_1, x_2, x_3; t) = \Theta(x_1^0, x_2^0, x_3^0; t) \tag{8.8}$$

Similarly, to the derivation of Eq. (8.7), the change of the temperature of a particle with time may be shown to be

$$\frac{DT}{Dt} \equiv \frac{\partial \Theta(\underline{x}^0; t)}{\partial t} = \frac{\partial T(\underline{x}; t)}{\partial t} + v_k(\underline{x}; t)\frac{\partial T(\underline{x}; t)}{\partial x_k} = \frac{\partial T}{\partial t} + \underline{v} \cdot (\text{grad } T) \tag{8.9}$$

The formation of the gradient on the right-hand side of this equation refers to the Euler-coordinates. The quantity *DT/Dt* is called the *convective time derivative* of the *temperature field.*

8.2 Deformation Gradient and Finite Strain Tensors

Consider two infinitesimally neighbored material points which are situated at the positions $\underline{x}^0$ and $\underline{x}^0 + \mathrm{d}\underline{x}^0$ at the time t_0. At the later time t they will still be infinitesimally neighbored but now at the positions $\underline{x}$ and $\underline{x} + \mathrm{d}\underline{x}$. Their distance has changed from the value $\mathrm{d}s_0 = |\mathrm{d}\underline{x}^0|$ to the value $\mathrm{d}s = |\mathrm{d}\underline{x}|$ as shown in Fig. 8.3.

By differentiating the Eq. (8.1) with respect to the Lagrange-coordinates x_i^0 we may express the components of the distance vector $\mathrm{d}\underline{x}$ at time t by the components of the distance vector $\mathrm{d}\underline{x}^0$at time t_0

$$\mathrm{d}x_i = F_{ik}\mathrm{d}x_k^0 \tag{8.10}$$

using the definition of a second rank tensor called the *deformation gradient*

$$F_{ik}(\underline{x}^0; t) = \frac{\partial f_i(\underline{x}^0; t)}{\partial x_k^0} \tag{8.11}$$

or we may express the components of the distance vector $\mathrm{d}\underline{x}^0$ by the components of the distance vector $\mathrm{d}\underline{x}$

$$\mathrm{d}x_i^0 = F_{ik}^{-1}\mathrm{d}x_k \tag{8.12}$$

using the *inverse deformation gradient*

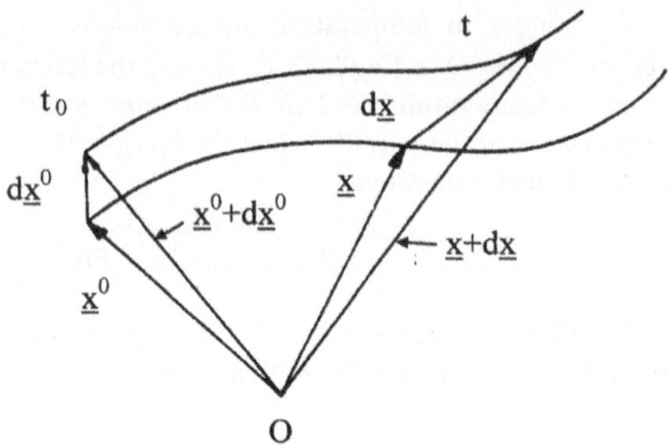

Fig. 8.3 Change of the distance of two neighbored points during the deformation. $|dx|$ and $|dx^0|$ are the distances of the two material points at the time t and t_0, respectively

$$F_{ik}^{-1}(\underline{x};t) = \frac{\partial g_i(\underline{x};t)}{\partial x_k} \tag{8.13}$$

By means of Eq. (8.10), we write the square of the distance ds in terms of the components of the distance vector $\mathrm{d}\underline{x}^0$

$$(\mathrm{d}s)^2 = \mathrm{d}x_m \mathrm{d}x_m = F_{mi}F_{mk}\mathrm{d}x_i^0\mathrm{d}x_k^0 = C_{ik}\mathrm{d}x_i^0\mathrm{d}x_k^0 \tag{8.14}$$

The symmetric tensor of the second rank $\underline{\underline{C}}$ is the *Cauchy strain tensor* already introduced in 7.2. Its components are given by

$$\boxed{C_{ik} = F_{mi}F_{mk} = \frac{\partial f_m(\underline{x}^0;t)}{\partial x_i^0}\frac{\partial f_m(\underline{x}^0;t)}{\partial x_k^0}} \tag{8.15}$$

In a similar way, the square of the distance ds_0 in terms of the components of the distance vector d$\underline{x}$ may be written as

$$(\mathrm{d}s_0)^2 = \mathrm{d}x_m^0\mathrm{d}x_m^0 = F_{mi}^{-1}F_{mk}^{-1}\mathrm{d}x_i\mathrm{d}x_k = A_{ik}\mathrm{d}x_i\mathrm{d}x_k \tag{8.16}$$

Here a symmetric tensor

$$\boxed{A_{ik} = F_{mi}^{-1}F_{mk}^{-1} = \frac{\partial g_m(\underline{x};t)}{\partial x_i}\frac{\partial g_m(\underline{x};t)}{\partial x_k}} \tag{8.17}$$

is defined, the *Piola strain tensor*. Two further strain tensors are sometimes used, the *Green-tensor* $\underline{\underline{G}} = \underline{\underline{A}}^{-1}$ and the *Finger tensor* $\underline{\underline{B}} = \underline{\underline{C}}^{-1}$, which are the inverse of the Piola and the Cauchy tensor. Their components are

$$\boxed{G_{ik} = F_{im}F_{km} = \frac{\partial f_i(\underline{x}^0;t)}{\partial x_m^0}\frac{\partial f_k(\underline{x}^0;t)}{\partial x_m^0}} \tag{8.18}$$

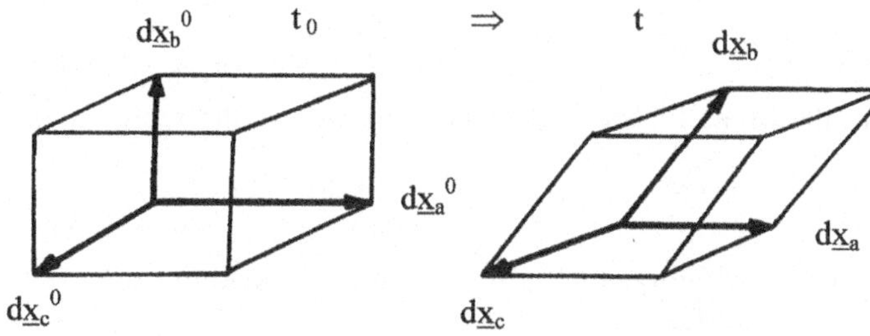

Fig. 8.4 Change of a volume element

and

$$\boxed{B_{ik} = F_{im}^{-1} F_{km}^{-1} = \frac{\partial g_i(\underline{x};t)}{\partial x_m} \frac{\partial g_k(\underline{x};t)}{\partial x_m}} \tag{8.19}$$

The four strain tensors can be represented as products of the deformation gradient $\underline{\underline{F}}$, its transposed $\underline{\underline{\tilde{F}}}$, its inverse $\underline{\underline{F}}^{-1}$ and its transposed inverse $\underline{\underline{\tilde{F}}}^{-1}$:

$$\underline{\underline{C}} = \underline{\underline{\tilde{F}}} \cdot \underline{\underline{F}} \qquad \underline{\underline{A}} = \underline{\underline{\tilde{F}}}^{-1} \cdot \underline{\underline{F}}^{-1} \qquad \underline{\underline{B}} = \underline{\underline{F}}^{-1} \cdot \underline{\underline{\tilde{F}}}^{-1} \qquad \underline{\underline{G}} = \underline{\underline{F}} \cdot \underline{\underline{\tilde{F}}} \tag{8.20}$$

Equations (8.10) and (8.12) describe the change of line elements by the deformation. We consider next the change of volume elements. As shown by Fig. 8.4, an infinitesimal cuboid is spanned by the three vectors $\mathrm{d}\underline{x}_a^0$, $\mathrm{d}\underline{x}_b^0$, $\mathrm{d}\underline{x}_c^0$ of the lengths $\mathrm{d}s_a$, $\mathrm{d}s_b$, $\mathrm{d}s_c$ parallel to the x, y, z-directions in the undeformed state.

The volume of the cuboid is $\mathrm{d}V_0 = \mathrm{d}s_a \cdot \mathrm{d}s_b \cdot \mathrm{d}s_c$. Due to the deformation it is transformed into an infinitesimal parallelepiped spanned by the three vectors $\mathrm{d}\underline{x}_a, \mathrm{d}\underline{x}_b, \mathrm{d}\underline{x}_c$. By the deformation the vector $\mathrm{d}\underline{x}_a^0$ is transformed into a vector $\mathrm{d}\underline{x}_a$

$$\mathrm{d}\underline{x}_a = \underline{\underline{F}} \cdot \mathrm{d}\underline{x}_a^0 = \begin{pmatrix} F_{11} & F_{12} & F_{13} \\ F_{21} & F_{22} & F_{23} \\ F_{31} & F_{32} & F_{33} \end{pmatrix} \cdot \begin{pmatrix} 1 \\ 0 \\ 0 \end{pmatrix} \cdot \mathrm{d}s_a = \begin{pmatrix} F_{11} \\ F_{21} \\ F_{31} \end{pmatrix} \cdot \mathrm{d}s_a$$

Corresponding equations are valid, of course, for the vectors $\mathrm{d}\underline{x}_b$ and $\mathrm{d}\underline{x}_c$. The volume of the transformed volume element is equal to the scalar product of the vector $\mathrm{d}\underline{x}_c$ with the vector product of $\mathrm{d}\underline{x}_a$ and $\mathrm{d}\underline{x}_b$, viz.,

$$\mathrm{d}V = (\mathrm{d}\underline{x}_a \times \mathrm{d}\underline{x}_b) \cdot \mathrm{d}\underline{x}_c = \det(\underline{\underline{F}}) \cdot \mathrm{d}V_0 = \sqrt{\det(\underline{\underline{C}})} \cdot \mathrm{d}V_0 \tag{8.21}$$

$\det\left(\underline{\underline{F}}\right)$ means the determinant of the matrix elements of $\underline{\underline{F}}$. Consequently, $\det\left(\underline{\underline{F}}\right)$ has to be positive. For the density we have the equation $\rho \mathrm{d}V = \mathrm{d}m$ and because of $\mathrm{d}m = \mathrm{d}m_0$ it follows

$$\rho = \rho_0 / \det(\underline{\underline{F}}) \tag{8.22}$$

To obtain the limits for small deformations we use the representation

$$x_i = x_i^0 + u_i(\underline{x}^0; t) \tag{7.9}$$

together with the definition (7.19) for ε_{ik} and the condition (7.18) for the displacement vector and find

$$\begin{aligned} C_{ik} &= \left(\delta_{im} + \frac{\partial u_m}{\partial x_i^0}\right)\left(\delta_{mk} + \frac{\partial u_m}{\partial x_k^0}\right) = \delta_{ik} + \frac{\partial u_i}{\partial x_k^0} + \frac{\partial u_k}{\partial x_i^0} + \frac{\partial u_m}{\partial x_i^0}\frac{\partial u_m}{\partial x_k^0} \\ &\cong \delta_{ik} + 2\varepsilon_{ik} + O\left[\varepsilon^2\right] \end{aligned} \tag{8.15a}$$

and

$$\begin{aligned} A_{ik} &= \left(\delta_{im} - \frac{\partial u_m}{\partial x_i}\right)\left(\delta_{mk} - \frac{\partial u_m}{\partial x_k}\right) \cong \left(\delta_{im} - \frac{\partial u_m}{\partial x_i^0}\right)\left(\delta_{mk} - \frac{\partial u_m}{\partial x_k^0}\right) \\ &\cong \delta_{ik} - 2\varepsilon_{ik} + O\left[\varepsilon^2\right] \end{aligned} \tag{8.17a}$$

and because of $\underline{\underline{G}} = \underline{\underline{A}}^{-1}$ and $\underline{\underline{B}} = \underline{\underline{C}}^{-1}$ and the condition (7.18)

$$G_{ik} \cong \delta_{ik} + 2\varepsilon_{ik} + O\left[\varepsilon^2\right] \tag{8.18a}$$

and

$$B_{ik} \cong \delta_{ik} - 2\varepsilon_{ik} + O\left[\varepsilon^2\right] \tag{8.19a}$$

With regard to the left-hand side of Eq. (8.17a), one should mention that differentiation after the variables x_i and x_i^0 give, of course, different results in the case of large deformations. The difference, however, is of the order ε^2 and may be neglected in the limit of small deformations.

8.3 Relative Deformation Gradient and Relative Strain Tensors

So far we only considered two states during the deformation, the undeformed state at the beginning of the experiment at the time t_0 and the deformed state at the time t at which the stress is to be calculated (*the state of observation*). The undeformed state has been chosen as a reference state. This view is justified for the description of the rheology of elastic solids which show a natural undeformed (stress free) state. For the description of fluids, however, this procedure is not appropriate as

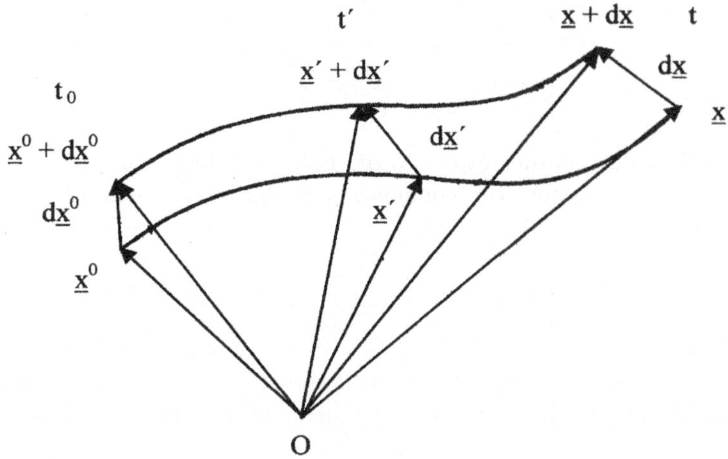

Fig. 8.5 The paths of two neighbored material particles and their distances, at t_0, the beginning of the experiment, $\mathrm{d}s_0 = \left|\mathrm{d}\underline{x}^0\right|$, at an intermediate time t' in the past history $\mathrm{d}s' = |\mathrm{d}\underline{x}'|$ and at the observation time t $\mathrm{d}s = |\mathrm{d}\underline{x}|$

fluids have the essential property of a fading memory and are, therefore, not able to remember an undeformed state of the past history.

Preferably, for the description of the rheological behavior of fluids the state of observation at the time t is chosen as the reference state. A third state in the past at the time t' has then to be considered for the comparison with the reference state. This is shown in Fig. 8.5, in which again the paths of two neighbored particles are sketched, but now at three times, at the beginning of the experiment t_0, at a time t' in the past history between t_0 and the observation time t and at the observation time t.

In order to avoid the occurrence of the fading time t_0 in the path-equations (8.1), the path of the particle $\underline{x}$ will be fixed now by equations which only contain the times t and t', viz.,

$$x_i' = p_i(x_1, x_2, x_3; t, t') = x_i'(x_1, x_2, x_3; t, t') \tag{8.23}$$

in which x_i' are the coordinates of a particle at time t' which will be found later at the observation time t at the coordinates x_i. The inverse equations are

$$x_i = q_i(x_1', x_2', x_3'; t, t') = x_i(x_1', x_2', x_3'; t, t') \tag{8.24}$$

The components of the distance vector at time t' are given by the expression

$$\mathrm{d}x_i' = F_{ik}\mathrm{d}x_k \tag{8.25}$$

with F_{ik} being the components of the *relative deformation gradient*

$$F_{ik}(\underline{x};t,t') = \frac{\partial p_i(\underline{x};t,t')}{\partial x_k} \equiv \frac{\partial x'_i}{\partial x_k} \tag{8.26}$$

For the differentiation with respect to the Euler-coordinates, both times t and t' have to be kept constant. The components of the *inverse relative deformation gradient* are

$$F_{ik}^{-1}(\underline{x}';t,t') = \frac{\partial q_i(\underline{x}';t,t')}{\partial x'_k} \equiv \frac{\partial x_i}{\partial x'_k} \tag{8.27}$$

In order to distinguish more clearly the observation time t from the current time t', Wagner [3] proposed the following designation for the relative deformation gradient

$$\underline{\underline{F}}_t(t') = \underline{\underline{F}}(\underline{x};t,t') \tag{8.28}$$

where the basis t (the time of observation) is indicated next to the underlining and t' is the current time in the past history. With this nomenclature Eq. (8.25) may be written in vector notation

$$d\underline{x}' = \underline{\underline{F}}_t(t') \cdot d\underline{x} \tag{8.25a}$$

The dot between the tensor $\underline{\underline{F}}$ and the column vector $d\underline{x}$ indicates their scalar product. By exchanging t and t' in Eq. (8.25a) one obtains

$$d\underline{x} = \underline{\underline{F}}_{t'}(t) \cdot d\underline{x}' \tag{8.25b}$$

from which it follows that the tensors $\underline{\underline{F}}_t(t')$and $\underline{\underline{F}}_{t'}(t)$ are inverse or

$$\underline{\underline{F}}_{t'}(t) = \underline{\underline{F}}_t^{-1}(t') \tag{8.29}$$

Exchanging the basis and the current time is equivalent with a change from the relative deformation gradient to its inverse.

The relative deformation gradient may be used to define relative deformation tensors. But not all of them turn out to be useful for the formulation of rheological equations of state. Only tensors whose components transform with a rotation of the coordinate system like the components of the stress tensor are admitted. Those are the absolute Green tensor (or its inverse) and the relative Cauchy tensor (or its inverse).

To describe the rheological behavior of cross-linked rubbers (purely elastic materials) under large deformations we may use either the *absolute Green tensor*

$$\underline{\underline{G}}_{t_0}(t) = \underline{\underline{F}}_{t_0}(t) \cdot \underline{\underline{\tilde{F}}}_{t_0}(t) \tag{8.30}$$

with the components

$$G_{ik}(\underline{x}^0; t) \equiv \left(\underline{\underline{G}}_{t_0}(t)\right)_{ik} = F_{im}(\underline{x}^0; t)F_{km}(\underline{x}^0; t) = \frac{\partial x_i}{\partial x_m^0}\frac{\partial x_k}{\partial x_m^0} \tag{8.18}$$

or its inverse, the *absolute Piola tensor*

$$\underline{\underline{A}}_{t_0}(t) = \underline{\underline{\tilde{F}}}_{t_0}^{-1}(t) \cdot \underline{\underline{F}}_{t_0}^{-1}(t) = \underline{\underline{\tilde{F}}}_{t}(t_0) \cdot \underline{\underline{F}}_{t}(t_0) \tag{8.31}$$

with the components

$$A_{ik}(\underline{x}, t) \equiv \left(\underline{\underline{A}}_{t_0}(t)\right)_{ik} = F_{mi}^{-1}(\underline{x}; t)F_{mk}^{-1}(\underline{x}; t) = \frac{\partial x_m^0}{\partial x_i}\frac{\partial x_m^0}{\partial x_k} \tag{8.17}$$

The right-hand side of Eq. (8.31) was found by using (8.29). To describe the rheological behavior of viscoelastic fluids we may use either the *relative Cauchy tensor*

$$\underline{\underline{C}}_{t}(t') = \underline{\underline{\tilde{F}}}_{t}(t') \cdot \underline{\underline{F}}_{t}(t') \tag{8.32}$$

with the components

$$C_{ik}(x, t, t') \equiv \left(\underline{\underline{C}}_{t}(t')\right)_{ik} = F_{mi}(\underline{x}; t, t')F_{mk}(\underline{x}; t, t') = \frac{\partial x'_m}{\partial x_i}\frac{\partial x'_m}{\partial x_k} \tag{8.33}$$

or its inverse, the *relative Finger tensor*

$$\underline{\underline{B}}_{t}(t') = \underline{\underline{F}}_{t}^{-1}(t') \cdot \underline{\underline{\tilde{F}}}_{t}^{-1}(t') = \underline{\underline{F}}_{t'}(t) \cdot \underline{\underline{\tilde{F}}}_{t'}(t) \tag{8.34}$$

with the components

$$B_{ik}(x', t, t') \equiv \left(\underline{\underline{B}}_{t}(t')\right)_{ik} = F_{im}^{-1}(\underline{x}'; t, t')F_{km}^{-1}(\underline{x}'; t, t') = \frac{\partial x_i}{\partial x'_m}\frac{\partial x_k}{\partial x'_m} \tag{8.35}$$

Using the relative Finger tensor and the relative Cauchy tensor, it is important to realize that the range for the intermediate time t' is $t_0 \leq t' \leq t$. For t' at the boundaries of this range or outside, special situations occur. If for instance $t' \leq t_0$, no deformation has taken place before $t = t_0$ and $\underline{x}'$ is equal to $\underline{x}^0$. The relative Finger tensor

$$B_{ik}(\underline{x}, t, t') = \frac{\partial x_i}{\partial x'_m}\frac{\partial x_k}{\partial x'_m} = \frac{\partial x_i}{\partial x^0_m}\frac{\partial x_k}{\partial x^0_m} = G_{ik}(\underline{x}, t) \quad \text{for } t' \leq t_0 \tag{8.36}$$

then reduces to the absolute Green tensor and the relative Cauchy tensor

$$C_{ik}(\underline{x}, t, t') = \frac{\partial x'_m}{\partial x_i}\frac{\partial x'_m}{\partial x_k} = \frac{\partial x^0_m}{\partial x_i}\frac{\partial x^0_m}{\partial x_k} = A_{ik}(\underline{x}, t) \quad \text{for } t' \leq t_0 \tag{8.36$'$}$$

to the absolute Piola tensor.

If on the other hand, $t' = t$, $\underline{x}'$ is equal to $\underline{x}$ and the relative Finger tensor and the relative Cauchy tensor reduce to unit tensors

$$B_{ik}(\underline{x}, t, t) = \frac{\partial x_i}{\partial x'_m}\frac{\partial x_k}{\partial x'_m} = \delta_{ik} \tag{8.37}$$

$$C_{ik}(\underline{x}, t, t) = \frac{\partial x'_m}{\partial x_i}\frac{\partial x'_m}{\partial x_k} = \delta_{ik} \tag{8.37$'$}$$

Finally, the absolute Green tensor and the absolute Piola tensor reduce to unit tensors for $t \leq t_0$

$$G_{ik}(\underline{x}, t) = \frac{\partial x_i}{\partial x^0_m}\frac{\partial x_k}{\partial x^0_m} = \delta_{ik} \tag{8.38}$$

$$A_{ik}(\underline{x}, t) = \frac{\partial x^0_m}{\partial x_i}\frac{\partial x^0_m}{\partial x_k} = \delta_{ik} \tag{8.38$'$}$$

In the limit of small deformations, $x_i(t) = x^0_i + u_i(\underline{x}^0; t)$ and $x'_i(t) = x^0_i + u_i(\underline{x}^0; t')$ according to Eq. (7.9) are used and with $x'_i(t) = x_i(t) + u_i(\underline{x}^0; t') - u_i(\underline{x}^0; t)$ one gets for the relative Cauchy tensor

$$C_{ik}(x; t, t') = \frac{\partial x'_m}{\partial x_i}\frac{\partial x'_m}{\partial x_k} \cong \delta_{ik} - 2[\varepsilon_{ik}(t) - \varepsilon_{ik}(t')] + O[\varepsilon^2] \tag{8.33a}$$

and for the relative Finger tensor

$$B_{ik}(x; t, t') = \frac{\partial x_i}{\partial x'_m}\frac{\partial x_k}{\partial x'_m} \cong \delta_{ik} + 2[\varepsilon_{ik}(t) - \varepsilon_{ik}(t')] + O[\varepsilon^2] \tag{8.35a}$$

8.4 The Rate of Strain Tensor

The state of stress in a purely viscous fluid does not depend on the distances of neighbored points of the matter but only on the difference of their velocities. Consequently, we consider the velocity of the change in distance of two neighbored points. The corresponding measure is obtained by differentiating $(\mathrm{d}s)^2$ with respect to the time while keeping the Lagrange-coordinates of both material particles constant. The so-called convective time derivative of $(\mathrm{d}s)^2$ reads

$$\frac{D(\mathrm{d}s)^2}{Dt} \equiv \left(\frac{\partial(\mathrm{d}s)^2}{\partial t}\right)_{x^0} = \frac{\partial}{\partial t}\left\{\frac{\partial x_m}{\partial x_i^0}\frac{\partial x_m}{\partial x_k^0}\mathrm{d}x_i^0\mathrm{d}x_k^0\right\} = \left\{\frac{\partial}{\partial t}\left[\frac{\partial x_m}{\partial x_i^0}\frac{\partial x_m}{\partial x_k^0}\right]\right\}\mathrm{d}x_i^0\mathrm{d}x_k^0$$

Because of (8.3) we find

$$\frac{\partial}{\partial t}\left(\frac{\partial x_m}{\partial x_i^0}\right)_{x^0} = \frac{\partial}{\partial x_i^0}\left(\frac{\partial x_m}{\partial t}\right)_{x^0} = \frac{\partial \varphi_m}{\partial x_i^0}$$

and may express the convective time derivative as

$$\frac{D(\mathrm{d}s)^2}{Dt} = 2\Delta_{ik}\mathrm{d}x_i^0\mathrm{d}x_k^0 \tag{8.39}$$

with

$$\Delta_{ik} = \frac{1}{2}\left(\frac{\partial \varphi_m}{\partial x_i^0}\frac{\partial x_m}{\partial x_k^0} + \frac{\partial \varphi_m}{\partial x_k^0}\frac{\partial x_m}{\partial x_i^0}\right) \tag{8.39a}$$

Equation (8.39) represents the convective time derivative of $(\mathrm{d}s)^2$ in Lagrange-coordinates. The components Δ_{ik} are the components of a symmetric tensor, the rate of strain tensor in Lagrange-coordinates. This tensor is not in use generally, but the corresponding tensor in Euler-coordinates.

To express the convective time derivative of $(\mathrm{d}s)^2$ in Euler-coordinates we insert

$$\mathrm{d}x_i^0 = \frac{\partial x_i^0}{\partial x_r}\mathrm{d}x_r \quad \text{and} \quad \mathrm{d}x_k^0 = \frac{\partial x_k^0}{\partial x_s}\mathrm{d}x_s$$

into (8.39) and (8.39a) and by using (8.5) we obtain

$$\begin{aligned} 2\Delta_{ik}\frac{\partial x_i^0}{\partial x_r}\frac{\partial x_k^0}{\partial x_s}\mathrm{d}x_r\mathrm{d}x_s &= \left(\frac{\partial \varphi_m}{\partial x_i^0}\frac{\partial x_i^0}{\partial x_r}\frac{\partial x_m}{\partial x_k^0}\frac{\partial x_k^0}{\partial x_s} + \frac{\partial \varphi_m}{\partial x_k^0}\frac{\partial x_k^0}{\partial x_s}\frac{\partial x_m}{\partial x_i^0}\frac{\partial x_i^0}{\partial x_r}\right)\mathrm{d}x_r\mathrm{d}x_s \\ &= \left(\frac{\partial v_m}{\partial x_r}\delta_{ms} + \frac{\partial v_m}{\partial x_s}\delta_{mr}\right)\mathrm{d}x_r\mathrm{d}x_s \end{aligned}$$

The result is the well-known formula

$$\frac{D(\mathrm{d}s)^2}{Dt} = 2d_{ik}\mathrm{d}x_i\mathrm{d}x_k \tag{8.40}$$

with

$$\boxed{d_{ik} = \frac{1}{2}\left(\frac{\partial v_i}{\partial x_k} + \frac{\partial v_k}{\partial x_i}\right)} \tag{8.41}$$

The d_{ik} form the components of a symmetric tensor $\underline{\underline{D}}$ the *rate of strain tensor* in *Euler-coordinates*. The rate of strain tensor is used for example to describe the flow behavior of Newtonian liquids by means of a linear relation between the stress tensor $\underline{\underline{S}}$ and the rate of strain tensor $\underline{\underline{D}}$. This theory forms the basis for classical hydrodynamics. Attempts to describe the flow behavior of polymer melts by means of a non-linear relation between the tensors $\underline{\underline{S}}$ and $\underline{\underline{D}}$ failed, as those theories cannot explain the elastic behavior of polymer melts and yield a wrong normal stress behavior, viz., a vanishing first normal stress coefficient [4, 5].

The rate of strain tensor in cylinder coordinates reads [6]

$$\begin{aligned}
d_{rr} &= \frac{\partial v_r}{\partial r} & 2d_{r\varphi} &= \frac{\partial v_\varphi}{\partial r} - \frac{v_\varphi}{r} + \frac{1}{r}\frac{\partial v_r}{\partial \varphi} & 2d_{rz} &= \frac{\partial v_r}{\partial z} + \frac{\partial v_z}{\partial r} \\
d_{\varphi\varphi} &= \frac{1}{r}\frac{\partial v_\varphi}{\partial \varphi} + \frac{v_r}{r} & 2d_{\varphi z} &= \frac{\partial v_\varphi}{\partial z} + \frac{1}{r}\frac{\partial v_z}{\partial \varphi} & d_{zz} &= \frac{\partial v_z}{\partial z}
\end{aligned} \tag{8.41a}$$

and in polar coordinates

$$\begin{aligned}
d_{rr} &= \frac{\partial v_r}{\partial r} \quad 2d_{r\vartheta} = \frac{\partial v_\vartheta}{\partial r} + \frac{1}{r}\frac{\partial v_r}{\partial \vartheta} - \frac{v_\vartheta}{r} \quad 2d_{r\varphi} = \frac{\partial v_\varphi}{\partial r} + \frac{1}{r\sin\vartheta}\frac{\partial v_r}{\partial \varphi} - \frac{v_\varphi}{r} \\
d_{\vartheta\vartheta} &= \frac{1}{r}\frac{\partial v_\vartheta}{\partial \vartheta} + \frac{v_r}{r} \qquad 2d_{\vartheta\varphi} = \frac{1}{r}\frac{\partial v_\varphi}{\partial \vartheta} + \frac{1}{r\sin\vartheta}\frac{\partial v_\vartheta}{\partial \varphi} - \frac{\cot\vartheta}{r}v_\varphi \\
d_{\varphi\varphi} &= \frac{1}{r\sin\vartheta}\frac{\partial v_\varphi}{\partial \varphi} + \frac{v_r}{r} + \frac{\cot\vartheta}{r}v_\vartheta
\end{aligned} \tag{8.41b}$$

8.5 Dynamics of Deformable Bodies

The equations to be discussed in the following are valid for all deformable materials independent of their consistency, viz., for fluids, gases, solids, polymeric substances in the non-molten and molten state. These equations are the laws of conservation of the mass, the momentum, the angular momentum and the energy. An extensive discussion and derivation may be found in [7]. Here, we restrict ourselves here to a summarizing presentation.

A substance is discussed which contains only one single component. Then the law of *conservation of the mass* reads if expressed by the density $\rho(\underline{x},t)$

$$\frac{\partial \rho}{\partial t} + \mathrm{div}(\rho \underline{v}) = \frac{\partial \rho}{\partial t} + \rho \mathrm{div}(\underline{v}) + \underline{v} \cdot \mathrm{grad}(\rho) = 0 \tag{8.42}$$

This equation is called the *equation of continuity*. The vector operations refer to Euler-coordinates and are valid also for curvilinear coordinate systems.

Mostly we deal with an *incompressible material*, for which the density does not change neither with time nor in space and therefore $\partial \rho / \partial t = 0$ and $\mathrm{grad}(\rho) = 0$. From (8.42) it may be seen that the divergence of the velocity field must then vanish

$$\mathrm{div}(\underline{v}) = 0 \tag{8.43}$$

This equation reads in Cartesian coordinates

$$\frac{\partial v_i}{\partial x_i} = 0 \tag{8.43a}$$

in cylinder coordinates

$$\frac{1}{r}\frac{\partial (r v_r)}{\partial r} + \frac{1}{r}\frac{\partial v_\varphi}{\partial \varphi} + \frac{\partial v_z}{\partial z} = 0 \tag{8.43b}$$

and in polar coordinates

$$\frac{1}{r^2}\frac{\partial (r^2 v_r)}{\partial r} + \frac{1}{r \sin\vartheta}\frac{\partial (v_\vartheta \sin\vartheta)}{\partial \vartheta} + \frac{1}{r \sin\vartheta}\frac{\partial (v_\varphi)}{\partial \varphi} = 0 \tag{8.43c}$$

The equations of motion were already discussed in Sect. 7.1 in tensor notation as

$$\frac{\partial \sigma_{ik}}{\partial x_k} + F_i = \rho a_i \tag{7.2}$$

and may be written in vector notation as

$$\underline{\nabla} \cdot \underline{\underline{S}} + \underline{F} = \rho \underline{a} \tag{8.44}$$

with $\underline{\nabla}$ being the *Nabla-operator*, which is a row vector with the components $\nabla_1 = \partial / \partial x_1$, $\nabla_2 = \partial / \partial x_2$, $\nabla_3 = \partial / \partial x_3$, $\underline{\underline{S}}$ the stress tensor, $\underline{F}$ the vector field of the volume forces and $\underline{a}$ the vector field of the local acceleration (cf. Eq. (8.7)).

The condition for the *conservation of the angular momentum* represented in Eq. (7.4) by the symmetry of the components of the stress tensor may be written as

$$\tilde{\underline{\underline{S}}} = \underline{\underline{S}} \tag{8.45}$$

Finally, the law of *conservation of the internal energy* may be written in the following form (cf. [7])

$$\rho c_v \frac{\mathrm{D}T}{\mathrm{D}t} - \underline{\nabla} \cdot (\lambda \underline{\nabla} T) = \underline{\underline{S}} : \underline{\underline{D}} \tag{8.46}$$

with c_v being the *specific heat at constant volume* of the material, $T = T(x,t)$ the *temperature field in Euler-coordinates*, DT/Dt its convective time derivative defined in Eq. (8.9), λ the heat conductivity and $\underline{\underline{S}} : \underline{\underline{D}}$ the double-scalar product of the stress tensor and the rate of strain tensor

$$\underline{\underline{S}} : \underline{\underline{D}} = \sigma_{ik} d_{ki} \tag{8.47}$$

Equation (8.46) is used for the calculation of the temperature distribution within a fluid due to heating by internal friction.

The rheological equation of state (constitutive equation) The equations discussed so far are not sufficient to completely describe the deformation and flow of viscoelastic materials. This statement becomes clear by comparing the number of unknown field functions with the number of independent equations available for calculations of:

the components of the stress tensor $\sigma_{ik}(\underline{x},t)$	9
the Lagrange-coordinates $g_i(\underline{x},t)$ or	
the velocity field $v_i(\underline{x},t)$	3
the temperature field $T(\underline{x},t)$	1
the density $\rho(\underline{x},t)$	1

For the determination of these 14 unknown field functions the following eight equations are available:

the equations of motion (8.44)	3
the conservation of the angular momentum (8.45)	3
the equation of continuity (8.42)	1
the conservation of energy (8.46)	1

The missing 6 equations for a complete description of deformation and flow are relationships between the symmetric stress tensor and the rate of strain tensor, the absolute Piola tensor (Eq. (8.17)), the relative Cauchy tensor (Eqs. (8.33)), or another appropriate measure of the deformation history. These relationships may have the form of differential or integral equations. They are symbolically designated as

$$\boxed{\sigma_{ik}(\underline{x},t) \Leftrightarrow d_{ik}(\underline{x},t) \quad \text{or} \quad A_{ik}(\underline{x},t) \quad \text{or} \quad C_{ik}(\underline{x};t,t')} \tag{8.48}$$

While the laws of conservation are the same for all materials, the relations (8.48) characterize their mechanical properties. Therefore, they are called the *rheological equations of state* or the *constitutive equations*. By those the differences between elastic, viscous, and viscoelastic materials show up.

It is not always necessary to use all the 14 equations for the solution of rheological problems. A special case which often occurs is the incompressibility of the material considered which make the changes in volume negligible in comparison to the changes in shape. In this case, the density is independent of time and position and the equation of continuity degenerates to (8.43) restricting the flow patterns to those of an incompressible medium.

A further simplification may originate from the assumption of isothermal conditions of the problem, if the temperature rise due to a local energy dissipation may be neglected. Then the temperature is constant everywhere and at every time and the equation for the conservation of energy (8.46) is fulfilled approximately in the sense that the dissipation term on the right-hand side of Eq. (8.46) is too small to disturb the temperature equilibrium.

8.6 Eigenvalues and Invariants of the Stress Tensor

Prior to the discussion of the various rheological equations of state, the transformation of symmetric tensors to their system of principal axis should be treated. This is done using the stress tensor as an example. Let $\underline{\underline{S}}$ be the stress tensor at the location $\underline{x}$ and σ_{ik} its Cartesian components. The stress vector $\underline{S}$ acting on an arbitrary plane with the normal vector $\underline{n}$ at the same position is given by Eq. (7.1) or in vector notation by $\underline{S} = \underline{\underline{S}} \cdot \underline{n}$. The question arises whether there exist planes at the position $\underline{x}$ at which only normal stresses, but no shear stresses act, or whether there exist planes for which the acting stresses are parallel to their normal vectors. Two vectors $\underline{S}$ and $\underline{n}$ are parallel if $\underline{S} = \lambda \cdot \underline{n}$ with λ being a scalar. By inserting $S_i = \lambda\, n_i$ in Eq. (7.1) one gets

$$\begin{aligned}(\sigma_{11} - \lambda)n_1 + \sigma_{12}n_2 + \sigma_{13}n_3 &= 0\\ \sigma_{21}n_1 + (\sigma_{22} - \lambda)n_2 + \sigma_{23}n_3 &= 0\\ \sigma_{31}n_1 + \sigma_{32}n_2 + (\sigma_{33} - \lambda)n_3 &= 0\end{aligned} \tag{8.49}$$

This is a system of three linear homogeneous equations for the three unknown components of the normal vector of the plane we look for. The coefficients σ_{ik} are given and the parameter λ may be chosen freely. The system of equations has a non-trivial solution then and only then, when the determinant of the coefficients vanishes:

$$\mathrm{Det}\left(\underline{\underline{S}} - \lambda\underline{\underline{E}}\right) = \begin{vmatrix} \sigma_{11} - \lambda & \sigma_{12} & \sigma_{13} \\ \sigma_{21} & \sigma_{22} - \lambda & \sigma_{23} \\ \sigma_{31} & \sigma_{32} & \sigma_{33} - \lambda \end{vmatrix} = 0 \tag{8.50}$$

This is an equation of the third degree for λ. Developing the determinant after powers of λ leads to the so-called *characteristic equation* of $\underline{\underline{S}}$.

$$-\lambda^3 + \mathrm{I}_\sigma\lambda^2 - \mathrm{II}_\sigma\lambda + \mathrm{III}_\sigma = 0 \tag{8.51}$$

The coefficients are abbreviations for the expressions:

$$\begin{aligned} \mathrm{I}_\sigma &= \sigma_{11} + \sigma_{22} + \sigma_{33} \\ \mathrm{II}_\sigma &= \sigma_{11}\sigma_{22} - \sigma_{12}\sigma_{21} + \sigma_{22}\sigma_{33} - \sigma_{23}\sigma_{32} + \sigma_{33}\sigma_{11} - \sigma_{31}\sigma_{13} \\ \mathrm{III}_\sigma &= \det\underline{\underline{S}} = \sigma_{11}\sigma_{22}\sigma_{33} + \sigma_{12}\sigma_{23}\sigma_{31} + \sigma_{13}\sigma_{32}\sigma_{21} \\ &\quad - \sigma_{11}\sigma_{23}\sigma_{32} - \sigma_{22}\sigma_{13}\sigma_{31} - \sigma_{33}\sigma_{12}\sigma_{21} \end{aligned} \tag{8.52}$$

and are called the *invariants of the stress tensor*. If the components of the stress tensor are known, the invariants may be calculated and the characteristic equation yields three roots for λ. If the tensor $\underline{\underline{S}}$ is symmetric and real, all three roots are real. We designate these roots with σ_1, σ_2 and σ_3. They are called *principal stresses* or *eigen-values of the stress tensor*. For each of these three roots there exists a non-trivial solution of Eq. (8.49):

Root	Normal vector (principal direction)	System of equations	
$\lambda = \sigma_1$	$\underline{n} = \underline{n}^{(1)} = (n_1^{(1)}, n_2^{(1)}, n_3^{(1)})$	$\sigma_{ki}n_k^{(1)} = \sigma_1 n_i^{(1)}$ (i = 1 to 3)	(8.53)
$\lambda = \sigma_2$	$\underline{n} = \underline{n}^{(2)} = (n_1^{(2)}, n_2^{(2)}, n_3^{(2)})$	$\sigma_{ki}n_k^{(2)} = \sigma_2 n_i^{(2)}$ (i = 1 to 3)	
$\lambda = \sigma_3$	$\underline{n} = \underline{n}^{(3)} = (n_1^{(3)}, n_2^{(3)}, n_3^{(3)})$	$\sigma_{ki}n_k^{(3)} = \sigma_3 n_i^{(3)}$ (i = 1 to 3)	

The three unit vectors $\underline{n}^{(1)}, \underline{n}^{(2)}, \underline{n}^{(3)}$ are called the *principal directions of the stress tensor*. If the three principle stresses differ from each other, the principal directions are perpendicular to each other and form the basis of a rectangular coordinate system, called the principal coordinate system as it is easily shown that from $\sigma_1 \neq \sigma_2$ follows $\underline{n}^{(1)} \perp \underline{n}^{(2)}$: Multiplying the first of Eqs. (8.53) with $n_i^{(2)}$, the second with $n_i^{(1)}$, forming the sum over all i, subtracting the two equations, and taking the symmetry of the stress tensor into account results in the equation

$$(\sigma_1 - \sigma_2) \cdot n_i^{(1)} n_i^{(2)} = 0 \text{ or because of } \sigma_1 \neq \sigma_2 \quad n_i^{(1)} n_i^{(2)} = 0 \quad \text{q.e.d.}$$

In the principal coordinate system the components of the stress tensor are

$$\underline{\underline{S}} = \begin{pmatrix} \sigma_1 & 0 & 0 \\ 0 & \sigma_2 & 0 \\ 0 & 0 & \sigma_3 \end{pmatrix} \tag{8.54}$$

i.e. its matrix is diagonal.

If $\sigma_1 = \sigma_2 \neq \sigma_3$ then each direction perpendicular to $\underline{n}^{(3)}$ is a principal direction and each coordinate system with one axis parallel to $\underline{n}^{(3)}$ will be a principal coordinate system. If $\sigma_1 = \sigma_2 = \sigma_3$ each direction will be a principal direction and each rectangular coordinate system will be a principal one.

The designation invariants for the three expressions (8.52) points to their independence of the choice of the coordinate system as shown in the following. Consider besides the original coordinate system with the basis vectors $\underline{e}_1$, $\underline{e}_2$, $\underline{e}_3$, a rotated coordinate system with the basis vectors $\underline{e}'_1, \underline{e}'_2, \underline{e}'_3$. If the components of a vector $\underline{x}$ in the original rectangular coordinate system are x_1, x_2, x_3, and those in the new coordinate system are x'_1, x'_2, x'_3, then the transformation

$$x'_i = a_{ij}x_j \tag{8.55}$$

relates the new components to the old ones. $a_{ij} = \left(\underline{e}'_j \cdot \underline{e}'_i\right)$ is the cosine of the angle between the new i-direction and the old j-direction. For the components of a tensor we have the corresponding transformation rule

$$\sigma'_{ik} = a_{ij}a_{kl}\sigma_{jl} \tag{8.56}$$

If the matrix of the components of the tensor $\underline{\underline{S}}$ is designated with S and the matrix of the coefficients a_{ik} with A then (8.56) may be written as the matrix equation

$$S' = A \cdot S \cdot \tilde{A} \tag{8.57}$$

whereby $\tilde{A}$ is the transpose of the matrix A. As A is a rotational matrix, it has to be orthogonal, obeying the conditions

$$A \cdot \tilde{A} = E \quad \text{and} \quad \tilde{A} \cdot A = E \tag{8.58}$$

or

$$a_{ik}a_{lk} = \delta_{il} \text{ and } a_{ik}a_{il} = \delta_{kl}$$

Though each component of the stress tensor transforms under a rotation due to Eq. (8.56), the invariants (8.52) remain unchanged, viz.,

$$\mathrm{I}'_\sigma = \mathrm{I}_\sigma \qquad \mathrm{II}'_\sigma = \mathrm{II}_\sigma \qquad \mathrm{III}'_\sigma = \mathrm{III}_\sigma \tag{8.59}$$

These equations may be proven by inserting (8.56) into the definitions (8.52) and making use of the orthogonal properties of the matrix A. An easier and more elegant proof makes use of the characteristic equation. As this equation determines the three principal stresses which are independent of the orientation of the coordinate system, this should be true for the characteristic equation and its coefficients, which are just the invariants.

As the characteristic equation has the three roots σ_1, σ_2, σ_3 it can be written in the form

$$-\lambda^3 + \mathrm{I}_\sigma\lambda^2 - \mathrm{II}_\sigma\lambda + \mathrm{III}_\sigma = -(\lambda - \sigma_1)(\lambda - \sigma_2)(\lambda - \sigma_3) \tag{8.60}$$

Comparing the coefficients of the powers of λ yields

$$\begin{aligned} \mathrm{I}_\sigma &= \sigma_1 + \sigma_2 + \sigma_3 \\ \mathrm{II}_\sigma &= \sigma_1\sigma_2 + \sigma_2\sigma_3 + \sigma_3\sigma_1 \\ \mathrm{III}_\sigma &= \sigma_1\sigma_2\sigma_3 \end{aligned} \tag{8.61}$$

which again proves the invariance of I_σ, II_σ, III_σ.

Finally, from the characteristic Eq. (8.51), we deduce an equation for the tensor $\underline{\underline{S}}$ and its matrix S. This equation is called the *Caley-Hamilton equation* and is of great importance for the formulation of rheological equations of state.

$$\boxed{-\underline{\underline{S}}^3 + \mathrm{I}_\sigma\underline{\underline{S}}^2 - \mathrm{II}_\sigma\underline{\underline{S}} + \mathrm{III}_\sigma\underline{\underline{E}} = \underline{\underline{0}}} \tag{8.62}$$

The tensor $\underline{\underline{S}}$ and its matrix S obey their own characteristic equation, in which now $\underline{\underline{E}}$ is the unit tensor and $\underline{\underline{O}}$ the zero tensor. The proof is simple. If this equation is valid in its principal coordinate system, it is also valid in each arbitrary coordinate system. $\underline{\underline{S}}, \underline{\underline{S}}^2, \underline{\underline{S}}^3$ have the same principal coordinate system. In this system, (8.62) is equivalent to the following equation for the corresponding matrices

$$-\begin{pmatrix} \sigma_1^3 & 0 & 0 \\ 0 & \sigma_2^3 & 0 \\ 0 & 0 & \sigma_3^3 \end{pmatrix} + \mathrm{I}_\sigma \begin{pmatrix} \sigma_1^2 & 0 & 0 \\ 0 & \sigma_2^2 & 0 \\ 0 & 0 & \sigma_3^2 \end{pmatrix} - \mathrm{II}_\sigma \begin{pmatrix} \sigma_1 & 0 & 0 \\ 0 & \sigma_2 & 0 \\ 0 & 0 & \sigma_3 \end{pmatrix} + \mathrm{III}_\sigma \begin{pmatrix} 1 & 0 & 0 \\ 0 & 1 & 0 \\ 0 & 0 & 1 \end{pmatrix}$$

$$= \begin{pmatrix} 0 & 0 & 0 \\ 0 & 0 & 0 \\ 0 & 0 & 0 \end{pmatrix}$$

For the non-diagonal elements this equation is simply $0 = 0$, for the diagonal elements it reads

$$-\sigma_i^3 + \mathrm{I}_\sigma \sigma_i^2 - \mathrm{II}_\sigma \sigma_i + \mathrm{III}_\sigma = 0 \quad \text{for} \quad i = 1, 2, 3$$

This equation is fulfilled for all i as the σ_i are the solutions of the characteristic equation.

Using the equation of Caley–Hamilton it is possible to express each integer (positive or negative) power of the stress tensor by a linear combination of the tensors $\underline{\underline{E}}$, $\underline{\underline{S}}$ and $\underline{\underline{S}}^2$. For instance

$$\underline{\underline{S}}^3 = \mathrm{III}_\sigma \underline{\underline{E}} - \mathrm{II}_\sigma \underline{\underline{S}} + \mathrm{I}_\sigma \underline{\underline{S}}^2$$
$$\underline{\underline{S}}^4 = \mathrm{III}_\sigma \underline{\underline{S}} - \mathrm{II}_\sigma \underline{\underline{S}}^2 + \mathrm{I}_\sigma \underline{\underline{S}}^3 = \mathrm{I}_\sigma \mathrm{III}_\sigma \underline{\underline{E}} + (\mathrm{III}_\sigma - \mathrm{I}_\sigma \mathrm{II}_\sigma)\underline{\underline{S}} + (\mathrm{I}_\sigma^2 - \mathrm{II}_\sigma)\underline{\underline{S}}^2$$
$$\underline{\underline{S}}^{-1} = \frac{\mathrm{II}_\sigma}{\mathrm{III}_\sigma}\underline{\underline{E}} - \frac{\mathrm{I}_\sigma}{\mathrm{III}_\sigma}\underline{\underline{S}} + \frac{1}{\mathrm{III}_\sigma}\underline{\underline{S}}^2$$

Generally, for each integer power of $\underline{\underline{S}}$ one gets

$$\underline{\underline{S}}^k = F_0(\mathrm{I}_\sigma, \mathrm{II}_\sigma, \mathrm{III}_\sigma)\underline{\underline{E}} + F_1(\mathrm{I}_\sigma, \mathrm{II}_\sigma, \mathrm{III}_\sigma)\underline{\underline{S}} + F_2(\mathrm{I}_\sigma, \mathrm{II}_\sigma, \mathrm{III}_\sigma)\underline{\underline{S}}^2 \tag{8.63}$$

whereby F_0, F_1, and F_2 are scalar functions of the invariants of the tensor.

As an example we consider the transformation to principal axes for a state of stress whose one direction is already a principal one.

Such a state of stress may be given by

$$\underline{\underline{S}} = \begin{pmatrix} \sigma_{11} & \sigma_{12} & 0 \\ \sigma_{21} & \sigma_{22} & 0 \\ 0 & 0 & \sigma_{33} \end{pmatrix} \tag{8.64}$$

with $\underline{e}_3$ being a principal direction with the eigenvalue $\sigma_3 = \sigma_{33}$. The other two principal directions are situated in the $\underline{e}_1$-$\underline{e}_2$ plane. Let $\underline{e}_I$ and $\underline{e}_{II}$ be the principal directions in this plane (compare Fig. 8.6). A point P with the coordinates x_1 and x_2 in the original coordinate system ($\underline{e}_1$, $\underline{e}_2$) has the coordinates x'_1 and x'_2 in the principal coordinate system spanned by the unit vectors $\underline{e}_{\mathrm{I}}$ and $\underline{e}_{\mathrm{II}}$. The two coordinate systems are rotated by an angle χ.

The transformation of the coordinates may be read directly from Fig. 8.6 as

$$\begin{aligned} x_1 &= x'_1 \cos\chi - x'_2 \sin\chi \\ x_2 &= x'_1 \sin\chi + x'_2 \cos\chi \\ x_3 &= x'_3 \end{aligned} \tag{8.65}$$

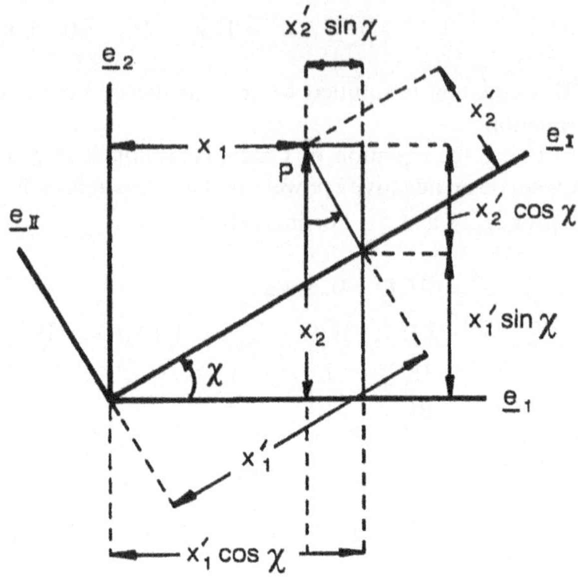

Fig. 8.6 Transformation to principal axes in a state of stress characterized by Eq. (8.64)

This transformation is effected by the following rotational matrix

$$x_i = b_{ik}x'_k \quad \text{with} \quad (b_{ik}) = \begin{pmatrix} \cos\chi & -\sin\chi & 0 \\ \sin\chi & \cos\chi & 0 \\ 0 & 0 & 1 \end{pmatrix}$$

Applying the same matrix to the transformation of the components of the stress tensor yields, because of $\sigma_{ik} = b_{il}b_{km}\sigma'_{lm}$ and $\sigma'_{11} = \sigma_1$, $\sigma'_{22} = \sigma_2$, $\sigma'_{12} = \sigma'_{21} = 0$, the equations

$$\begin{aligned} \sigma_{11} &= \sigma_1\cos^2\chi + \sigma_2\sin^2\chi \\ \sigma_{22} &= \sigma_1\sin^2\chi + \sigma_2\cos^2\chi \\ \sigma_{12} = \sigma_{21} &= (\sigma_1 - \sigma_2)\sin\chi\cos\chi \end{aligned} \tag{8.66}$$

The inversion of these equations is obtained as follows

$$\begin{aligned} &\sigma_{11} - \sigma_{22} = (\sigma_1 - \sigma_2)\cos 2\chi \\ &2\sigma_{12} = (\sigma_1 - \sigma_2)\sin 2\chi \\ &\sigma_{11} + \sigma_{22} = \sigma_1 + \sigma_2 \end{aligned} \tag{8.67}$$

Summing up the squares of the first and second equations of (8.67) yields $(\sigma_1 - \sigma_2)^2$ and dividing the second equation by the first an expression for tan 2χ:

$$\begin{aligned} \sigma_1 - \sigma_2 &= \sqrt{(\sigma_{11} - \sigma_{22})^2 + 4\sigma_{12}^2} \\ \tan 2\chi &= 2\sigma_{12}/(\sigma_{11} - \sigma_{22}) \\ \sigma_1 + \sigma_2 &= \sigma_{11} + \sigma_{22} \end{aligned} \tag{8.68}$$

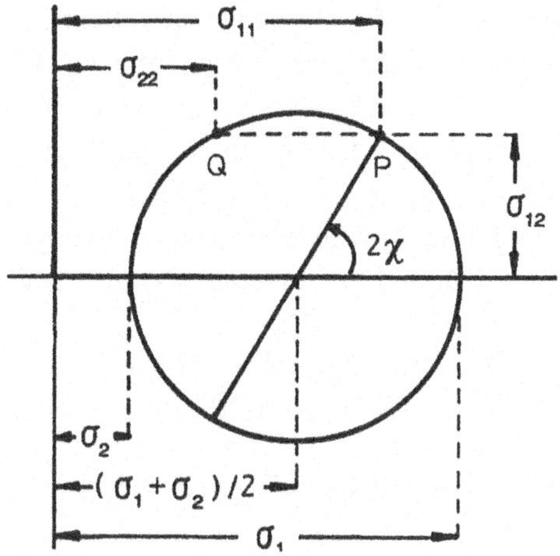

Fig. 8.7 Mohr's stress circle [8]

These equations solve the problem of the transformation of the stress tensor to principal axis. The second equation of (8.68) yields the position of the principal coordinate system (the angle χ). From the first and third equations the two principal stresses can be obtained.

Equations (8.68) permit the following interpretation as shown in Fig. 8.7 (*Mohr's stress circle*). A circle of the radius $(\sigma_1 - \sigma_2)/2$ with its center on the abscissa at the position $(\sigma_1 + \sigma_2)/2$ describes the positions of all possible combinations of normal stresses and shear stresses in coordinate systems which are rotated by an angle χ against the principal coordinate system.

A diagonal of the circle which includes the angle 2χ with the abscissa intersects the circle at the point P. The coordinates of this point are the normal stress σ_{11} and the shear stress σ_{12}. A line parallel to the abscissa through P intersects the circle at the point Q. The coordinate of Q yields the second normal stress σ_{22}. In the drawing, σ_1 was assumed to be the larger of the two principal stresses. The figure can be used to solve the problem graphically. A circle with its center on the abscissa is drawn through the points P and Q. This circle intersects the abscissa at the points σ_1 and σ_2.

8.7 Transformation of the Strain Tensors and the Rate of Strain Tensor to Principal Axis

The derivations given in Sect. 8.6 with respect to the stress tensor can be applied, of course, to each second order tensor with real and symmetric components. The finite strain tensors $\underline{\underline{C}}$, $\underline{\underline{A}}$, $\underline{\underline{G}}$,and $\underline{\underline{B}}$ are real and symmetric and, therefore, admit a transformation to principal axis. Being inverse to each other, the tensors $\underline{\underline{A}}$ and $\underline{\underline{G}}$ have the same principal coordinate system as have the tensors $\underline{\underline{B}}$ and $\underline{\underline{C}}$. The principal values of the tensors $\underline{\underline{G}}$ and $\underline{\underline{C}}$ and those of the tensors $\underline{\underline{A}}$ and $\underline{\underline{B}}$ are the same. We call the eigenvalues of $\underline{\underline{A}}$ and $\underline{\underline{B}}$ b_1, b_2, b_3, and their invariants I_B, II_B, III_B, the eigenvalues of $\underline{\underline{C}}$ and $\underline{\underline{G}}$ c_1, c_2, c_3 and their invariants I_C, II_C, III_C. Then the following equations result directly from Eqs. (8.51), (8.52), and (8.62) of Sect. 8.6, and the relations $\underline{\underline{G}} = \underline{\underline{A}}^{-1}$ and $\underline{\underline{B}} = \underline{\underline{C}}^{-1}$ [3]:

$$\begin{aligned}
&\mathrm{I}_B = \mathrm{I}_A = B_{11} + B_{22} + B_{33} = A_{11} + A_{22} + A_{33} = b_1 + b_2 + b_3\\
&\mathrm{II}_B = \mathrm{II}_A = (B_{11}B_{22} - B_{12}B_{21}) + \cdots = (A_{11}A_{22} - A_{12}A_{21}) + \cdots = b_1b_2 + b_2b_3 + b_3b_1\\
&\mathrm{III}_B = \mathrm{III}_A = \det \underline{\underline{B}} = \det \underline{\underline{A}} = b_1b_2b_3 = \left(\frac{\mathrm{d}V_0}{\mathrm{d}V}\right)^2
\end{aligned} \tag{8.69}$$

and

$$\begin{aligned}
&\mathrm{I}_C = \mathrm{I}_G = C_{11} + C_{22} + C_{33} = G_{11} + G_{22} + G_{33} = c_1 + c_2 + c_3\\
&\mathrm{II}_C = \mathrm{II}_G = (C_{11}C_{22} - C_{12}C_{21}) + \cdots = (G_{11}G_{22} - G_{12}G_{21}) + \cdots = c_1c_2 + c_2c_3 + c_3c_1\\
&\mathrm{III}_C = \mathrm{III}_G = \det \underline{\underline{C}} = \det \underline{\underline{G}} = c_1c_2c_3 = \left(\frac{\mathrm{d}V}{\mathrm{d}V_0}\right)^2
\end{aligned} \tag{8.70}$$

For the significance of the eigenvalues c_1, c_2, c_3 we refer to Eq. (8.14) which in the system of principal axes reads

$$(\mathrm{d}s)^2 = c_1(\mathrm{d}x_1^0)^2 + c_2(\mathrm{d}x_2^0)^2 + c_3(\mathrm{d}x_3^0)^2$$

A line element originally parallel to the x_1^0-direction ($\mathrm{d}x_2^0 = 0$, $\mathrm{d}x_3^0 = 0$), changes to a line element of the length $\mathrm{d}s$ according to

$$(\mathrm{d}s)^2 = c_1(\mathrm{d}x_1^0)^2 = \lambda_1^2(\mathrm{d}x_1^0)^2$$

Line elements which are oriented in the principal directions of the strain tensor $\underline{\underline{C}}$ are extended to the λ_1, λ_2, λ_3–fold of their original length, whereby

$$\lambda_1 = \sqrt{c_1}, \quad \lambda_2 = \sqrt{c_2}, \quad \lambda_3 = \sqrt{c_3}$$

Therefore, λ_1, λ_2, λ_3 are called the *principal extension ratios*. The basic vectors of the principle axes are transformed by the deformation into perpendicular line elements. A cuboid with the edges $\mathrm{d}x_1^0$, $\mathrm{d}x_2^0$, $\mathrm{d}x_3^0$ parallel to the principal axis is

[3] A plus-sign followed by two dots in a formula means its completion by cyclic substitution.

transformed into a cuboid with the edges $\lambda_1 dx_1^0$, $\lambda_2 dx_2^0$, $\lambda_3 dx_3^0$. Because of $\underline{\underline{C}} = \underline{\underline{B}}^{-1}$ a corresponding relation between the eigenvalues of $\underline{\underline{B}}$ and λ follows

$$c_1 = g_1 = \frac{1}{b_1} = \lambda_1^2, \quad c_2 = g_2 = \frac{1}{b_2} = \lambda_2^2, \quad c_3 = g_3 = \frac{1}{b_3} = \lambda_3^2 \tag{8.71}$$

Inserting $c_i = 1/b_i$ and $a_i = 1/g_i$ into Eq. (8.69) and (8.70) leads to the following relationships between the invariants

$$\begin{aligned} \mathrm{I}_C &= \mathrm{II}_B/\mathrm{III}_B \quad & \mathrm{II}_C &= \mathrm{I}_B/\mathrm{III}_B \quad & \mathrm{III}_C &= 1/\mathrm{III}_B \\ \mathrm{I}_A &= \mathrm{II}_G/\mathrm{III}_G \quad & \mathrm{II}_A &= \mathrm{I}_G/\mathrm{III}_G \quad & \mathrm{III}_A &= 1/\mathrm{III}_G \end{aligned} \tag{8.72}$$

Using the equations of Caley-Hamilton and $\underline{\underline{G}} = \underline{\underline{A}}^{-1}$ and $\underline{\underline{B}} = \underline{\underline{C}}^{-1}$, one obtains:

$$\begin{aligned} \underline{\underline{C}} &= \mathrm{I}_C\underline{\underline{E}} - \mathrm{II}_C\underline{\underline{B}} + \mathrm{III}_C\underline{\underline{B}}^2 \qquad & \underline{\underline{A}} &= \mathrm{I}_A\underline{\underline{E}} - \mathrm{II}_A\underline{\underline{G}} + \mathrm{III}_A\underline{\underline{G}}^2 \\ \underline{\underline{G}} &= \mathrm{I}_G\underline{\underline{E}} - \mathrm{II}_G\underline{\underline{A}} + \mathrm{III}_G\underline{\underline{A}}^2 \qquad & \underline{\underline{B}} &= \mathrm{I}_B\underline{\underline{E}} - \mathrm{II}_B\underline{\underline{C}} + \mathrm{III}_B\underline{\underline{C}}^2 \end{aligned} \tag{8.73}$$

Similar considerations can be applied to the rate of strain tensor $\underline{\underline{D}}$. Notice, however, that the principal axes of the rate of strain tensor do not coincide with the principal axes of the strain tensors. The tensors $\underline{\underline{D}}$ and $\underline{\underline{G}}$ are in diagonal form in the same coordinate system only in special cases. Let d_1, d_2, d_3 be the eigenvalues of $\underline{\underline{D}}$ and I_D, II_D, III_D its invariants. Then the relations between the invariants and the eigenvalues are

$$\begin{aligned} \mathrm{I}_D &= d_{11} + d_{22} + d_{33} = d_1 + d_2 + d_3 = \mathrm{div}(\underline{v}) \\ \mathrm{II}_D &= (d_{11}d_{22} - d_{12}d_{21}) + \cdots = d_1d_2 + d_2d_3 + d_3d_1 \\ \mathrm{III}_D &= \det \underline{\underline{D}} = d_1d_2d_3 \end{aligned} \tag{8.74}$$

As an example we discuss the transformation of the tensors $\underline{\underline{C}}$ and $\underline{\underline{D}}$ to principal axes in *planar deformation*. In planar deformation all material particles move in the 1–2 plane, i.e., their x_3-coordinate remains constant. The path equations then are

$$x_1 = f_1(x_1^0, x_2^0; t) \quad x_2 = f_2(x_1^0, x_2^0; t) \quad x_3 = x_3^0 \tag{8.75}$$

and the velocity field in Euler-coordinates reads

$$v_1 = v_1(x_1, x_2; t) \quad v_2 = v_2(x_1, x_2; t) \quad v_3 = 0 \tag{8.76}$$

The matrices of the tensors $\underline{\underline{C}}$ and $\underline{\underline{D}}$

$$C_{ik} = \begin{pmatrix} C_{11} & C_{12} & 0 \\ C_{21} & C_{22} & 0 \\ 0 & 0 & 1 \end{pmatrix} \tag{8.77}$$

and

$$d_{ik} = \begin{pmatrix} d_{11} & d_{12} & 0 \\ d_{21} & d_{22} & 0 \\ 0 & 0 & 0 \end{pmatrix} \tag{8.78}$$

are of similar structure as those of the stress tensor in Eq. (8.64). The 3-direction is a principal direction with eigenvalues $c_3 = 1$ and $d_3 = 0$, respectively. The transformation to principle axes proceeds in the same manner as for the stress tensor (8.64). For the eigenvalues of $\underline{\underline{C}}$ we find the equations

$$\begin{aligned} C_{11} - C_{22} &= (c_1 - c_2)\cos 2\chi \\ 2C_{12} &= (c_1 - c_2)\sin 2\chi \\ C_{11} + C_{22} &= c_1 + c_2 \end{aligned} \tag{8.79}$$

$$\begin{aligned} c_1 - c_2 &= \sqrt{(C_{11} - C_{22})^2 + 4C_{12}^2} \\ c_1 + c_2 &= C_{11} + C_{22} \\ \tan 2\chi &= 2C_{12}/(C_{11} - C_{22}) \end{aligned} \tag{8.80}$$

and for those of $\underline{\underline{D}}$

$$\begin{aligned} d_1 - d_2 &= \sqrt{(d_{11} - d_{22})^2 + 4d_{12}^2} \\ d_1 + d_2 &= d_{11} + d_{22} \\ \tan 2\chi &= 2d_{12}/(d_{11} - d_{22}) \end{aligned} \tag{8.81}$$

8.7.1 Time-Dependent Simple Shear

Time-dependent simple shear is a special planar deformation for which

$$\begin{aligned} &x_1 = f_1 = x_1^0 + \gamma(t)x_2^0 \quad x_2 = f_2 = x_2^0 \quad x_3 = f_3 = x_3^0 \\ &x_1^0 = g_1 = x_1 - \gamma(t)x_2 \quad x_2^0 = g_2 = x_2 \quad x_3^0 = g_3 = x_3 \end{aligned} \tag{8.82}$$

$\gamma(t)$ is an arbitrary piecewise differentiable function of the time with the condition that $\gamma(t) = 0$ for $t < 0$. From (8.11) and (8.13) it follows

$$F_{ik} = \begin{pmatrix} 1 & \gamma(t) & 0 \\ 0 & 1 & 0 \\ 0 & 0 & 1 \end{pmatrix} \quad \text{and} \quad F_{ik}^{-1} = \begin{pmatrix} 1 & -\gamma(t) & 0 \\ 0 & 1 & 0 \\ 0 & 0 & 1 \end{pmatrix} \tag{8.83}$$

and from (8.20)

$$C_{ik} = \begin{pmatrix} 1 & \gamma & 0 \\ \gamma & 1+\gamma^2 & 0 \\ 0 & 0 & 1 \end{pmatrix} \quad A_{ik} = \begin{pmatrix} 1 & -\gamma & 0 \\ -\gamma & 1+\gamma^2 & 0 \\ 0 & 0 & 1 \end{pmatrix}$$

$$B_{ik} = \begin{pmatrix} 1+\gamma^2 & -\gamma & 0 \\ -\gamma & 1 & 0 \\ 0 & 0 & 1 \end{pmatrix} \quad G_{ik} = \begin{pmatrix} 1+\gamma^2 & \gamma & 0 \\ \gamma & 1 & 0 \\ 0 & 0 & 1 \end{pmatrix} \tag{8.84}$$

Inserting (8.82) into (8.3) and (8.5) yields

$$\varphi_1 = \dot{\gamma}(t)x_2^0 \qquad \varphi_2 = \varphi_3 = 0 \quad \text{and} \quad v_1 = \dot{\gamma}(t)x_2 \qquad v_2 = v_3 = 0 \tag{8.85}$$

and, therefore, from (8.39)

$$d_{12} = d_{21} = \dot{\gamma}(t)/2 \qquad d_{11} = d_{22} = d_{33} = d_{13} = d_{23} = 0 \tag{8.86}$$

A comparison of (8.85) with (3.23) shows that time-dependent simple shear is equivalent with laminar shear flow whereby the *velocity gradient* $q(t)$ equals the *shear rate* $\dot{\gamma}(t)$.

In order to calculate the principal coordinate system of the Cauchy tensor for simple shear we insert (8.84) into (8.79) and (8.80) and obtain the equations

$$\gamma^2 = (c_2 - c_1)\cos 2\chi \quad 2\gamma = (c_1 - c_2)\sin 2\chi \quad 2 + \gamma^2 = c_1 + c_2 \tag{8.87}$$

$$c_1 = 1 + \gamma^2/2 + \gamma\sqrt{1+\gamma^2/4} \qquad c_2 = 1 + \gamma^2/2 - \gamma\sqrt{1+\gamma^2/4} \qquad \tan 2\chi = -2/\gamma \tag{8.88}$$

From the third of Eq. (8.88) we conclude that tan 2χ is negative and has two solutions for the angle 2χ. One solution is that cos $2\chi > 0$ and sin $2\chi < 0$ and 2χ is negative and situated between zero and $-90°$, the other that cos $2\chi < 0$ and sin $2\chi > 0$ and 2χ is positive and situated between 90° and 180°.

We choose the latter solution and find from the first two equations of (8.88) that c_1 is the larger of the two eigenvalues. As the product $c_1 \cdot c_2$ is equal to unity, we may designate c_1 and c_2 by

$$c_1 = \lambda^2 \quad c_2 = 1/\lambda^2 \tag{8.89}$$

with $\lambda \geq 1$. Due to Eq. (8.71), in the system of principle axis of the Cauchy tensor the length of line elements in the 1-direction is increased to the λ-fold, in the 2-direction decreased by $1/\lambda$, in the 3-direction it is left unchanged.

A simple relation between λ and γ may be obtained as follows. From (8.88) and (8.89) we find

$$c_1 - c_2 = \lambda^2 - 1/\lambda^2 = 2\gamma\sqrt{1+\gamma^2/4}$$

and inserting this relation into the first of Eq. (8.88) yields

$$\lambda^2 = 1 + \gamma^2/2 + \gamma\sqrt{1+\gamma^2/4} = 1 + \gamma^2/2 + (\lambda^2 - 1/\lambda^2)/2$$

which is equivalent to $\gamma^2 = \lambda^2 - 2 + 1/\lambda^2$ or

$$\gamma = \lambda - 1/\lambda \tag{8.90}$$

The angle χ determines the orientation of the principal coordinate system of the tensor $\underline{\underline{C}}$. From the third of Eq. (8.88) we find that χ is positive and a function of the magnitude of the shear γ. For small values of γ, $-2/\gamma = \tan 2\chi$ tends to minus infinity and 2χ to 90°. Therefore, the principal axis starts at an angle of 45° for small values of the shear. With increasing values of γ, 2χ becomes less negative and the angle χ increases. The type of deformation in principal axes is called *pure shear.*

The transformation of the Cauchy tensor for simple shear into its principal coordinate system written in form of matrices reads

$$R \cdot \begin{pmatrix} 1 & \gamma & 0 \\ \gamma & 1+\gamma^2 & 0 \\ 0 & 0 & 1 \end{pmatrix} \cdot \tilde{R} = \begin{pmatrix} \lambda^2 & 0 & 0 \\ 0 & 1/\lambda^2 & 0 \\ 0 & 0 & 1 \end{pmatrix} \quad \text{with} \quad R = \begin{pmatrix} \cos\chi & \sin\chi & 0 \\ -\sin\chi & \cos\chi & 0 \\ 0 & 0 & 1 \end{pmatrix} \tag{8.91}$$

R is the orthogonal matrix which transforms the coordinates from the original system into the principal system of $\underline{\underline{C}}$ and χ is given by the positive solution of Eq. (8.88).

The transformation to principal axes is illustrated in Fig. 8.8, which shows a square in the x_1–x_2-plane (full black lines). This square is transformed into a parallelogram (broken line) by a simple shear strain of magnitude $\gamma = 1$ resulting in the angle $\chi = 58°$ for the inclination of the principal coordinate system (red full lines) against the direction of shear x_1. The rotated square drawn in red changes by the transformation (8.91) into a rectangle with edges magnified by the ratio $\lambda = 1.62$ in the x'_1-direction and compressed by the ratio $1/\lambda = 0.62$ in the x'_2-direction compared to the edges of the square.

Equations (8.81) and (8.86) yield for the principal axes of the rate of strain tensor $\underline{\underline{D}}$

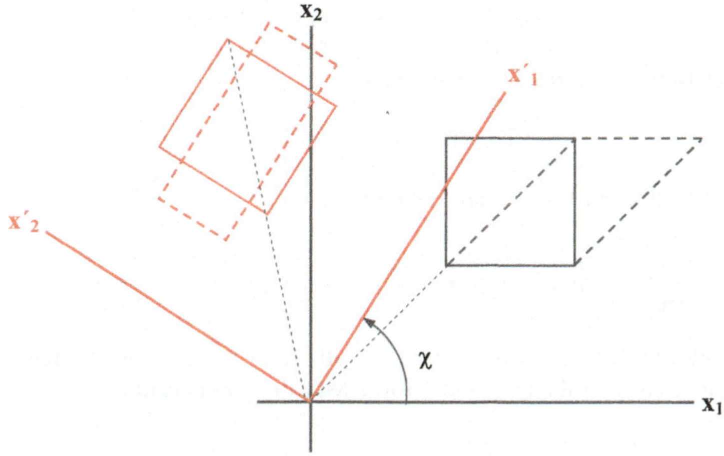

Fig. 8.8 Illustrating the transformation to principal axis in simple shear of magnitude $\gamma = 1$ and an inclination of the principal axis of $\chi = 58°$

$$d_1 = -d_2 = \dot{\gamma}(t) \qquad \tan 2\chi = \infty \quad \chi = 45° \tag{8.92}$$

For the rate of strain tensor, the principal direction is always inclined by 45° against the direction of motion, independent of the magnitude of the shear rate.

8.7.2 Multidimensional Time-Dependent Incompressible (Isochoric) Extension

A cube with edges parallel to the coordinate axes is extended or compressed into a cuboid without changing its volume. This type of deformation is described by the path equations

$$x_1 = \lambda_1(t)x_1^0 \quad x_2 = \lambda_2(t)x_2^0 \quad x_3 = \lambda_3(t)x_3^0 \tag{8.93}$$

with $\lambda_1(t)$, $\lambda_2(t)$, $\lambda_3(t)$ being arbitrary piecewise differentiable functions of the time subjected to the conditions $\lambda_i(t) = 1$ for $t < 0$ and to the incompressibility condition

$$\lambda_1(t) \cdot \lambda_2(t) \cdot \lambda_3(t) = 1 \tag{8.94}$$

The λ_i are called the *principal extension ratios*. Often the so-called *Hencky-strains* $\varepsilon_{H,i}$ (t) are introduced defined by

$$\varepsilon_{H,i}(t) = \ln \lambda_i(t) \quad \text{for } i = 1 \text{ to } 3 \tag{8.95}$$

Using the Hencky-strains Eq. (8.94) reads

$$\varepsilon_{H_1}(t) + \varepsilon_{H_2}(t) + \varepsilon_{H_3}(t) = 0 \tag{8.94'}$$

The velocity field in Euler-representation becomes

$$v_1(\underline{x}, t) = \frac{\dot{\lambda}_1}{\lambda_1} x_1 = \dot{\varepsilon}_{H_1} x_1 \quad v_2(\underline{x}, t) = \frac{\dot{\lambda}_2}{\lambda_2} x_2 = \dot{\varepsilon}_{H_2} x_2 \quad v_3(\underline{x}, t) = \frac{\dot{\lambda}_3}{\lambda_3} x_3 = \dot{\varepsilon}_{H_3} x_3 \tag{8.96}$$

The deformation gradient, its inverse, all strain tensors and the rate of strain tensor are already in their diagonal form with the components

$$F_{ik} = \begin{pmatrix} \lambda_1 & 0 & 0 \\ 0 & \lambda_2 & 0 \\ 0 & 0 & \lambda_3 \end{pmatrix} \quad \text{and} \quad F_{ik}^{-1} = \begin{pmatrix} \lambda_1^{-1} & 0 & 0 \\ 0 & \lambda_2^{-1} & 0 \\ 0 & 0 & \lambda_3^{-1} \end{pmatrix} \tag{8.97}$$

$$C_{ik} = \begin{pmatrix} \lambda_1^2 & 0 & 0 \\ 0 & \lambda_2^2 & 0 \\ 0 & 0 & \lambda_3^2 \end{pmatrix} = G_{ik} \quad A_{ik} = \begin{pmatrix} \lambda_1^{-2} & 0 & 0 \\ 0 & \lambda_2^{-2} & 0 \\ 0 & 0 & \lambda_3^{-2} \end{pmatrix} = B_{ik} \tag{8.98}$$

$$d_{ik} = \begin{pmatrix} \dot{\varepsilon}_{H_1} & 0 & 0 \\ 0 & \dot{\varepsilon}_{H_2} & 0 \\ 0 & 0 & \dot{\varepsilon}_{H_3} \end{pmatrix} \tag{8.99}$$

Multidimensional incompressible extension may be classified by introducing a parameter m relating the extension in the 1-direction to that in 2-direction.

$$\lambda_1 = \lambda \qquad \lambda_2 = \lambda^m \qquad \lambda_3 = \lambda^{-(1+m)} \tag{8.100}$$

with values of m between $-1/2$ and +1. Table 8.1 lists some typical modes of extensional deformation and the first and second invariants of their strain tensors $\underline{\underline{C}}$ and $\underline{\underline{G}}$.

As the comparison of (8.91) with table 8.1 shows, pure shear and planar extension have the same principle values.

Table 8.1 Special modes of isochoric multiaxial extension, the values of the parameter m and their invariants

Type of extension	m	Principal extension ratios	$I_C = I_G$	$II_C = II_G$
Uniaxial extension in the 1-direction	−1/2	$\lambda_1 = \lambda, \lambda_2 = \lambda_3 = \lambda^{-1/2}$	$\lambda^2 + 2/\lambda$	$1/\lambda^2 + 2\lambda$
Planar extension in the 1–3 plane	0	$\lambda_1 = \lambda, \lambda_2 = 1, \lambda_3 = 1/\lambda$	$1 + \lambda^2 + 1/\lambda^2$	$1 + \lambda^2 + 1/\lambda^2$
Equibiaxial extension in the 1–2 plane	1	$\lambda_1 = \lambda, \lambda_2 = \lambda, \lambda_3 = 1/\lambda^2$	$2\lambda^2 + 1/\lambda^4$	$2/\lambda^2 + \lambda^4$

References

1. Green AE, Zerna W (1954) Theoretical elasticity. Clarendon, Oxford
2. Bird RB, Armstrong RC, Hassager O (1987) Dynamics of polymeric liquids. John Wiley, New York
3. Wagner MH (1976) Rheol Acta 15:133, 136 (1977) Rheol Acta 16:43
4. Reiner M (1945) Am J Math 67:350
5. Rivlin RS (1948) Proc Roy Soc A 193:260
6. Happel J, Brenner H (1965) Low Reynolds number hydrodynamics. Prentiss Hall, Englewood Cliffs, NY
7. Hohnerkamp J, Römer H (1986) Grundlagen der klassischen theoretischen Physik. Springer, Heidelberg, New York
8. Mohr O (1882) Der Zivilingenieur, p 113

Chapter 9
Large Deformations of Polymers

9.1 Stress-Strain Behavior of Polymeric Materials

For the mechanical characterization of materials, particularly in the nonlinear range of deformation, the tensile behavior characterized by the stress–strain curve plays an important role. In technical applications the stress σ is defined by relating the force F to the initial cross section of the sample A_0, i.e.,

$$\sigma = F/A_0 \tag{9.1}$$

and the strain ε by the ratio of the change in length $\Delta l = l - l_0$ to the initial length, l_0 i.e.,

$$\varepsilon = \Delta l/l_0 \tag{9.2}$$

with l being the actual length of the sample.

The so-called engineering stress and engineering strain are physically meaningful as long as the deformations applied are small as it is usually the case for investigations on metals, glass, and ceramics. For polymer materials these conditions are a more or less coarse approximation in the glassy state, but not at all fulfilled in the rubber-elastic or molten state. In these cases, physically well-defined quantities should be used which take the large changes of the sample length and the cross section during elongation into account. The "natural" deformation often expressed by the so-called Hencky strain and the "true" stress are described and discussed in Sect. 12.1.

Due to the importance for the mechanical characterization of materials, the technical equipment and the methods to measure stress–strain curves are standardized. The standard [1] describes the test equipment for polymer materials, in others, e.g., [2] and [3], the test procedures are given.

The characteristic quantities derived from stress–strain curves are the same for all materials but their values can be very different, of course. These facts may be briefly discussed by comparing the schematic curves of a metal like steel and a polymer.

H. Münstedt and F. R. Schwarzl, *Deformation and Flow of Polymeric Materials*,
DOI: 10.1007/978-3-642-55409-4_9, © Springer-Verlag Berlin Heidelberg 2014

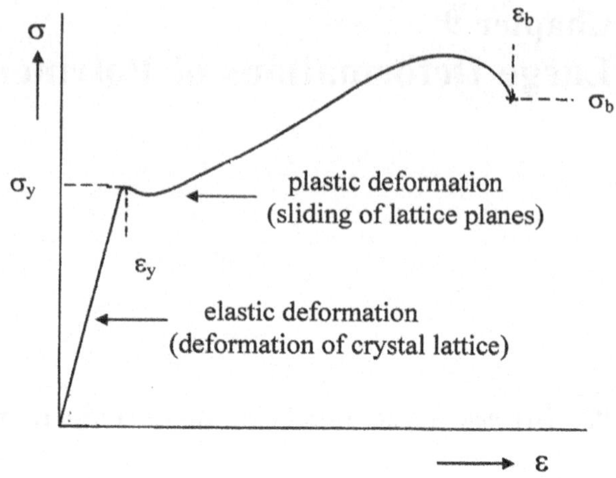

Fig. 9.1 Schematic diagram of the engineering stress σ as a function of the engineering strain ε for steel. σ_y is the yield stress, ε_y the elongation at yield, σ_b the stress at break, ε_b the elongation at break

In Fig. 9.1 the stress as a function of the strain is schematically plotted for steel. After a steep and approximately linear increase of the stress with strain a small maximum is reached which is followed by a further increase of the stress which is not as pronounced, however, as in the first section. Then the curve runs through another broad maximum before it breaks. As indicated in Fig. 9.1, the first maximum due to a yielding of the material is described by the elongation at yield ε_y and the corresponding stress σ_y. The elongation and the stress at break are denominated ε_b and σ_b, respectively. The first part of the deformation is characterized by the slope of the curve which represents the tensile or Young's modulus E. As long as the material-specific modulus is independent of strain or stress, respectively, i.e., the slope is constant, the deformation behavior is linear. The underlying microscopic processes are deformations of the crystal lattice that are fully reversible within certain limits of atomic displacements. The elongation after the yield point is irreversible and goes back to lattice planes slipping along each other. At too high stresses bonds break and the material fails. Typical data for steel are given in Table 9.1.

For polymeric materials different states of deformation have to be distinguished which go back to the molecular mobility changing with temperature. In the glassy state at temperatures not too close to the glass transition the material can behave very brittle. A stress–strain curve typical of such a state is schematically presented in Fig. 9.2. Similar to Fig. 9.1, the stress σ increases with the elongation ε but the sample breaks without having reached a yield point before. The failure is characterized by the critical elongation to break ε_b and the corresponding stress σ_b. Typical values for these quantities are given in Table 9.1. The highest moduli of around 10,000 MPa are found for commercial not reinforced polymers at cryogenic temperatures [4]. In contrast to steel, $\sigma(\varepsilon)$ is not linear up to the point of failure, however, i.e., a constant modulus only exists for small deformations.

Table 9.1 Comparison between typical mechanical data of steel and a polymeric material in different states of deformation (The definitions of the quantities are given in Fig. 9.1)

	E [MPa]	ε_y	σ_y [MPa]	ε_b	σ_b [MPa]
Steel	210,000	0.1–0.2	250	≈0.3	400–1400
Polymer					
Brittle	3,000–10,000			0.002–0.03	10–200
Cold drawn	≈3000	0.02–0.03	10–60	2–5	$>\sigma_y$
Rubber-elastic	1–10			2–10	10–50

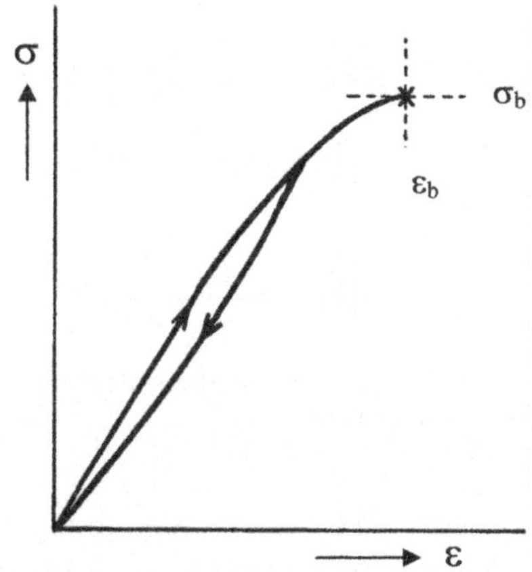

Fig. 9.2 Schematics of a stress–strain diagram for a brittle polymer material

The fact that the deformation is not totally elastic is expressed by the hysteresis in Fig. 9.2. The area between the two curves is proportional to the mechanical energy dissipated during one cycle.

Due to the pronounced curvature of the stress–strain curve, in case of polymeric materials the tangent modulus is replaced by the secant modulus. The latter requires the specification of two strains for its defined determination.

Preferably in the *leather-like state* of semicrystalline polymers and at about 30 °C below the glass temperature in the case of amorphous polymers, a stress–strain behavior can be observed as sketched in Fig. 9.3. After a steep increase of the stress as a function of strain a yield point characterized by σ_y and ε_y occurs. Increasing the strain further leaves the stress nearly constant which rises again just before the failure of the sample. The stress–strain curve specific for polymers under certain conditions in the solid state is accompanied by a remarkable change in the geometry of the sample also shown in Fig. 9.3. The yielding falls together with the formation of a notch (1) which extends along the sample with increasing stretching. Shoulders are formed which separate the drawn section from the undrawn ones (2). After the whole sample is elongated, the stress slightly increases again before failure occurs (3). This phenomenon is called cold-drawing as it leads to the thinning of a sample far below temperatures usual for drawing processes from the molten state. Most of the deformation is irreversible as indicated by the

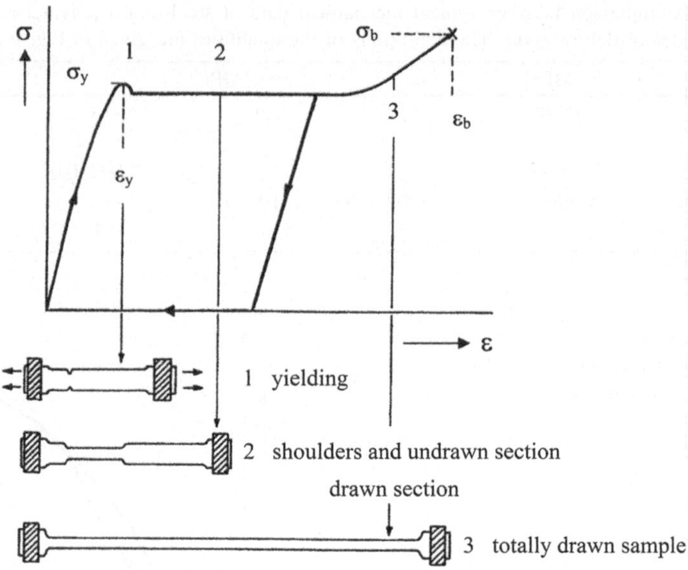

Fig. 9.3 Schematics of the stress–strain behavior of an amorphous polymer cold drawn some few degrees below T_g

pronounced hysteresis loop in Fig. 9.3. For amorphous polymers the cold-drawing effect has been found to get reversed by heating the sample distinctly above T_g in the unloaded state.

The molecular process underlying cold-drawing of amorphous polymers is the orientation of the molecule chains under the large mechanical stresses applied. This mechanism is more complicated, however, in the case of semicrystalline polymers as the change of the orientation of crystallites additionally comes into play.

The stress–strain curve in the rubber-elastic state is sketched in Fig. 9.4. A constant modulus is only found in a very small range of deformation and yielding and cold-drawing are absent. The stress increases more or less continuously up to the failure of the sample which is characterized by the stress σ_b and elongation ε_b at break. As can be seen from Table 9.1 these quantities cover a wide range. ε_b can assume rather large values comparable to those of rubbers whereas σ_b is not too far away from the range found for polymers in the other deformation states. As it becomes obvious from Fig. 9.4, nearly up to the point of failure most of the elongation is reversible. This reversibility is understandable from the entanglement network still existing in the rubber-elastic regime (cf. Sect. 9.4).

All the three types of the stress–strain behavior described can be obtained for amorphous polymers if the corresponding temperatures are chosen. They are summarized in Fig. 9.5. At temperatures far below T_g one finds a pronounced brittleness. After an approximately linear increase of the stress with strain the samples break at relatively small elongations. Around 30 °C below T_g first

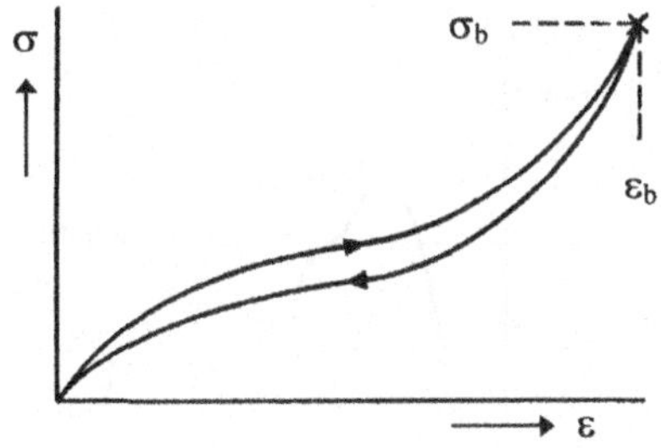

Fig. 9.4 Schematics of a stress–strain diagram for a polymer in the rubber-elastic state

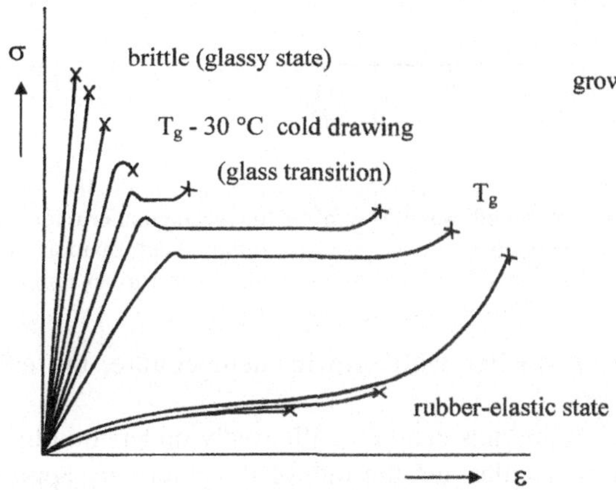

Fig. 9.5 Schematic stress–strain curves of an amorphous polymer at various temperatures below and above T_g

indications of yielding occur which extends into wide ranges of cold-drawing around T_g. Above T_g in the rubber-elastic state the inhomogeneous deformation of cold-drawing is not found anymore. The modulus becomes comparatively low and the stress remains nearly constant over a wide range of elongation before it increases up to the failure of the sample. In this deformation regime a growing temperature effects a decrease of the stress and elongation to break but the corresponding parts of the stress–strain curves remain nearly unchanged.

Besides the temperature, the deformation rate can have a strong influence on the stress–strain curve. This is typical of viscoelastic materials. Such behavior is presented in Fig. 9.6 for a rubber-modified polystyrene. For the sake of simplicity the forces F are plotted as functions of the change in sample length Δl. At the constant temperature of 23 °C the yield point shifts to higher stresses and slightly larger deformations with growing deformation rates. The same tendency is seen for the force at break, but the opposite is observed for the deformation at break. The lower the deformation rate the larger becomes this quantity and seems to approach a constant value (cf. Fig. 9.6a). In Fig. 9.6b the dependence on temperature is presented. At the constant deformation rate of 500 mm/min a higher temperature decreases the force and the deformation at the yield point. The force at break

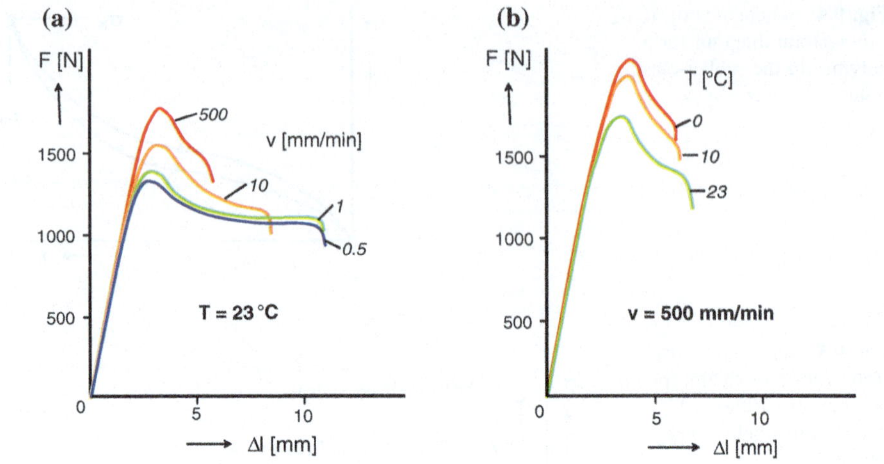

Fig. 9.6 Force F as a function of change in length Δl for a rubber-modified polystyrene. The initial sample length was 100 mm and the rectangular cross section $3 \times 10\ \text{mm}^2$. **a** Measurements at a constant temperature of 23 °C, but different deformation rates v, **b** Measurements at a constant deformation rate of $v = 500$ mm/min and three various temperatures [5]

becomes smaller with growing temperature, the critical deformation increases, however.

These findings can be qualitatively understood by the interplay of the mobility of the molecules and the rate of the externally applied deformation. If the deformation is slow enough to be followed by the molecules, the acting stress may be so small that they are not broken. Increasing the deformation rate leads to a stress enhancement that may be reduced, however, by a growth in molecular mobility which can be achieved by higher temperatures.

9.2 Rheological Equation of State for Isotropic Elastic Materials

For a purely elastic material the stress tensor is a unique function of either the absolute Green tensor $\underline{\underline{G}}$ (cf. Eq. 8.18) or of the absolute Piola tensor $\underline{\underline{A}}$ (cf. Eq. 8. 17). If moreover the material is isotropic, the stress tensor has to be an isotropic tensor function of $\underline{\underline{A}}$ or $\underline{\underline{G}}$. For reasons which will become clear in Sect. 9.3 we prefer the presentation in terms of $\underline{\underline{G}}$ which due to Eq. (8.63) can be written as

$$\underline{\underline{S}} = f_0(I_G, II_G, III_G)\underline{\underline{E}} + f_1(I_G, II_G, III_G)\underline{\underline{G}} + f_2(I_G, II_G, III_G)\underline{\underline{G}}^2 \qquad (9.3)$$

f_0, f_1, and f_2 are three scalar functions of the three invariants of the Green tensor. These functions characterize the elastic properties and are to be determined by experiments. The functions f_0, f_1, and f_2 are not independent of each other, but related by the elastic deformation energy.

Large elastic deformations are observed, for example, for well cross-linked rubbers. They are incompressible with a shear modulus many orders of magnitude smaller than their bulk modulus. Therefore, under non isotropic pressures, the volume changes are much smaller than the changes in shape and may be neglected As only isochoric deformations are admitted, i.e.,

$$III_G = \left(\frac{dV}{dV_0}\right)^2 = 1 \tag{9.4}$$

according to Eq. (8.70), the third invariant does not occur as an independent variable in the scalar functions f_0, f_1, and f_2. Moreover, an arbitrary hydrostatic pressure may always be added to the stress tensor without changing the state of deformation. Consequently, the stress tensor can only be determined apart from an arbitrary hydrostatic pressure p. The first term in Eq. (9.3) may represent this quantity and the *rheological equation of state of a purely elastic isotropic incompressible material* then becomes

$$\underline{\underline{S}} = -p\underline{\underline{E}} + f_1(I_G, II_G)\underline{\underline{G}} + f_2(I_G, II_G)\underline{\underline{G}}^2 \tag{9.5}$$

The functions f_1 and f_2 depend on the two invariants I_G and II_G, only.

A further simplification results from the introduction of the elastic deformation energy. In the principal coordinate system of the tensor $\underline{\underline{G}}$ also $\underline{\underline{G}}^2$ and $\underline{\underline{S}}$ will be in their diagonal forms and Eq. (9.5) reads as

$$\sigma_i = -p + f_1(I_G, II_G)\lambda_i^2 + f_2(I_G, II_G)\lambda_i^4 \quad \text{for } i = 1,2,3 \tag{9.6}$$

with λ being the extension ratio. Then σ_1 is the first principal stress, i.e., the normal stress acting on the 1-plane related to the area of this plane in the deformed state. To calculate the deformation energy it is necessary to introduce the engineering stress, since energy is defined as the path integral of the forces, not of the stresses. Let s_1 be the engineering stress acting on the 1-plane related to the area of this plane in the undeformed state then because of the incompressibility condition $\lambda_1\lambda_2\lambda_3 = 1$, the following relation between the *engineering stress* and the true stress holds:

$$s_1 = \lambda_2\lambda_3\sigma_1 = \sigma_1/\lambda_1; \tag{9.7}$$

The deformation energy per unit volume necessary to deform a small cube parallel to the axes of the coordinate system to the λ_1, λ_2, λ_3-fold of its original lengths is designated by w. For an incompressible elastic isotropic material this *deformation energy density* can only depend on the two invariants I_G and II_G of the strain tensor

$$w = w(I_G, II_G) \tag{9.8}$$

By changing the principle extension ratios from λ_1, λ_2, λ_3 to $\lambda_1 + \mathrm{d}\lambda_1$, $\lambda_2 + \mathrm{d}\lambda_2$, $\lambda_3 + \mathrm{d}\lambda_3$, while considering the incompressibility condition

$$\mathrm{d}\lambda_1/\lambda_1 + \mathrm{d}\lambda_2/\lambda_2 + \mathrm{d}\lambda_3/\lambda_3 = 0 \tag{9.9}$$

the change in the deformation energy density is

$$\mathrm{d}w = s_1\mathrm{d}\lambda_1 + s_2\mathrm{d}\lambda_2 + s_3\mathrm{d}\lambda_3 \tag{9.10}$$

Applying the chain rule of differentiation, the change of the energy may also be represented by differentiation of w with respect to λ_1, λ_2, λ_3.

$$\mathrm{d}w = \frac{\partial w}{\partial I_G}\frac{\partial I_G}{\partial \lambda_i}\mathrm{d}\lambda_i + \frac{\partial w}{\partial II_G}\frac{\partial II_G}{\partial \lambda_i}\mathrm{d}\lambda_i \tag{9.11}$$

Using (8.70) and (8.71) I_G and II_G may be expressed explicitly by the λ and the derivatives become

$$\frac{\partial I_G}{\partial \lambda_i} = 2\lambda_i \quad \frac{\partial II_G}{\partial \lambda_i} = 2\lambda_i(I_G - \lambda_i^2)$$

The second equation may be understood by writing $\partial II_G/\partial\lambda_i$ for the index 1, i.e., $2\lambda_1\,(\lambda_2^2 + \lambda_3^2) = 2\lambda_1\,(I_G - \lambda_1^2)$ and then generalize it to the index i. Inserting these relations into (9.11) yields

$$\mathrm{d}w = 2\left[\frac{\partial w}{\partial I_G} + I_G\frac{\partial w}{\partial II_G}\right]\lambda_i\mathrm{d}\lambda_i - 2\frac{\partial w}{\partial II_G}\lambda_i^3\mathrm{d}\lambda_i \tag{9.12}$$

The variations $d\lambda_i$ are not independent of each other because of the incompressibility condition (9.9). Therefore, before comparing the expressions (9.10) and (9.12) $\mathrm{d}\lambda_3$ has to be eliminated from both expressions using Eq. (9.9). After elimination of $\mathrm{d}\lambda_3$, $\mathrm{d}\lambda_1$ and $\mathrm{d}\lambda_2$ are independent of each other and setting (9.10) equal (9.12) results in two equations, one of them being

$$\lambda_1 s_1 - \lambda_3 s_3 = 2\left[\frac{\partial w}{\partial I_G} + I_G\frac{\partial w}{\partial II_G}\right](\lambda_1^2 - \lambda_3^2) - 2\frac{\partial w}{\partial II_G}(\lambda_1^4 - \lambda_3^4) \tag{9.13}$$

The other follows from (9.6) as

$$\lambda_1 s_1 - \lambda_3 s_3 = \sigma_1 - \sigma_3 = f_1(I_G, II_G)(\lambda_1^2 - \lambda_3^2) + f_2(I_G, II_G)(\lambda_1^4 - \lambda_3^4) \tag{9.14}$$

Considering that λ_1 and λ_3 are independent of each other, from a comparison of the coefficients in (9.13) and (9.14) follows for the functions f_1 and f_2 of the constitutive Eq. (9.3)

$$f_1(I_G, II_G) = 2\left[\frac{\partial w}{\partial I_G} + I_G \frac{\partial w}{\partial II_G}\right] \qquad f_2(I_G, II_G) = -2\frac{\partial w}{\partial II_G} \tag{9.15}$$

A purely elastic isotropic incompressible material is completely characterized by one function depending on two independent variables, the deformation energy density $w(I_G, II_G)$.

Using the equation of Caley-Hamilton (Eq. 8.62) $\underline{\underline{G}}^2$ can be substituted by a linear combination of $\underline{\underline{G}}$ and $\underline{\underline{G}}^{-1}$ which results in a still simpler form for the rheological equation of state.

$$\boxed{\underline{\underline{S}} = -p\underline{\underline{E}} + 2\frac{\partial w}{\partial I_G}\underline{\underline{G}} - 2\frac{\partial w}{\partial II_G}\underline{\underline{G}}^{-1}} \tag{9.16}$$

As an example we consider the case of simple shear. Inserting (8.82) and (8.84) into (9.16) yields

$$\begin{aligned} \sigma_{12} &= G\gamma \\ \sigma_{11} - \sigma_{22} &= G\gamma^2 \\ \sigma_{22} - \sigma_{33} &= -2\gamma^2\frac{\partial w}{\partial II_G} \end{aligned} \tag{9.17}$$

with G being an abbreviation for the expression

$$G = 2\left[\frac{\partial w}{\partial I_G} + \frac{\partial w}{\partial II_G}\right] \tag{9.18}$$

From these equations it follows that simple shear cannot be realized by a shear stress only, but also requires the application of normal stress differences, a *first normal stress difference* σ_{11}–σ_{22} and a *second normal stress difference* σ_{22}–σ_{33}. G is the ratio of the shear stress to the shear strain and is called the *shear modulus* of the material which is not necessarily a constant, but may depend on the magnitude of the shear γ. Only in the case that the deformation energy density depends linearly on the first and second invariants, the shear modulus is a constant. The first normal stress difference is always positive, as G and γ^2 are positive. The ratio of the first normal stress difference to the square of the shear, the first normal stress coefficient, is equal to the shear modulus. As a consequence, for values of the shear larger than one the first normal stress difference will exceed the shear stress. The second normal stress difference may be positive or negative. It is different from zero only, if the deformation energy depends on the second invariant.

An experiment in simple shear does not admit the complete determination of the deformation energy. This may be seen from the equality of the two invariants in simple shear which are

$$I_G = II_G = 3 + \gamma^2 \tag{9.19}$$

In order to determine $w(I_G, II_G)$ experiments should be performed in which both invariants can be changed independently of each other within a range as wide as possible.

An example for an experiment in which I_G and II_G differ is uniaxial extension which is described by

$$\lambda_1 = \lambda \quad \lambda_2 = \lambda_3 = 1/\sqrt{\lambda} \tag{9.20}$$

and

$$I_G = \lambda^2 + 2/\lambda \quad II_G = 1/\lambda^2 + 2\lambda \tag{9.21}$$

As the planes perpendicular to the direction of extension are free of stress, the stresses for an isochoric uniaxial extension become

$$\sigma_{11} = 2\left(\frac{\partial w}{\partial I_G} + \frac{1}{\lambda}\frac{\partial w}{\partial II_G}\right)\left(\lambda^2 - \frac{1}{\lambda}\right) \qquad \sigma_{12} = \sigma_{23} = \sigma_{31} = \sigma_{22} = \sigma_{33} = 0 \tag{9.22}$$

Further combinations of I_G and II_G may be realized by other types of multi-axial extensions as listed in Table 8.1. For instance equibiaxial extension of the 1-2 plane is characterized by

$$\lambda_1 = \lambda \quad \lambda_2 = \lambda \quad \lambda_3 = 1/\lambda^2 \tag{9.23}$$

and the stresses in the 1 and 2-directions are

$$\sigma_{11} = \sigma_{22} = 2\left(\frac{\partial w}{\partial I_G} + \lambda^2\frac{\partial w}{\partial II_G}\right)\left(\lambda^2 - \frac{1}{\lambda^4}\right) \qquad \sigma_{33} = 0 \tag{9.24}$$

In uniaxial extension for $\lambda \gg 1$ we find $I_G > II_G$, while in equibiaxial extension for $\lambda \gg 1$ $II_G \gg I_G$ is valid.

9.3 Rheological Equation of State for the Ideal Rubber

An ideal rubber is defined as a purely elastic isotropic incompressible material with a very simple rheological equation of state:

$$\boxed{\underline{\underline{S}} = -p\underline{\underline{E}} + G \cdot \underline{\underline{G}}} \tag{9.25}$$

G is the *constant shear modulus*. The corresponding elastic deformation energy

$$w(I_G, II_G) = \frac{1}{2} G(I_G - 3) \tag{9.26}$$

may be derived from a simple model using the Gaussian chain statistics (compare Sect. 9.4). As the validity of Gaussian chain statistics is limited to those cases in which macromolecules are not stretched up to the region of their finite length, one should expect the applicability of Eq. (9.25) to be limited to moderate deformations. Experimentally, (9.25) turns out to give a reasonable description of the rheological behavior of rubbers in shear up to $\gamma = 1$ and in uniaxial extension up to $\lambda = 2$ ([6, 7]).

In the undeformed state, i.e., $\lambda_1 = \lambda_2 = \lambda_3 = 1$ and $I_G = 3$, w vanishes. Furthermore, from (9.17) and (9.18) it follows that G equals the constant shear modulus and the *constant first normal stress coefficient*, while the second normal stress coefficient is zero. If we had used the tensor $\underline{\underline{A}}$ instead of $\underline{\underline{G}}$ for the description of the equation of state, a more complicated expression would have been the result for the rheological law of the ideal rubber. This was the reason for preferring $\underline{\underline{G}}$ as the deformation tensor for the constitutive Eq. (9.25).

The linear relation (9.25) between the stress tensor $\underline{\underline{S}}$ and the strain tensor $\underline{\underline{G}}$ may be inverted to

$$\underline{\underline{G}} = J\left(\underline{\underline{S}} + p\underline{\underline{E}}\right) \tag{9.27}$$

with $J = 1/G$ being the *constant shear compliance*. The term $Jp\delta_{ik}$ in Eq. (9.27) remains undetermined because the hydrostatic pressure p is unknown. It may, however, be derived from the condition of incompressibility which reads as

$$III_G = Det(G_{ik}) = 1 \tag{9.4}$$

or because of $G_{ik} = J(\sigma_{ik} + p\delta_{ik})$

$$Det(G_{ik}) = J^3 \begin{vmatrix} \sigma_{11} + p & \sigma_{12} & \sigma_{13} \\ \sigma_{21} & \sigma_{22} + p & \sigma_{23} \\ \sigma_{31} & \sigma_{32} & \sigma_{33} + p \end{vmatrix} = 1$$

Developing the determinant yields just the left-hand side of the characteristic equation of the stress tensor (8.60) with p = – λ. Therefore the condition (9.4) reads as

$$p^3 + I_\sigma p^2 + II_\sigma p + III_\sigma = 1/J^3 = G^3 \tag{9.28}$$

This equation has at least one real root for p which can be inserted into (9.27). This determines the strain tensor uniquely.

In conclusion we may distinguish two cases:

(a) The deformations are given; it follows the state of stress apart from an arbitrary additional hydrostatic pressure, which remains undetermined.
(b) The stresses are given; the state of deformation follows apart from the hydrostatic component $Jp\delta_{ik}$ which follows from Eq. (9.28).

For a given state of stress, only one state of strain exists, for a given state of strain an infinite number of states of stress is possible which differ by their hydrostatic component.

The predictions of the theory of rubberlike elasticity are compared with some experimental results taken from [8].

Simple shear:
For the case of simple shear one obtains by inserting (8.84) into (9.25)

$$\begin{aligned} \sigma_{12} &= G\gamma \\ \sigma_{11} - \sigma_{22} &= G\gamma^2 \\ \sigma_{22} - \sigma_{33} &= 0 \end{aligned} \tag{9.29}$$

The shear stress is proportional to the shear strain, the first normal stress difference is positive, the second one is zero.

In Fig. 9.7 the deformation in simple shear of a vulcanized natural rubber is shown after Treloar [8] and compared with the prediction of (9.29). Up to values of $\gamma = 1$ theory and experiment agree well. At higher values of the shear, the shear stress measured lies below the prediction. Even in the range between $\gamma = 1$ and $\gamma = 5$ the differences between experiment and Eq. (9.29) do not exceed 10 %, however.

Uniaxial extension:
For uniaxial extension in the 1-direction we insert $\lambda_1 = \lambda$, $\lambda_2 = \lambda_3 = 1/\lambda^{1/2}$ into (8.98) and find $G_{11} = \lambda^2$, $G_{22} = G_{33} = 1/\lambda$, $G_{ik} = 0$ for $i \neq k$. From Eq. (9.25) the principal stresses follow as

$$\sigma_1 = -p + G\lambda^2 \quad \sigma_2 = \sigma_3 = -p + G/\lambda \tag{9.30}$$

The transversal 2- and 3-planes of the sample are free of stress ($\sigma_2 = \sigma_3 = 0$). On the 1-plane a force acts which, related to the cross section in the undeformed state, is designated the *engineering stress f*.

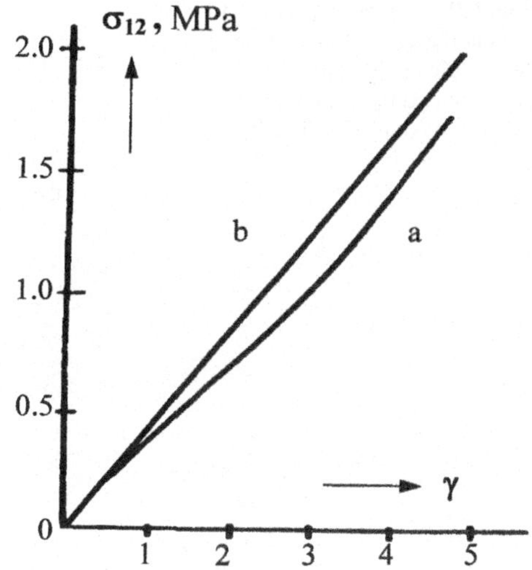

Fig. 9.7 Shear stress versus shear for a vulcanized rubber
(*a*) experimental
(*b*) Eq. (9.29) [8]

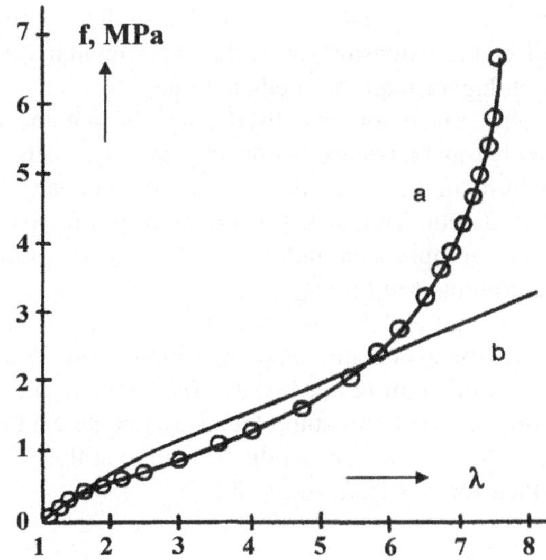

Fig. 9.8 Engineering stress f versus the extension ratio λ for a vulcanized natural rubber in uniaxial extension
(*a*) experimental
(*b*) Eq. (9.32) [8]

$$\sigma_1 = \lambda f \tag{9.31}$$

Eliminating p from Eq. (9.30) by using the condition $\sigma_2 = 0$ results in the following relation between the engineering stress f and the extension ratio λ

$$f = G(\lambda - 1/\lambda^2) \tag{9.32}$$

The engineering stress-extension diagram of a vulcanized natural rubber is compared with the prediction (9.32) in Figs. 9.8 and 9.9. Figure 9.8 shows a good

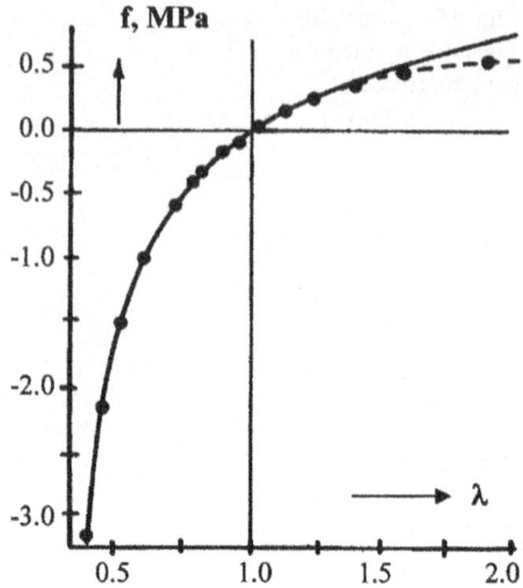

Fig. 9.9 Engineering stress f versus the extension ratio λ for a vulcanized natural rubber in compression and small uniaxial extension [8]

agreement in the range $\lambda < 1.4$ between Eq. (9.32) and the experiment. In the extension range $1.4 < \lambda < 5.6$ the prediction surpasses the experiment, while at still higher extension ratios the experimental values of the engineering stress are much higher than the predicted ones.

The reason for this discrepancy is that the macromolecules are so strongly oriented at higher extension that the Gaussian chain statistics loses its validity. By introducing a statistics which approximately takes into consideration the finite chain length, Treloar [8] could describe the upswing in the curve in Fig. 9.8.

The compression and the uniaxial stress–strain behavior at small deformations are presented in Fig. 9.9.

The figure shows the excellent agreement between theory and experiment at small extension ratios and in compression. It can be seen that the stress-deformation diagram is not linear, even close to the undeformed state, but exhibits a strong negative curvature. One way to define a modulus in uniaxial extension is to determine the tangent on the $f(\lambda)$ curve at the point $\lambda = 1$. For the *tensile modulus* E then follows from Eq. (9.32)

$$E = \lim_{\lambda \to 1} \frac{df}{d\lambda} = 3G \tag{9.33}$$

However, even at small deformations, Eqs. (9.25) and (9.27) do not present a generally valid description of the rheological behavior. This deficiency becomes obvious when considering the behavior under two-dimensional stresses, for which Eq. (9.25) fails.

A much better phenomenological description of various deformation modes has been given by Mooney [9] proposing the following deformation energy density

$$w = C_1(I_G - 3) + C_2(II_G - 3) \tag{9.34}$$

which according to Eq. (9.16) results in the stress–strain relation

$$\underline{\underline{S}} = -p\underline{\underline{E}} + 2C_1\underline{\underline{G}} - 2C_2\underline{\underline{G}}^{-1} \tag{9.35}$$

with C_1 and C_2 being two positive constants. From Eq. (9.18) it follows that $2(C_1 + C_2)$ equals the *constant shear modulus* and the *constant first normal stress coefficient*, while according to Eq. (9.17) C_2 gives rise to a *negative second normal stress difference* in shear.

The influence of the C_2-term in the deformation energy density does not show up in simple shear, but already appears in uniaxial extension. For a rubber with the stress–strain relation (9.35), the relationship between engineering stress and extension ratio becomes instead of (9.32)

$$f = 2(C_1 + C_2/\lambda)(\lambda - 1/\lambda^2) \tag{9.36}$$

By plotting $f/(\lambda - 1/\lambda^2)$ versus $1/\lambda$ the relevance of the C_2-term is easily proved. In this way, C_2 has been found to be about 10 % of C_1 [9]. Consequently, the second normal stress difference is much smaller—in absolute measure—than the first one. Moreover, the third term in Eq. (9.35) improves especially the description of the behavior under two-dimensional stresses. This has been shown by Rivlin and Saunders [10] who investigated pure shear, pure shear with superimposed extension, pure torsion and torsion with superimposed extension. They concluded that within wide limits, the behavior of vulcanized rubbers could be described by the *Mooney-Equation* (9.34). They proved that C_1 is a material constant independent of I_G and II_G and that C_2 is independent of I_G but slightly dependent on II_G. The ratio C_2/C_1 decreases from 0.25 to 0.04 with increasing II_G.

9.4 Statistical Theory of Rubber Elasticity

The statistical theory of the conformations of macromolecules as described in Sect. 2.3 has often been used to predict the properties of macromolecular networks. We restrict our considerations to the discussion of a simple model for rubber elasticity as described by *Flory and Rehner* [11].

As elementary cell of a cross-linked rubber a regular tetraeder is chosen consisting of four chains of equal length and one four-functional cross-linking point P (cf. Fig. 9.10). The four chains are kept fixed at their end points A, B, C, D while their other endpoints which are bundled in P have to move together with the point P, which is assumed to be free in space. The probability of this conformation is calculated from which its entropy and its free energy follow.

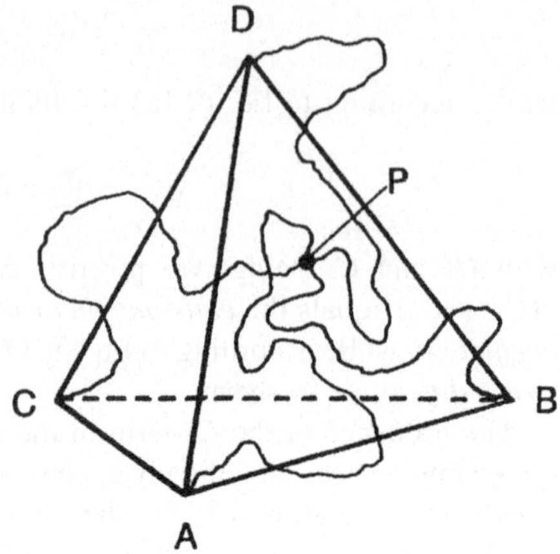

Fig. 9.10 Elementary cell for a cross-linked rubber

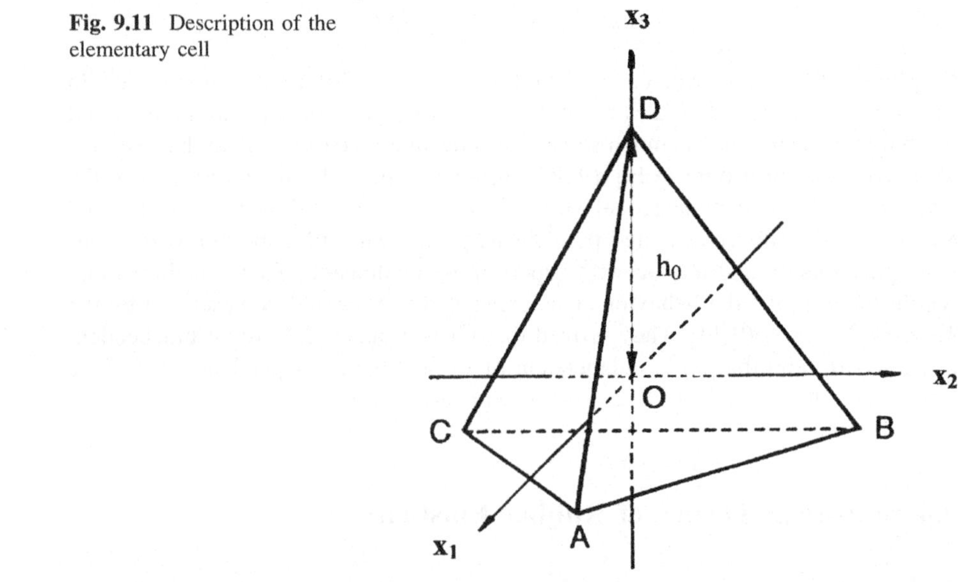

Fig. 9.11 Description of the elementary cell

A linear homogeneous deformation is applied characterized by the principal extension ratios λ_1, λ_2, λ_3 which is to be followed by the points A, B, C, and D. By this deformation, the probability of the conformation is changed as is its entropy and free energy. From the latter the deformation energy of the tetraeder is obtained as function of the extension ratios.

The first step is the mathematical description of the tetraeder referring to Fig. 9.11. The origin of the coordinate system is placed in the center O of the tetraeder.

Point D is assumed to lie on the x_3-axis at a distance h_0 from the origin; the points A, B, and C then come to lie in a plane parallel to the 1-2 plane. Under the assumption that A lies in the 1-3 plane, the coordinates of all four corners of the tetraeder are fixed as follows:

$$
\begin{aligned}
A &= \left[\frac{\sqrt{8}}{3}, 0, -\frac{1}{3}\right] \cdot h_0 \quad B = \left[-\frac{\sqrt{2}}{3}, \frac{\sqrt{2}}{\sqrt{3}}, -\frac{1}{3}\right] \cdot h_0 \\
C &= \left[-\frac{\sqrt{2}}{3}, -\frac{\sqrt{2}}{\sqrt{3}}, -\frac{1}{3}\right] \cdot h_0 \quad D = [0, 0, 1] \cdot h_0
\end{aligned} \tag{9.37}
$$

Applying a linear homogeneous deformation, the coordinates x_1, x_2, x_3 change to $x_1{}' = \lambda_1 \cdot x_1$, $x_2{}' = \lambda_2 \cdot x_2$, $x_3{}' = \lambda_3 \cdot x_3$, and the initial tetraeder changes to a tetraeder which is not regular any longer and has the corners A′, B′, C′, and D′ with the coordinates

$$
\begin{aligned}
A' &= \left[\frac{\sqrt{8}}{3}\lambda_1, 0, -\frac{1}{3}\lambda_3\right] \cdot h_0 \quad B' = \left[-\frac{\sqrt{2}}{3}\lambda_1, \frac{\sqrt{2}}{\sqrt{3}}\lambda_2, -\frac{1}{3}\lambda_3\right] \cdot h_0 \\
C' &= \left[-\frac{\sqrt{2}}{3}\lambda_1, -\frac{\sqrt{2}}{\sqrt{3}}\lambda_2, -\frac{1}{3}\lambda_3\right] \cdot h_0 \quad D' = [0, 0, \lambda_3] \cdot h_0
\end{aligned} \tag{9.38}
$$

We designate the vectors, $\underline{\mathrm{OA}'}$, $\underline{\mathrm{OB}'}$, $\underline{\mathrm{OC}'}$, $\underline{\mathrm{OD}'}$ as $\underline{R}_1$, $\underline{R}_2$, $\underline{R}_3$, $\underline{R}_4$. Of two of the properties of these vectors is made use: The one is that their vector sum vanishes

$$
\underline{R}_1 + \underline{R}_2 + \underline{R}_3 + \underline{R}_4 = \underline{0} \tag{9.39}
$$

The verification follows directly from Eq. (9.38). The other is the relation between the sum of square lengths of the vectors and the elongation ratios which follows as

$$
R_1^2 + R_2^2 + R_3^2 + R_4^2 = \frac{4h_0^2}{3}[\lambda_1^2 + \lambda_2^2 + \lambda_3^2] \tag{9.40}
$$

For the calculation of the entropy, we consider the tetraeder in its deformed state. In this case the cross-linking point P of the four chains not necessarily stays at the origin of the coordinate system. The vector $\underline{\mathrm{OP}}$ is designated by $\underline{s}$. The probability of the conformation for a fixed value of P will be equal to the product of four probabilities

$$
w(\lambda_1, \lambda_2, \lambda_3, P) = w_1 \cdot w_2 \cdot w_3 \cdot w_4 \tag{9.41}
$$

whereby w_1 is the probability that the endpoint of chain number 1 is situated at P if its starting point is at A′, w_2 the probability that the endpoint of chain number 2 is situated at P if its starting point is at B′, etc. Equation (2.36) describes the probability for a random flight chain with its starting point in O to have its end point in x as follows:

$$w(x,y,z) \equiv w(\underline{r}) = \left(\frac{3}{2\pi\langle r^2\rangle}\right)^{3/2} e^{-3r^2/2\langle r^2\rangle} \tag{2.36}$$

with $\langle r^2\rangle$ being the expectation value of the squared end to end distance of the chain in its undisturbed state.

From (2.36) the probability w_1 for a random flight chain with the starting point R_1 to have its end point at the position $\underline{s}$ follows as

$$w_1 = w(\underline{R}_1 - \underline{s}) = \left(\frac{3}{2\pi\langle r^2\rangle}\right)^{3/2} e^{-3(\underline{R}_1-\underline{s})^2/2\langle r^2\rangle} \tag{9.42}$$

and analog expressions are valid for w_2, w_3, and w_4. Inserting these relations into (9.41) results in

$$w(\lambda_1,\lambda_2,\lambda_3,P) = \left(\frac{3}{2\pi\langle r^2\rangle}\right)^{6} e^{-3[(\underline{R}_1-\underline{s})^2+(\underline{R}_2-\underline{s})^2+(\underline{R}_3-\underline{s})^2+(\underline{R}_4-\underline{s})^2]/2\langle r^2\rangle} \tag{9.43}$$

The expression in the brackets of Eq. (9.43) simplifies by using (9.39) and (9.40) to

$$(\underline{R}_1-\underline{s})^2+(\underline{R}_2-\underline{s})^2+(\underline{R}_3-\underline{s})^2+(\underline{R}_4-\underline{s})^2 = 4s^2+4h_0^2(\lambda_1^2+\lambda_2^2+\lambda_3^2)/3 \tag{9.44}$$

Then the probability reads as

$$w(\lambda_1,\lambda_2,\lambda_3,P) = \left(\frac{3}{2\pi\langle r^2\rangle}\right)^{6} e^{-6s^2/\langle r^2\rangle} e^{-2(\lambda_1^2+\lambda_2^2+\lambda_3^2)h_0^2/\langle r^2\rangle} \tag{9.43'}$$

Under a first assumption, the points A′, B′, C′ and D′ are kept fixed, while the cross-linking point P may be anywhere in space. Consequently, the probability of the conformation is found by integrating $w(\lambda_1, \lambda_2, \lambda_3, P)$ over all positions of P in the three-dimensional space with the result

$$W(\lambda_1,\lambda_2,\lambda_3) = \int_0^\infty 4\pi s^2 w(\lambda_1,\lambda_2,\lambda_3,P)ds$$
$$= \left(\frac{3}{2\pi\langle r^2\rangle}\right)^{6} e^{-2(\lambda_1^2+\lambda_2^2+\lambda_3^2)h_0^2/\langle r^2\rangle} \int_0^\infty 4\pi s^2 e^{-6s^2/\langle r^2\rangle} ds$$

The integral over the variable s does not depend on the values of λ_1, λ_2, λ_3, and constitutes a constant, being without any influence on the result, which reads as

$$W(\lambda_1, \lambda_2, \lambda_3) = \text{Const} \cdot e^{-2(\lambda_1^2+\lambda_2^2+\lambda_3^2)h_0^2/\langle r^2\rangle} \tag{9.45}$$

From the Boltzmann equation for the entropy

$$S = k \ln W(\lambda_1, \lambda_2, \lambda_3) \tag{9.46}$$

it follows that

$$S = \text{Const} - 2k\frac{h_0^2}{\langle r^2\rangle}(\lambda_1^2 + \lambda_2^2 + \lambda_3^2)$$

for the entropy of the unit cell in the deformed state. Deducting the entropy S_0 in the undeformed state, which is obtained by setting $\lambda_1 = \lambda_2 = \lambda_3 = 1$ in S, the change in entropy of the unit cell due to the deformation is found as

$$S - S_0 = -2k\frac{h_0^2}{\langle r^2\rangle}[(\lambda_1^2 + \lambda_2^2 + \lambda_3^2) - 3] \tag{9.47}$$

A difficult problem arises now from the question how to deduce the entropy change of the complete network from the entropy change of the unit cell.

The simplest assumption is the existence of n_c chains per unit volume between two cross-linking points, which are distributed over $n_c/4$ tetraeders. In this case Eq. (9.45) should be multiplied by a factor of $n_c/4$ to obtain the change in entropy per unit volume, the *specific deformation entropy* $s(\lambda_1,\lambda_2,\lambda_3)$,

$$s(\lambda_1, \lambda_2, \lambda_3) = -\frac{G}{2T}[(\lambda_1^2 + \lambda_2^2 + \lambda_3^2) - 3] = -\frac{G}{2T}[I_G - 3] \tag{9.48}$$

with T being the absolute temperature and G being an abbreviation for the expression

$$G = n_c kT\frac{h_0^2}{\langle r^2\rangle} \tag{9.49}$$

Under isothermal conditions the free energy density

$$w = u - Ts \tag{9.50}$$

determines the deformation. Assuming that the *internal energy density* u does not change with deformation, the *deformation energy density* is given by

$$w(I_G, II_G) = \frac{1}{2}G[I_G - 3] \tag{9.51}$$

which is equal to (9.26), the expression for the elastic deformation energy of an ideal rubber with the shear modulus G. In reality, the deformation energy density also changes somewhat with the deformation, but these effects are small compared with those of the entropy changes.

The shear modulus G is proportional to the absolute temperature and the number of chains between cross-links. This result is restricted to the following requirements

(1) The deformation does not change the internal energy of the material.
(2) All chains between cross-links have the same length and their statistics follow that in a theta-condition.
(3) The chains are long enough to be described by the Gaussian chain statistics.
(4) The deformations are limited to a range, in which the finite lengths of the chains do not affect the repulsive forces.

If these requirements are met, the Flory-Rehner model describes an ideal rubber with a shear modulus G.

The expression (9.49) for the shear modulus overestimates the repulsive forces of the network in the deformed state. By distributing the n_c chains over the $n_c/4$ tetraeders the assumption was made that also points in the interior of the material (the points A, B, C, and D of the inner tetraeders) are forced to follow the external deformation affinely. In reality those network points have more degrees of freedom. They are able to move freely, whereby however, the motion of those chain ends which are connected with chain ends of other tetraeders has to take place in compliance with the latter. Only chain ends which are on the surface of the specimen have to move affinely with the deformation.

Considering these facts Duiser and Staverman [12, 13] have refined the calculation of the shear modulus of the network with the following result. Let n_c be the number of chains between adjacent cross-links per unit volume, f the functionality of the cross-links and n the number of cross-links per unit volume, then the equation

$$n = 2n_c/f \tag{9.52}$$

expresses the fact that each chain between adjacent cross-links is fixed to exactly two cross-linking points. For the shear modulus the equation

$$\boxed{G = (n_c - n)kT\frac{h_0^2}{\langle r^2\rangle} = n(\frac{f}{2} - 1)kT\frac{h_0^2}{\langle r^2\rangle} = n_c(1 - \frac{2}{f})kT\frac{h_0^2}{\langle r^2\rangle}} \tag{9.53}$$

was derived instead of (9.49). h_0 is the mean distance between two cross-linking points, $\langle r^2\rangle$ the mean square end to end distance of the random walk chain. If the cross-linking takes place in the undeformed state, one may expect

$$h_0^2 = \langle r^2 \rangle \tag{9.54}$$

and the quotient $h_0^2/\langle r^2 \rangle$ does not play any role.

For *tri-functional cross-linking points* it follows from (9.52) and (9.53)

$$f = 3 \qquad G = \frac{1}{2} nkT = \frac{1}{3} n_c kT \tag{9.55}$$

and for *tetra-functional cross-linking points*

$$f = 4 \qquad G = nkT = \frac{1}{2} n_c kT \tag{9.56}$$

Equation (9.56) is identical to Eq. (4.9) for the shear modulus of a rubber with *n* tetra-functional cross-links per unit volume.

It is also identical with Eq. (4.6) for the shear modulus of an uncross-linked polymer in the rubber-elastic plateau. In this case n equals n_e, the *number of entanglements* per unit volume and the entanglements are considered as temporary tetra-functional cross-links.

Equation (9.56) is the base for the description of rheological properties of polymer melts by the Lodge theory of elastic liquids.

References

1. ISO 5893
2. ISO 527
3. ASTM D 638
4. Hartwig G (1978) Prog Coll Pol Sci 64:56
5. Ramsteiner F, private communication
6. Treloar LRG (1975) The physics of rubber elasticity. Clarendon Press, Oxford
7. Treloar LRG (1944) Trans Faraday Soc 40:59
8. Treloar LRG, (1956) The structure and mechanical properties of rubberlike materials. In: Stuart HA (ed) Die Physik der Hochpolymeren Bd. IV. Springer Berlin, Heidelberg, New York
9. Mooney M ((1940) J Appl Phys 11:582
10. Rivlin RS, Saunders DW, Phil Trans Roy Soc London (1951) A 243:251
11. Flory PJ, Rehner J Jr (1943) J Chem Phys 11:512
12. Duiser JA (1965) Doctoral thesis Leiden
13. Duiser JA, Staverman AJ (1964) Physics of non-crystalline solids. In: Prins JA (ed). North Holland, Amsterdam S. p 376

Chapter 10
Equations of State for Polymer Melts

10.1 Introduction

For the description of the rheological behavior of polymer melts, no general phenomenological framework exists up to now. However, very significant progress has been made in the development of special molecular theories in this field during the last 20 years. As the intention of this book is to give an overview of mechanical properties of polymers, but not to go into the details of molecular theories, we restrict ourselves to discuss a number of equations of state which have been proposed for the description of rheological properties of polymers.

For the reader who would like to get deeper into theoretical matters, we recommend a number of monographs dealing with molecular theories of the rheology of polymer melts as for example [1–5]. Especially the survey by Dealy and Larson of recent developments in theory [5] is to be mentioned in this connection.

The majority of the various constitutive equations which have been proposed for polymer melts contain the time dependence in the form of a simple factor in the kernel of the integral equation which describes the stress tensor at the observation time t as the sum of the contributions during the deformation history. This factor—the memory function $m(t)$—is related to the linear time-dependent relaxation modulus $G(t)$ of the melt.

An important result of the theoretical studies by Dealy and Larson concerns the *time dependence* of the *relaxation modulus* $G(t)$ of polymer melts. Due to their investigations one has to distinguish between linear and branched polymers, while the former also include short-chain branched structures.

For linear, nearly monodisperse polymers, the storage and loss moduli as functions of the angular frequency could be described almost quantitatively for a large number of polymer melts by general equations involving only three material-specific parameters, which are the molar mass M, the entanglement molar mass M_e, and G_e^0, the modulus of the melt before the processes of *reptation* and *primitive path fluctuations* come into action. The polymers discussed by Dealy and Larson were linear PE, atactic, isotactic, and syndiotactic PP, PIB, cis-PI, PMMA, atactic and isotactic PS, PVA, and PDMS. The values for G_e^0, M_e, and M_c were

H. Münstedt and F. R. Schwarzl, *Deformation and Flow of Polymeric Materials*,
DOI: 10.1007/978-3-642-55409-4_10,

tabulated in Appendix A of [5]. The conclusion can be drawn that for linear monodisperse polymer melts, the linear viscoelastic properties may be approximately predicted.

They also could explain the existence of the empirical relation between molar mass and viscosity of linear polymer melts, which is observed with overwhelming experimental evidence stating that the zero-shear viscosity of linear polymer melts is proportional to M^{α} above a certain *critical molar mass* M_c.

$$\begin{aligned} \eta_0 &= a(M/M_c)^{\alpha} \quad \text{for} \quad M \geq M_c \text{ and} \\ \eta_0 &= a(M/M_c) \quad \text{for} \quad M \leq M_c \end{aligned} \tag{10.1}$$

with $\alpha \cong 3.4$–3.6 and a being a constant depending on the type of the polymer and the temperature. The ratio M_c/M_e varies from polymer to polymer showing values between 2.0 and 3.5.

10.2 Rheological Equation of State for the Elastic Liquid After Lodge

10.2.1 The Constitutive Equation

A model which has been useful for the description of some aspects of the rheological behavior of polymer melts, is that of the elastic liquid proposed by Lodge [6]. This model is based on the assumption that entanglements in a polymer melt are created and annihilated continuously by the Brownian motion. The polymer melt is regarded as a temporary network of entanglements with a finite life time only depending on the temperature. A certain part of the entanglement network created at the time t' in the past, is deformed during the following period by the flow process, and contributes to the stress tensor at the observation time t. Its contribution to the stress tensor is calculated using the theory of the elasticity of an ideal rubber as described in the Sects. 9.3 and 9.4.

Consider an entanglement network which had existed from the time $t_0 = 0$ at which the deformation history started until the observation time t and which consisted during this time of n entanglements per unit volume. According to Eqs. (9.25) and (9.56), its contribution to the stress tensor at the observation time t would be

$$\sigma_{ik}(t) + p\delta_{ik} = G \cdot G_{ik} \tag{9.25}$$

with G being the shear modulus G = nkT and G_{ik} the components of the absolute Green tensor.

$$G_{ik} = \frac{\partial f_i}{\partial x_m^0}\frac{\partial f_k}{\partial x_m^0} = \frac{\partial x_i}{\partial x_m^0}\frac{\partial x_k}{\partial x_m^0} \tag{8.18}$$

The functions f_i are defined by Eq. (8.1). The absolute Green tensor $\underline{\underline{G}}$ measures the strain at the time t relative to the undeformed state at the time t_0.

Consider next a network which was created at the past time t' and which still exists at the observation time t. Its contribution to the stress tensor at time t is given by Eqs. (9.25) and (9.56), whereby the absolute Green tensor is to be replaced now by the relative Finger tensor (8.35)

$$B_{ik}(x, t, t') = \frac{\partial x_i}{\partial x'_m}\frac{\partial x_k}{\partial x'_m} \tag{8.35}$$

which measures the strain at the time t' relative to the state at time t.

The number of entanglements per unit volume which were created in the time interval between t' and $t' + \mathrm{d}t'$, and which still exist at the observation time t, only depends on the time difference $t - t'$ and is proportional to $\mathrm{d}t'$. Let this number be $n(t - t')\mathrm{d}t'$, then the contribution of these entanglements to the stress tensor at the observation time t is $m(t - t')\cdot B_{ik}(x,t,t')\mathrm{d}t'$ with $m(t - t') = n(t - t')kT$. Summation over the contributions of all entanglements which were created in the past before the observation time t yields the *constitutive equation* of *Lodge's elastic liquid*

$$\sigma_{ik}(t) + p\delta_{ik} = \int_{-\infty}^{t} m(t - t')B_{ik}(\underline{x}; t, t')\mathrm{d}t' \tag{10.2}$$

The kernel $m(t)$ is called the *memory function*. It will be shown in (10.12) to be equal to the negative time derivative of the *time-dependent shear relaxation modulus* of the liquid $G(t)$ at small strains

$$m(t) = -\frac{\mathrm{d}G(t)}{\mathrm{d}t} \tag{10.3}$$

The long-time limit of the relaxation modulus vanishes because liquids show *fading memory* expressed by the equation

$$G_\infty = \lim_{t\to\infty} G(t) = 0 \tag{10.4}$$

and therefore (10.3) may be integrated to

$$G(t) = G_0 - \int_0^t m(\tau)\mathrm{d}\tau = G_\infty + \int_t^\infty m(\tau)\mathrm{d}\tau = \int_t^\infty m(\tau)\mathrm{d}\tau \tag{10.5}$$

and to

$$G_0 = \lim_{t \to +0} G(t) \tag{10.6}$$

Let us assume that the strain history started at $t = 0$. Then before $t = 0$ no deformation was applied to the sample and due to (8.36) and (8.38) the relative Finger tensor reduces in this time interval to the absolute Green tensor

$$B_{ik}(\underline{x}; t, t') = \frac{\partial x_i}{\partial x'_m} \frac{\partial x_k}{\partial x'_m} = G_{ik}(\underline{x}; t) \quad \text{for} \quad t' < 0 \tag{8.36}$$

which does not depend on t'. Using this result and (10.5), the constitutive Eq. (10.2) may be written as

$$\sigma_{ik} + p\delta_{ik} = G(t) \cdot G_{ik}(\underline{x}; t) + \int_0^t m(t - t') B_{ik}(\underline{x}; t, t') \mathrm{d}t' \tag{10.7}$$

This constitutive equation is the most simple one for the description of an elastic liquid under large deformations. It is a linear integral equation expressing the stress tensor by the relative Finger tensor with the time-dependent function $m(t)$ as kernel.

Another form of the Lodge equation is found by partial integration of (10.7) and using (8.36) and (8.37)

$$\sigma_{ik} + p\delta_{ik} = -\int_0^t G(t - t') \frac{\partial B_{ik}(\underline{x}; t, t')}{\partial t'} \mathrm{d}t' \tag{10.8}$$

which is valid, if $B_{ik}(x,t,t')$ is continuous and piecewise differentiable, with respect to t'.

For small deformations Eq. (10.7) becomes

$$\sigma_{ik}(t) + p\delta_{ik} \cong 2G_0 \varepsilon_{ik}(t) - 2\int_0^t m(t - t') \varepsilon_{ik}(t') \mathrm{d}t' \tag{7.27}$$

Making use of Eqs. (8.18a) and (8.19a), (10.8) changes into

$$\sigma_{ik}(t) + p\delta_{ik} \cong 2\int_0^t G(t - t') \dot{\varepsilon}_{ik}(t') \mathrm{d}t' \tag{7.28}$$

which are identical with the equations of state for a linear viscoelastic incompressible material [(cf. (7.27)). (7.28) is found from (7.27) by partial integration.]

10.2.2 The Lodge Liquid in Time-Dependent Simple Shear

The Lodge liquid is discussed in the following for time-dependent simple shear. *Time-dependent simple shear* is characterized by the path equations

$$x_1 = x_1^0 + x_2^0\gamma(t) \quad x_2 = x_2^0 \quad x_3 = x_3^0 \quad \text{for} \quad t \geq 0 \tag{10.9}$$

and by the relative path equations

$$\begin{aligned} x_1 &= x_1' + x_2'[\gamma(t) - \gamma(t')] \quad \text{for} \quad 0 < t' < t \\ x_2 &= x_2' \quad x_3 = x_3' \\ &\text{and} \quad \gamma(t') = 0 \quad \text{for} \quad t' < 0 \end{aligned} \tag{10.10}$$

The non-zero components of the relative Finger tensor and of the absolute Green tensor result by inserting (10.9) and (10.10) into (8.35) and (8.18) as

$$\begin{aligned} B_{11}(\underline{x};t,t') &= 1 + [\gamma(t) - \gamma(t')]^2 & G_{11}(\underline{x};t) &= 1 + \gamma^2(t) \\ B_{12}(\underline{x};t,t') &= \gamma(t) - \gamma(t') & G_{12}(\underline{x};t) &= \gamma(t) \\ B_{22}(\underline{x};t,t') &= B_{33}(\underline{x};t,t') = 1 & G_{22}(\underline{x};t) &= G_{33}(\underline{x};t) = 1 \end{aligned} \tag{10.11}$$

Inserting these tensor components into Eqs. (10.7) yields the following equations for the stresses

$$\begin{aligned} \sigma_{12}(t) &= G_0\gamma(t) - \int_0^t m(t-t')\gamma(t')\mathrm{d}t' = G_0\gamma(t) + \int_0^t \dot{G}(t-t')\gamma(t')\mathrm{d}t' \\ \sigma_{11}(t) - \sigma_{22}(t) &= G(t)\gamma^2(t) + \int_0^t m(t-t')[\gamma(t) - \gamma(t')]^2\mathrm{d}t' \geq 0 \\ \sigma_{22}(t) - \sigma_{33}(t) &= 0 \end{aligned} \tag{10.12}$$

The first of Eqs. (10.12) represents the *superposition principle* known for the linear behavior of viscoelastic materials (cf. Eq. 5.17), but in this case it is valid for large shear strains in the non-linear regime, too. It follows that $G(t)$ of Eq. (10.12) is equal to the shear modulus of the theory of linear viscoelasticity. Contrary to this theory, however, the Lodge theory yields a first normal stress difference typical of polymer melts which according to Eq. (10.12) is positive.

What the Lodge theory is not able to describe is the occurrence of a second normal stress difference which according to Eq. (10.12) is zero and the shear rate dependencies of the shear stress and the first normal stress difference as well, which are found experimentally for polymer melts (cf. Fig. 11.4). Therefore, the Lodge theory represents a rudimentary description of the viscoelastic properties of polymer melts, only, which however, shows some merits in the modelling of the time behavior.

10.2.3 The Lodge Liquid in a Stressing Experiment in Simple Shear

A *stressing experiment* is a special case of time-dependent simple shear with a constant shear rate $\dot{\gamma}_0$. Its time history may be set as

$$\begin{aligned} \gamma(t) &= \dot{\gamma}_0 t \quad \text{for} \quad t \geq 0 \quad \text{and} \\ \gamma(t) &= 0 \quad \text{for} \quad t < 0 \end{aligned} \tag{10.13}$$

For the calculation of the stresses, Eq. (10.8) is used. After inserting (10.11) and (10.13) it follows with $\xi = t - t'$

$$\sigma_{12}(t) = \eta^0(t)\,\dot{\gamma}_0 \quad \text{with} \quad \eta^0(t) = \int_0^t G(\xi)\mathrm{d}\xi \tag{10.14}$$

$$\sigma_{11}(t) - \sigma_{22}(t) = n_1^0(t)\dot{\gamma}_0^2 \quad \text{with} \quad n_1^0(t) = 2\int_0^t G(\xi)\xi\mathrm{d}\xi \tag{10.15}$$

The *viscosity* $\eta^0(t)$ is time dependent, but independent of the shear rate. The flow behavior of the Lodge liquid is, therefore, Newtonian. The *first normal stress coefficient* $n_1^0(t)$ is time dependent, but independent of the shear rate, too. The *second normal stress coefficient* $n_2(t)$ is zero.

Both functions $\eta^0(t)$ and $n_1^0(t)$ increase monotonously with time and reach a plateau value after sufficiently long times as $G_\infty = 0$. They may be expressed in terms of the relaxation spectrum $g(\tau)$. For this purpose, the stress relaxation modulus is written as integral transform of the spectrum using Eq. (5.24) with $G_\infty = 0$, then the order of the integrals is changed and the inner integral is evaluated. The following expressions result for the viscosity and the first normal stress coefficient

$$\eta^0(t) = \int_0^\infty g(\tau)\tau[1 - e^{-t/\tau}]\mathrm{d}\tau \tag{10.16}$$

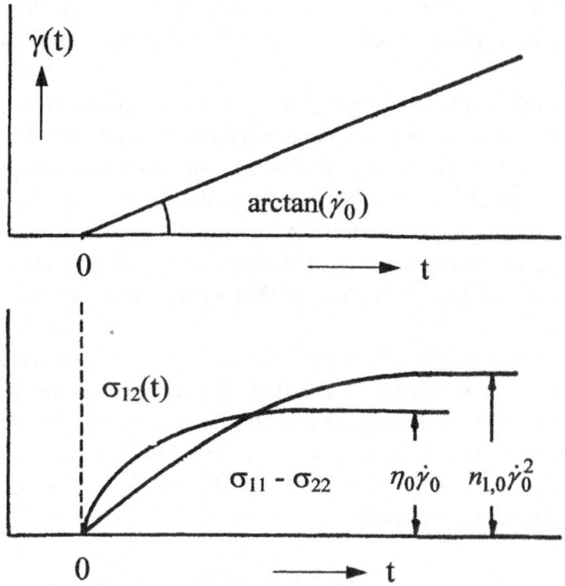

Fig. 10.1 The Lodge liquid in a shear stressing experiment: shear strain (*above*) and shear stress and first normal stress difference (*below*)

$$n_1^0(t) = 2\int_0^\infty g(\tau)\tau[\tau(1 - e^{-t/\tau}) - te^{-t/\tau}]\mathrm{d}\tau \tag{10.17}$$

The plateau value of the function $\eta^0(t)$ is obtained for $t \to \infty$ from (10.14) or (10.16) and that of $n_1^0(t)$ from (10.15) or (10.17) for $t \to \infty$. The plateau values are

$$\eta^0(t \to \infty) = \eta_0 = \int_0^\infty G(\xi)\mathrm{d}\xi = \int_0^\infty g(\tau)\tau\mathrm{d}\tau \tag{10.18}$$

$$n_1^0(t \to \infty) = n_{1,0} = 2\int_0^\infty G(\xi)\xi\mathrm{d}\xi = 2\int_0^\infty g(\tau)\tau^2\mathrm{d}\tau \tag{10.19}$$

The time dependence of the shear strain, the shear stress, and the first normal stress difference is sketched in Fig. 10.1. From this figure it may be seen that $N_1(t) = \sigma_{11} - \sigma_{22}$ reaches its stationary value later than σ_{12} as for $n_{1,0}$ the longer relaxation times have more importance than for η^0. It may also be seen that the Lodge theory is not able to describe a stress overshoot as experimentally found in Fig. 11.2, for example.

As the stationary values of η^0 and n_1^0 are independent of the shear rate, Lodge's liquid exhibits a shear stress and a first normal stress difference proportional to the first and second power of the shear rate, respectively, i.e., it exhibits Newtonian flow behavior and a quadratic dependence of the first normal stress difference on the shear rate.

For polymer melts, this is only true for small values of the shear rate, as may be seen from Fig. 11.4, where the stationary values of η and n_1 are shown as functions of the shear rate for LDPE. Outside the Newtonian regime both quantities show a strong decrease with the shear rate, reaching finally the non-Newtonian regime where $\eta^0(\infty)$ decreases approximately linearly with $\dot{\gamma}_0$ and $n_1(\infty)$ even decreases approximately quadratically with $\dot{\gamma}_0$. Consequently, Lodge's theory is unable to describe the non-Newtonian behavior neither for the viscosity nor for the normal stress coefficient.

The Eqs. (10.16) and (10.17) allow the derivation of numerical formulae to convert dynamic shear moduli into the time dependence of the viscosity and the first normal stress coefficient as predicted by the Lodge theory. A comparison of the representation of $G'(\omega)$ and $G''(\omega)$ as functions of the angular frequency [formulae (5.101) and (5.102)] with (10.16) yields the following approximation for the time-dependent shear viscosity

$$\eta^0(t)/t \cong [G'(\omega) + 0.27G''(2\omega) + 0.115G''(4\omega)]_{\omega=1/t} \tag{10.20}$$

with bounds for the relative error equal to +12 % and −12 %. Meissner confirmed the validity of this approximation on data for an LDPE melt [7].

More accurate approximation formulae have been derived and checked for PS-melts by Gortemaker et al. [8]. For the viscosity they read

$$\begin{aligned} & && \text{error bounds} \\ \eta^0(t)/t &\cong [G'(\omega) - 0.322G''(\omega/2) + 0.580G''(2\omega)]_{\omega=1/t} && \pm 5.5\% \\ \eta^0(t)/t &\cong [G'(\omega) - 1.142G''(\omega) + 1.428G''(3\omega/2)]_{\omega=1/t} && \pm 7.0\% \\ \eta^0(t)/t &\cong [G'(\omega) - 0.540G''(3\omega/2) + 0.906G''(2\omega)]_{\omega=1/t} && \pm 9.4\% \end{aligned} \tag{10.21}$$

and for the first normal stress coefficient

$$\begin{aligned} & && \text{error bounds} \\ n_1^0(t)/t^2 &\cong [0.939G'(\omega/2) - 2.41G'(\omega) + 2.47G'(1.30\omega)]_{\omega=1/t} && \pm 4.8\% \\ n_1^0(t)/t^2 &\cong [1.68G'(0.6\omega) - 5.41G'(\omega) + 4.72G'(1.20\omega)]_{\omega=1/t} && \pm 4.8\% \\ n_1^0(t)/t^2 &\cong [0.50G'(\omega) - 0.415G'(2\omega) - 0.010G'(4\omega)]_{\omega=1/t} && \pm 9.8\% \end{aligned} \tag{10.22}$$

As far as the Lodge theory may be applied, those formulae are very useful as they connect the time-dependent viscosity and the first normal stress coefficient with the dynamic shear moduli. Of course, they are not appropriate for use in the non-Newtonian regime.

10.2.4 The Lodge Liquid in Shear Creep and Creep Recovery

Next we compare the stressing experiment in simple shear with a creep experiment in simple shear, both followed by recovery, for the Lodge liquid. The creep experiment is defined as follows: At time $t = 0$, the shear stress σ_{12} is instantaneously increased from zero to the value σ_0 and then kept constant. At the same time, a normal stress difference $\sigma_{11} - \sigma_{22}$ has to be applied in order to maintain the deformation of simple shear. This may be realized by using an instrument with constant gap width between the plates.

The first of Eqs. (10.12) expresses the principle of superposition valid between the shear stress $\sigma_{12}(t)$ and the shear strain $\gamma(t)$. According to Sect. 5.3 an inverse equation exists which can be written as

$$\gamma(t) = J_0\sigma_{12}(t) + \int_{-\infty}^{t} \dot{J}(t-t')\sigma_{12}(t')\mathrm{d}t' \tag{5.15}$$

with $J(t)$ being the shear creep compliance which is related to the shear modulus by the equation

$$\int_{0}^{t} G(t-t')J(t')\mathrm{d}t' = t \quad \text{for all} \quad t \geq 0 \tag{5.48}$$

The creep experiment in simple shear is defined by

$$\begin{aligned} \sigma_{12}(t) &= 0 && \text{for} \quad t<0 \\ \sigma_{12}(t) &= \sigma_0 && \text{for} \quad t\geq 0 \end{aligned} \tag{10.23}$$

and

$$\gamma(t) = \sigma_0 J(t) \quad \text{for} \quad t \geq 0 \tag{10.24}$$

A Lodge liquid behaves in a creep experiment in simple shear like a linear viscoelastic incompressible material. But additionally to the shear stress, a first normal stress difference occurs in case of the Lodge liquid which follows by inserting (10.24) into the second of Eqs. (10.12).

In Fig. 10.2, the stressing experiment (broken line) is compared with the creep experiment (full line) for a Lodge liquid in simple shear. The constant stress in the creep experiment is chosen such that $\sigma_0 = \eta_0\dot{\gamma}_0$ where $\dot{\gamma}_0$ equals the constant shear rate in the stressing experiment and η_0 the viscosity of the Lodge liquid.

The upper part of the figure shows the shear strain $\gamma(t)$ divided by the constant shear stress σ_0, the lower part shows the shear stress. In the stressing experiment, the shear stress starts at zero and increases monotonously with time until it reaches

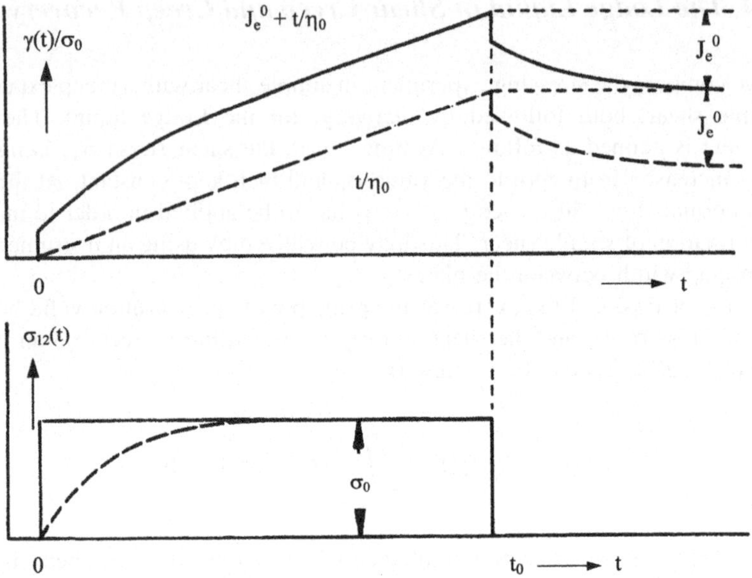

Fig. 10.2 Shear stressing experiment (*broken line*) and shear creep experiment (*full line*) of the Lodge liquid with recovery: shear strain divided by σ_0 (*above*) and shear stress (*below*)

finally its stationary value $\sigma_0 = \eta_0 \dot{\gamma}_0$. As $\dot{\gamma}_0$ was chosen such that $\eta_0 \dot{\gamma}_0$ equals the constant stress σ_o in the creep experiment, the shear stress becomes equal in both experiments in the stationary state as indicated in the lower part of Fig. 10.2.

In the creep experiment, the compliance may be decomposed as

$$\gamma(t)/\sigma_0 = J_0 + \psi(t) + t/\eta_0 \tag{5.27}$$

and in the stationary state one can write

$$\gamma(t)/\sigma_0 = J_e^0 + t/\eta_0 \tag{5.46}$$

with J_e^0 being the steady-state recoverable compliance and t/η_0 the flow term.

In the stressing experiment

$$\frac{\gamma(t)}{\sigma_0} = \frac{\dot{\gamma}_0}{\sigma_0} t = \frac{t}{\eta_0} \tag{10.25}$$

is valid. Hence the shear strain in the stressing experiment equals the flow term in the creep experiment. In the stationary state, the strain in both experiments differs only by the constant term $J_e^0 \sigma_0$ and therefore the molecular state in the melt will be the same in both the experiments as the effect of the additive component in strain has reached its equilibrium value in the stationary state.

At the time $t = t_0$ the shear stress is released, but a first normal stress difference has further to be applied to maintain during the recovery period a simple shear. If the recovery experiment starts after reaching the stationary state, it will proceed in the same way in both experiments. The recovery curve after a creep experiment is known from Sect. 5.5 and has been drawn in Fig. 10.2. The *recovery* after the *stressing experiment* proceeds in parallel to that of the *creep recovery*, but $\sigma_0 J_e^0$ below. The maximum recoverable strain is the same in both cases, viz., $\gamma_e = \sigma_0 J_e^0 = \eta_0 \dot{\gamma}_0 J_e^0$.

The following expression has been derived in Sect. 5.8 for the second moment of the relaxation spectrum

$$J_e^0 \eta_0^2 = \int_0^\infty \tau^2 g(\tau) \mathrm{d}\tau \tag{5.106}$$

Comparing this with Eq. (10.19), we find

$$\lim n_1^0(t \to \infty) \equiv n_{1,0} = 2\eta_0^2 J_e^0 \tag{10.26}$$

valid for the Lodge liquid.

10.2.5 The Lodge Liquid in Harmonic Oscillation in Simple Shear

In analogy to Sect. 5.8 the *oscillatory* experiment defined by prescribing the shear strain as

$$\begin{aligned} \gamma(t) &= 0 & \text{for} \quad t<0 \\ \gamma(t) &= \gamma_0 \sin(\omega t) & \text{for} \quad t \geq 0 \end{aligned} \tag{5.70}$$

is discussed in the following. Inserting these conditions into the first of Eqs. (10.12), the shear stress in the stationary state of the oscillation follows like in Sect. 5.8 as

$$\sigma_{12}(t) = \gamma_0 [G'(\omega) \sin(\omega t) + G''(\omega) \cos(\omega t)] \tag{10.27}$$

$G'(\omega)$ and $G''(\omega)$ are the *storage* and *loss shear modulus*, defined by Eqs. (5.74) and (5.75). For the Lodge liquid, the shear stress is the same as for a linear viscoelastic material, while the strain amplitude γ_0 does not need to be small now. Contrary to the linear viscoelastic case, besides the shear stress an oscillating first normal stress difference occurs. The normal stress difference is found by inserting (5.70) into the second of Eqs. (10.12) as

$$\sigma_{11}(t) - \sigma_{22}(t) = \frac{1}{2}\gamma_0^2\{2G'(\omega) - \cos(2\omega t)[2G'(\omega) - G'(2\omega)]\} + \\ + \frac{1}{2}\gamma_0^2\{\sin(2\omega t)[2G''(\omega) - G''(2\omega)]\} \tag{10.28}$$

The first normal stress difference is harmonic too, but with the double frequency.

If $G'(\omega)$ and $G''(\omega)$ are expressed as integral transforms of the relaxation spectrum $g(\tau)$, Eqs. (5.102) and (5.104), but without the term G_∞, are found, viz.,

$$G''(\omega) = \int_0^\infty g(\tau)\frac{\omega\tau}{1+\omega^2\tau^2}\,d\tau \tag{10.29}$$

$$G'(\omega) = \int_0^\infty g(\tau)\frac{\omega^2\tau^2}{1+\omega^2\tau^2}\,d\tau \tag{10.30}$$

From these equations it is seen that $G''(\omega)/\omega$ and $G'(\omega)/\omega^2$ stay finite with $\omega \to 0$ and these limits are identical with those in (10.18) and (10.19). Consequently, for the *Lodge liquid,* the following equations are valid

$$\lim_{\omega\to 0}\frac{G'(\omega)}{\omega^2} = \lim_{t\to\infty}\frac{[\sigma_{11}(t) - \sigma_{22}(t)]}{2\dot{\gamma}_0^2} = \eta_0^2 J_e^0 = \frac{1}{2}n_{1,0} \tag{10.31}$$

$$\lim_{\omega\to 0}\frac{G''(\omega)}{\omega} = \lim_{t\to\infty}\frac{\sigma_{12}(t)}{\dot{\gamma}_0} = \eta_0 \tag{10.32}$$

In the region of small ω ($\omega \to 0$) it follows:

$$G'(\omega) \cong \frac{n_{1,0}}{2}\omega^2 \tag{10.31a}$$

$$G''(\omega) \cong \eta_0\omega \tag{10.32a}$$

and in the region of small $\dot{\gamma}_0$ ($\dot{\gamma}_0 \to 0$)

$$\frac{\sigma_{11}(t\to\infty) - \sigma_{22}(t\to\infty)}{2} \cong \frac{n_{1,0}\dot{\gamma}_0^2}{2} \tag{10.31b}$$

$$\sigma_{12}(t\to\infty) \cong \eta_0\dot{\gamma}_0 \tag{10.32b}$$

These relations mean that if $G''(\omega)$ is plotted versus ω and σ_{12} $(t \to \infty)$ is plotted versus $\dot{\gamma}_0$ whereby ω is set equal to $\dot{\gamma}_0$, both quantities should coincide and start increasing proportional to $\dot{\gamma}_0$ in the region of small ω and $\dot{\gamma}_0$, respectively. Correspondingly, if $G'(\omega)$ is plotted versus ω and $[\sigma_{11}(t \to \infty) - \sigma_{22}(t \to \infty)]/2$

versus $\dot{\gamma}_0$ whereby ω is set equal to $\dot{\gamma}_0$, these two quantities should coincide and start increasing proportional to $\dot{\gamma}_0^2$ in the region of small ω and $\dot{\gamma}_0$, respectively.

These equations follow from the Lodge theory but are part of another more general analogy found experimentally between the quantities characterizing a dynamic experiment and a stressing experiment in simple shear, the *analogy of* Cox and Merz [9].

This analogy which is discussed in Chap. 11 [formulae (11.6) and (11.9)] relates the characteristic quantities found in *oscillatory shear* in the limit of *small strain amplitudes* to those in a *stressing experiment* in simple shear in the limit of *small shear rates*. It may be summarized as follows:

$G'(\omega)$ as a function of ω runs similar to $(\sigma_{11} - \sigma_{22})/2$ as a function of $\dot{\gamma}_0$ in the stationary state of a stressing experiment in simple shear, and $G''(\omega)$ as a function of ω runs similar to σ_{12} as a function of $\dot{\gamma}_0$, if ω is set equal to $\dot{\gamma}_0$. In the region of small ω resp. $\dot{\gamma}_0$ the corresponding functions coincide. For larger values of ω resp. $\dot{\gamma}_0$, $G''(\omega)$ bends horizontally and σ_{12} $(t \to \infty)$ exceeds $G''(\omega)$ and $G'(\omega)$ bends horizontally and $(\sigma_{11} - \sigma_{22})/2$ exceeds $G'(\omega)$.

It will be shown in Chap. 11 that the following analogies are found experimentally:

$$\eta(\dot{\gamma}_0) = \eta(\dot{\gamma}_0, t \to \infty) \cong |\eta^*(\omega)| \geq \frac{G''}{\omega} \quad \text{for} \quad \dot{\gamma}_0 = \omega \tag{11.6}$$

and

$$n_1(\dot{\gamma}_0, t \to \infty) \cong 2G' \frac{(1 + (G'/G'')^2)^{3/2}}{\omega^2} \geq 2\frac{G'}{\omega^2} \quad \text{for} \quad \dot{\gamma}_0 = \omega \tag{11.9}$$

In the limit $\omega \to 0$ the loss modulus is considerably larger than the storage modulus $G'' \gg G'$ and therefore the equality sign on the right-hand side of Eqs. (11.6) and (11.7) will be valid. These are Eqs. (10.32) and (10.31) following from the Lodge theory. If on the other hand ω becomes larger, $G''(\omega)$ and $G'(\omega)$ bend horizontally and even cross each other. In this region, $G''(\omega) \cong G'(\omega)$ and the > sign in Eqs. (11.6) and (11.7) will be valid. This agrees with the last sentence of the statement above.

Figure 10.3 illustrates these statements schematically.

Note that due to the Lodge theory σ_{12} and $\sigma_{11} - \sigma_{22}$ should follow straight lines in this double logarithmic plot, while experimentally both are well known to bend horizontally in the non-Newtonian region. Only at very low values of ω and $\dot{\gamma}_0$, the result of Lodge's theory is found experimentally, viz., that the two corresponding quantities coincide and follow straight lines with the double-logarithmic slopes 2 and 1, respectively. Later, all curves bend horizontally and spread, while the moduli run below the shear stress and the first normal stress difference.

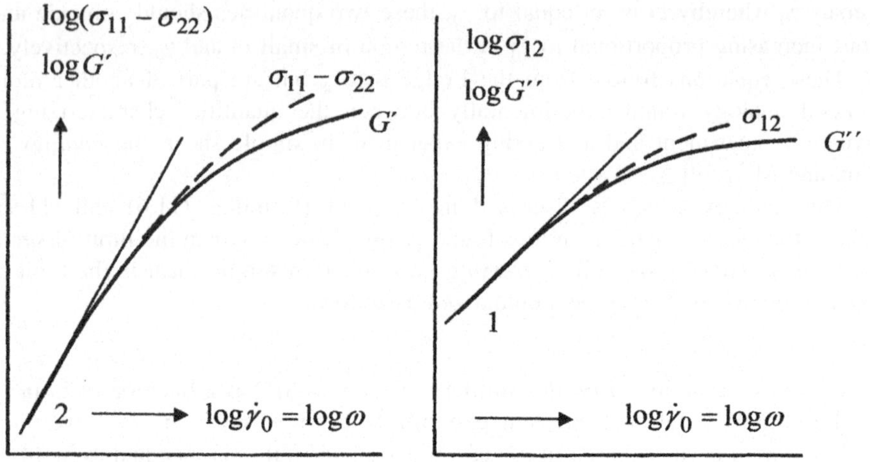

Fig. 10.3 Illustration of the analogy of Cox and Merz: The left-hand side shows schematically in double logarithmic representation the storage modulus $G'(\omega)$ in oscillating shear and the first normal stress difference $\sigma_{11} - \sigma_{22}$ in a stressing experiment as functions of the angular frequency ω or the shear rate $\dot{\gamma}_0$, the right-hand side shows the loss modulus $G''(\omega)$ in oscillating shear and the shear stress σ_{12} in a stressing experiment. Only at very low values of ω or $\dot{\gamma}_0$ the *curves* run as *straight lines* with double logarithmic slopes 2 and 1, respectively

10.2.6 The Calculation of the Relaxation Spectrum from Dynamic Data

Storage and loss moduli will be mostly determined at small oscillation amplitudes and, therefore, the formulae of the linear viscoelastic theory are applicable to evaluate the measurements of dynamic moduli of polymer melts. Two problems which often occur in this context are the calculation of the relaxation spectrum and the retardation spectrum of the polymer. Expressed in terms of a *discrete relaxation spectrum* of *n relaxation times*, the stress relaxation modulus, and the storage and loss moduli read (cf. Sect. 5.8)

$$G(t) = \sum_{k=1}^{n} a_k e^{-t/\tau_k} \tag{10.33}$$

$$G'(\omega) = \sum_{k=1}^{n} a_k \frac{\omega^2 \tau_k^2}{1 + \omega^2 \tau_k^2} \tag{10.34}$$

$$G''(\omega) = \sum_{k=1}^{n} a_k \frac{\omega \tau_k}{1 + \omega^2 \tau_k^2} \tag{10.35}$$

where ($\tau_1 < \tau_2 < \ldots < \tau_n$) are the *n relaxation times* arranged after their magnitude and ($a_1, a_2, \ldots a_n$) the corresponding *relaxation strengths*.

Applying these equations to derive the relaxation spectrum from dynamic data measured on a polymer melt, a number of precautions have to be taken into account.

1. From experimental data for $G'(\omega)$ and $G''(\omega)$, the relaxation spectrum *cannot* be *derived in a unique way*, but an infinite number of different spectral representations is always possible to describe the experimental data with equal accuracy. Therefore, a limitation of the number of possible descriptions is necessary to simplify the mathematical problem. This is attained by predetermining the number and the values of the relaxation times and by calculating the corresponding relaxation strengths which yield the best description of the given data.
2. Moduli and angular frequency mostly extend over several orders of magnitude. In those cases the experimental error is generally proportional to the magnitude of the measured quantity, and the property which is to be minimized in determining the appropriate approximation is not the sum of the squares of the differences between theoretical and experimental values of G' and G'' but the sum of the squares of their ratios minus unity. The quantity which has to be minimized will therefore be the variance

$$V = \frac{1}{2n_m - 1} \sum_{i=1}^{n_m} \left\{ \left(\frac{G_{1i}}{g_{1i}} - 1 \right)^2 + \left(\frac{G_{2i}}{g_{2i}} - 1 \right)^2 \right\} \tag{10.36}$$

with g_{1i} and g_{2i} being the n_m values of the storage moduli and the loss moduli measured at the angular frequencies ω_i with $i = 1, 2, \ldots n_m$ and G_{1i} and G_{2i} the corresponding values obtained by inserting the angular frequencies of the measurements into Eqs. (10.34) and (10.35).

$$G_{1i} = G'(\omega_i) = \sum_{k=1}^{n} a_k \frac{\omega_i^2 \tau_k^2}{1 + \omega_i^2 \tau_k^2} \tag{10.37}$$

and

$$G_{2i} = G''(\omega_i) = \sum_{k=1}^{n} a_k \frac{\omega_i \tau_k}{1 + \omega_i^2 \tau_k^2} \tag{10.38}$$

3. All relaxation strengths have to be positive, which can be achieved by an appropriate choice for the relaxation times. They are chosen logarithmically equidistant corresponding to a factor $\sqrt{10}$ between two successive times as

$$\tau_n = \tau_{\max} \qquad \tau_{n-1} = \kappa\tau_n \qquad \tau_{n-2} = \kappa^2\tau_n \ldots \tau_2 = \kappa^{n-2}\tau_n \qquad \tau_1 = \kappa^{n-1}\tau_n \, .$$

with $\kappa = 1/\sqrt{10} \cong 0.316$. Furthermore, the longest relaxation time may not be chosen too large. Then in most cases a positive relaxation spectrum will be obtained. If not, the choice of a shorter value for τ_n may help. If this procedure does not yield the desired result, a positive spectrum can be obtained as follows:

4. *Determination of the spectrum in two successive steps*: First a set of relaxation times is chosen with a logarithmic distance corresponding to a factor ten between two successive relaxation times. This always results in a positive spectrum, which, however, gives a wavy approximation of the experimental data. Then a second set of relaxation times is chosen which is shifted against the first one by a factor $\sqrt{10}$ to shorter times and again the relaxation strengths are calculated. Then both spectra are added and the common set of relaxation strength is renormalized by multiplication with a common factor a determined to minimize the variance V_2

$$V_2 = \frac{1}{2n_m - 1} \sum_{i=1}^{n_m} \left\{ \left(a \frac{G'(\omega_i)}{g_{1i}} - 1 \right)^2 + \left(a \frac{G''(\omega_i)}{g_{2i}} - 1 \right)^2 \right\} \tag{10.39}$$

with respect to the choice of a by the condition $\partial V_2/\partial a = 0$. This procedure always results in a positive spectrum with a logarithmic distance corresponding to the factor κ. If both methods yield a positive spectrum, then the calculation in one step generally gives a slightly more accurate description of the data (a slightly lower value of the variance.)

The method is illustrated on data measured by Resch for an LLDPE [10], which are shown in Fig. 10.4. These data were measured during the preparation of the doctoral thesis [11] but have not been included in it. The authors are much obliged to Dr. J. A. Resch, Schwanenstadt, Austria for her permission to use these data for the construction and calculation of the Figs. 10.4, 10.5, 10.6 and 10.8. Figure 10.4 shows, in a double logarithmic representation, the storage moduli (squares) and the loss moduli (circles) as function of the angular frequency at the temperature $T = 170$ °C. Open symbols represent the results of dynamic measurements, while filled symbols indicate the results of calculations from creep and recovery measurements in shear converted by means of the formulae of Tables (5.3) and (5.4) to the values of G' and G''.

From the figure the high accuracy of the oscillatory measurements may be seen, as well as the excellent fit of the data derived from creep and recovery with those of the dynamic experiments. By adding the results of creep and recovery measurements, the experimental window is considerably enlarged into the direction of

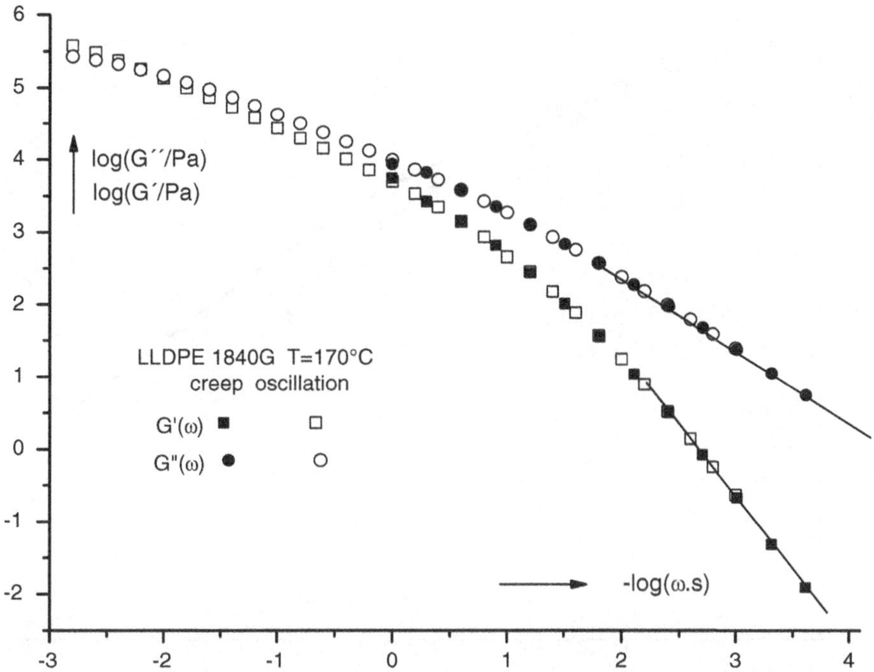

Fig. 10.4 Storage and loss moduli versus reciprocal angular frequency in a double logarithmic representation for a long-chain branched LLDPE at 170 °C, measured by Resch [10] by forced oscillations (*open symbols*) and calculated from torsional creep and recovery measurements (*closed symbols*)

lower frequencies. The two drawn lines indicate the limiting slopes of $G'(\omega)$ and $G''(\omega)$ for $\omega \to 0$ which are -2 and -1, respectively.

For the data shown in Fig. 10.4, the relaxation spectrum has been calculated in one step and the 12 relaxation strengths are presented as red stars in the Fig. 10.5 together with $G'(\omega)$ (black drawn line) and $G''(\omega)$ (red drawn line) as calculated from the spectrum. The accuracy of the representation is very good as may be seen from the figure and from the value of the square root of the variance which equals $\sqrt{V} = 2.8\,\%$.

The same data as shown in Fig. 10.5, were represented by a relaxation spectrum calculated in two steps, while simultaneously the longest relaxation time was shifted to longer times as far as possible to obtain still a positive value for a_n. In this case, 13 relaxation times were obtained, the longest relaxation time was $\tau_{max} = 506$ s and a slightly worse value for the square root of the variance ($\sqrt{V} = 3.1\,\%$) was found. (Fig. 10.6)

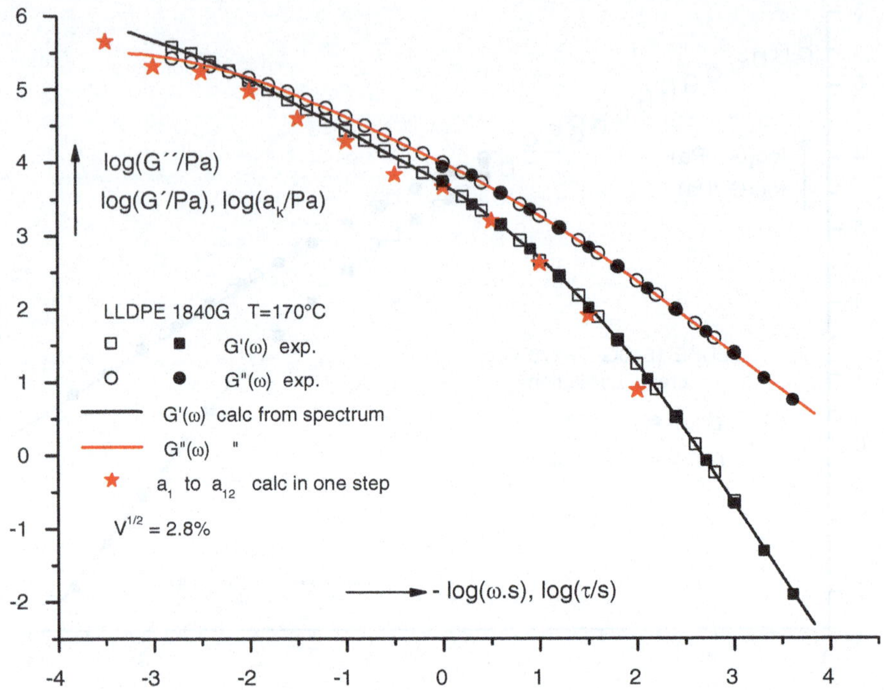

Fig. 10.5 Storage and loss moduli versus angular frequency in a double logarithmic representation for a long-chain branched LLDPE at 170 °C, measured by Resch [10] by forced oscillations (*open symbols*) and calculated from torsional creep and recovery measurements (*closed symbols*). Further shown are the relaxation spectrum calculated in one step (*red stars*) and G′ and G″ calculated from the spectrum (*lines drawn in black and red*)

10.2.7 Calculation of the Retardation Spectrum from the Relaxation Spectrum

If the relaxation spectrum is known for a linear viscoelastic material, the retardation spectrum may be calculated from it and vice versa. This problem often occurs in discussing the flow properties of polymer melts. Though the corresponding formulae seem to be simple, their numerical use can be very troublesome. Therefore we consider it to be useful to discuss this problem in some detail here.

The relation between the spectra is found by the following equation valid between the *Laplace transform* of the *relaxation modulus*, the *relaxance*

$$\hat{G}(p) = p \int_{0}^{\infty} e^{-pt} G(t) \mathrm{d}t \qquad (10.40)$$

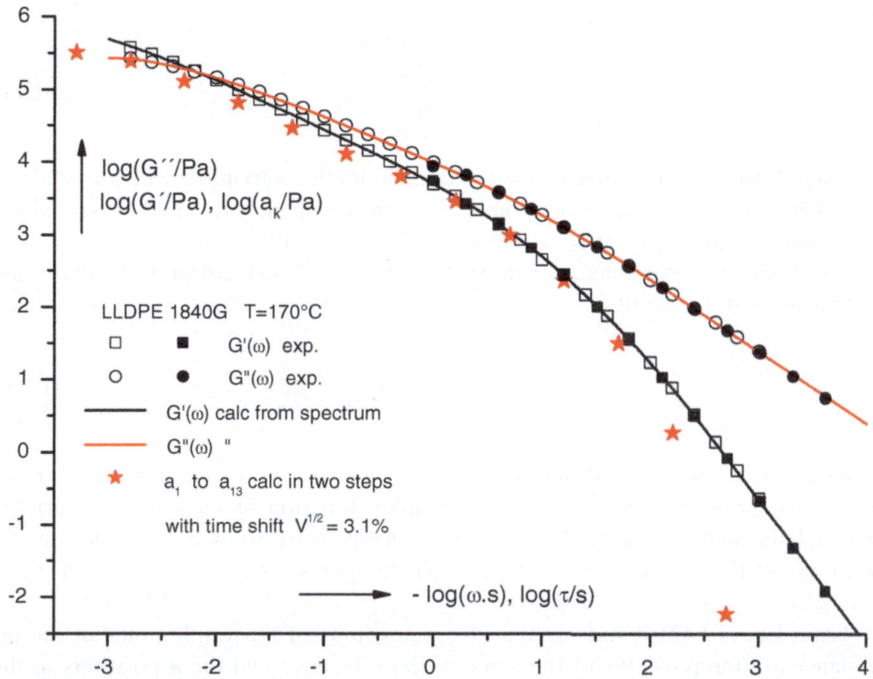

Fig. 10.6 Storage and loss moduli versus the reciprocal angular frequency in a double logarithmic representation for a long-chain branched LLDPE at 170 °C, measured by Resch [10] and the relaxation spectrum calculated in two steps with time shift (*red stars*) and G′ and G″ calculated from the spectrum (*lines drawn in black and red*)

and the Laplace transform of the creep compliance, the *retardance*

$$\hat{J}(p) = p\int_0^\infty e^{-pt} J(t)\mathrm{d}t \tag{10.41}$$

which reads

$$\hat{G}(p)\cdot\hat{J}(p) = 1 \tag{10.42}$$

with p being a complex variable. For the derivation of these and the following equations, we refer the reader to an article of Baumgärtel and Winter [12].

For a viscoelastic liquid represented by a discrete relaxation spectrum (10.33) is inserted into (10.40) and the transformation is performed with the result

$$\hat{G}(p) = \sum_{i=1}^{n}\frac{a_i p}{p + 1/\tau_i} = G_0 - \sum_{i=1}^{n}\frac{a_i/\tau_i}{p + 1/\tau_i} \tag{10.43}$$

with

$$G_0 = \sum_{i=1}^{n} a_i \tag{10.44}$$

The relaxance is a complex function of the complex variable p which is analytic everywhere in the p-plane apart from n positions on the negative real p-axis, where it has single poles, viz., at $p = -1/\tau_n, -1/\tau_{n-1}, \ldots -1/\tau_2, -1/\tau_1$.

To obtain the retardance, we insert (5.26) into (10.41) and perform the transformation with the result

$$\hat{J}(p) = J_0 + \sum_{i=1}^{n-1} \frac{b_i/\bar{\tau}_i}{p + 1/\bar{\tau}_i} + \frac{1}{p\eta_0} \tag{10.45}$$

$\bar{\tau}_1, \bar{\tau}_2, \ldots, \bar{\tau}_{n-1}$ are the $n-1$ *retardation times* and $b_1, b_2, \ldots b_{n-1}$ the $n-1$ *retardation strengths*. The retardance is a complex function of the complex variable p which is analytic everywhere in the p-plane apart from n positions on the negative real p-axis, where it has single poles, viz., at $p = 0, -1/\bar{\tau}_{n-1}, -1/\bar{\tau}_{n-2}, \ldots - 1/\bar{\tau}_1$.

From Eq. (10.42) it follows that the n positions of the single poles of the retardance are the positions of the zeros of the relaxance and the n positions of the single poles of the relaxance are the positions of the zeros of the retardance. Hence the relaxance is a complex function of the variable p with zeros at $p = 0$, $-1/\bar{\tau}_{n-1}, -1/\bar{\tau}_{n-2}, \ldots - 1/\bar{\tau}_1$ and single poles at $p = -1/\tau_n, -1/\tau_{n-1}, \ldots - 1/\tau_1$. Such a function may be written as the product

$$\hat{G}(p) = G_0 \cdot \frac{p(p + 1/\bar{\tau}_1)(p + 1/\bar{\tau}_2)\ldots(p + 1/\bar{\tau}_{n-1})}{(p + 1/\tau_1)(p + 1/\tau_2)\ldots(p + 1/\tau_n)} \tag{10.46}$$

The prefactor G_0 is found as the limit of $\hat{G}(p)$ for $p \to +\infty$ in (10.43) and (10.46).

Using (5.13) it follows from (10.42) for the retardance

$$\hat{J}(p) = J_0 \cdot \frac{(p + 1/\tau_1)(p + 1/\tau_2)\ldots(p + 1/\tau_n)}{p(p + 1/\bar{\tau}_1)(p + 1/\bar{\tau}_2)\ldots(p + 1/\bar{\tau}_{n-1})} \tag{10.47}$$

Baumgärtel and Winter [12] have shown that the expression (10.47) may be transformed by partial decomposition into the form of Eq. (10.45), whereby

$$J_0 = 1/G_0 \tag{5.13}$$

$$\eta_0 = G_0 \tau_n \prod_{i=1}^{n-1} \frac{\tau_i}{\bar{\tau}_i} = \sum_{i=1}^{n} a_i \tau_i \tag{10.48}$$

and

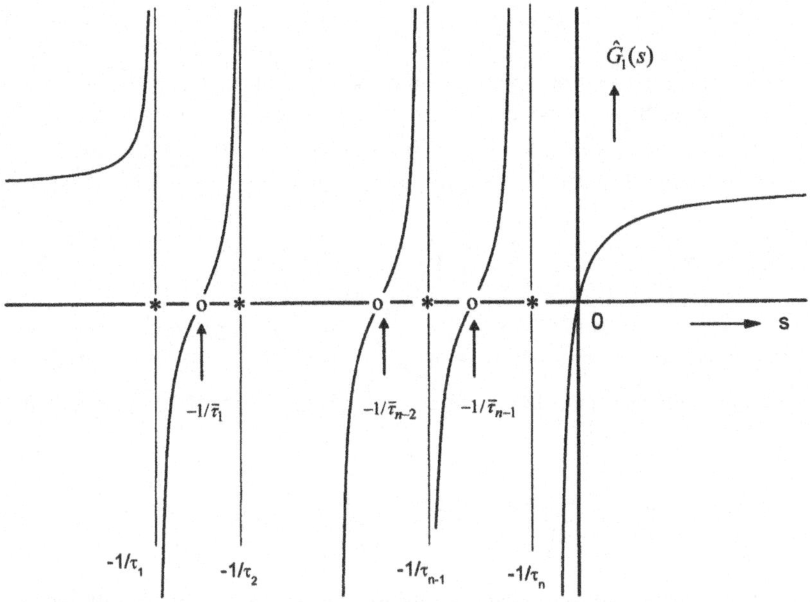

Fig. 10.7 The real part of the relaxance $\hat{G}_1(s)$ as a function of s, the real part of p. Zeros of $\hat{G}_1(s)$ are indicated by *circles*, poles by *crosses*

$$b_i = -\frac{1}{G_0}\prod_{k=1}^{n}(1-\bar{\tau}_i/\tau_k)\Bigg/\prod_{k=1,i}^{n}(1-\bar{\tau}_i/\bar{\tau}_k) \quad \text{for} \quad i = 1 \text{ to } n-1 \tag{10.49}$$

The subscript $k = 1$, *i to n* indicates that the product has to be performed for all values of the index k between 1 and n excluding the value $k = i$.

The relaxance $\hat{G}(p) = \hat{G}_1(s,u) + i\hat{G}_2(s,u)$ is a complex function of the complex variable $p = s + iu$. Along the real axis of p, $\hat{G}(s,0) = \hat{G}_1(s)$ is real. As a function of s the real part of $\hat{G}(p)$ viz. $\hat{G}_1(s)$ is sketched in Fig. 10.7.

For $s = +\infty$, $\hat{G}_1(s) = G_0$. With decreasing s, $\hat{G}_1(s)$ decreases monotonously and passes zero at the origin. Along the negative s-axis it becomes negative and tends to $-\infty$ if s approaches the first pole $s = -1/\tau_n$. On the other side of the first pole, $\hat{G}_1(s)$ starts from $+\infty$ and decreases with decreasing value of s, passes zero at the second root of $\hat{G}_1(s)$, becomes negative then and tends to $-\infty$ if s approaches the second pole $s = -1/\tau_{n-1}$. On the negative side of the last pole $s = -1/\tau_1$, $\hat{G}_1(s)$ starts from $+\infty$, decreases with decreasing s, but stays positive and tends to G_0 if s approaches $-\infty$. The *order* of the *magnitudes* of *relaxation and retardation times* follows from Fig. 10.7 as

$$\tau_1 < \bar{\tau}_1 < \tau_2 < \bar{\tau}_2 < \ldots \tau_{n-1} < \bar{\tau}_{n-1} < \tau_n \qquad (10.50)$$

To convert the relaxation spectrum into the retardation spectrum, first the $n-1$ retardation times have to be calculated. Considering that the retardation times correspond to the roots of the function $\hat{G}_1(s)/s$ and taking into account Eq. (10.43), this is effected by determination of the $n-1$ roots of the expression

$$\frac{\hat{G}_1(s)}{s} = \sum_{i=1}^{n} \frac{a_i}{s + 1/\tau_i} = \sum_{i=1}^{n} a_i \left(\prod_{j=1,i}^{n} (s + 1/\tau_j) \right) \Bigg/ \prod_{i=1}^{n} (s + 1/\tau_i) \qquad (10.51)$$

The right-hand side of this equation was obtained by multiplying nominator and denominator with the same factor $\prod_{j=1}^{n} (s + 1/\tau_j)$. Equation (10.51) is rewritten as

$$\frac{\hat{G}_1(s)}{s} = G_0 P(s) \Bigg/ \prod_{i=1}^{n} (s + 1/\tau_i) \qquad (10.52)$$

where $P(s)$ is a polynomial of the degree $n-1$, for which the coefficient of the highest power of s is unity:

$$G_0 P(s) \equiv \sum_{i=1}^{n} a_i \left(\prod_{j=1,i}^{n} (s + 1/\tau_j) \right) = \alpha_{n-1} s^{n-1} + \alpha_{n-2} s^{n-2} + \cdots + \alpha_1 s + \alpha_0 \qquad (10.53)$$

As the denominator in Eq. (10.52) does not vanish at the roots of the nominator, Eq. (10.52) and the polynomial $P(s)$ have the same roots. Next we have to calculate the coefficients of the polynomial $P(s)$, $\alpha_i = \alpha_i(a_1, a_2, \ldots a_n, \tau_1, \tau_2, \ldots \tau_n)$ in Eq. (10.53). This is easy for small values of n. For larger values of n, a recursive method for computational purposes has been developed.

The most difficult step is to find the roots of (10.53). From (10.50), the interval is known, where e.g., $\bar{\tau}_1$ has to come to lie, viz., the interval ($\tau_1 < s < \tau_2$). This interval is divided into 4000 logarithmically equidistant subintervals and within the subinterval the s-value has to be found where $P(s)$ changes its sign. Then this subinterval is divided again into 4000 logarithmically equidistant subintervals and the procedure is repeated. In this manner it is possible to determine the retardation times with an accuracy of ±0.00001. This high accuracy is necessary to obtain the retardation strengths from Eq. (10.49) with reasonable accuracy. Once the retardation times are known, they are inserted into (10.48) and (10.49) yielding the retardation spectrum and the viscosity.

As an example, Fig. 10.8 shows the *relaxation spectrum* of an LLDPE as calculated from measurements by Resch at 170 °C [10]. The curve decreasing with the relaxation time represents the fit to the points obtained by three different

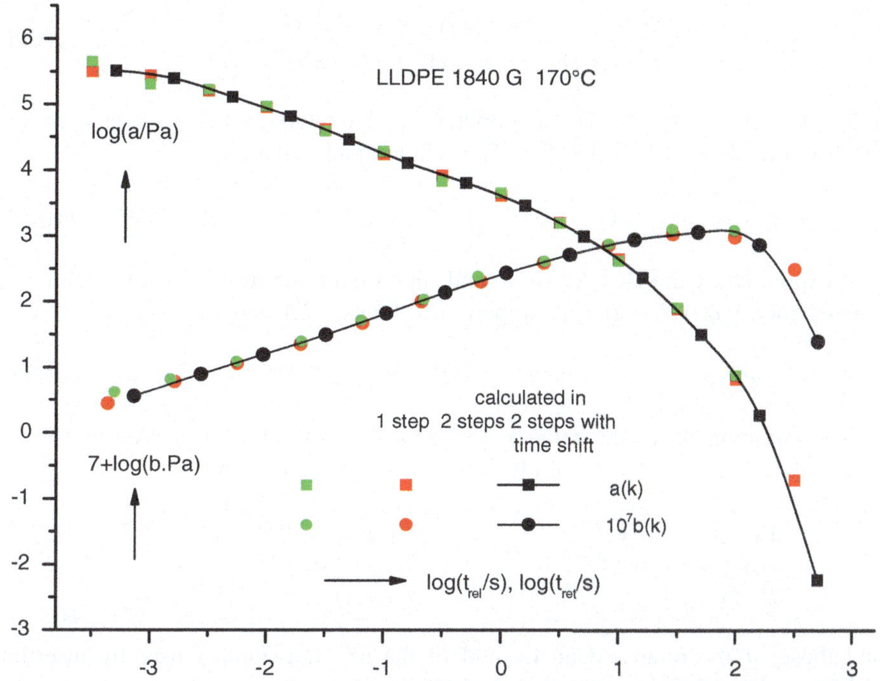

Fig. 10.8 Discrete relaxation spectra a(k) as functions of the relaxation times t_{rel} and discrete retardation spectra b(k) as functions of the retardation times t_{ret} calculated in one step, in two steps, and in two steps with time shift for a long-chain branched LLDPE measured by Resch at 170 °C [10]

methods: Calculation in one step (green squares), in two steps (red squares), and in two steps with time shift (black squares). The various discrete relaxation strengths all lie very well on the same smooth curve representing—apart from a normalization factor—the *continuous relaxation spectrum* $g(\tau)$.

From the relaxation spectra, the corresponding discrete retardation strengths were calculated together with the retardation times. They are shown as green, red, and black circles. The various discrete spectra all come to lie very well on the same smooth curve representing—apart from a normalization factor—the *continuous retardation spectrum* $f(\tau)$.

10.2.8 The Lodge Liquid in Isochorous Multidimensional Extension

Consider next the Lodge liquid in multi-dimensional isochorous extension. *Multi-dimensional isochorous extension* is described by the path equations

$$\begin{array}{lll} x_1 = \lambda_1 x_1^0 & x_2 = \lambda_2 x_2^0 & x_3 = \lambda_3 x_3^0 \\ \lambda_1 = \lambda(t) & \lambda_2 = \lambda(t)^m & \lambda_3 = \lambda(t)^{-(1+m)} \end{array} \tag{10.54}$$

and $\lambda(t) = 1$ for $t < 0$. The parameter m characterizes the geometry of the extension as defined in Table 8.1. The relative path equations are

$$x_1 = [\lambda(t)/\lambda(t')]\, x_1' \quad x_2 = [\lambda(t)/\lambda(t')]^m x_2' \quad x_3 = [\lambda(t)/\lambda(t')]^{-(1+m)} x_3' \tag{10.55}$$

or if expressed by the Hencky strain and suppressing the index H for the matter of convenience $[\varepsilon_H(t) \equiv \varepsilon\,(t)]$. (Compare also Sects. 12.1 and 12.2.)

$$x_1 = e^{\varepsilon(t)-\varepsilon(t')} x_1' \quad x_2 = e^{m[\varepsilon(t)-\varepsilon(t')]} x_2' \quad x_3 = e^{-(1+m)[\varepsilon(t)-\varepsilon(t')]} x_3' \tag{10.55'}$$

The components of the relative Finger and of the absolute Green tensor are found by inserting (10.54) and (10.55′) into (8.35) and (8.18)

$$\begin{array}{ll} B_{11}(\underline{x}; t, t') = e^{2[\varepsilon(t)-\varepsilon(t')]} & G_{11}(\underline{x}; t) = e^{2\varepsilon(t)} \\ B_{22}(\underline{x}; t, t') = e^{2m[\varepsilon(t)-\varepsilon(t')]} & G_{22}(\underline{x}; t) = e^{2m\varepsilon(t)} \\ B_{33}(\underline{x}; t, t') = e^{-2(1+m)[\varepsilon(t)-\varepsilon(t')]} & G_{33}(\underline{x}; t) = e^{-2(1+m)\varepsilon(t)} \end{array} \tag{10.56}$$

and those of the relative Cauchy and of the absolute Piola tensor by inserting (10.54) and (10.55′) into (8.33) and (8.17).

$$\begin{array}{ll} C_{11}(\underline{x}; t, t') = e^{-2[\varepsilon(t)-\varepsilon(t')]} & A_{11}(\underline{x}; t) = e^{-2\varepsilon(t)} \\ C_{22}(\underline{x}; t, t') = e^{-2m[\varepsilon(t)-\varepsilon(t')]} & A_{22}(\underline{x}; t) = e^{-2m\varepsilon(t)} \\ C_{33}(\underline{x}; t, t') = e^{2(1+m)[\varepsilon(t)-\varepsilon(t')]} & A_{33}(\underline{x}; t) = e^{2(1+m)\varepsilon(t)} \end{array} \tag{10.57}$$

The tensor components are independent of the spatial coordinates. From (10.7) the tensile stresses result as

$$\begin{aligned} \sigma_{11}(t) + p &= G(t) e^{2\varepsilon(t)} + \int_0^t m(t-t') e^{2[\varepsilon(t)-\varepsilon(t')]} \mathrm{d}t' \\ \sigma_{22}(t) + p &= G(t) e^{2m\varepsilon(t)} + \int_0^t m(t-t') e^{2m[\varepsilon(t)-\varepsilon(t')]} \mathrm{d}t' \\ \sigma_{33}(t) + p &= G(t) e^{-2(1+m)\varepsilon(t)} + \int_0^t m(t-t') e^{-2(1+m)[\varepsilon(t)-\varepsilon(t')]} \mathrm{d}t' \end{aligned} \tag{10.58}$$

In these equations $m(t - t')$ is the memory function, which of course differs from the geometry parameter m.

The formulae (10.58) become much easier to handle, if the considerations are restricted to *stressing experiments*, in which the Hencky strain is a linearly increasing function of the time, i.e.,

$$\varepsilon(t) = 0 \text{ for } t<0 \quad \text{and} \quad \varepsilon(t) = \dot{\varepsilon}_0 t \quad \text{for} \quad t \geq 0 \tag{10.59}$$

with $\dot{\varepsilon}_0$ being the constant Hencky strain rate in the 1-direction. As this strain is continuous and piecewise differentiable, Eqs. (10.8) may be used to calculate the stresses which after introducing the new integration variable $\xi = t - t'$ become

$$\begin{aligned} \sigma_{11}(t) + p &= 2\dot{\varepsilon}_0 \int_0^t G(\xi) e^{2\dot{\varepsilon}_0 \xi} \mathrm{d}\xi \\ \sigma_{22}(t) + p &= 2m\dot{\varepsilon}_0 \int_0^t G(\xi) e^{2\dot{\varepsilon}_0 m \xi} \mathrm{d}\xi \\ \sigma_{33}(t) + p &= -2(1+m)\dot{\varepsilon}_0 \int_0^t G(\xi) e^{-2(1+m)\dot{\varepsilon}_0 \xi} \mathrm{d}\xi \end{aligned} \tag{10.60}$$

Three distinguished cases are considered:

Linear extension in the 1*-direction*: m = −½ is inserted into Eq. (10.60), in the second of the equations $\sigma_{22}(t) + p = 0$ is set, and the value of p obtained is inserted into the first and third of the equations with the result

$$\sigma_{11}(t) = \dot{\varepsilon}_0 \int_0^t G(\xi)[2e^{2\dot{\varepsilon}_0 \xi} + e^{-\dot{\varepsilon}_0 \xi}] \mathrm{d}\xi \qquad \sigma_{22}(t) = \sigma_{33}(t) = 0 \tag{10.61}$$

The *tensile viscosity* $\mu(t)$ defined as the stress divided by the Hencky strain rate

$$\mu(t, \dot{\varepsilon}_0) = \frac{\sigma_{11}}{\dot{\varepsilon}_0} = \int_0^t G(\xi)[2e^{2\dot{\varepsilon}_0 \xi} + e^{-\dot{\varepsilon}_0 \xi}] \mathrm{d}\xi \tag{10.62}$$

is a monotonously increasing function of the time t and depends on the strain rate $\dot{\varepsilon}_0$ as a parameter. In the limit of small strain rates, this viscosity tends to three times the linear shear viscosity

$$\lim_{\dot{\varepsilon}_0 \to 0} \mu(t, \dot{\varepsilon}_0) = \mu^0(t) = 3 \int_0^t G(\xi) \mathrm{d}\xi = 3\eta^0(t) \tag{10.63}$$

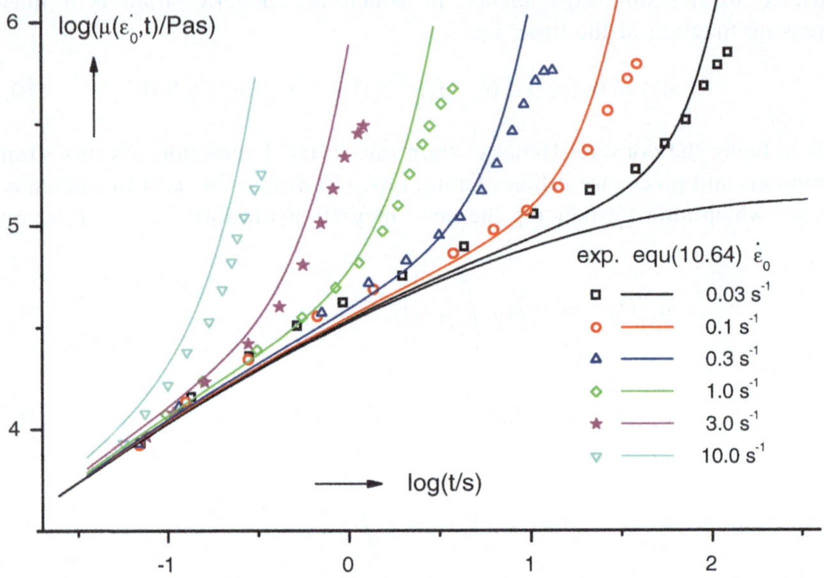

Fig. 10.9 Uniaxial tensile viscosities measured by Stadler [13] as function of the time under various constant Hencky strain rates for a Lupolen (1840 H) at 150 °C (various symbols) and the predicitions of Lodge's theory (*drawn lines*)

The dependence of μ on t and $\dot{\varepsilon}_0$ is obtained by inserting the discrete spectral representation of the shear modulus (5.23) into (10.62) and integration after the time variable as

$$\mu(t,\dot{\varepsilon}_0) = \sum_{i=1}^{n} a_i \tau_i \left[\frac{2\left[1 - e^{-(1-2\dot{\varepsilon}_0\tau_i)t/\tau_i}\right]}{1 - 2\dot{\varepsilon}_0\tau_i} + \frac{1 - e^{-(1+\dot{\varepsilon}_0\tau_i)t/\tau_i}}{1 + \dot{\varepsilon}_0\tau_i} \right] \tag{10.64}$$

This function shows a finite plateau value only if the product of $\dot{\varepsilon}_0$ and the largest relaxation time τ_e does not exceed 0.5, i.e., if $\dot{\varepsilon}_0\tau_e < 1/2$. Otherwise, if $\dot{\varepsilon}_0\tau_e > 1/2$, the first term in Eq. (10.64) will increase exponentially with increasing time and μ will become divergent.

If the relaxation spectrum is known from oscillation experiments, $\mu(t,\dot{\varepsilon}_0)$ may be calculated and compared with the result of extensional measurements. As an example, Fig. 10.9 shows the uniaxial tensile viscosities measured by Stadler [13] on LDPE at 150 °C under different constant Hencky strain rates as indicated by the experimental points in various colors. As Stadler also determined the relaxation spectrum of this material at the same temperature [13], the tensile viscosities as predicted by Lodge's theory could be calculated by inserting the spectrum into Eq. (10.64). The result is given by the drawn lines in the corresponding colors.

The experiment shows considerable strain hardening which starts the later, the lower applied the strain rate. The Lodge theory predicts the onset of strain hardening at about the right stressing times. It is, however, not able to predict a finite viscosity level, as in Lodge's theory the molecules are assumed to follow the exponential increase in extension. In reality, the molecules when stretched to their limit of extensibility will slip out of their entanglements which will lead to a finite viscosity level. This effect can be described only when the life time of the entanglements is assumed to depend on the magnitude of the strain or the strain rate. This aspect is taken into account in Wagner's theory and treated in the next section.

Equibiaxial extension in the 1–2 *plane*: m = 1 is inserted into Eq. (10.60), $\sigma_{33}(t) + p = 0$ is set in the third of the equations, and the value of p obtained is introduced to the first and second of the equations with the result

$$\sigma_{11}(t) = \sigma_{22}(t) = 2\dot{\varepsilon}_0 \int\limits_0^t G(\xi)[e^{2\dot{\varepsilon}_0\xi} + 2e^{-4\dot{\varepsilon}_0\xi}]\mathrm{d}\xi \qquad \sigma_{33}(t) = 0 \tag{10.65}$$

The *equibiaxial elongational viscosity* $\mu_e(t)$ is defined as the stress divided by the Hencky strain rate

$$\mu_e(t, \dot{\varepsilon}_0) = \frac{\sigma_{11}}{\dot{\varepsilon}_0} = 2\int\limits_0^t G(\xi)[e^{2\dot{\varepsilon}_0\xi} + 2e^{-4\dot{\varepsilon}_0\xi}]\mathrm{d}\xi \tag{10.66}$$

In the limit of small strain rates, this viscosity tends to six times the linear shear viscosity.

Planar extension in the 1–3 *plane*: m = 0 is introduced to Eq. (10.60), $\sigma_{33}(t) + p = 0$ is set in the third of the equations, and the value of p obtained is introduced to the first and second of the equations with the result

$$\sigma_{11}(t) = 2\dot{\varepsilon}_0 \int\limits_0^t G(\xi)[e^{2\dot{\varepsilon}_0\xi} + e^{-2\dot{\varepsilon}_0\xi}]\mathrm{d}\xi \qquad \dot{\sigma}_{22}(t) = 2\dot{\varepsilon}_0 \int\limits_0^t G(\xi)e^{-2\dot{\varepsilon}_0\xi}\mathrm{d}\xi \qquad \sigma_{33}(t) = 0 \tag{10.67}$$

This case may be realized by applying tensile stresses in the 1- and 2-direction of different magnitudes and two viscosities may be defined according to

$$\mu_{p1}(t, \dot{\varepsilon}_0) = \sigma_{11}/\dot{\varepsilon}_0 \quad \text{and} \quad \mu_{p2}(t, \dot{\varepsilon}_0) = \sigma_{22}/\dot{\varepsilon}_0$$

10.3 Rheological Equations of State After Wagner

10.3.1 The Constitutive Equation

As shown in the last chapter, the model of Lodge for the elastic liquid is able to describe the visco-elastic properties of polymer melts in their Newtonian regime as well as the occurrence of a first normal stress coefficient and the onset of strain hardening in uniaxial stressing experiments. It fails, however, for the description of the dependence of the viscosity and the first and second normal stress coefficients on the shear rate and for modeling the existence of a constant elongational viscosity after strain hardening. To describe these essential features of the rheology of polymer melts, various authors proposed the introduction of damping functions into the constitutive equation of Lodge. A historical review over various proposals has been given by Wagner [14], who introduced the following generalization of Lodge's equation [15].

$$\sigma_{ik}(t) + p\delta_{ik} = \int_{-\infty}^{t} m(t-t')B_{ik}(\underline{x};t,t')\mathrm{d}t' \tag{10.2}$$

The linearity of this equation is based on the assumption that at a constant temperature the memory function $m(t-t')$ only depends on the time difference $t-t'$, but neither on the magnitude of the strain reached during the time interval $[t',t]$, nor on the strain rate to which the network was exposed. In the molecular interpretation of Lodge, the life time of entanglements is determined by the Brownian motion only, but not by the deformation history within the time interval $[t',t]$.

This assumption is improbable from the molecular point of view as one should expect the entanglements to become disentangled at high stretch or rate of stretching of the molecules connecting them. Therefore, the deformation history between t' and t should influence the lifetime of the entanglement network. As the function $m(t-t')$ is a scalar, it can depend only on the first and second invariant of the deformation tensor, but not on the third invariant which is equal to unity. These considerations led Wagner to the following constitutive equation

$$\sigma_{ik}(t) + p\delta_{ik} = \int_{-\infty}^{t} m(t-t';I_B,II_B)B_{ik}(\underline{x};t,t')\mathrm{d}t' \tag{10.68}$$

In this equation, $I_B = I_B(x,t,t')$ and $II_B = II_B(x,t,t')$ are the first and the second invariants of the relative Finger tensor (8.35). By

$$\frac{1}{kT}m(t-t';I_B,II_B)\mathrm{d}t' = n(t-t';I_B,II_B)\mathrm{d}t' \tag{10.69}$$

the kernel m is related to the number of entanglements per volume which were created between t' and $t' + dt'$ and still exist at the time t, after the deformation history within the interval $[t't]$, as described by I_B and II_B.

Wagner further assumed that the kernel may be factorized into two functions, one depending only on the elapsed time $t - t'$, the other on the deformation of the network:

$$m(t - t'; I_B, II_B) = m(t - t') \cdot h(I_B, II_B) \tag{10.70}$$

$m(t - t')$ is the *memory function* introduced in Lodge's theory describing the survival probability of the network under the action of the Brownian motion. The function $h(I_B, II_B)$ is called the *damping function*, varying between unity for moderate deformations in the region of the validity of Lodge's theory and zero for very large deformations. It describes the probability of survival of the network during the deformation history characterized by I_B and II_B. The factorization (10.70) is equivalent to the assumption that the annihilation of entanglements due to the Brownian motion and the stretching of molecules are independent processes.

For small deformations, the damping function tends to unity and Eq. (10.68) to Lodge's equation. Consequently, Eqs. (10.3)–(10.6) remain valid, under the condition that $G(t)$ equals the time-dependent shear modulus of the viscoelastic liquid in the linear region. Combining Eq. (10.68) and (10.70), the *constitutive equation* of the *special Wagner theory* follows.

$$\sigma_{ik}(t) + p\delta_{ik} = \int\limits_{-\infty}^{t} m(t - t') \cdot h[I_B(\underline{x}; t, t'), II_B(\underline{x}; t, t')] B_{ik}(\underline{x}; t, t') \mathrm{d}t' \tag{10.71}$$

Let us assume that the strain history started at $t = 0$. Then before $t = 0$ no deformation was applied to the sample and in the time interval $(-\infty < t' < 0]$ $B_{ik}(\underline{x} : t, t')$, $I_B(\underline{x}; t, t')$ and $II_B(\underline{x}; t, t')$ degenerate to $G_{ik}(\underline{x}; t)$, $I_G(\underline{x}; t)$ and $II_G(\underline{x}; t)$ which are independent of the integration variable t' and may be placed outside the integral. Dividing the integral into two parts one with the limits $t' = -\infty$ and 0 and one with the limits $t' = 0$ and $t' = t$ yields with (10.5)

$$\begin{aligned}\sigma_{ik}(t) + p\delta_{ik} =& G(t)[I_G(\underline{x}; t), II_G(\underline{x}; t)] G_{ik}(\underline{x}; t) + \\ &+ \int\limits_{0}^{t} m(t - t') \cdot h[I_B(\underline{x}; t, t'), II_B(\underline{x}; t, t')] B_{ik}(\underline{x}; t, t') \mathrm{d}t'\end{aligned} \tag{10.72}$$

Another form is found by partial integration of (10.72), if the necessary conditions of continuity and differentiability are met as

$$\sigma_{ik} + p\delta_{ik} = -\int_0^t G(t-t')\frac{\partial\{h[I_B(\underline{x},t,t'),II_B(\underline{x},t,t')]B_{ik}(\underline{x};t,t')\}}{\partial t'}\mathrm{d}t' \qquad (10.73)$$

This equation may be applied if $h[I_B(\underline{x};t,t'),II_B(\underline{x};t,t')]\cdot B_{ik}(\underline{x};t,t')$ is continuous and piecewise differentiable as a function of t'.

It will be shown in Sect. 10.3.2 that the constitutive Eq. (10.71) is not able to describe non-zero second normal stress coefficients. To remove this deficiency, Wagner and Demarmels [16] generalized the theory and completed the integrand of (10.71) with a second term proportional to the inverse relative Finger tensor and a second damping function h_{-1} . In this version, the theory is known as the *constitutive equation* of the *general Wagner theory*

$$\boxed{\sigma_{ik}(t) + p\delta_{ik} = \int_{-\infty}^t m(t-t')\cdot[h_1(I_B,II_B)B_{ik}(\underline{x};t,t') + h_{-1}(I_B,II_B)B_{ik}^{-1}(\underline{x};t,t')]\mathrm{d}t'} \qquad (10.74)$$

If the deformation history started at the time $t = 0$, this equation may be written as

$$\boxed{\begin{aligned}\sigma_{ik}(t) + p\delta_{ik} = {} & G(t)\cdot[h_1(I_G,II_G)G_{ik}(\underline{x};t) + h_{-1}(I_G,II_G)A_{ik}(\underline{x};t)] + \\ & + \int_0^t m(t-t')\cdot[h_1(I_B,II_B)B_{ik}(\underline{x};t,t') + h_{-1}(I_B,II_B)B_{ik}^{-1}(\underline{x};t,t')]\mathrm{d}t'\end{aligned}} \qquad (10.75)$$

and if the necessary conditions of continuity and differentiability are met, as

$$\boxed{\sigma_{ik}(t) + p\delta_{ik} = \int_0^t G(t-t')\cdot\frac{\partial[h_1(I_B,II_B)B_{ik}(\underline{x};t,t') + h_{-1}(I_B,II_B)B_{ik}^{-1}(\underline{x};t,t')]}{\partial t'}\mathrm{d}t'} \qquad (10.76)$$

These are the most general equations of state, which we will consider. They contain as special cases all rheological laws to be discussed further.

10.3.2 The Wagner Theories in Time-Dependent Simple Shear

We apply the special Wagner theory to time-dependent simple shear characterized by the path Eqs. (10.9), the relative path Eqs. (10.10), and the components of the relative Finger tensor and the absolute Green tensor (10.11). Before inserting these relations into Eqs. (10.71), we remark that for simple shear the first and second invariant for both tensors are equal $I_G = II_G = 3 + \gamma^2(t)$ *and* $I_B = II_B = 3+[\gamma(t) - \gamma(t')]^2$ and the function $h(I_G, II_G)$ reduces to a function of one independent variable which we will designate by the letter h_s and call the *damping function in shear.*

$$h[I_G(\underline{x}, t), II_G(\underline{x}, t)] \equiv h_s(\gamma(t)) = h_s(3 + \gamma^2(t)) \tag{10.77}$$

$$h[I_B(x, t, t'), II_B(x, t, t')] \equiv h_s((\gamma(t), \gamma(t')) = h_s(3 + [\gamma(t) - \gamma(t')]^2) \tag{10.78}$$

Consequently, the damping function in simple shear can depend only on the absolute value of the difference between $\gamma(t)$ and $\gamma(t')$. Inserting these equations into (10.71), yields for the stresses in simple shear

$$\begin{aligned}
&\sigma_{12}(t) = G(t)h_s(\gamma(t)) \cdot \gamma(t) + \int_0^t m(t - t') \cdot h_s(|\gamma(t) - \gamma(t')|) \cdot [\gamma(t) - \gamma(t')]\mathrm{d}t' \\
&\sigma_{11}(t) - \sigma_{22}(t) = G(t)h_s(\gamma(t)) \cdot \gamma^2(t) + \int_0^t m(t - t') \cdot h_s(|\gamma(t) - \gamma(t')|) \cdot [\gamma(t) - \gamma(t')]^2\mathrm{d}t' \\
&\sigma_{22}(t) - \sigma_{33}(t) = 0
\end{aligned} \tag{10.79}$$

For the shear stress and the shear strain, the *superposition principle* is *not valid* any longer, as the integral transforms become non-linear due to the presence of the damping function. A shear stress is not sufficient to produce a shear strain. In addition, normal stress differences are necessary. Due to the special Wagner theory, a first normal stress difference is sufficient, which is always positive, while the second normal stress difference vanishes.

Equations (10.79) are used to describe the *stress relaxation experiment in simple shear*, in which

$$\begin{aligned}
\gamma(t) &= 0 \quad \text{for} \quad t<0 \quad \text{and} \\
\gamma(t) &= \gamma_0 \quad \text{for} \quad t\geq 0
\end{aligned} \tag{10.80}$$

The integrals in the first two equations of (10.79) vanish and the equations for stress relaxation due to the special Wagner theory follow as

$$\begin{aligned}
&\sigma_{12}(t) = G(t)h_s(\gamma_0) \cdot \gamma_0 \equiv G(t; \gamma_0)\gamma_0 \\
&\sigma_{11}(t) - \sigma_{22}(t) = G(t)h_s(\gamma_0) \cdot \gamma_0^2 = G(t; \gamma_0)\gamma_0^2 \\
&\sigma_{22}(t) - \sigma_{33}(t) = 0
\end{aligned} \tag{10.81}$$

Due to the special Wagner theory, the *non-linear stress relaxation modulus* $G(t,\gamma_0)$ may be factorized into a time-dependent function, the *stress relaxation modulus in the linear range* and a function depending on the magnitude of the applied shear strain, the *damping function in shear.* It may be determined as the ratio of the shear stress to the shear strain or as the ratio of the first normal stress

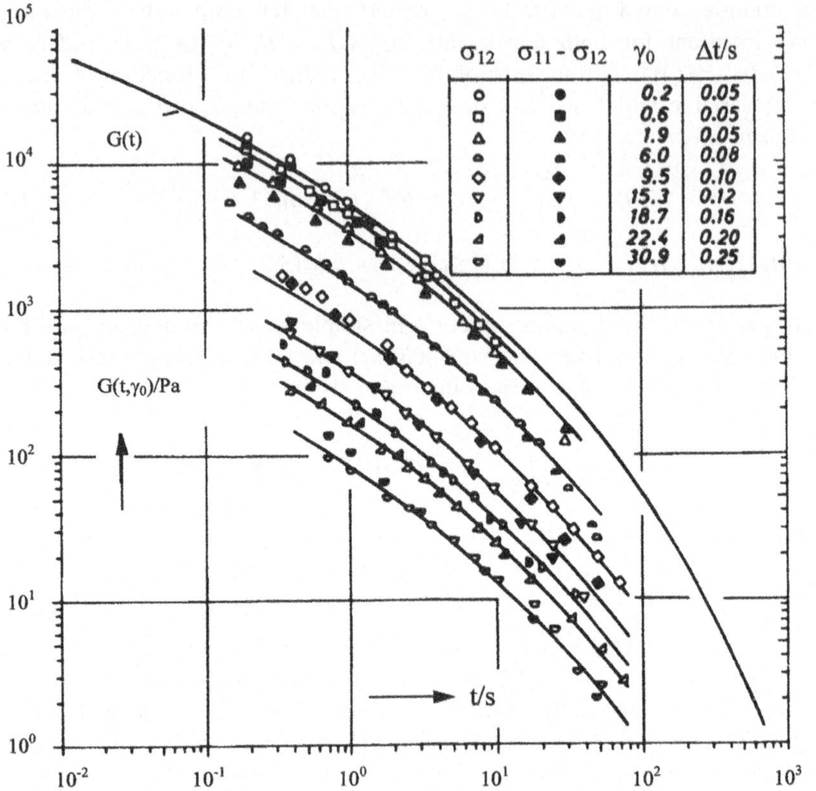

Fig. 10.10 Time-dependent non-linear relaxation modulus in simple shear of IUPAC I melt determined by Laun [17] at 150 °C (M_w = 284 kg/mol and M_w/M_n = 28) as function of the time and the constant shear strain γ_0. Δt was the time necessary to attain the set value of the constant shear strain

difference to the square of the shear strain. According to (10.81) the second normal stress difference vanishes.

The validity of the two first Eqs. (10.81) has been checked by Laun [17] on an LDPE melt whose molecular data are given by Meissner [18]. The non-linear time-dependent shear modulus of this melt is shown in Fig. 10.10 in double logarithmic representation for different values of the shear strain γ_0 at 150 °C. Points indicated by open symbols were obtained from measurements of the shear stress, those indicated by closed symbols by measurements of the first normal stress difference. As to be seen from the figure, the values obtained for $G(t,\gamma_0)$ are identical in both cases.

The representation of $G(t,\gamma_0)$ as a product of the two functions $G(t)$ and $h_s(\gamma_0)$ is justified as the curves in double logarithmic plot may be brought to coincidence by

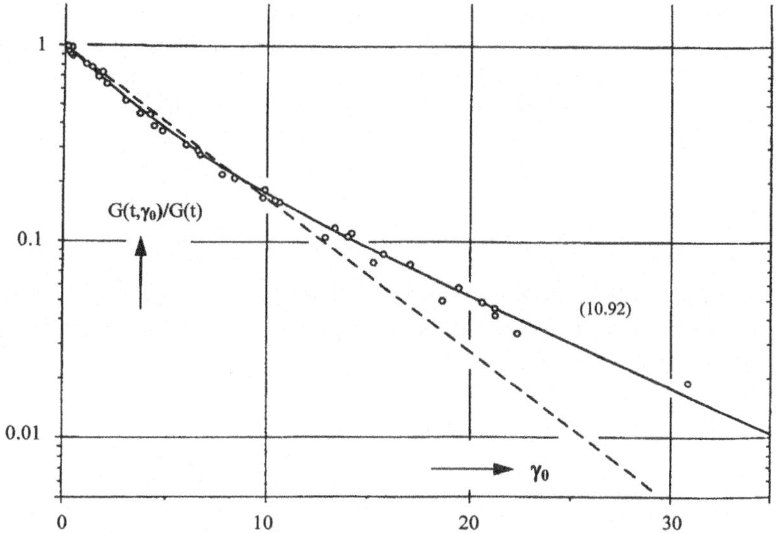

Fig. 10.11 Damping function in shear for the IUPAC I melt after Laun [17]

parallel shifts in vertical direction. The drawn line, designated by $G(t)$, was calculated from oscillation measurements by Zosel and represents the linear case. The points obtained at the smallest strains are in good agreement with these data (see [18]).

From the vertical shifts of the curves in the figure, the damping function in shear is obtained, which is shown in Fig. 10.11 and which may be represented by the sum of two exponential terms

$$h_s(\gamma_0) \cong f_1 e^{-\mu_1 \gamma_0} + f_2 e^{-\mu_2 \gamma_0} \qquad (10.82)$$

with $f_1 = 0.57$, $\mu_1 = 0.31$, $f_2 = 0.43$, $\mu_2 = 0.106$. This simple expression which was first proposed by Osaki [19] is very appropriate for further calculations in connection with the spectral representation of $G(t)$.

Measurements on branched and linear polystyrenes were reported by Hepperle [20], however, within a smaller range of shear strains ($0 < \gamma_0 < 5$). Within this range, the damping functions could be described by one single exponential term. The decrease of the damping function depends on the degree of branching. The higher the degree of branching, the smaller is the influence of the shear strain. That means μ_1 decreases with increasing branching content. By the way, all figures by Hepperle show a horizontal dependence of $h_s(\gamma_0)$ on γ_0 in the range of very small γ_0-values which is not described by Eq. (10.82). This result is in agreement with the general statement that $h_s(\gamma_0)$ has to be an *even function of* γ_0 as shear experiments under the conditions γ_0 and $-\gamma_0$ are physically identical.

If the *general Wagner theory* is applied to the case of simple shear, besides the components of the relative Finger tensor and the absolute Green tensor, also the components of the relative Cauchy tensor and the absolute Piola tensor are needed. For the case of time-dependent simple shear those are given by

$$\begin{aligned} &C_{22}(\underline{x};t,t') = 1 + [\gamma(t) - \gamma(t')]^2 && A_{22}(\underline{x};t) = 1 + \gamma^2(t) \\ &C_{12}(\underline{x};t,t') = -[\gamma(t) - \gamma(t')] && A_{12}(\underline{x};t) = -\gamma(t) \\ &C_{11}(\underline{x};t,t') = C_{33}(\underline{x};t,t') = 1 && A_{11}(\underline{x};t) = A_{33}(\underline{x};t) = 1 \end{aligned} \tag{10.83}$$

If these components together with (10.11) are inserted into (10.75) and if the damping functions in shear of the first and second term in (10.74) are designated by $h_{1,s}(\gamma(t))$ and $h_{2,s}(\gamma(t))$ Eqs. (10.79) change into

$$\begin{aligned} &\sigma_{12}(t) = G(t)[h_{1,s}(\gamma(t)) - h_{2,s}(\gamma(t))]\cdot\gamma(t) + \\ &+\int_0^t m(t-t')\cdot[h_{1,s}(|\gamma(t)-\gamma(t')|) - h_{2,s}(|\gamma(t)-\gamma(t')|)]\cdot[|\gamma(t)-\gamma(t')|]\mathrm{d}t' \\ &\sigma_{11}(t) - \sigma_{22}(t) = G(t)[h_{1,s}(\gamma(t)) - h_{2,s}(\gamma(t))]\cdot\gamma^2(t) + \\ &+\int_0^t m(t-t')\cdot[h_{1,s}(|\gamma(t)-\gamma(t')|) - h_{2,s}(|\gamma(t)-\gamma(t')|)]\cdot[\gamma(t)-\gamma(t')]^2\mathrm{d}t' \\ &\sigma_{22}(t) - \sigma_{33}(t) = G(t)\cdot h_{2,s}(\gamma(t))\cdot\gamma^2(t) + \\ &+\int_0^t m(t-t')\cdot[h_{2,s}(|\gamma(t)-\gamma(t')|)]\cdot[\gamma(t)-\gamma(t')]^2\mathrm{d}t' \end{aligned} \tag{10.84}$$

For *stress relaxation in simple shear* instead of (10.81), the equations

$$\begin{aligned} &\sigma_{12}(t) = G(t)[h_{1,s}(\gamma_0) - h_{2,s}(\gamma_0)]\cdot\gamma_0 = G(t;\gamma_0)\gamma_0 \\ &\sigma_{11}(t) - \sigma_{22}(t) = G(t)[h_{1,s}(\gamma_0) - h_{2,s}(\gamma_0)]\cdot\gamma_0^2 = G(t;\gamma_0)\gamma_0^2 \\ &\sigma_{22}(t) - \sigma_{33}(t) = G(t)h_{2,s}(\gamma_0)\cdot\gamma_0^2 \end{aligned} \tag{10.85}$$

result. Generally, $|h_{2,s}| \ll |h_{1,s}|$ and $|h_{2,s}| < 0$ are found experimentally.

Due to the general Wagner theory, the *non-linear time-dependent shear modulus* is the product of the linear shear modulus and the *sum of the absolute values of the two damping functions in shear* and may be found as the shear stress divided by the constant shear strain or by the quotient of the first normal stress difference and the square of the shear strain. The *first normal stress coefficient* is *positive*. The *second normal stress coefficient* is *negative*. This shows that within the scope of the general Wagner theory second normal stresses and second normal stress coefficients might be fitted to experimental data.

Finally, the predictions of the general Wagner theory for a stressing experiment in simple shear are discussed. A stressing experiment is defined by the time-dependent simple shear as prescribed in Eq. (10.13). Inserting Eqs. (10.11), (10.77), (10.78), and (10.83) into Eqs. (10.76), we find after the introduction of the new integration variable $\xi = t - t'$

$$\sigma_{12}(t) = \eta(t, \dot{\gamma}_0)\, \dot{\gamma}_0 \quad \text{with}$$

$$\eta(t, \dot{\gamma}_0) = \int_0^t G(\xi) \frac{\partial\{\xi \cdot [h_{1,s}(\dot{\gamma}_0\xi) - h_{2,s}(\dot{\gamma}_0\xi)]\}}{\partial\xi} \mathrm{d}\xi \tag{10.86a}$$

$$\sigma_{11}(t) - \sigma_{22}(t) = n_1(t, \dot{\gamma}_0)\dot{\gamma}_0^2 \quad \text{with}$$

$$n_1(t, \dot{\gamma}_0) = \int_0^t G(\xi) \frac{\partial\{\xi^2 \cdot [h_{1,s}(\dot{\gamma}_0\xi) - h_{1,s}(\dot{\gamma}_0\xi)]\}}{\partial\xi} \mathrm{d}\xi \tag{10.86b}$$

$$\sigma_{22}(t) - \sigma_{33}(t) = n_2(t, \dot{\gamma}_0)\dot{\gamma}_0^2 \quad \text{with}$$

$$n_2(t, \dot{\gamma}_0) = \int_0^t G(\xi) \frac{\partial[\xi^2 \cdot h_{2,s}(\dot{\gamma}_0\xi)]}{\partial\xi} \mathrm{d}\xi \tag{10.86c}$$

The *viscosity* $\eta(t, \dot{\gamma}_0)$ and the *first and second normal stress coefficients* $n_1(t, \dot{\gamma}_0)$ and $n_2(t, \dot{\gamma}_0)$ depend on time and shear rate. The first normal stress coefficient $n_1(t, \dot{\gamma}_0)$ is positive and the second normal stress coefficient $n_2(t, \dot{\gamma}_0)$ is negative. The general Wagner theory should therefore be able to describe non-Newtonian flow behavior.

The two functions $\eta(t, \dot{\gamma}_0)$ and $n_1(t, \dot{\gamma}_0)$ are no longer monotonously increasing functions of time, but may exhibit a maximum before reaching a plateau value after sufficiently long times. This may be seen by the following consideration. Assuming that the damping function has an exponential decrease similar to Eq. (10.82), the expressions in the brackets in (10.86a) and (10.86b) will be zero at very small arguments and at very large arguments. As they are positive everywhere, they will run through a positive maximum in between. Therefore their derivative after ξ will show a negative part which will lead to a decrease in the integral and a maximum in the viscosity and the first normal stress coefficient. Therefore a picture as given in Fig. 12.12 might qualitatively be described by means of the Wagner-Demarmels theory.

The functions $\eta(t, \dot{\gamma}_0)$ and $n_1(t, \dot{\gamma}_0)$ tend to finite limits for $t \to \infty$, the *stationary non-Newtonian viscosity* and the *stationary non-Newtonian first normal stress coefficient*. These may be calculated by partial integration of (10.86a) and (10.86b) and performing the limiting process to infinite time and using the fact that both products $t \cdot G(t)$ and $t^2 \cdot G(t)$ vanish for $t \to \infty$ for liquids with fading memory.

$$\eta(t \to \infty, \dot{\gamma}_0) = -\int_0^\infty \dot{G}(\xi)\xi \cdot [h_{1,s}(\dot{\gamma}_0\xi) - h_{2,s}(\dot{\gamma}_0\xi)]\mathrm{d}\xi \tag{10.87}$$

$$n_1(t \to \infty, \dot{\gamma}_0) = -\int_0^\infty \dot{G}(\xi)\xi^2 \cdot [h_{1,s}(\dot{\gamma}_0\xi) - h_{2,s}(\dot{\gamma}_0\xi)]\mathrm{d}\xi \tag{10.88}$$

If damping functions and spectrum are known, the stationary non-Newtonian viscosity and the first normal stress coefficient may be calculated under the assumption of the validity of Wagner's constitutive equation. This has been verified by Laun on the LDPE IUPAC A melt [17].

10.3.3 The Wagner Theories for Uniaxial Extension

Finally we consider the *general Wagner theory for uniaxial extension*, assuming the relative path Eqs. (10.55′) with $m = -1/2$

$$x_1 = e^{\varepsilon(t)-\varepsilon(t')}\, x_1' \quad x_2 = e^{-[\varepsilon(t)-\varepsilon(t')]/2} x_2' \quad x_3 = e^{-[\varepsilon(t)-\varepsilon(t')]/2} x_3' \tag{10.55′}$$

The components of the relative Finger and the absolute Green tensor are given by (10.56) with $m = -1/2$ and the components of the relative Cauchy and the absolute Piola tensor by (10.57) with $m = -1/2$. The invariants are

$$I_B(\underline{x}; t, t') = e^{2[\varepsilon(t)-\varepsilon(t')]} + 2e^{-[\varepsilon(t)-\varepsilon(t')]} \qquad I_C(\underline{x}; t, t') = e^{-2[\varepsilon(t)-\varepsilon(t')]} + 2e^{[\varepsilon(t)-\varepsilon(t')]}$$
$$II_B(\underline{x}; t, t') = 2e^{[\varepsilon(t)-\varepsilon(t')]} + e^{-2[\varepsilon(t)-\varepsilon(t')]} \qquad II_C(\underline{x}; t, t') = 2e^{-[\varepsilon(t)-\varepsilon(t')]} + e^{2[\varepsilon(t)-\varepsilon(t')]}$$

with the consequence that the damping functions can depend only on the difference $\varepsilon(t) - \varepsilon(t')$.

$$h_1(I_B, II_B) \equiv h_{1,e}((\varepsilon(t) - \varepsilon(t'))) \qquad h_{-1}(I_B, II_B) \equiv h_{2,e}((\varepsilon(t) - \varepsilon(t'))) \tag{10.89}$$

The functions $h_{1,e}$ and $h_{2,e}$ are designated as *damping functions in uniaxial extension.* Inserting these equations into (10.75) results into the following equations for the tensile stresses in uniaxial extension regarding that $\sigma_{22}(t)$ and $\sigma_{33}(t)$ vanish:

$$\sigma(t) = \sigma_{11}(t) - \sigma_{22}(t) = G(t)\{h_{1,e}(\varepsilon(t))[e^{2\varepsilon(t)} - e^{-\varepsilon(t)}] + h_{2,e}(\varepsilon(t))[e^{-2\varepsilon(t)} - e^{\varepsilon(t)}]\} -$$
$$- \int_0^t \dot{G}(t-t') \cdot \{h_{1,e} \cdot [e^{2[\varepsilon(t)-\varepsilon(t')]} - e^{-[\varepsilon(t)-\varepsilon(t')]}] + h_{2,e} \cdot [e^{-2[\varepsilon(t)-\varepsilon(t')]} - e^{[\varepsilon(t)-\varepsilon(t')]}]\}dt' \tag{10.90}$$

Here again the most simple case is *stress relaxation in uniaxial extension* defined by

$$\begin{aligned} \varepsilon(t) &= 0 \quad \text{for} \quad t<0 \\ \varepsilon(t) &= \varepsilon_0 \quad \text{for} \quad t \geq 0 \end{aligned} \tag{10.91}$$

where the terms under the integral in Eq. (10.90) vanish with the result

$$\begin{aligned} \sigma(t) &= \varepsilon_0 \cdot E(t, \varepsilon_0) \quad \text{whereas} \\ E(t, \varepsilon_0) &= G(t) \cdot \{h_{1,e}(\varepsilon_0) \cdot [e^{2\varepsilon_0} - e^{-\varepsilon_0}] + h_{2,e}(\varepsilon_0) \cdot [e^{-2\varepsilon_0} - e^{\varepsilon_0}]\}/\varepsilon_0 \end{aligned} \tag{10.92}$$

The *non-linear relaxation modulus* in *uniaxial extension* $E(t,\varepsilon_0)$ may be written as the product of a function depending on the constant Hencky strain with the time-dependent shear modulus $G(t)$ of the linear region. From the quotient of $E(t,\varepsilon_0)$ to $G(t)$ the damping function in extension may be derived, in principle. Measurements at high steps in elongation are difficult to perform, however, for several experimental reasons.

The *stressing experiment in uniaxial extension* was defined by

$$\varepsilon(t) = 0 \quad \text{for} \quad t<0 \quad \varepsilon(t) = \dot{\varepsilon}_0 t \quad \text{for} \quad t \geq 0 \tag{10.59}$$

with $\dot{\varepsilon}_0$ being the constant Hencky strain rate. For the tensile stress, the following expression is obtained from Eq. (10.76)

$$\sigma(t) = \int_0^t G(\xi) \frac{d[h_{1,e}(\dot{\varepsilon}_0\xi)(e^{2\dot{\varepsilon}_0\xi} - e^{-\dot{\varepsilon}_0\xi}) + h_{2,e}(\dot{\varepsilon}_0\xi)(e^{-2\dot{\varepsilon}_0\xi} - e^{\dot{\varepsilon}_0\xi})]}{d\xi} d\xi \tag{10.93}$$

If the special Wagner theory is applied instead of the general one, Eq. (10.93) may be used with the conditions $h_{1,e} = h_e$ and $h_{2,e} = 0$ resulting into:

$$\sigma(t) = \int_0^t G(\xi) \frac{d[h_e(\dot{\varepsilon}_0\xi)(e^{2\dot{\varepsilon}_0\xi} - e^{-\dot{\varepsilon}_0\xi})]}{d\xi} d\xi \tag{10.94}$$

Wagner [21] analyzed experiments performed by Meissner [18] under various elongational rates on LDPE (IUPAC I) at 150 °C shown in Fig. 10.12 as the tensile viscosity [1])

$$\mu(t, \dot{\varepsilon}_0) = \sigma(t)/\dot{\varepsilon}_0 \tag{10.95}$$

versus time for various strain rates in a double logarithmic plot. As no second normal stress differences were measured, he restricted his evaluation to the use of the special Wagner theory and Eq. (10.94). He has shown how the damping function may be derived from this equation if the tensile stress has been measured and the linear relaxation modulus is known.

The predictions of Eq. (10.94) have been indicated by drawn lines in Fig. 10.12. The higher the extensional rate was, the earlier experiment and theory deviate from the limiting curve with the lowest extensional rate. The onset of the strain hardening is predicted in good agreement with the experiment, the magnitude of strain hardening seems to be overestimated by the theory particularly at smaller $\dot{\varepsilon}_0$. Concerning the experimental and theoretical challenges, the agreement in Fig. 10.12 might be considered as satisfactory.

The damping function shown versus the Hencky strain in Fig. 10.13, can be described by the expression

$$h_e(\varepsilon) = \left[ae^{2\varepsilon} + (1-a)e^{r\varepsilon}\right]^{-1} \tag{10.96}$$

with $a = \exp(-6) \approx 0.0025$ and $r = 0.30$ being the two fitting parameters. Different from the case of simple shear, Eq. (10.96) is not a sum of exponential functions. As a consequence, the integral in Eq. (10.94) cannot be solved analytically, but has to be treated numerically.

Concerning the description of *recovery experiments* the following remark should be made. The damping function may be considered as a unique function of the invariants only in time intervals, in which the deformation increases, i.e., where h decreases. During periods of decreasing deformation (periods of recovery) h is to be interpreted as a functional of the history of the invariants. Entanglements may be loosened or even disappear by increasing deformation, but are unlikely to be restored again within periods of decreasing deformation. Broken entanglements will remain in this state during recovery.

Due to a proposal by Raible et al. [22] in such a case the value of the damping function should be set equal to the minimum value which was attained during the deformation history:

$$h \cong \mathrm{H}(I_B, II_B) = \min_{\tau=t'}^{\tau=t}\left\{h[I_B(\tau,t'), II_B(\tau,t')]\right\} \tag{10.97}$$

---

[1] With respect to the term tensile viscosity we want to refer the reader to the principle remarks given in Sect. 12.1.

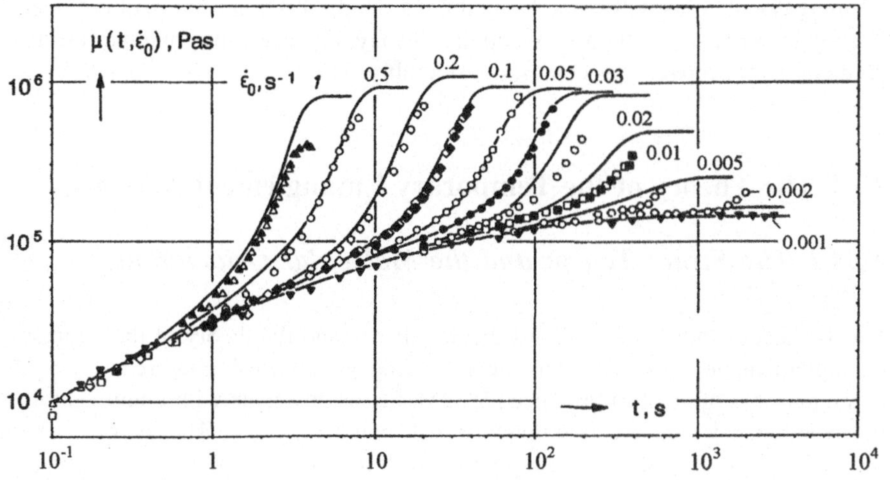

Fig. 10.12 Uniaxial tensile viscosities measured by Meissner [18] as function of the time at various constant Hencky elongational rates for an LDPE (IUPAC I) at 150 °C and the predicitions of Wagner's theory as drawn lines. Reproduced from [21] by permission from Elsevier

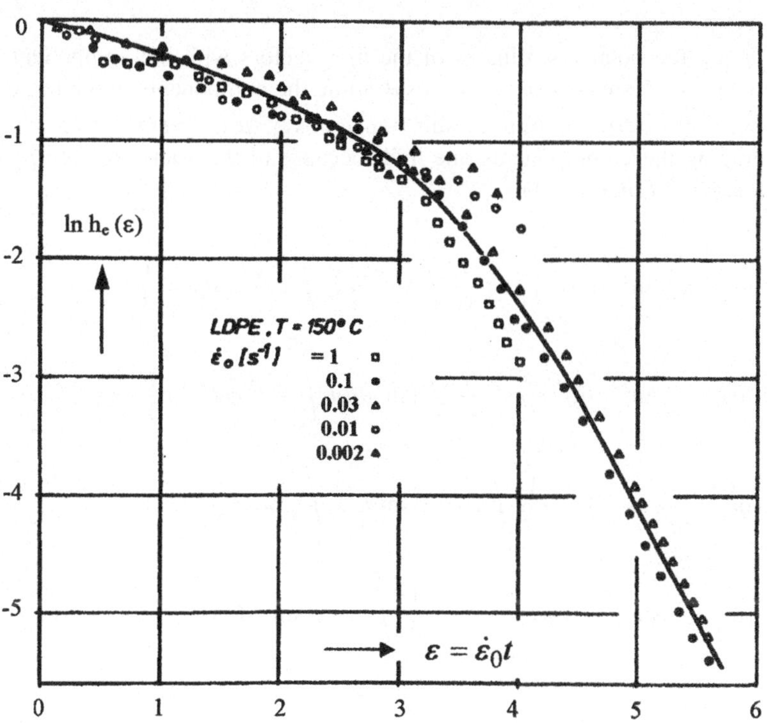

Fig. 10.13 Damping function in uniaxial extension calculated from the data of Fig. 10.12 by Wagner. Reproduced from [21] by permission from Elsevier

As most calculations will be performed numerically this condition will not restrict the applicability of the theory essentially. It may, however, improve the results of predictions of recovery processes, considerably.

10.4 The Theory of the Temporary Entanglement Network

10.4.1 The Strain Tensor and the Molecular Orientation

For the understanding of the Doi-Edwards theory and the theory of the temporary entanglement network with slip the following geometrical interpretation of the strain tensor is of great importance [23–25]. Let us designate with $\underline{u}$ unit vectors at the position $\underline{x}$ which are isotropically distributed in space. The average of the tensor $\langle \underline{u}\,\underline{u}\rangle$ is defined as the average over all directions of the vectors $\underline{u}$ using a distribution function $f(\vartheta, \varphi)$ for their directions

$$\langle \underline{u}\,\underline{u}\rangle = \frac{1}{4\pi}\int_0^{\pi} \sin\vartheta d\vartheta \int_0^{2\pi} d\varphi(\underline{u}\,\underline{u})f(\vartheta, \varphi) \tag{10.98}$$

ϑ and φ are the polar coordinates of the unit vectors with the components $u_1 = \sin\vartheta\cos\varphi,\ u_2 = \sin\vartheta\sin\varphi,\ u_3 = \cos\vartheta$ while the components of the tensor $(\underline{u}\,\underline{u})$ are $u_1u_1 = \sin^2\vartheta\cos^2\varphi,\ u_1u_2 = \sin^2\vartheta\sin\varphi\cos\varphi,\ u_1u_3 = \sin\vartheta\cos\vartheta\cos\varphi$ etc. On averaging these components one finds because of the assumed isotropy of the vectors $\underline{u}$ (viz., $f(\vartheta, \varphi) = 1$)

$$\langle u_1u_1\rangle = \langle \sin^2\vartheta\cos^2\varphi\rangle = \frac{1}{4\pi}\int_0^{\pi} \sin^3\vartheta d\vartheta \int_0^{2\pi} \cos^2\varphi d\varphi = \frac{1}{3}$$

$$\langle u_2u_2\rangle = \langle \sin^2\vartheta\sin^2\varphi\rangle = \frac{1}{4\pi}\int_0^{\pi} \sin^3\vartheta d\vartheta \int_0^{2\pi} \sin^2\varphi d\varphi = \frac{1}{3}$$

$$\langle u_3u_3\rangle = \langle \cos^2\vartheta\rangle = \frac{1}{4\pi}\int_0^{\pi} \cos^2\vartheta \sin\vartheta d\vartheta \int_0^{2\pi} d\varphi = \frac{1}{3}$$

$$\langle u_1u_2\rangle = \langle \sin^2\vartheta\sin\varphi\cos\varphi\rangle = \frac{1}{4\pi}\int_0^{\pi} \sin^2\vartheta\sin\vartheta d\vartheta \int_0^{2\pi} \sin\varphi\cos\varphi d\varphi = 0$$

Therefore, the dyadic average of isotropically distributed unit vectors equals

$$\langle u_i u_k \rangle = \tfrac{1}{3}\delta_{ik} \quad \text{or} \quad \langle \underline{u}\underline{u} \rangle = \tfrac{1}{3}\underline{\underline{E}} \tag{10.99}$$

Consider now the state at the position $\underline{x}$ at the past time t'. There were isotropically distributed unit vectors $\underline{u}'$ which obeyed the conditions $\langle \underline{u}'_i \underline{u}'_k \rangle = \delta_{ik}/3$. By the deformation imposed on the material from the time t' until the time t the unit vectors $\underline{u}'$ are transformed into vectors $\underline{u}$ with the components $u_i = [\underline{\underline{F}}_t^{-1}(t')]_{im} u'_m$. For calculating the average $\langle u_i u_k \rangle$ use is made of the distributive property of the symbol $\langle\rangle$ and of the fact that the components of the inverse deformation gradient $[\underline{\underline{F}}_t^{-1}(t')]_{ik}$ do not depend on ϑ or φ and can be drawn outside the symbol $\langle\rangle$. Therefore, the average can be written as

$$\begin{aligned}\langle u_i u_k \rangle &= \left\langle [\underline{\underline{F}}_t^{-1}(t')]_{im} u'_m [\underline{\underline{F}}_t^{-1}(t')]_{kl} u'_l \right\rangle = [\underline{\underline{F}}_t^{-1}(t')]_{im} \langle u'_m u'_l \rangle [\underline{\underline{F}}_t^{-1}(t')]_{kl} = \\ &= [\underline{\underline{F}}_t^{-1}(t')]_{im} \frac{1}{3}\delta_{ml} [\underline{\underline{F}}_t^{-1}(t')]_{kl} = \frac{1}{3}[\underline{\underline{F}}_t^{-1}(t')]_{im} [\tilde{\underline{\underline{F}}}_t^{-1}(t')]_{mk} = \frac{1}{3}[\underline{\underline{C}}_t^{-1}(t')]_{ik}\end{aligned}$$

or $\langle \underline{u}\underline{u} \rangle = \underline{\underline{C}}_t^{-1}(t')/3$. Wagner designates this average by $\langle \underline{u}'\underline{u}' \rangle$, because it describes the change of the dyadic product from the state at the past time t' to the state at the observation time t and therefore constitutes a measure for the state of the deformation at the past time t'.

$$\langle \underline{u}'\underline{u}' \rangle = \frac{1}{3}\underline{\underline{C}}_t^{-1}(t') = \frac{1}{3}\underline{\underline{B}}_t(t') \tag{10.100}$$

$\langle \underline{u}'\underline{u}' \rangle$ is the average of the dyadic product of those unit vectors at the position $\underline{x}$ at the observation time t, which have been isotropically distributed at the past time t'. Due to Eq. (8.35) this average equals one third of the relative Finger tensor.

The average of the square of the length of the vectors is found as the trace of the dyadic $\langle \underline{u}'\underline{u}' \rangle$ i.e.

$$\langle u'^2 \rangle = I_B/3 \tag{10.101}$$

Besides this quantity, a further average will play a role in the following, which is much more difficult to calculate than $\langle u'^2 \rangle$, viz., the *average* of the *length of the vectors* themselves, $\langle u' \rangle$. This average differs somewhat from the square root of $\langle u'^2 \rangle$, and may be calculated by taking the square root of the expression

$$\begin{aligned}u'^2 &= u'_m u'_m = [\underline{\underline{F}}_t^{-1}(t')]_{mi} u_i [\underline{\underline{F}}_t^{-1}(t')]_{mk} u_k = [\tilde{\underline{\underline{F}}}_t^{-1}(t')]_{im} [\underline{\underline{F}}_t^{-1}(t')]_{mk} u_i u_k \\ &= [\underline{\underline{B}}_t^{-1}(t')]_{ik} u_i u_k\end{aligned}$$

and averaging over the angles which results in the integral

$$\langle u' \rangle = \left\langle \sqrt{[\underline{\underline{B}}_t^{-1}(t')]_{ik} u_i u_k} \right\rangle \tag{10.102}$$

Because of the square root under the integral the evaluation of Eq. (10.102) leads to elliptic integrals and is possible only in simple cases and by numerical integration. For deformations in which the tensor $\underline{\underline{B}}^{-1}$ is in its principle coordinate system, the integrals can be solved for unidirectional extension in the 1-direction, for equibiaxial extension in the 1 and 2-direction and by numerical integration for planar extension in the 1 and 2-direction [26].

The difference between the averages $\langle u' \rangle$ and $\sqrt{\langle u'^2 \rangle}$ is rather small, as may be seen from Fig. A.1 in the appendix on page 359 in which both quantities are plotted versus the extension ratio λ for the three cases of deformation considered. Apart from the case of planar extension for which the averages practically coincide, $\langle u' \rangle$ is only slightly higher than $\sqrt{\langle u'^2 \rangle}$. Explicit formulae may be found in Appendix A and in the cited literature.

10.4.2 The Temporary Entanglement Network with and Without Slip

According to Doi and Edwards [27] and de Gennes [28] linear macromolecules are not able to penetrate each other, and can move only within a tube formed by the neighboring molecules. According to an illustrative picture by Janeschitz-Kriegl [1], they behave as snakes creeping through a dense bamboo wood, and can only proceed into the direction of their head or tail, creating hereby a new part of their own tube and leaving behind a corresponding part of their old tube. This motion called *reptation* is thought to be one of the principal mechanisms for stress relaxation in the deformed melt. The *tube model* as proposed by Doi and Edwards [27] and the *temporary entanglement network model* as proposed by Wagner and Schaeffer [24, 29, 30] are illustrated in Fig. 10.14.

The entanglement model contains slip links. These are parts of rings formed at the entanglements where the polymer molecule can slide through (cf. Fig. 4.7c). We discuss the slip link model following the presentation by Wagner and Schaeffer [29].

The distance vectors $\underline{r}$ of neighbored entanglements are isotropically distributed in the undeformed (stress free) state. Their distances r differ, showing a Gaussian distribution with $\langle r^2 \rangle = a_0^2$ where a_0^2 is the mean square distance between neighbored entanglements. a_0 is also assumed to be the *diameter* of the *tubes* in the undeformed state.

Polymer molecules in the melt may be considered as statistically equivalent random walk chains with *n Kuhn's statistical segments of length b*. In the Gaussian approximation ($/r/ \ll$ nb), the probability $w(\underline{r})d^3\underline{r}$ to find a chain with an end-to-end vector between $\underline{r}$ and $\underline{r} + d\underline{r}$ is independent of the orientation of the vector $\underline{r}$.

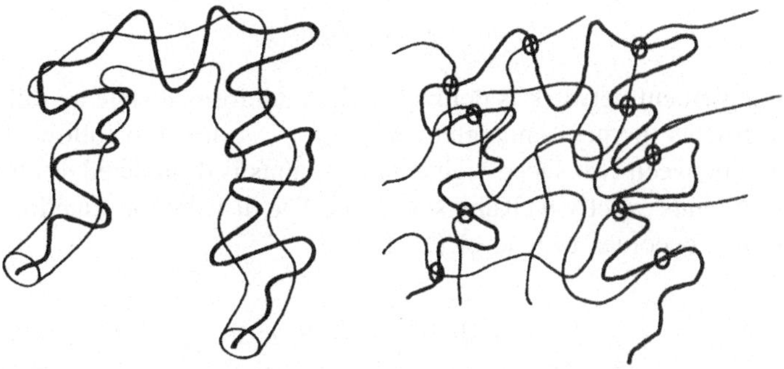

Fig. 10.14 Tube model after Doi and Edwards (*left*) and temporary entanglement network model after Wagner and Schaeffer (*right*) in comparison, reproduced from [24] by permission from the Society of Rheology (U.S.)

The distribution was already introduced in Chap. 2 as Eq. (2.30) and is written here again in the following form:

$$\begin{aligned} w(\underline{r})d^3\underline{r} &= \frac{a^3}{\pi^{3/2}}\exp(-a^2r^2)\mathrm{d}x_1\mathrm{d}x_2\mathrm{d}x_3 = \\ &= \frac{a^3}{\pi^{3/2}}\exp(-a^2r^2)\cdot r^2\mathrm{d}r\sin\vartheta\mathrm{d}\vartheta\mathrm{d}\varphi = \psi_0^A(r)\psi_0^O(\underline{u})\mathrm{d}r\mathrm{d}\Omega \end{aligned} \quad (10.103)$$

with a^2 being proportional to the reciprocal *mean square end to end distance* of the *entire molecule* $\langle r^2\rangle_{\mathrm{mol}}$.

$$a = \sqrt{\frac{3}{2nb^2}} = \sqrt{\frac{3}{2\langle r^2\rangle_{\mathrm{mol}}}} \quad (10.104)$$

and $\mathrm{d}\Omega = \sin\vartheta\mathrm{d}\vartheta\mathrm{d}\varphi$. Expressed in polar coordinates, the function $w(\underline{r})d^3\underline{r}$ is the product of a *distribution function for the distance*

$$\psi_0^A(r)\mathrm{d}r = \frac{4a^3}{\pi^{1/2}}e^{-a^2r^2}r^2\mathrm{d}r \quad (10.103a)$$

and one for the *orientation*

$$\psi_0^O(\underline{u})\mathrm{d}\Omega = \frac{1}{4\pi}\sin\vartheta\mathrm{d}\vartheta\mathrm{d}\varphi \quad (10.103b)$$

For parts of the molecule of sufficient length, the same statistics is valid. For a part of the molecule with n_0 Kuhn segments, Eqs. (10.103), (10.103a), and (10.103b) remain valid, if n is replaced by n_0 and a by a_0

$$a_0 = \sqrt{(3/(2n_0b^2)} \tag{10.105}$$

The entanglement network is defined in the undeformed state as follows: After n_0 Kuhn segments the next entanglement point is assumed to follow. The part of the molecule between two successive entanglements is designated as *chain bow*. A chain bow contains exactly n_0 Kuhn segments. The distribution function for its end to end distance becomes instead of (10.103a)

$$\psi_0^A(r)\mathrm{d}r = \frac{4a_0^3}{\pi^{1/2}} e^{-a_0^2r^2} r^2 \mathrm{d}r \tag{10.106}$$

and its end to end vector has the mean square distance

$$\langle r^2 \rangle_{\mathrm{ch}} = 3/(2a_0^2) = n_0b^2 \tag{10.107}$$

which differs of course from $\langle r^2 \rangle_{\mathrm{mol}}$. The entropy of one chain bow with the end to end distance r equals apart from an additive constant $\Delta S = k \ln \psi_0^A(r) = -ka_0^2r^2 - 2k \ln r \cong -ka_0^2r^2$ and, assuming free rotation of the Kuhn segments, its free energy ΔF equals

$$\Delta F = \Delta U - T\Delta S = -T\Delta S = kTa_0^2r^2 \tag{10.108}$$

The force to be exerted by the vicinity on the chain bow to hold its end-to-end vector in position is found by taking the gradient of the free energy and using Eq. (10.105) as

$$\underline{K}(\underline{r}) = \mathrm{grad}(\Delta F) = \frac{\mathrm{d}\Delta F}{\mathrm{d}r}\mathrm{grad}(r) = 2kTa_0^2\underline{r} = \frac{3kT}{n_o b^2} \cdot \underline{r} \tag{10.109}$$

This force acts in the direction of the end-to-end vector of the chain bow, is proportional to its end-to-end distance and the absolute temperature, and inversely proportional to its number of Kuhn segments.

If interaction between polymer molecules and entanglements only occurs at the entanglements, the contribution of the chain bow to the stress tensor will be the dyadic product of the force on the chain bow and its end-to-end vector, averaged over the distribution of the chains in the deformed state $w(r)$. If for the force and the end vector the values in the deformed state $\underline{K}'$ and $\underline{r}'$ are taken, averaging can be performed over the distribution of the chains $w_0(r)$ in the undeformed state. The contribution of all chain bows to the stress tensor may be expressed as

$$\underline{\underline{\sigma}} + p\underline{\underline{E}} = cN_0 \langle \underline{K}\,\underline{r} \rangle = cN_0 \int w(\underline{r})\underline{K}(\underline{r})\underline{r}d^3\underline{r} = cN_0 \int w_0(\underline{r})\underline{K}'(\underline{r})\underline{r}'d^3\underline{r} \tag{10.110}$$

where c equals the *number of macromolecules per volume*, N_0 the *number of chain bows per macromolecule*, cN_0 the *number of chain bows per volume*.

The temporary entanglement network without slip, the Lodge theory:
We first discuss the network under the assumptions that

(1) The shift of the entanglements is affine to the macroscopic deformation
(2) No slip of the chain bows occurs in the tubes, after the deformation

If the entanglement network has been formed in the undeformed state at the past time t', the distribution of the end-to-end vectors $\underline{r} = r\underline{u}$ will be isotropic, with $\underline{u}$ being isotropically distributed unit vectors. By the deformation the vectors $\underline{r}$ are transformed into vectors $\underline{r}'$ according to

$$\underline{r}' = r\underline{\underline{F}}_t^{-1} \cdot \underline{u} = r \cdot \underline{u}' \tag{10.111}$$

and the force of the vicinity on the deformed chain bow equals

$$\underline{K}'(\underline{r}) = \frac{3kT}{n_0 b^2} \cdot \underline{r}' = \frac{3kT}{n_0 b^2} r \cdot \underline{u}' \tag{10.112}$$

The dyadic in (10.110) becomes $\underline{K}'\,\underline{r}' = (3kT/n_0b^2)r^2\underline{u}'\,\underline{u}'$ and the stress tensor (10.110) reads considering (10.107) and (10.100).

$$\begin{aligned}\underline{\underline{\sigma}} + p\underline{\underline{E}} &= cN_0\frac{3kT}{n_0b^2}\int w_0(\underline{r})r^2\underline{u}'\,\underline{u}'d^3\underline{r} = \frac{3kTcN_0}{n_0b^2}\int \psi_0^A(r)r^2\mathrm{d}r\int \psi_0^O(\underline{u})\underline{u}'\,\underline{u}'d\Omega = \\ &= 3kTcN_0\int \psi_0^O(\underline{u})\,\underline{u}'\,\underline{u}'\mathrm{d}\Omega = 3kTcN_0\langle\underline{u}'\,\underline{u}'\rangle = kTcN_0\underline{\underline{B}}_t(t')\end{aligned} \tag{10.113}$$

Taking into account finally the relaxation of the number of chain bows by replacing $kTcN_0$ by $kTn(t - t') = m(t - t')$ and adding the contributions of all networks formed in the past by integration after the time t', results in the equation of state of the *Lodge theory*

$$\underline{\underline{\sigma}} + p\underline{\underline{E}} = \int_{-\infty}^{t} kTn(t - t')\underline{\underline{B}}_t(t')\mathrm{d}t' = \int_{-\infty}^{t} m(t - t')\underline{\underline{B}}_t(t')\mathrm{d}t' \tag{10.2}$$

The assumptions that the tubes are deformed affine to the deformation and no slip of chain bows occurs within the tubes lead to the equation of state of Lodge.

The temporary entanglement network with slip:
In an extension of the model, the occurrence of slip of the chain bows directly after the orientation of the tubes is admitted now. After Wagner and Schaeffer [29] slip will always be accompanied by disentanglement, as parts of the chain which move from one chain bow into the next, are missing at the end of the molecules for an entanglement. Therefore, slip will be characterized by two functions depending on the deformation, the *slip function* $S(\underline{r}')$ and the *disentanglement function* $D(\underline{r}')$.

$S = n''/n_0$ is the ratio of the number of Kuhn segments between neighbored entanglements (of a chain bow considered) after the deformation (n'') and before the deformation (n_0), $D = N''/N_0$ is the ratio of the number of chain bows of the molecule after (N'') and before (N_0) the deformation.

n'' and N'' will generally depend on the orientation and the length of the vector $\underline{r}'$. Wagner and Schaeffer assumed that n'' and N'' only depend on the stretch of the chain bow, not on its orientation, that means on the absolute value of the vector $\underline{u}'$, viz., on u': $n'' = n''(u')$ and $N'' = N''(u')$ with the consequence

$$S(\underline{r}') = \frac{n''}{n_0} = S(u') \tag{10.114}$$

$$D(\underline{r}') = \frac{N''}{N_0} = D(u') \tag{10.115}$$

Instead of (10.109), the force exerted by the vicinity on the chain bow now reads

$$\underline{K}(\underline{r}) = \frac{3kT}{n''b^2} \cdot \underline{r} = \frac{1}{S}\frac{3kT}{n_0 b^2} \cdot \underline{r} \tag{10.116}$$

and instead of (10.110), the average of the dyadic becomes

$$\begin{aligned} \underline{\underline{\sigma}} + p\underline{\underline{E}} &= c\langle N''\underline{K}'(\underline{r}')\underline{r}'\rangle = cN_0 \frac{3kT}{n_0 b^2}\left\langle \frac{D(u')}{S(u')} r^2 \underline{u}'\underline{u}' \right\rangle = \\ &= \frac{3kTcN_0}{n_0 b^2} \int \psi_0^A(r) r^2 dr \int \psi_0^O(\underline{u}) \frac{D(u')}{S(u')} \underline{u}'\underline{u}' d\Omega = 3kTcN_0 \left\langle \frac{D(u')}{S(u')} \underline{u}'\underline{u}' \right\rangle \end{aligned}$$

which is equivalent to the equation of state

$$\boxed{\underline{\underline{\sigma}} + p\underline{\underline{E}} = 3kTcN_0 \left\langle \frac{D(u')}{S(u')} \underline{u}'\underline{u}' \right\rangle} \tag{10.117}$$

In this equation, the averaging procedure relates only to the directional coordinates ϑ and φ. $D(u')$ and $S(u')$ are not independent from each other, but are related by the condition

$$\langle D(u') \cdot S(u') \rangle = 1 \tag{10.118}$$

which may be proved as follows: Adding the product $n'' \cdot N''$ over all orientations of the vectors $\underline{r}$ in the undeformed state results into $\langle n'' \cdot N'' \rangle$ and this equals the complete number of Kuhn segments pro molecule, viz., ($n_0 N_0$). Dividing $\langle n'' \cdot N'' \rangle$ by $n_0 N_0$ yields (10.118).

10.5 The Doi-Edwards Theory

10.5.1 The Constitutive Equations of the Doi-Edwards Theory

The theory of Doi and Edwards [27] is a special case of the temporary entanglement network with slip. Doi and Edwards proposed that slip occurs in such a manner that the force between two adjacent entanglements after slip is equal to the force between them before the deformation started. The macromolecule would then retract in the deformed and oriented tube to its original contour length. The absolute value of the force on the chain bow before the deformation is found from (10.109) as $|K(r)| = (3kT/n_0 b^2) \cdot r$ and after the deformation as $|K(r')| = (3kT/n'' b^2) \cdot r'$. Equating the two forces results in the condition $r'/n' = r/\, n_0$. Because of $r' = u' r$ the slip function becomes

$$S(u') \equiv \frac{n''}{n_0} = u' \tag{10.119}$$

Regarding the disentanglement, Doi and Edwards distinguish between two possibilities:

(1) Disentanglement occurs only at the chain ends: *Doi-Edwards theory* 1

That means the *disentanglement* is *isotropic* and N'' is independent of the magnitude of the stretch of the chains, or the disentanglement function $D(u')$ is a constant independent of u'. From (10.118), then follows $\langle D(u')S(u')\rangle = D(u') \cdot \langle S(u')\rangle = 1$ or using (10.119)

$$D(u') = \frac{1}{\langle S(u')\rangle} = \frac{1}{\langle u'\rangle} \tag{10.120}$$

After inserting (10.119) and (10.120) into Eq. (10.117), it follows for the stress tensor in the Doi-Edwards theory 1

$$\underline{\underline{\sigma}} + p\underline{\underline{E}} = 3kTcN_0 \frac{1}{\langle u'\rangle} \left\langle \frac{\underline{u}'\,\underline{u}'}{u'} \right\rangle \tag{10.121}$$

In the Doi–Edwards theory 1, the deformed unit vectors are first divided by the square root of their length and then multiplied dyadically with each other.

(2) Disentanglement occurs in the different sections of the tube independently. (*Independent alignment assumption*): *Doi-Edwards theory* 2

Another simple solution of Eq. (10.118) is found by the assumption

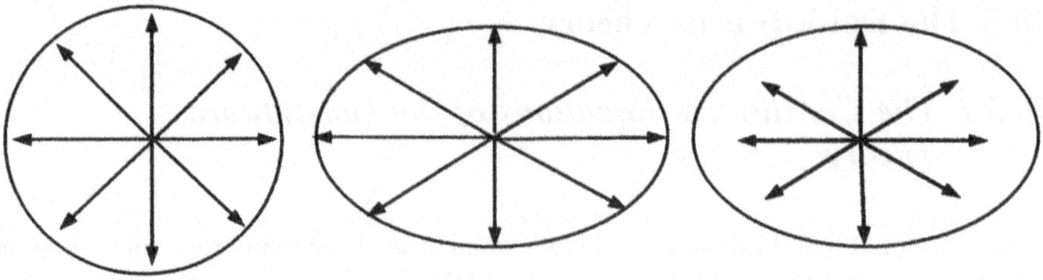

Fig. 10.15 Illustration of the orientation and length of isotropic unit vectors (*left*) by deformation (*middle*) and subsequent slip (*right*) according to the Doi–Edwards theory 2

$$D(u') = \frac{1}{S(u')} = \frac{1}{u'} \tag{10.122}$$

which implies that *anisotropic disentanglement* occurs which is inversely proportional to the average stretch of the chain bow under consideration. Inserting of (10.122) and (10.119) into (10.117) yields for the stress tensor in the *Doi–Edwards theory* 2

$$\boxed{\underline{\underline{\sigma}} + p\underline{\underline{E}} = 3kTcN_0\left\langle \frac{\underline{u}'\,\underline{u}'}{u'\,u'} \right\rangle} \tag{10.123}$$

In the Doi–Edwards theory 2, the deformed unit vectors are first divided by their length and then multiplied dyadically with each other. The tensor in (10.123) is often used in the literature, it is called the *DE-strain tensor* and designated by $\underline{\underline{S}}$

$$\underline{\underline{S}} = \underline{\underline{S}}_t(t') = \left\langle \frac{\underline{u}'\,\underline{u}'}{u'\,u'} \right\rangle \tag{10.123'}$$

The DE-strain tensor is illustrated in Fig. 10.15, which consists of three parts. In its left part the arrows show the isotropically distributed unit vectors in the undeformed state within a material element of a circular cross section. During the deformation the material is compressed in vertical and extended in horizontal direction. This is indicated by the change of the circular cross-section into an ellipsoidal one in the middle part of the figure. There the arrows show the orientation and length of the chainbows after the deformation, but before the slipping process. In the right part of Fig. 10.15 the arrows indicate the orientation and length of the chainbows after the deformation and the slipping process. According to Eq. (10.123), the vectors of the right part of the figure are to be multiplied dyadically with each other.

The dyadic expressions occurring in Eqs. (10.121) and (10.123) are too complicated to admit integration over the directional angles in closed form. The integration has to be performed numerically in each special case of deformation separately, similarly to the calculation of the averages $\langle u' \rangle$ in Sect. 10.4.1 This problem is more serious for Eq. (10.121) than for (10.123) as in the former a square root in the denominator of the integrand occurs, which is absent in the dyadic of (10.123).

Equation (10.123) has a further very essential advantage over Eq. (10.121). Currie [31] has proven the existence of a very good approximation for the dyadic $\underline{\underline{S}}$ in (10.123) consisting of a linear combination of the tensors $\underline{\underline{B}}_t(t')$ and $\underline{\underline{B}}_t^{-1}(t')$ which are easy to calculate from the given deformation history. The *Currie approximation* reads

$$\boxed{\left\langle \frac{\underline{u}'\,\underline{u}'}{u'\,u'} \right\rangle \cong \frac{1}{J}\underline{\underline{B}}_t(t') - \frac{1}{J\sqrt{II_B + 13/4}}\underline{\underline{B}}_t^{-1}(t') + \frac{1}{3}\left[1 - \frac{I_B}{J} + \frac{II_B}{J\sqrt{II_B + 13/4}}\right]\underline{\underline{E}}} \tag{10.124}$$

with

$$J = I_B - 1 + 2\sqrt{II_B + 13/4} \tag{10.124a}$$

The isotropic term in (10.124) gives rise to an isotropic term in the equation for the stress tensor which might be comprehended in the hydrostatic pressure and need not to be regarded further.

For the case of stress relaxation in simple shear and in the limit of small deformations $I_B = 0$ $II_B = 3$, $J = 7$ and

$$\sigma_{12} \cong \frac{3kTcN_0}{J}\gamma\left[1 + \frac{1}{\sqrt{II_B + 13/4}}\right] = \frac{3kTcN_0}{5}\gamma = \gamma \cdot G(t)$$

is found with $G(t)$ being the linear time-dependent shear modulus of the melt.

Hence, for the Doi-Edwards theory 2, the factor $3kTcN_0$ has to be related to the shear modulus by the equation

$$3kTcN_0 = 5G(t) \tag{10.125}$$

and the exact *constitutive equation of the Doi-Edwards theory* 2 becomes

$$\underline{\underline{\sigma}} + p\underline{\underline{E}} = \int_{-\infty}^{t} 5m(t-t')\left\langle \frac{\underline{u}'\,\underline{u}'}{u'\,u'} \right\rangle \mathrm{d}t' = \int_{-\infty}^{t} 5m(t-t')\underline{\underline{S}}_t(t')\mathrm{d}t' \tag{10.126}$$

with m being the memory function, related to the time-dependent shear modulus by Eq. (10.3).

If use is made of the approximation of Currie, the constitutive equation of the Doi-Edwards theory 2 takes the form of the general Wagner theory

$$\underline{\underline{\sigma}}(t) + p\underline{\underline{E}} = \int_{-\infty}^{t} m(t-t')[h_1(I_B, II_B)\underline{\underline{B}}_t(t') + h_{-1}(I_B, II_B)\underline{\underline{B}}_t^{-1}(t')]\mathrm{d}t' \quad (10.127)$$

with the two *damping functions*

$$h_1(I_B, II_B) \cong \frac{5}{J} = \frac{5}{I_B - 1 + 2\sqrt{II_B + 13/4}} \quad (10.127a)$$

and

$$h_{-1}(I_B, II_B) \cong \frac{-5}{J\sqrt{II_B + 13/4}} \quad (10.127b)$$

with J given by Eq. (10.124a). The discussion of the consequences of the Doi–Edwards theory 2 will be based on these equations which are easy to specify for the various types of deformation.

In proving the validity of the approximation (10.124), Currie further derived the following exact equations for the two damping functions of the Doi–Edwards theory 2 which may be obtained by partial differentiation of the scalar function

$$U(I_B, II_B) = 5\langle \ln(u'^2) \rangle = 10\langle \ln(u') \rangle \quad (10.128)$$

after the two invariants I_B and II_B

$$h_{-1}(I_B, II_B) = \frac{\partial U(I_B, II_B)}{\partial I_B} \quad (10.128a)$$

$$h_1(I_B, II_B) = -\frac{\partial U(I_B, II_B)}{\partial II_B} \quad (10.128b)$$

If the two damping functions in a constitutive equation can be derived from a potential, the equation is called a *separable B.K.Z.-equation* [32]. Thus Currie proved that the constitutive equation of the Doi–Edwards theory 2 is a separable BKZ-equation related to the potential given in (10.128).

The Doi–Edwards theory 2 (Eq. 10.123) is a special case of the general Wagner theory (Eq. 10.74). But those theories are by no means identical and have a totally different background. In the Doi–Edwards theory 2, the shape of the two damping functions is given by the molecular picture, in the general Wagner theory the damping functions may be chosen to adapt the theory to the experiment.

A corresponding analytical representation of the damping functions of the Doi–Edwards theory 1 is not known. Doi and Edwards, however, deduced from Eq. (10.121) the following rheological equation of state [2]

$$\underline{\underline{\sigma}} + p\underline{\underline{E}} = \int_{-\infty}^{t} \frac{15}{4} m(t-t') \frac{1}{\langle u' \rangle} \left\langle \frac{\underline{\underline{u'u'}}}{u'} \right\rangle \mathrm{d}t' \tag{10.129}$$

The calculation of the dyadic in Eq. (10.129) is much more difficult than that of Eq. (10.126) as already mentioned. Due to Wagner [22] the differences between both theories in calculating the tensile viscosities are only small. He therefore compared the consequences of Eq. (10.126) with the experiment.

In the theory of Doi and Edwards [27] and in the slip link model of Wagner and Schaeffer [24], the volume of the tubes is assumed to increase proportional to u′ and their diameter to remain unchanged with the deformation. Already more than 10 years before the treatment of Wagner, Marrucci and de Cindio [33] proposed a slip link model with the different assumption that the volume of the tubes remains constant while their diameter decreases proportional to $1/\sqrt{u'}$. Using the independent alignment assumption, the result is a constitutive equation which is still more complicated than (10.129) viz.:

$$\underline{\underline{\sigma}} + p\underline{\underline{E}} = \int_{-\infty}^{t} \frac{15}{4} \frac{m(t-t')}{\sqrt{n_0 b^2}} \frac{1}{\langle \sqrt{u'} \rangle} \left\langle \frac{\underline{\underline{u'u'}}}{\sqrt{u'}} \right\rangle \mathrm{d}t' \tag{10.130}$$

In the *Marucci–de Cindio equation* the deformed unit vectors $\underline{u}'$ are first to be divided by the fourth root of their length and then multiplied dyadically with each other.

10.5.2 The Doi–Edwards Theory 2 for Simple Shear and Multiaxial Extension

Though it is well established today that the Doi–Edwards theory is not able to describe the rheological behavior of polymer melts, it often serves as starting point for the development of more advanced theories. Because it is free of parameters, it is furthermore used for comparison with experimental results to determine the molecular stress functions of Wagner's theory experimentally. Therefore, it may be useful for the reader to summarize some of the most important results of this theory in the following.

For the description of *time-dependent simple shear* by the Doi-Edwards theory 2, the corresponding equations of the general Wagner theory (10.74) are specialized to the Doi-Edwards theory 2 by introducing the expressions (10.127a) and (10.127b) as damping functions. The latter contain the invariants of the relative Finger tensor, which are for simple shear equal and equal to

$$I_B = II_B = 3 + (\gamma(t) - \gamma(t'))^2 \tag{10.131}$$

The following equations for the stresses result

$$\begin{aligned}
\sigma_{12}(t) &= G(t)n_1(t,0)\gamma(t) + \int_0^t m(t-t')n_1(t,t')\cdot[\gamma(t)-\gamma(t')]\mathrm{d}t' \\
\sigma_{11}(t) - \sigma_{22}(t) &= G(t)n_1(t,0)\gamma^2(t) + \int_0^t m(t-t')n_1(t,t')\cdot[(\gamma(t)-\gamma(t')]^2\mathrm{d}t' \\
\sigma_{22}(t) - \sigma_{33}(t) &= G(t)n_2(t,0)\gamma^2(t) + \int_0^t m(t-t')n_2(t,t')\cdot[\gamma(t)-\gamma(t')]^2\mathrm{d}t'
\end{aligned} \tag{10.132}$$

where use is made of the abbreviations

$$\begin{aligned}
n_1(t,t') &= \frac{5(1+\sqrt{I_B+13/4})}{J(t,t')\sqrt{I_B+13/4}} & n_1(t,0) &= \frac{5(1+\sqrt{\gamma^2(t)+25/4})}{J(t,0)\sqrt{\gamma^2(t)+25/4}} \\
n_2(t,t') &= -\frac{5}{J(t,t')\sqrt{I_B+13/4}} & n_2(t,0) &= -\frac{5}{J(t,0)\sqrt{\gamma^2(t)+25/4}} \\
J(t,t') &= I_B - 1 + 2\sqrt{I_B+13/4} & J(t,0) &= 2+\gamma^2(t)+2\sqrt{\gamma^2(t)+25/4}
\end{aligned} \tag{10.133}$$

For *stress relaxation in simple shear,* the terms under the integrals in (10.132) vanish and the stresses become

$$\begin{aligned}
\sigma_{12}(t) &= G(t)n_1(\gamma_0)\gamma_0 \\
\sigma_{11}(t) - \sigma_{22}(t) &= G(t)n_1(\gamma_0)\gamma_0^2 \\
\sigma_{22}(t) - \sigma_{33}(t) &= G(t)n_2(\gamma_0)\gamma_0^2
\end{aligned} \tag{10.134}$$

with the *non-linearity factor* $n_1(\gamma_0)$ and the *second normal stress coefficient* (in stress relaxation) $n_2(\gamma_0)$ given by

$$n_1(\gamma_0) = \frac{5(1+\sqrt{\gamma_0^2+25/4})}{J(\gamma_0)\sqrt{\gamma_0^2+25/4}} \tag{10.134a}$$

$$n_2(\gamma_0) = -\frac{5}{J(\gamma_0)\sqrt{\gamma_0^2+25/4}} \tag{10.134b}$$

and with

$$J(\gamma_0) = 2+\gamma_0^2+2\sqrt{\gamma_0^2+25/4} \tag{10.134c}$$

The non-linearity factor $n_1(\gamma_0)$ starts with unity for $\gamma_0 = 0$ and decreases as $5/\gamma_0^2$ for high γ_0. The ratio of the second to the first normal stress difference

$$\frac{\sigma_{22}-\sigma_{33}}{\sigma_{11}-\sigma_{22}} = \frac{n_2(\gamma_0)}{n_1(\gamma_0)} = -\frac{1}{1+\sqrt{\gamma_0^2+25/4}} \tag{10.135}$$

is negative, its absolute value approaches 2/7 at small shear strains and decreases with increasing shear strain. According to the Doi–Edwards theory 2 in a relaxation experiment in simple shear, a positive first and a negative second normal stress difference should occur besides the shear stress.

The *stressing experiment in simple shear* may be calculated by Eqs. (10.76), inserting (10.127a) and (10.127b) as damping functions with the result

$$\begin{aligned}
&\sigma_{12}(t) = \eta(t,\dot{\gamma}_0)\,\dot{\gamma}_0 && \text{with} \quad \eta(t,\dot{\gamma}_0) = \int_0^t G(\xi)\frac{\partial[\xi\cdot n_1(\dot{\gamma}_0\xi)]}{\partial\xi}\,\mathrm{d}\xi \\
&\sigma_{11}(t)-\sigma_{22}(t) = n_1(t,\dot{\gamma}_0)\dot{\gamma}_0^2 && \text{with} \quad n_1(t,\dot{\gamma}_0) = \int_0^t G(\xi)\frac{\partial[\xi^2\cdot n_1(\dot{\gamma}_0\xi)]}{\partial\xi}\mathrm{d}\xi \\
&\sigma_{22}(t)-\sigma_{33}(t) = n_2(t,\dot{\gamma}_0)\dot{\gamma}_0^2 && \text{with} \quad n_2(t,\dot{\gamma}_0) = \int_0^t G(\xi)\frac{\partial[\xi^2\cdot n_2(\dot{\gamma}_0\xi)]}{\partial\xi}\mathrm{d}\xi
\end{aligned} \tag{10.136}$$

with n_1 and n_2 given by (10.134a) and (10.134b).

For the stressing experiment in simple shear, the Doi–Edwards theory 2 predicts a *time-dependent non-linear viscosity* $\eta(t,\dot{\gamma}_0)$, a *time-dependent non-linear positive first normal stress coefficient* $n_1(t,\dot{\gamma}_0)$ and a *time-dependent non-linear negative second normal stress coefficient* $n_2(t,\dot{\gamma}_0)$. The dependence of those quantities on $\dot{\gamma}_0$ is free of parameters, and can be calculated if the stress relaxation modulus in shear under small deformations is known.

In the description of *stressing experiments in multiaxial extension* by the Doi–Edwards theory 2, the following invariants

$$\begin{aligned}
I_B &= e^{2\dot{\varepsilon}_0\xi} + e^{2m\dot{\varepsilon}_0\xi} + e^{-2(1+m)\dot{\varepsilon}_0\xi} \\
II_B &= e^{-2\dot{\varepsilon}_0\xi} + e^{-2m\dot{\varepsilon}_0\xi} + e^{2(1+m)\dot{\varepsilon}_0\xi}
\end{aligned} \tag{10.137}$$

with $\xi = t - t'$ occur. They depend on the value chosen for the geometry parameter m. If these are inserted into (10.127a) and (10.127b) to obtain the damping functions for multiaxial extension, these also will depend on the value of m and are therefore designated by

$$h_{1,m}(\dot{\varepsilon}_0\xi) \equiv h_1(I_B, II_B) = \frac{5}{J} \tag{10.137a}$$

$$h_{-1,m}(\dot{\varepsilon}_0\xi) \equiv h_{-1}(I_B, II_B) = \frac{-5}{J\sqrt{II_B + 13/4]}} \tag{10.137b}$$

and J given by (10.124a). For the invariants I_B and II_B, Eqs. (10.137) with the corresponding value of m should be inserted. The equations for the tensile stresses become

$$\begin{aligned}
\sigma_{11}(t)+p &= \int_0^t G(\xi)\frac{\partial}{\partial\xi}[h_{1,m}(\dot{\varepsilon}_0\xi)e^{2\dot{\varepsilon}_0\xi}+h_{-1,m}(\dot{\varepsilon}_0\xi)e^{-2\dot{\varepsilon}_0\xi}]\mathrm{d}\xi \\
\sigma_{22}(t)+p &= \int_0^t G(\xi)\frac{\partial}{\partial\xi}[h_{1,m}(\dot{\varepsilon}_0\xi)e^{2m\dot{\varepsilon}_0\xi}+h_{-1,m}(\dot{\varepsilon}_0\xi)e^{-2m\dot{\varepsilon}_0\xi}]\mathrm{d}\xi \\
\sigma_{33}(t)+p &= \int_0^t G(\xi)\frac{\partial}{\partial\xi}[h_{1,m}(\dot{\varepsilon}_0\xi)e^{-2(1+m)\dot{\varepsilon}_0\xi}+h_{-1,m}(\dot{\varepsilon}_0\xi)e^{2(1+m)\dot{\varepsilon}_0\xi}]\mathrm{d}\xi
\end{aligned} \tag{10.138}$$

These are to be specified for the three cases $m=-1/2$, 1, and 0.

Uniaxial extension is found by inserting $m=-½$ into Eqs. (10.138), subtracting the second of the equations obtained from the first one and putting $\sigma_{22}=0$. If we use the abbreviation $\dot{\varepsilon}_0\xi\equiv z$, the result may be written as

$$\begin{aligned}
\sigma_{11}(t) &= \dot{\varepsilon}_0\int_0^t G(\xi)\tfrac{d}{dz}\{h_{1,-1/2}(z)[e^{2z}-e^{-z}]+h_{-1,-1/2}(z)[e^{-2z}-e^{z}]\}\mathrm{d}\xi \\
\sigma_{22}(t) &= \sigma_{33}(t)=0
\end{aligned} \tag{10.139}$$

$h_{1,-1/2}$ and $h_{-1,-1/2}$ are found by inserting $m=-1/2$ in Eqs. (10.137), (10.137a), (10.137b), and (10.124a). The *uniaxial extensional viscosity* is given then as

$$\mu(t,\dot{\varepsilon}_0)=\sigma_{11}(t)/\dot{\varepsilon}_0 \tag{10.139a}$$

Developing (10.139) in powers of z, differentiating the expression between curved brackets after z and forming the limit of $\dot{\varepsilon}_0$ to zero, results in

$$\mu(t,0)=3\eta^0(t) \tag{10.139b}$$

which corresponds to Eq. (10.63) with $\eta^0(t)$ given by Eq. (10.14) or (10.16).

Equibiaxial extension is found by inserting $m=1$ into Eqs. (10.138), subtracting the third of the equations obtained from the first one and putting $\sigma_{33}=0$. This results in

$$\begin{aligned}
\sigma_{11}(t) &= \int_0^t G(\xi)\tfrac{\partial}{\partial\xi}\{h_{1,1}(\dot{\varepsilon}_0\xi)[e^{2\dot{\varepsilon}_0\xi}-e^{-4\dot{\varepsilon}_0\xi}]+h_{-1,1}(\dot{\varepsilon}_0\xi)[e^{-2\dot{\varepsilon}_0\xi}-e^{4\dot{\varepsilon}_0\xi}]\}d\xi \\
\sigma_{22}(t) &= \sigma_{11}(t) \\
\sigma_{33}(t) &= 0
\end{aligned} \tag{10.140}$$

$h_{1,1}$ and $h_{-1,1}$ are found by inserting $m = 1$ in Eqs. (10.137), (10.137a), (10.137b), and (10.124a). The *equibiaxial viscosity* follows then as

$$\mu_e(t, \dot{\varepsilon}_0) = \sigma_{11}(t)/\dot{\varepsilon}_0 \tag{10.140a}$$

with the limiting curve for small $\dot{\varepsilon}_0$ equal to

$$\mu_e(t, 0) = 6\eta^0(t) \tag{10.140b}$$

Planar extension in the 1–3 *plane* is found by inserting $m = 0$ into Eqs. (10.138). From the equations obtained, the third one is subtracted from the first and then from the second one. Putting $\sigma_{33} = 0$ results into

$$\begin{aligned}
\sigma_{11}(t) &= \dot{\varepsilon}_0 \int_0^t G(\xi) \frac{d}{dz} \{h_{1,0}(z)[e^{2z} - e^{-2z}] + h_{-1,0}(z)[e^{-2z} - e^{2z}]\} d\xi \\
\sigma_{22}(t) &= \dot{\varepsilon}_0 \int_0^t G(\xi) \frac{d}{dz} \{h_{1,0}(z)[1 - e^{-2z}] + h_{-1,0}(z)[1 - e^{2z}]\} d\xi \\
\sigma_{33}(t) &= 0
\end{aligned} \tag{10.141}$$

$h_{1,0}$ and $h_{-1,0}$ are found by inserting $m = 0$ in Eqs. (10.137), (10.137a), (10.137b), and in (10.124a). This deformation may be realized by loading the 1-plane and the 2-plane with stresses of different magnitude and leaving the 3-plane stress free. Consequently, two *planar extensional viscosities* may be defined as

$$\mu_{p1}(t, \dot{\varepsilon}_0) = \sigma_{11}(t)/\dot{\varepsilon}_0 \tag{10.141a}$$

$$\mu_{p2}(t, \dot{\varepsilon}_0) = \sigma_{22}(t)/\dot{\varepsilon}_0 \tag{10.141b}$$

with the limiting curves for small $\dot{\varepsilon}_0$

$$\mu_{p1}(t, 0) = 4\eta^0(t) \tag{10.141c}$$

$$\mu_{p2}(t, 0) = 2\eta^0(t) \tag{10.141d}$$

In conclusion it may be stated that the Doi-Edwards theory 2 describes qualitatively a number of characteristic properties which are found for polymer melts, as a non-Newtonian time-dependent shear viscosity, a positive non-linear time-dependent first normal stress coefficient, a negative non-linear time-dependent second normal stress coefficient, and non-Newtonian extensional viscosities in uniaxial, equibiaxial, and planar extension. As the two damping functions $h_{1,m}(I_B, II_B)$ and $h_{-1,m}(I_B, II_B)$ of the Doi-Edwards theory are free of parameters, the applicability of the theory can be checked by the experimental determination of those properties and their comparison with the predictions.

This had been done by Wagner [23], who compared measurements of the damping function in shear obtained in stress relaxation for the IUPAC-melt at 150 °C by Laun [17] with the Doi-Edwards prediction (10.134), measurements of the damping function obtained from stressing experiments in linear extension on the same melt at 150 °C by Laun and Münstedt [34] with the prediction (10.139), measurements of the damping functions in uniaxial, equibiaxial, ellipsoidal, and planar extension in stressing experiments on PIB at room temperature by Demarmels and Meissner [35] with (10.139) (10.140), and (10.141).

His conclusion was that the Doi-Edwards theory 2 predicts damping functions which are much stronger than those obtained by the experiment. In some cases, the difference was as large as one half of a decade. He concluded that the stresses predicted by the Doi-Edwards theory 2 are too small, the one predicted by the Lodge theory are too high in comparison with the experiment. The most promising description should be searched for in a theory situated in between those two extremes. This is the M.S.F.theory using the molecular stress function, which will be described in Sect. 10.6.1.

10.6 Theory of the Molecular Stress Function (MSF)

10.6.1 The Constitutive Equation of the MSF-Theory

For the reasons mentioned at the end of Sect. 10.5.2, Wagner and Schaeffer changed the theory of Doi and Edwards in a way that the predictions come to lie between those of the Lodge theory and the Doi-Edwards theory 2 [23, 29]. Both theories may be considered as based on temporary entanglement networks treated in Sect. 10.4.2. The Lodge theory was shown to be identical with the temporary entanglement network without any slip. The force on the deformed chain bows is then given by Eq. (10.109). In the Doi-Edwards theory, complete slip is assumed meaning that the absolute value of the force on the deformed chain bows $|\underline{K}|$ is the same as the force before the deformation:

$$|\underline{K}| = |\underline{K}_{\text{equi}}| = \frac{3kT}{n_o b^2} r \qquad (10.142)$$

The assumption of partial slip means that the force on the deformed chain bow containing n'' Kuhn segments is

$$\underline{K}'(\underline{r}) = \frac{3kT}{n'' b^2} \cdot \underline{r}' = \frac{1}{S} \frac{3kT}{n_o b^2} \cdot \underline{r}' = \frac{1}{S} \frac{3kT}{n_o b^2} \cdot \underline{r} u' \qquad (10.143)$$

with the slip function $S(u')$ defined by (10.114). The absolute value of the force $\underline{K}'$ becomes

$$|\underline{K}'| = \frac{1}{S}\frac{3kT}{n_o b^2} \cdot ru' \tag{10.144}$$

The assumption is made that the force on the deformed chain bow is larger than $\underline{K}_{equi}$. The ratio $|K'|/|\underline{K}_{equi}|$ is assumed to be independent of the orientation of the chain bow, but dependent on an average of the extension of all chain bows. This ratio is designated by f and is called the *molecular stress function.* From (10.144) and (10.142) it follows

$$\frac{|K'|}{|K_{equi}|} = \frac{u'}{S(u')} = f(\langle \ln u' \rangle) \tag{10.145}$$

Being the ratio of two scalars, f is a scalar function, too. First, Wagner assumed that f is a function of the average $\langle u' \rangle$, viz., $f = f(\langle u' \rangle)$, later he proposed $\langle ln\ u' \rangle$ to be the argument of the molecular stress function as indicated in Eq. (10.145). The two assumptions are not equivalent, as $\langle u' \rangle$ and $\langle ln\ u' \rangle$ are not connected in a unique way with each other. Their relation depends to some extent on the type of the deformation considered (cf. Appendix A page 358).

From (10.145) the slip function is found as

$$S(u') = \frac{u'}{f} \tag{10.146}$$

If the independent alignment assumption (10.122) is further accepted to be valid, the disentanglement function follows as

$$D(u') = \frac{f}{u'} \tag{10.147}$$

Inserting these functions into the Eq. (10.117) for the temporary network with slip and considering that the scalar f may be drawn in front of the acute brackets, the dyadic becomes

$$\underline{\underline{\sigma}} + p\underline{\underline{E}} = 3kTcN_0\left\langle \frac{f}{u'}\frac{f}{u'}\,\underline{u'}\,\underline{u'} \right\rangle = 3kTcN_0 f^2\left\langle \frac{\underline{u'}\,\underline{u'}}{u'\,u'} \right\rangle$$

For the shear modulus in the linear regime (10.125) is valid and the *constitutive equation of the MSF theory* follows as

$$\underline{\underline{\sigma}}(t) + p\underline{\underline{E}} = \int_{-\infty}^{t} 5m(t-u')f^2\left\langle \frac{\underline{u'}\,\underline{u'}}{u'\,u'} \right\rangle dt' \tag{10.148}$$

m is the memory function, related to the time-dependent shear modulus by Eq. (10.3). If use is made of the approximation of Currie (Eq. 10.124), the constitutive equation takes the form of (10.127), i.e.,

$$\underline{\underline{\sigma}}(t) + p\underline{\underline{E}} = \int_{-\infty}^{t} m(t-t')[h_1(I_B, II_B)\underline{\underline{B}}_t(t') + h_{-1}(I_B, II_B)\underline{\underline{B}}_t^{-1}(t')]\mathrm{d}t' \quad (10.149)$$

with the damping functions

$$h_1(I_B, II_B) \cong \frac{5f^2}{J} \quad (10.149\text{a})$$

and

$$h_{-1}(I_B, II_B) \cong -\frac{5f^2}{J\sqrt{II_B + 13/4}} \quad (10.149\text{b})$$

instead of (10.127a) and (10.127b). The molecular stress function will be larger than or equal to unity and can be used to fit the theory to experimental results. As f is a scalar, changes of the first normal stress coefficient will automatically effect the second normal stress coefficient in the same way and their ratio (10.135) will remain unchanged.

As the molecular stress function is assumed to depend on the argument $\langle ln\ u' \rangle$, the relation between $\langle ln\ u' \rangle$ and the extension ratio λ, the Hencky strain ε or the shear strain γ is essential. This problem is similar to that of the relation between $\langle u' \rangle$ and λ, which was shortly addressed in Sect. 10.4.1. It is also treated in Appendix A.

Of course, only if one succeeds to describe the different experiments for a certain polymer by means of the same molecular stress function, the *MSF* theory can be considered to be successful. At present, such a conclusion seems to be far away.

10.6.2 The MSF Theory for Simple Shear and Multiaxial Extension

In the constitutive equation of the *MSF* theory (10.149), the molecular stress function f is assumed to be a function of the argument $\langle (ln\ u') \rangle$ which should be a scalar depending on the two invariants I_B and II_B. The dependence of $\langle (ln\ u') \rangle$ on I_B and II_B is complicated and has to be calculated numerically. As shown in Appendix A, in the cases of simple shear and multiaxial extension, this dependence is somewhat easier to handle as the following expressions were derived

$$\langle \ln u' \rangle = \Phi_s(\gamma) \quad \text{for simple shear} \tag{10.150}$$

$$\langle \ln u' \rangle = \Theta_u(\varepsilon) \quad \text{for uniaxial extension} \tag{10.151}$$

$$\langle \ln u' \rangle = \Theta_p(\varepsilon) \quad \text{for planar extension} \tag{10.152}$$

$$\langle \ln u' \rangle = \Theta_e(\varepsilon) \quad \text{for equibiaxial extension} \tag{10.153}$$

These functions all depend on one independent variable but only (10.151) can be described explicitly. The others still involve one numerical integration. All functions start with the value zero, then increase with increasing argument and finally approximate their asymptotes which are $\ln \gamma - 1$, $\varepsilon - 1$, $\varepsilon - 1$ and $\varepsilon + \ln 2 - 1$, respectively.

The calculation of time-dependent simple shear or multiaxial extensions proceeds in the same way as for the Doi-Edwards theory 2. The corresponding Eqs. (10.132)–(10.136) and (10.137)–(10.141) may be used if the values of Eqs. (10.149a) and (10.149b) are inserted for the damping functions instead of (10.127a) and (10.127b).

For simple shear Eqs. (10.132) remain valid if the non-linearity factor and the second normal stress coefficient are multiplied by

$$f^2(t,t') = f^2[\Phi_s(\gamma(t) - \gamma(t'))] \tag{10.154}$$

and

$$f^2(t,0) = f^2[\Phi_s(\gamma(t)] \tag{10.154a}$$

respectively. For *stress relaxation in simple shear*, Eqs. (10.134) change to

$$\begin{aligned} \sigma_{12}(t) &= G(t) n_1(\gamma_0) F^2(\gamma_0) \cdot \gamma_0 \\ \sigma_{11}(t) - \sigma_{22}(t) &= G(t) n_1(\gamma_0) F^2(\gamma_0) \cdot \gamma_0^2 \\ \sigma_{22}(t) - \sigma_{33}(t) &= G(t) n_2(\gamma_0) F^2(\gamma_0) \cdot \gamma_0^2 \end{aligned} \tag{10.155}$$

with $F^2(\gamma_0) = f^2[\Phi_s(\gamma_0)]$. These equations offer a simple way to determine the molecular stress function experimentally by measuring the three stress combinations in simple shear and dividing the result by the corresponding equations of the set (10.134). The quotient represents the square of the molecular stress function as a function of γ_0.

For a *stressing experiment in simple shear* Eqs. (10.136) may be used if $n_1(\dot{\gamma}_0\xi)$ and $n_2(\dot{\gamma}_0\xi)$ are replaced by $n_1(\dot{\gamma}_0\xi) f^2(\dot{\gamma}_0\xi)$ and $n_2(\dot{\gamma}_0\xi) f^2(\dot{\gamma}_0\xi)$. As the molecular stress function f depends on the elapsed time $\xi = t - t'$, the time dependence of the non-Newtonian viscosity and the normal stress coefficients may change considerably due to the introduction of the molecular stress function.

To derive the equations for *multiaxial extensions in stressing experiments* under constant Hencky strain rate $\dot{\varepsilon}_0$, the damping functions (10.137a, b) have to be multiplied by $f_m^2(\dot{\varepsilon}_0\xi)$. The molecular stress function f_m to be used depends on the type of deformation, viz.,

$$f_{-1/2}(\dot{\varepsilon}_0\xi) = f[\Theta_u(\dot{\varepsilon}_0\xi)] \quad \text{for uniaxial extension} \tag{10.156a}$$

$$f_0(\dot{\varepsilon}_0\xi) = f[\Theta_p(\dot{\varepsilon}_0\xi)] \quad \text{for planar extension} \tag{10.156b}$$

$$f_1(\dot{\varepsilon}_0\xi) = f[\Theta_e(\dot{\varepsilon}_0\xi)] \quad \text{for equibiaxial extension} \tag{10.156c}$$

Note that in the limit $\dot{\varepsilon}_0 \to 0$ all three functions assume the value unity. The further procedure is equal to that for the Doi-Edwards theory 2. We again use the abbreviation $\dot{\varepsilon}_0\xi = z$.

The *uniaxial tensile viscosity* becomes equal to

$$\mu(t,\dot{\varepsilon}_0) = \int_0^t G(\xi)\frac{d}{dz}\{f_{-1/2}^2(z)\cdot[h_{1,-1/2}(z)(e^{2z}-e^{-z}) + h_{-1,-1/2}(z)(e^{-2z}-e^{z})]\}d\xi \tag{10.157}$$

Its limiting curve for $\dot{\varepsilon}_0 \to 0$ is given by (10.139b).

The *equibiaxial viscosity* becomes

$$\mu_e(t,\dot{\varepsilon}_0) = \int_0^t G(\xi)\frac{d}{dz}\{f_1^2(z)\cdot[h_{1,1}(z)(e^{2z}-e^{-4z}) + h_{-1,1}(z)(e^{-2z}-e^{4z})]\}d\xi \tag{10.158}$$

with its limiting curve for $\dot{\varepsilon}_0 \to 0$ given by (10.140b)

The two *planar viscosities* are

$$\begin{aligned}\mu_{p,1}(t,\dot{\varepsilon}_0) &= \int_0^t G(\xi)\frac{\mathrm{d}}{\mathrm{d}z}\{f_0^2(z)\cdot[h_{1,0}(z)(e^{2z}-e^{-2z}) + h_{-1,0}(z)(e^{-2z}-e^{2z})]\}\mathrm{d}\xi \\ \mu_{p,2}(t,\dot{\varepsilon}_0) &= \int_0^t G(\xi)\frac{\mathrm{d}}{\mathrm{d}z}\{f_0^2(z)\cdot[h_{1,0}(z)(1-e^{-2z}) + h_{-1,0}(z)(1-e^{2z})]\}\mathrm{d}\xi\end{aligned} \tag{10.159}$$

and their limiting curves for $\dot{\varepsilon}_0 \to 0$ are given by (10.141c) and (10.141d).

For linear polymers Wagner et al. proposed the following form for the molecular stress function [36]

$$f^2 = 1 + (f_{max}^2 - 1)\left[1 - e^{-(x-1)/(f_{max}^2 - 1)}\right] \quad \text{with} \quad x = e^{\langle \ln(u') \rangle} \tag{10.160}$$

f_{max} determines the maximum stretch of the molecules in uniaxial extension resulting in the plateau values of the elongational viscosity. For $x = 1, f$ starts with unity and increases with increasing x until its saturation value f_{max}.

An example for the use of the molecular stress function (10.160) in the description of stressing experiments in uniaxial, planar, and biaxial extension for a linear polyethylene and a linear polystyrene melt is given in a publication of Wagner et al. [36].

If all experiments at one temperature performed on one polymer could be described with satisfactory accuracy by one molecular stress function which were a function of the logarithmic average orientation of the chain bows $\langle(\ln u'\rangle)$ only, formulating a descriptive theory of the rheological behavior would be straightforward. With the knowledge of the molecular stress function, determined in one experiment and the linear shear modulus measured over a sufficiently broad frequency range all rheological experiments should be predictable.

Unfortunately, this situation is by far not realized. First, the molecular stress function is *not* a unique function of the average $\langle \ln u' \rangle$, but differs for *rotational* and *rotation free flows*. Simple shear (rotational flow) and planar extension (rotation free flow) have the same value of $\langle \ln u' \rangle$ but show different molecular stress functions. In planar flow $f^2(\langle \ln u' \rangle)$ starts with unity at small elongations and increases monotonously until it reaches the saturation value f_{max}^2. In shear flow, $f^2(\langle \ln u' \rangle)$ first follows the function in planar flow, but soon deviates from it, passes through a maximum at $\langle \ln u' \rangle \cong 0.5$ and then decreases to unity again. Wagner et al. [37] described this behavior theoretically and applied it to measurements on a HDPE and an LDPE melt in extensional and shear flow. They could describe the occurrence of maxima in the shear viscosity and the first normal stress coefficient for both, linear and long-chain branched melts.

Secondly, the molecular stress function is *not* a *unique function of the average stretch* but also depends on strain rate and strain history. This is accounted for by the introduction of an *evolution equation for the molecular stress function*. This is a first-order differential equation for f^2 expressing $\partial f^2/\partial t$ as a function of the components of the Doi-Edwards strain tensor $\underline{\underline{S}}$ and certain combinations with the rate of strain tensor $\underline{\underline{D}}$ (cf. Eq. 8.41) as, e.g., a term expressing the *dissipated energy density* as the double scalar product $(\underline{\underline{D}} : \underline{\underline{S}}) = D_{ik}S_{ik}$.

In simple cases, the molecular stress function may be calculated by solving the corresponding evolution equation without the knowledge of the time dependence of the strain tensor. In those cases, the function $f^2(\langle \ln u' \rangle)$ obtained can be inserted into Eq. (10.148) and the latter may be solved.

In the majority of cases, however, the integration of the evolution equation involves the time dependence of the strain tensor $\underline{\underline{S}}$. In those cases, the *evolution equation* and the *constitutive Eq.* (10.148) are *coupled* equations and are to be solved simultaneously which is possible only numerically. This makes the procedure difficult and non-transparent. For branched polymers, Wagner et al. [38] and Rolon-Garrido and Wagner [39] proposed various different forms for the molecular stress function, but a real breakthrough seems to be still far away.

Appendix A: The Averages $\langle(u')\rangle$ and $\langle(\ln u')\rangle$ and the Components of the Strain Tensor

We first consider the evaluation of the average

$$\langle u' \rangle = \left\langle \sqrt{[\underline{\underline{B}}_t^{-1}(t')]_{ik} u_i u_k} \right\rangle \tag{10.102}$$

which was introduced in Sect. 10.4.1. Because of the square root in (10.102) its evaluation leads to an elliptic integral and is possible only in simple cases. For deformations in which the tensor $\underline{\underline{B}}^{-1}$ is in its principle coordinate system, the integrals can be solved analytically for unidirectional extension in the 1-direction, for biaxial extension in the 1 and 2-direction and, by numerical integration, for planar extension in the 1 and 2-direction (pure shear) [25]. The results and the integral to be solved numerically for planar extension are summarized in Table A.1. The same table also shows the formulae for the corresponding values of the averages $\sqrt{\langle u'^2 \rangle}$ which are obtained from Eq. (10.101) as follows. From the definition of the tensor $\langle(\underline{u}'\ \underline{u}')\rangle$ in (10.100), we find for multiaxial extension that

$$x_i = \lambda_i x'_i \quad \text{for } i = 1, 2, 3 \tag{A.1}$$

with x'_i and x_i being the coordinates of a material point at the past time t' and at the observation time t, respectively. From the definition (8.35), the components of the Finger tensor follow and from (10.101) its trace and its square root divided by $\sqrt{3}$ as listed in the third column of the table.

The difference between the averages $\langle u' \rangle$ and $\sqrt{\langle u'^2 \rangle}$ is rather small, as may be seen from Fig. A.1 in which both quantities are plotted versus the extension ratio λ for the three cases of deformation. Values calculated for $\langle u' \rangle$ are indicated by filled squares, circles, and triangles, those for $\sqrt{\langle u'^2 \rangle}$ by drawn lines. Apart from the case of planar extension (pure shear), for which the averages practically coincide, $\langle u' \rangle$ is only slightly higher than $\sqrt{\langle u'^2 \rangle}$.

The asymptotic equations for $\lambda \gg 1$ are straight lines with different slopes according to

$$\begin{aligned} \langle u' \rangle &\approx \tfrac{\pi}{4}\lambda && \text{for biaxial extension} \\ \langle u' \rangle &\approx \tfrac{1}{\sqrt{3}}\lambda && \text{for planar extension} \\ \langle u' \rangle &\approx \tfrac{1}{2}\lambda && \text{for uniaxial extension} \end{aligned} \tag{A.2}$$

Table A.1 Formulae for $\langle u' \rangle$ and $\sqrt{\langle u'^2 \rangle}$ as functions of the extension ratio λ for biaxial extension, planar extension (pure shear), and uniaxial extension, for $\lambda \geq 1$

Type of deformation	$\langle u' \rangle$	$\sqrt{\langle u'^2 \rangle}$
Biaxial extension	$\frac{1}{2\lambda^2} + \frac{\lambda}{2\sqrt{1-1/\lambda^6}} \arctan(\sqrt{\lambda^6 - 1})$	$\sqrt{\frac{2}{3}}\lambda\sqrt{1 + \frac{1}{2\lambda^6}}$
Planar extension	$\frac{1}{2\lambda} + \frac{1}{\pi}\int\limits_0^{\pi/2} \frac{1+(\lambda^2-1)\cos^2\varphi}{\sqrt{(\lambda^2-1)(\cos^2\varphi+1/\lambda^2)}} \arcsin\sqrt{\frac{(\lambda^2-1)(\cos^2\varphi+1/\lambda^2)}{1+(\lambda^2-1)\cos^2\varphi}}d\varphi$	$\frac{\lambda}{\sqrt{3}}\sqrt{1 + \frac{1}{\lambda^2} + \frac{1}{\lambda^4}}$
Uniaxial extension	$\frac{\lambda}{2}\left(1 + \frac{1}{\sqrt{\lambda^3}\sqrt{\lambda^3-1}} \ln(\sqrt{\lambda^3 - 1} + \sqrt{\lambda^3})\right)$	$\frac{\lambda}{\sqrt{3}}\sqrt{1 + \frac{2}{\lambda^3}}$

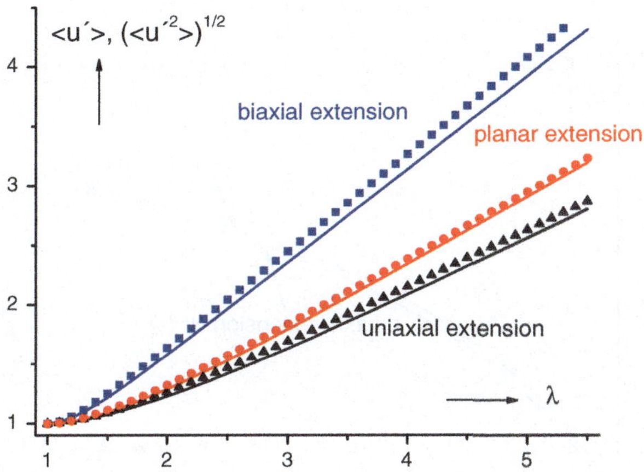

Fig. A.1 Comparison of the averages $\langle u' \rangle$ (*triangles, circles*, and *squares*) and $\sqrt{\langle u'^2 \rangle}$ (*drawn lines*) as functions of the extension ratio λ for uniaxial extension, planar extension (pure shear), and biaxial extension

As the molecular stress function introduced in Sect. 10.6.1 was assumed to depend on the argument $\langle(\ln u')\rangle$, the question concerning the relation between $\langle(\ln u')\rangle$ and the extension ratio λ, the Hencky strain ε or the shear strain γ arises. This problem is treated next.

For uniaxial extension the equation for $\langle(\ln u')\rangle$ is obtained in explicit form; for the other cases, one numerical integration is involved. The formulae valid for $\lambda \geq 1$ are given in Table A.2 where a and c are used as abbreviations for the terms in the integrand of the integral as presented in the first row. For uniaxial extension $a = 1$ and $c = \lambda^2$ and the integral may be solved analytically.

The solutions of the integrals are designated as $\Psi_b(\lambda)$, $\Psi_p(\lambda)$, and $\Psi_e(\lambda)$ in the case of biaxial, planar, and uniaxial extension. These functions are shown in Fig. A.2 versus the extension ratio λ. All curves start with $\langle(\ln u')\rangle = 0$ for $\lambda = 1$, increase with a positive curvature in the region of very small extension ratios (only visible under large magnification), then further increase with a negative curvature and finally run parallel to ln λ. Their asymptotic behavior for large values of λ is indicated in the figure.

Table A.2 Formulae for $\langle(\ln u')\rangle$ as functions of the extension ratio λ for uniaxial extension, planar extension (pure shear), and biaxial extension, for λ ≥ 1 [25]

	$\langle \ln u' \rangle = \Psi(\lambda) = \frac{1}{2}\ln c - 1 + \frac{2}{\pi}\int_0^{\pi/2} \sqrt{\frac{a}{c-a}} \arctan \sqrt{\frac{c-a}{a}} d\varphi$
Biaxial extension	$\langle \ln u' \rangle = \Psi_b(\lambda) \quad a = \lambda^2 \cos^2 \varphi + \frac{\sin^2 \varphi}{\lambda^4} \quad c = \lambda^2$
Planar extension	$\langle \ln u' \rangle = \Psi_p(\lambda) \quad a = \cos^2 \varphi + \frac{\sin^2 \varphi}{\lambda^2} \quad c = \lambda^2$
Uniaxial extension	$\langle \ln u' \rangle = \Psi_e(\lambda) = \ln \lambda - 1 + \frac{1}{\sqrt{\lambda^3 - 1}} \arctan \sqrt{\lambda^3 - 1}$

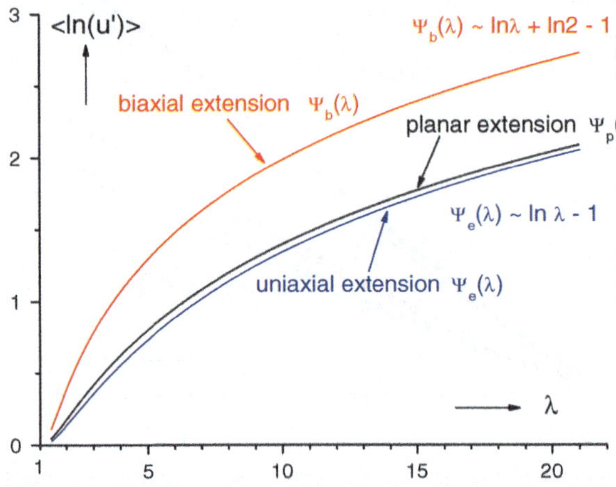

Fig. A.2 The average $\langle(\ln u')\rangle$ versus the extension ratio λ for uniaxial, planar, and equibiaxial extension for λ ≥ 1. Indicated are the formulae for the asymptotic behavior at large values of λ

As $\langle(\ln u')\rangle$ is a scalar, it is the same for simple shear and pure shear (planar extension). Therefore, the relation between $\langle(\ln u')\rangle$ and γ is found by inserting into $\Psi_p(\lambda)$ the value for λ which results from (8.88) and (8.89) as function of γ

$$\lambda = \sqrt{c_1} = \sqrt{1 + \gamma^2/2 + \gamma\sqrt{1 + \gamma^2/4}} \tag{A.3}$$

This function is designated by $\Phi_s(\gamma) \equiv \Psi_p(\lambda(\gamma))$ and shown in Fig. A.3. $\Phi_s(\gamma)$ starts with zero at γ = 0, then increases with a positive curvature until γ = 1, changes to a negative curvature, and finally increases parallel to ln γ −1. It runs above its asymptote which is shown as a broken line.

To obtain the averages $\langle(\ln u')\rangle$ as functions of the Hencky strain ε = ln λ, the extension ratio λ = exp(ε) is to be inserted into the formulae of Table A.2 prior to the numerical integration. The functions obtained are designated by

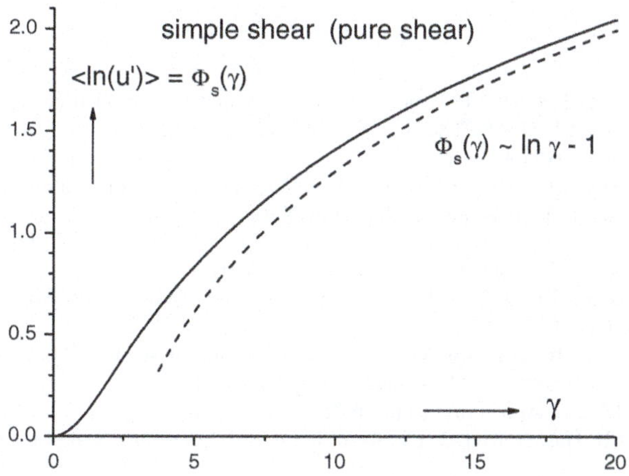

Fig. A.3 The average $\langle(\ln u')\rangle$ versus the shear strain γ for simple shear (pure shear). Indicated is the asymptotic behavior at large values of γ as a *broken line*

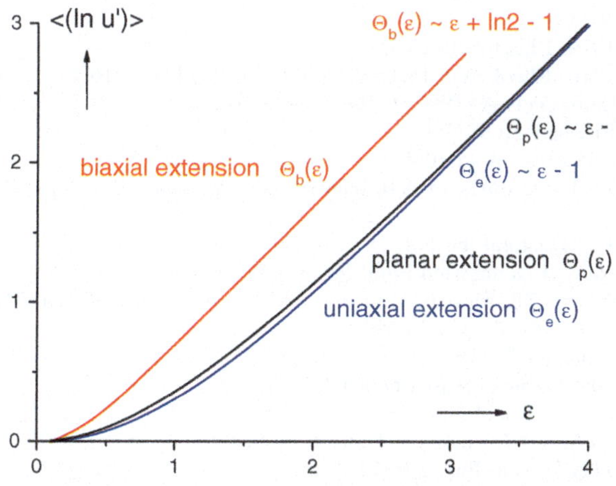

Fig. A.4 The average $\langle(\ln u')\rangle$ versus the Hencky strain ε for uniaxial, planar, and biaxial extension for $\varepsilon \geq 0$. Indicated are the formulae for the asymptotic behavior at large values of ε

$\Theta_b(\varepsilon) = \psi_b\,(\exp(\varepsilon))$, $\Theta_p(\varepsilon) = \psi_p\,(\exp(\varepsilon))$ and $\Theta_e(\varepsilon) = \psi_e\,(\exp(\varepsilon))$ for biaxial, planar, and uniaxial extension, respectively, and are shown in Fig. A.4. These functions start with zero at $\varepsilon = 0$, show a short increase with positive curvature and finally change into straight lines with the slope unity.

References

1. Janeschitz-Kriegl H (1983) Polymer melt rheology and flow birefringence. Springer, Berlin
2. Doi M, Edwards SF (1986) The theory of polymer dynamics. Oxford Science Publications Clarendon press, Oxford
3. Bird RB, Armstrong RC, Hassager O (1987) Dynamics of polymeric liquids. Fluid mechanics, Vol. 1. John Wiley & Sons, NewYork
4. Bird RB, Curtiss CF, Armstrong RC, Hassager O (1987) Dynamics of polymeric liquids. Kinetic theory, Vol 2. John Wiley & Sons, New York
5. Dealy JM, Larson RG, (2006) Structure and rheology of molten polymers. Hans Gardener Publications, Ohio
6. Lodge AS (1954) In: Harrison VGW (ed) Proceedings of second international congress of Rheology Oxford, Butterworth Scientific, London, p 229
7. Meissner J (1972) J Appl Polym Sci 16:2877
8. Gortemaker FH, Hansen MG, de Cindio B, Laun MH, Janeschitz-Kriegl H (1976) Rheol Acta 15:256
9. Cox WP, Merz EH (1958) J Polym Sci 28:619
10. Resch JA (2010) Private communication
11. Resch JA (2010) Doctoral thesis, University Erlangen-Nürnberg
12. Baumgärtel M, Winter HH (1989) Rheol Acta 28:511
13. Stadler FJ (2007) Doctoral Thesis. University Erlangen-Nürnberg, Sierke Verlag, Göttingen. ISBN 978-3-940333-24-7
14. Wagner MH (1979) Rheol Acta 18:33
15. Wagner MH (1976) Rheol Acta 15:133,136, (1977) Rheol Acta 16:43
16. Wagner MH, Demarmels A (1990) J Rheology 36:943
17. Laun HM(1978) Rheol Acta 17:1
18. Meissner J (1971) Rheol Acta 10:230
19. Osaki K (1976) Proceedings of VII international congress. Rheology 104 (Gothenburg Sweden)
20. Hepperle J (2002) Doctoral thesis University Erlangen-Nürnberg
21. Wagner MH (1978) J Nonnewton Fluid Mech 4:39
22. Raible T, Stephenson SE, Meissner J, Wagner MH (1982) J Nonnewton Fluid Mech 11:239
23. Wagner MH (1990) Rheol Acta 29:594
24. Wagner MH, Schaeffer J (1992) J Rheol 36:1
25. Wagner MH (1999) Korean-Aust Rheol J 11:293
26. Wagner MH (1994) J Rheol 38:655
27. Doi M, Edwards SF (1978) Faraday Trans II 74(1789):1802
28. Gennes De (1971) J Chem Phys 55:572
29. Wagner MH, Schaeffer J (1992) Rheol Acta 31:22
30. Schaeffer J (1994) Doctoral thesis University, Stuttgart
31. Currie PK (1982) J Nonnewton Fluid Mech 11:53
32. Bernstein B, Kearsley E, Zapas I (1963) Trans Soc Rheol 7:39
33. Marrucci G, de Cindio B (1980) Rheol Acta 19:68
34. Laun HM, Münstedt H (1976) Rheol Acta 15:517
35. Demarmels A, Meissner J (1986) J Coll Polym Sci 264:829
36. Wagner MH, Bastian H, Hachmann P, Meissner J, Kurzbeck S, Münstedt H, Langouche F (2000) Rheol Acta 39:97
37. Wagner MH, Rubio P, Bastian H (2001) J Rheol 45:1387
38. Wagner MH, Yamagushi M, Takahashi M (2003) J Rheol 47:779
39. Rolon-Garrido VH, Wagner MH (2007) Rheol Acta 46:583

Chapter 11
Shear Rheology

Measurements in shear are widely used to characterize rheological properties of polymer melts. Rotational rheometers as described in Sect. 15.3 are the instrumental base for rheological experiments at small shear rates or shear stresses, respectively, capillary rheometers are used for characterizations at higher shear rates (cf. Sects. 15.1 and 15.2). The advantages of rotational rheometers are that different experimental modes can be applied. In stressing experiments the viscous and elastic components of polymer melts as a function of time can be determined in parallel at a given constant shear rate, in the dynamic-mechanical mode these properties are measured in dependence on frequency. Creep recovery is another mode interesting for investigating the viscoelastic behavior of polymeric materials, particularly of those with long retardation times. In the linear range the viscoelastic functions obtained by various experiments are related to each other according to the theory of linear viscoelasticity as shown in Fig. 5.30. The mathematical procedure can be complicated in some cases, however. Therefore, approximate relations between measurable viscoelastic functions were developed (cf. Sect. 5.9).

At the beginning of the experimental rotational rheology, stressing experiments with cone-plate geometries were favored as the shear rate is nearly constant within the gap (cf. Sect. 15.3). Due to an easier sample preparation parallel plates are preferred today. Besides viscosity functions, the steady-state first normal stress differences as a function of the shear rate can be obtained from stressing experiments. Nowadays, dynamic-mechanical measurements dominate the field of rotational rheometry. The real and imaginary parts of the complex modulus or compliance are the quantities which are determined. But more and more the advantages of creep and creep recovery are realized if slow molecular processes have to be investigated. From these modes the viscous and elastic parts of deformation follow directly.

H. Münstedt and F. R. Schwarzl, *Deformation and Flow of Polymeric Materials*,
DOI: 10.1007/978-3-642-55409-4_11,

11.1 Experiments at Constant Shear Rate (Stressing Experiments)

A stressing experiment in shear is defined by the following specifications of the time dependence of the shear strain γ

$$\begin{aligned} \gamma(t) &= 0 && \text{for} \quad t<0 \\ \gamma(t) &= \dot{\gamma}_0 \cdot t && \text{for} \quad t \geq 0 \end{aligned} \qquad (10.13)$$

with $\dot{\gamma}_0$ being the constant shear rate. The resulting shear stress σ in general depends on both parameters, the shear rate $\dot{\gamma}_0$ and the time t, i.e., $\sigma = \sigma\ (\dot{\gamma}_0, t)$. From the given shear rate and the resulting shear stress σ, a time-dependent viscosity can directly be determined as

$$\eta(\dot{\gamma}_0, t) = \sigma(\dot{\gamma}_0, t)/\dot{\gamma}_0 \qquad (11.1)$$

Strictly taken, the shear stress is the component σ_{12} of the stress tensor. Whenever a shear deformation is discussed in the following, the indices are omitted for the matter of convenience, however.

Pioneering work in the field of stressing experiments with cone- plate geometry was performed by Meißner [1] using a rotational rheometer the principle of which goes back to Weissenberg [2]. Typical of such kind of experiments on polymer melts is the occurrence of a normal force F_N acting on the surface of the plate A in addition to the torque exerted on the rotational axis (cf. Fig. 15.13). The *first normal stress difference* is defined as

$$N_1 = F_N/A = \sigma_{11}(\dot{\gamma}_0, t) - \sigma_{22}(\dot{\gamma}_0, t) \qquad (11.2)$$

with σ_{11} being the normal stress in the flow direction and σ_{22} the normal stress in the direction of the flow gradient. From the first normal stress difference the *first normal stress coefficient* may be calculated as

$$n_1(\dot{\gamma}_0, t) = (\sigma_{11}(\dot{\gamma}_0, t) - \sigma_{22}(\dot{\gamma}_0, t))/\dot{\gamma}_0^2 \qquad (11.3)$$

In Fig. 11.1, the shear stress σ and the first normal stress difference N_1 measured in parallel are presented as functions of time in a linear plot for the LDPE IUPAC A at a temperature of 150 °C and a shear rate of $\dot{\gamma}_0 = 10\ \text{s}^{-1}$ [1]. A complete characterization of this material can be found in [3]. N_1 is significantly larger than σ over the time range measured. Both quantities exhibit a steep decay at smaller times and seem to approach plateau values at longer ones.

Plotted on a logarithmic timescale, in Fig. 11.2 two features of the normal stress difference become obvious. For $\dot{\gamma}_0 = 10\ \text{s}^{-1}$, N_1 runs through a steep maximum at around 0.6 s. Due to the timescale chosen for the linear plot this maximum cannot be seen in Fig. 11.1. A steady-state is difficult to determine, however, as the measuring time was not chosen long enough. In addition, in Fig. 11.2 the normal

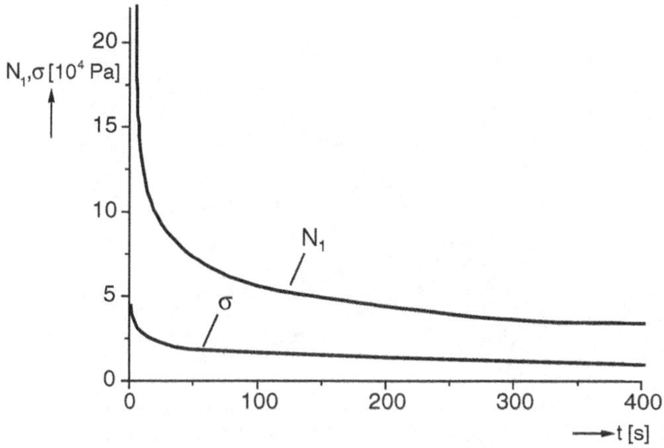

Fig. 11.1 Shear stress σ and first normal stress difference N_1 as functions of time in a linear plot for the LDPE IUPAC A at the shear rate $\dot{\gamma}_0 = 10\ \mathrm{s}^{-1}$ and $T = 150$ °C. (Reprinted from [1] with permission from John Wiley and Sons)

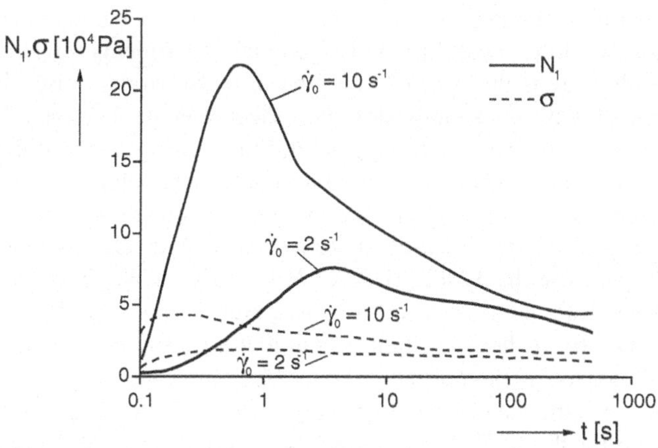

Fig. 11.2 Shear stress σ and first normal stress difference N_1 as functions of time t for the LDPE IUPAC A at $T = 150$ °C and two constant shear rates $\dot{\gamma}_0$. (Reprinted from [1] with permission from John Wiley and Sons)

stress differences obtained at a smaller shear rate of $\dot{\gamma}_0 = 2\ \mathrm{s}^{-1}$ is presented. The maximum becomes lower and is shifted to longer times.

The maxima for the shear stresses are smaller but broader than those for the normal stress differences. Like for N_1, a steady-state is not clearly seen, however, in the plot of Fig. 11.2 due to the limited experimental time.

Figure 11.3 shows the viscosities as a function of time from stressing experiments in a wide range of constant shear rates. The curves at lower shear rates do not exhibit a maximum any more and reach stationary states which become independent of the shear rate if it is chosen small enough. The rate-independent curve

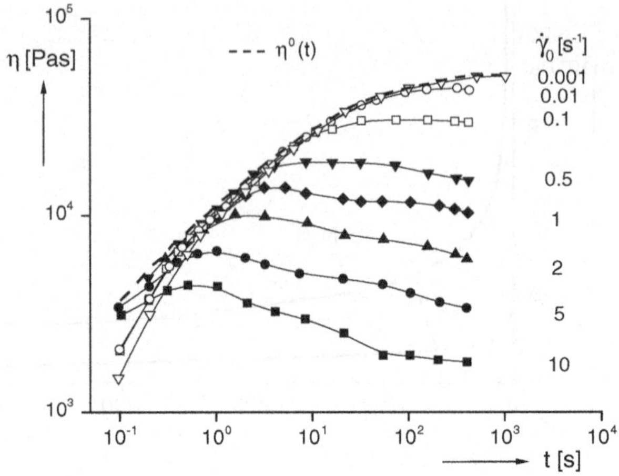

Fig. 11.3 Shear viscosities η as functions of the time t for the LDPE IUPAC A at 150 °C and various constant shear rates $\dot{\gamma}_0$. (Reprinted from [1] with permission from John Wiley and Sons)

represents the time-dependent linear shear viscosity $\eta^0(t)$ which is of some interest for an assessment of the reliability of elongational experiments (cf. Sects. 12.2 and 12.5). Its steady-state is the so-called zero-shear or Newtonian viscosity η_0.

Measurements with a specially designed rheometer made it possible to attain stationary values of the shear stress σ and the first normal stress difference N_1 as well in a wide range of shear rates. From the steady-state values, the steady-state or stationary viscosities $\eta(\dot{\gamma}) = \sigma/\dot{\gamma}$ and the first normal stress coefficients $n_1 = N_1/\dot{\gamma}^2$ can be obtained. These two quantities are plotted in Fig. 11.4 as functions of the shifted shear rate for the LDPE IUPAC A [4]. The wide range presented was achieved by measurements at various temperatures and by shifting them to master curves (cf. Chap. 6). It should be pointed out that n_1 has to be related to the square of the shift factor a_T, as the normal stress coefficient scales with $\dot{\gamma}^2$.

As can be seen, the data come to lie on one curve for the shifted viscosities and on another one for the shifted first normal stress coefficients. At small shear rates the viscosity reaches the zero-shear viscosity η_0 which is typical of polymer melts (cf. Sect. 13.1.2). The first normal stress coefficient approaches a constant value, too, for small shear rates. According to Eq. (10.26), the steady-state linear recoverable compliance J_e^0 can be determined from this value making use of η_0 measured in parallel. Both material functions in Fig. 11.4 show a decrease in dependence on the shear rate which is more pronounced for the first normal stress coefficient than for the viscosity.

Reliable measurements of the first normal stress differences need a lot of experimental care due to their sensitivity to already small temperature variations and flow instabilities which may occur close to the rim of a sample at certain shear deformations [5]. That is the reason why only few reliable data on the first normal stress coefficient of polymer melts can be found in the literature. The experimental requirements are still more challenging for measurements of the second normal

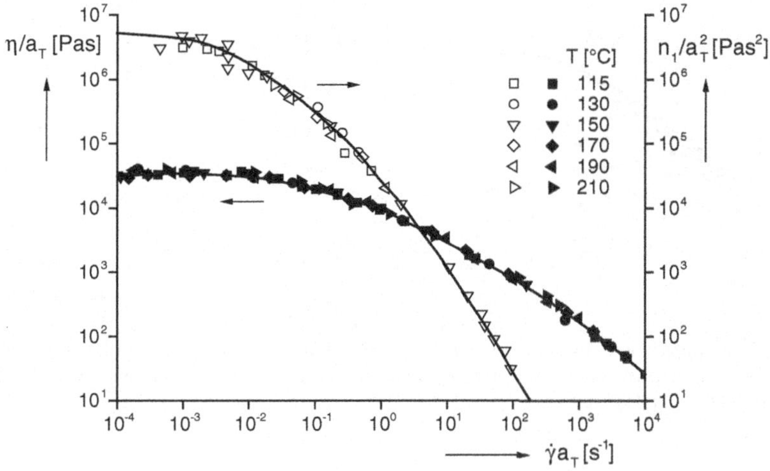

Fig. 11.4 Steady-state values of the shifted viscosity η/a_T and the normal stress coefficient n_1/a_T^2 as functions of the shifted shear rate $a_T\dot{\gamma}$ for the LDPE IUPAC A. The curves were obtained from measurements at different temperatures making use of the time-temperature superposition with the shift factor a_T. The reference temperature was chosen as $T_0 = 150$ °C. The symbols mark the different measuring temperatures indicated in the figure [4]

stress coefficient [5] and, therefore, only its negative sign and its order of magnitude being around 10 % of the first normal stress difference can be taken for granted up to now.

11.2 Relation Between the Time and Shear-Rate Dependence of the Viscosity

An interesting relationship between measurements of the time-dependent linear viscosity in a stressing experiment

$$\eta^0(t) = \lim_{\dot{\gamma}_0 \to 0} \eta(\dot{\gamma}_0, t) \tag{11.4}$$

and the viscosity function, i.e., the dependence of the steady-state viscosity on the shear rate, was found first by Gleissle on a silicone oil and a polyethylene [6]. Further experiments showed the validity of this relation for other polymer melts, too. The essence of this so-called "*mirror relation*" is depicted in Fig. 11.5 making use of the results shown in Fig. 11.3 on a polyethylene. The symbols in the right part of Fig. 11.5 mark the viscosities at the longest times from the left part of this figure. They are plotted as a function of the shear rates at which they were attained. Surprisingly, they fall together with the time-dependent linear viscosity function $\eta^0(t)$ if $\dot{\gamma} = 1/t$ is chosen. This curve is represented by the full line in the

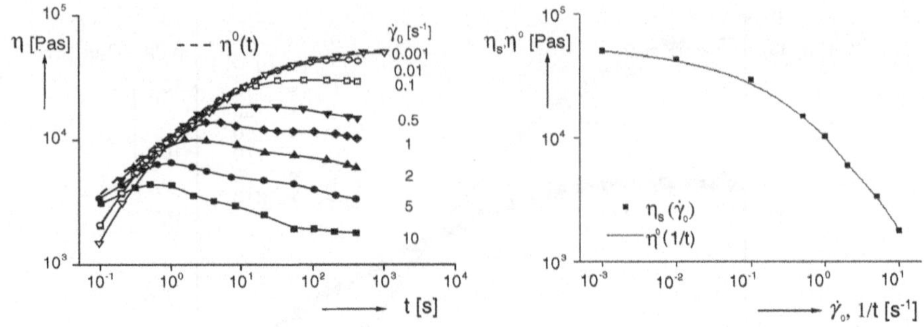

Fig. 11.5 Comparison between the linear time-dependent viscosity η^0 (t) plotted as a function of $\dot{\gamma} \equiv 1/t$ and the steady-state viscosities η_s measured at different constant shear rates $\dot{\gamma}_0$ for the LDPE IUPAC A at 150 °C

right part of Fig. 11.5, i.e., $\eta^0(t)$ from the left diagram is mirrored with respect to the viscosity axis. This "*mirror relation*" is purely empirical and, therefore, in principle its validity has to be shown for each material separately.

11.3 Dynamic-Mechanical Experiments

Dynamic-mechanical experiments in the linear range of deformation are very versatile for the rheological characterization of polymer melts. The basics of dynamic-mechanical experiments are presented in Sect. 5.8. They provide the zero-shear viscosity η_0 and the linear steady-state elastic compliance J_e^0 at frequencies small enough, i.e., in the so-called terminal regime which is distinguished by the validity of Eq. (5.105) in the case of G'' and of Eq. (5.106) for G'. To make use of these relationships the amplitude of the oscillatory strain or stress has to be chosen in a range in which the corresponding material function is independent of the amplitude. The most straightforward way to obtain η_0 and J_e^0 from dynamic-mechanical measurements is the plot of the real part J' and the imaginary part J'' of the complex compliance $J^* = J' - iJ''$ as functions of the angular frequency ω. From Eqs. (5.99) and (5.100) it is obvious that J_e^0 corresponds to the value of J' for $\omega \to 0$ and η_0 can be determined from the linear part of $J''(\omega)$ in a double-logarithmic plot (cf. Fig. 11.12).

More often, however, the storage modulus G' and the loss modulus G'' are discussed as functions of the angular frequency ω. These functions reflect molecular processes which often become more transparent by the calculation of relaxation spectra. According to Eq. (5.105), η_0 can be obtained in the terminal regime from a double-logarithmic plot of G'' as a function of ω. Knowing η_0, J_e^0 can experimentally be determined from $G'(\omega)$ according to Eq. (5.106) at low enough frequencies.

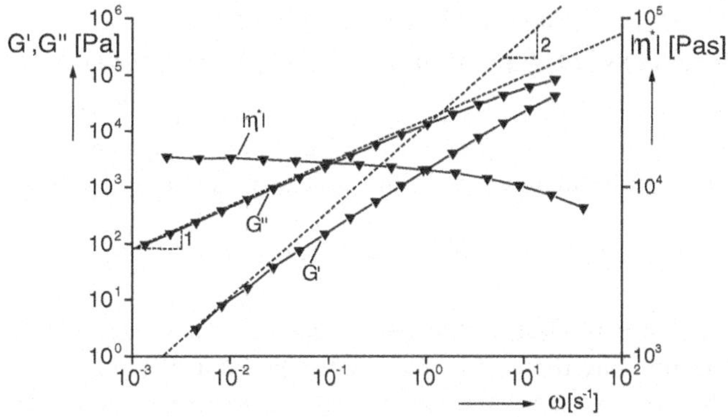

Fig. 11.6 Typical results for the storage modulus G', loss modulus G'', and magnitude of the complex viscosity $|\eta^*|$ as functions of the angular frequency ω for a linear low density polyethylene (mLLDPE) of $M_w = 100$ kg/mol, and a narrow molar mass distribution of $M_w/M_n \cong 2$ at $T = 150$ °C [7]

Figure 11.6 shows measurements of $G'(\omega)$ and $G''(\omega)$ in a double-logarithmic plot of a metallocene-catalyzed linear low density polyethylene (mLLDPE) with a narrow molar mass distribution. The slope of 1 for $G''(\omega)$ and that of 2 for $G'(\omega)$, i.e., the terminal regime, is attained as indicated by the broken lines. The deviations from the terminal regime occur at smaller frequencies for G' than for G''.

From dynamic-mechanical experiments further information on the molecular structure of a polymer can be obtained. At high enough frequencies, G' may reach the so-called rubber elastic plateau described by G_N^0 from which the average molar mass M_e of chains between entanglements can be calculated according to Eq. (4.6) based on the theory of rubber elasticity. This state is by far not reached by the sample shown in Fig. 11.6, however.

Formally, a complex viscosity can be defined according to Eq. (5.107). In Fig. 11.6, its magnitude

$$|\eta^*| = |G^*|/\omega = \frac{\sqrt{G'^2 + G''^2}}{\omega} \tag{11.5}$$

as a function of ω is presented, too. After a regime of approximately constant viscosity, a distinct decrease of the viscosity is observed at frequencies higher than $\omega \approx 10^{-1}$ s^{-1}. The linear range of $|\eta^*(\omega)|$ is much wider than the terminal regime of G' but in agreement with that of G''. This finding is due to the fact that at lower frequencies, G'' is distinctly larger than G' and, therefore, dominates the complex viscosity according to Eq. (11.5).

11.3.1 Relations Between Dynamic-Mechanical and Stressing Experiments (Cox-Merz Rule)

The function $|\eta^*(\omega)|$ is of importance insofar as for a great number of polymer melts it is found to be identical with the viscosity function $\eta(\dot{\gamma})$, i.e.,

$$\eta(\dot{\gamma}) = \lim_{t\to\infty} \eta(\dot{\gamma}, t) \cong |\eta^*(\omega)| \quad for\ \dot{\gamma} \equiv \omega \tag{11.6}$$

The validity of this relationship called the *Cox-Merz rule* [8] is depicted in Fig. 11.7. The full line represents averaged data from dynamic-mechanical measurements, the dots stand for the steady-state viscosities obtained from stressing experiments at definitely chosen shear rates. The material was a long-chain branched polyethylene very similar to the IUPAC A. The agreement over a shear-rate range of six decades is excellent. For very small ω, i.e., $G' << G''$, the magnitude of the complex viscosity becomes (cf. Eq. 11.5)

$$\lim_{\omega\to 0} |\eta^*| = \lim_{\omega\to 0} G''/\omega = \eta_0 \tag{11.7}$$

The Cox-Merz rule has an interesting consequence. As $|\eta^*| = |G^*|/\omega$ in a dynamic-mechanical experiment and $\eta = \sigma/\dot{\gamma}$ in a stressing experiment it follows

$$\lim_{t\to\infty} \sigma(\dot{\gamma}, t) \cong |G^*(\omega)| \quad \text{for } \dot{\gamma} \equiv \omega \tag{11.8}$$

The validity of the *Cox-Merz rule* is reported in the literature for many polymer melts, but it is purely empirical and that means its validity has to be shown, in principle, for each class of materials before it can be applied. From the wide scope of data which were obtained over the years the probability is very high, however,

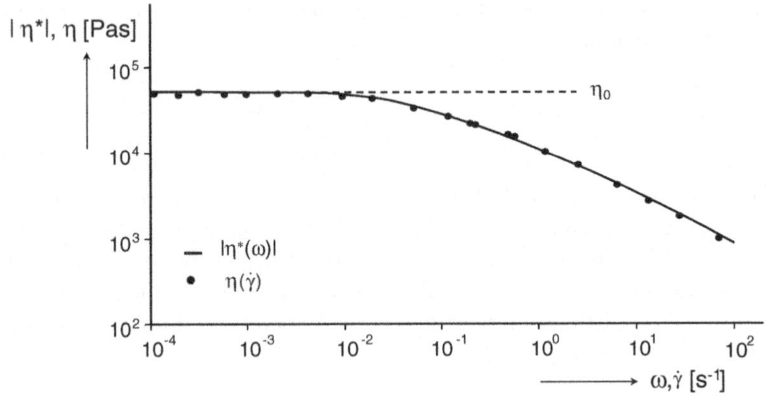

Fig. 11.7 Comparison of the magnitude of the complex viscosity $|\eta^*|$ as a function of the angular frequency ω and the stationary viscosity η as a function of the shear rate $\dot{\gamma}$ obtained in stressing experiments for an LDPE similar to LDPE IUPAC A at $T = 150$ °C

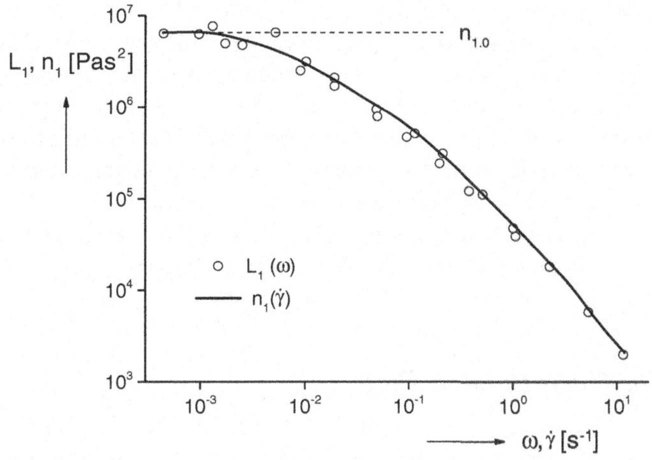

Fig. 11.8 Comparison of the first normal stress coefficient n_1 as a function of the shear rate $\dot{\gamma}$ with the quantity derived from dynamic-mechanical measurements according to Eq. (11.9). (Reprinted from [9] with permission from the Society of Rheology)

that the Cox-Merz rule can be used for the melts of unfilled linear polymers without any restriction.

A relationship between the first normal stress coefficient n_1 and the dynamic-mechanical quantities G' and G'' as pendant to the Cox-Merz rule for the viscosity was proposed in [9] as

$$n_1(\dot{\gamma}) \cong L_1(\omega) \equiv 2G' \frac{\left(1 + (G'/G'')^2\right)^{3/2}}{\omega^2} \quad \text{for } \dot{\gamma} \equiv \omega \tag{11.9}$$

For small ω, i.e. $G' << G''$, and taking Eq. (5.106) into account, Eq. (11.9) reads

$$\lim_{\dot{\gamma}\to 0} n_1(\dot{\gamma}) = n_{1.0} = \lim_{\omega\to 0} 2G'/\omega^2 = 2J_e^0 \eta_0^2 \tag{11.10}$$

In Fig. 11.8, it is shown for an LDPE similar to IUPAC A that, indeed, normal stress and dynamic-mechanical measurements can be related to each other if $n_1(\dot{\gamma}) = L_1(\omega)$ and $\dot{\gamma} \equiv \omega$ are chosen. This result is an empirical one and was not confirmed by investigations on other polymeric materials up to now, probably because of the experimental difficulties to reliably determine steady-state normal stress differences over a wide range of shear rates.

Taking the equivalence given by Eq. (11.9) for granted, this relationship represents a convenient way to get some insight into the normal stresses of polymer melts making use of dynamic-mechanical measurements which are comparably easy to perform.

The *Cox-Merz rule* together with the "*mirror relation*" is of some interest insofar as it allows to determine $\eta^0(t)$ from dynamic-mechanical experiments, which can be carried out over a wide frequency range. $|\eta^*(\omega)|$ can be set approximately equal to $\eta(\dot{\gamma})$ and $\eta^0(t)$ can then be obtained by applying the "*mirror relation*". Such a procedure is of particular advantage if viscosity data at short times are required, as measurements at higher frequencies corresponding to shorter times can reliably be performed up to $\omega = 100\ \text{s}^{-1}$.

It has to be stressed once more, however, that the two relations which are the base of such a procedure are of a purely empirical nature.

11.3.2 Van Gurp-Palmen Plot

Another way of evaluating dynamic-mechanical data is the so-called *van Gurp-Palmen plot* [10]. In this presentation, the phase angle δ is plotted as a function of $|G^*|$. Such an evaluation of dynamic-mechanical measurements seems to be straightforward at the first glance as δ and $|G^*| = \sigma_0/\gamma_0$ are directly measured quantities. σ_0 and γ_0 are the stress and strain amplitudes, respectively (cf. Sect. 5.8). The obvious disadvantage, however, is that G' and G'' and, therefore, material properties as a function of time or frequency cannot be discussed by using this plot. Its elegance can be seen in a quick and sensitive assessment of the thermorheological behavior of a material. In the case of a thermorheologically simple material, *van Gurp-Palmen plots* at different temperatures fall on a master curve. This may be seen as follows. A *van Gurp-Palmen plot* is a parameter presentation of the quantities $\delta = f_1(\omega)$ versus $|G^*| = f_2(\omega)$ originating from an elimination of the parameter ω. In case of the validity of the principle of time-temperature superposition for a thermorheologically simple material, δ and $|G^*|$ become equal to $f_1(a_T\omega)$ and $f_2(a_T\omega)$ at other temperatures. After the elimination of the parameter $a_T\omega$, the same curves are found. However, time-temperature shift factors obviously cannot be determined from the *van Gurp-Palmen plot* and, therefore, its significance for thermorheology is limited. In the case of a thermorheologically complex material, the curves of $\delta(|G^*|)$ measured at different temperatures do not fall together.

Although the *van Gurp-Palmen plot* provides less information on the viscoelastic behavior of a polymer melt than the usual analysis of the complex modulus or compliance and lacks the possibility of a spectral analysis, it has been used during the last years for a rheological characterization of long-chain branched materials [7, 11–13]. Its merits lie in a first hint to the existence of longer relaxation times if compared to the curve of a linear standard. Longer relaxation times are typically induced by long-chain branches, but it has to be taken into account that the conclusions may not be without some ambiguity as the spectrum of a polymer melt is determined by relaxation times the molecular origin of which can be different.

11.3.3 Large Amplitude Oscillations (LAOS)

At large amplitudes, i.e., in the non-linear range of deformation, a sinusoidal mechanical excitation leads to a periodic answer which is more or less distorted. The determination of a complex modulus is not possible then any more. In publications between the 1960s and 1970s it was tried to analyze the non-linear behavior and correlate it with molecular or morphological structures of soft materials. An overview of these first experiments and the further development of LAOS is given in a review article [14].

The breakthrough of the LAOS technique has its roots in the design of monochromatic mechanical excitation devices and sensitive special torque transducers. The adoption of the signal processing technique developed to a high standard in the field of nuclear magnetic resonance was another decisive step for making LAOS more easy to use. The pioneering work regarding this experimental progress is documented, for example, in [15] and [16]. The base of the evaluation of the large amplitude oscillations is the Fourier Transformation Analysis. The distorted signal is decomposed into the harmonics of the exciting frequency and the corresponding amplitudes are determined. The ratio of the first to the third harmonic and quantities derived from that are usually assessed to find correlations with the structure of a sample. Thus, it could be shown, for example, that the answer of a linear polypropylene to a non-linear excitation was different from that of a long-chain branched sample [14]. A deeper understanding of these results and their description by existing theories is still in a rudimentary state, however.

11.4 Creep Recovery Experiments

The creep recovery experiment schematically presented in Fig. 5.12 is defined by

$$\begin{aligned} \sigma &= 0 \quad \text{for} \quad t<0 \\ \sigma &= \sigma_0 \quad \text{for} \quad 0 \leq t < t_0 \\ \sigma &= 0 \quad \text{for} \quad t \geq t_0 \end{aligned} \tag{5.1}$$

At the constant stress σ_0 the shear deformation γ is measured as a function of the creep time t.

For polymer melts with a pronounced non-linear range of deformation, the creep compliance J which presents the material function depends on stress and time, i.e.

$$J(\sigma_0, t) = \gamma(\sigma_0, t)/\sigma_0$$

which in the linear range of deformation becomes

$$J(t) = \gamma(\sigma_0, t)/\sigma_0 \tag{5.6}$$

When σ_0 is set to zero after the creep time t_0, a viscoelastic material recovers and the recoverable strain γ_r, is obtained as a function of the recovery time t_r. Generally, the recoverable compliance J_r is determined as

$$J_r(\sigma_0, t_0, t_r) = \gamma_r(\sigma_0, t_0, t_r)/\sigma_0$$

For long enough creep times t_0 and recovery times t_r one gets the steady-state recoverable compliance J_e defined as

$$\lim_{\substack{t_0 \to \infty \\ t_r \to \infty}} J_r(\sigma_0, t_0, t_r) = J_e(\sigma_0) \tag{11.11}$$

which in the linear range, i.e. $\sigma \to 0$, becomes the linear steady-state recoverable compliance J_e^0.

The linear regime of a creep recovery experiment is of particular interest in so far as from the theory of linear viscoelasticity presented in Sect. 5.5 the creep compliance can be decomposed according to

$$J(t) = J_0 + \psi(t) + t/\eta_0 = J_r(t) + t/\eta_0 \tag{5.27}$$

with J_0 being the instantaneous elastic compliance, $\psi(t)$ the retarded viscoelastic compliance, and η_0 the zero-shear viscosity.

The zero-shear viscosity η_0 and the linear steady-state recoverable compliance J_e^0 are of special importance for relationships between rheological properties and molecular parameters (cf. Chap. 13).

Stress-controlled plate-plate and cone-plate rotational rheometers are principally suited for creep recovery experiments. Three experimental conditions have to be fullfilled, however, to get reliable results for η_0 and J_e^0. First, the stress has to be chosen in the linear range of deformation, second, stationary states have to be attained in the creep and the subsequent recovery experiment and third, the influences of the bearing of the rheometer during recovery have to be kept negligibly small or get corrected.

The last requirement is the most demanding one as due to the mechanical positioning of the sample during shear, bearings cannot be avoided. In [17] it was shown that, in principle, bearings can have two negative effects on creep recovery measurements. Friction leads to a recoverable shear which is lower than the real one and some unbalanced torques within the bearing may superimpose a drift on the recovery. To minimize these effects special rheometers were constructed [18]. One example is an apparatus described in [19] which is based on a magnetic bearing with a very small amount of friction and a low residual torque. While the resulting stress is small in comparison to the stresses applied in a creep experiment and can be neglected, it may give rise to a drift, however, when the recovery is

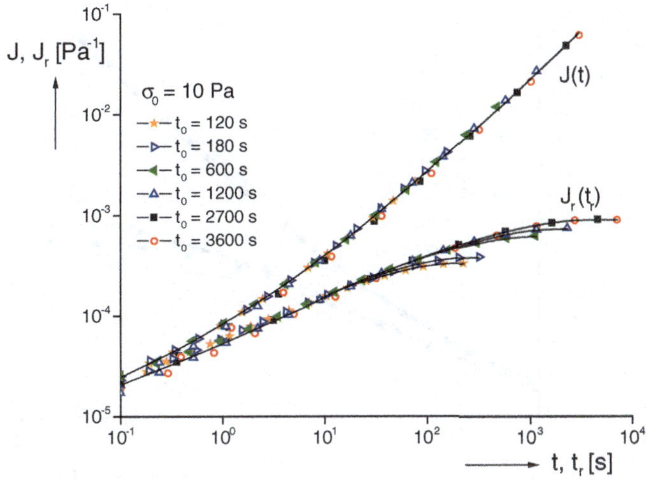

Fig. 11.9 Determination of the stationary state of a creep recovery experiment at 180 °C on a polypropylene with the molar mass $M_w = 560$ kg/mol and the polydispersity factor $M_w/M_n = 6.3$. J is the creep compliance, J_r the recoverable creep compliance, t the creep time, t_r the recovery time and t_0 the creep time after which the recovery was started. σ_0 is the constant stress of the creep experiment [21]

determined. The recoverable compliance measured can be corrected for this effect to give the real material property as was shown in [17]. A more detailed discussion of creep-recovery experiments and an example for the procedure correcting the influence of the residual torque is given in [20].

An example for an experimental determination of the stationarity of a creep and creep recovery experiment is shown in Fig. 11.9 on a polypropylene. At a shear stress of $\sigma_0 = 10$ Pa it is clearly seen that the steady-state recoverable creep compliance becomes higher with longer preceding creep times t_0. For $t_0 = 2{,}700$ and 3,600 s the data come to lie on one curve indicating that the preceding creep was large enough to reach the maximum recoverable deformation. Furthermore, Fig. 11.9 demonstrates that the time necessary to attain the steady state of the recovery process becomes the smaller the shorter t_0. The creep curves $J(t)$ up to different creep times $t = t_0$ only show small deviations from each other demonstrating the good reproducibility of the creep experiments.

An example for testing the range of linearity of a creep and recovery experiment is shown in Fig. 11.10 for the PP of Fig. 11.9 [21]. Up to the applied shear stress of 20 Pa, the time-dependent recoverable compliances $J_r(t)$ measured come to lie on one curve representing the linear recoverable compliance. For all stresses the preceding creep times t_0 were chosen long enough to reach the stationary regime (cf. [21]).

From the creep and recovery curves in Fig. 11.10 it becomes evident that $J_r(t)$ is much more sensitive to the stress applied than $J(t)$. Up to a stress of 200 Pa, the creep curves are indistinguishable from each other. For 5000 Pa, $J(t)$ is slightly

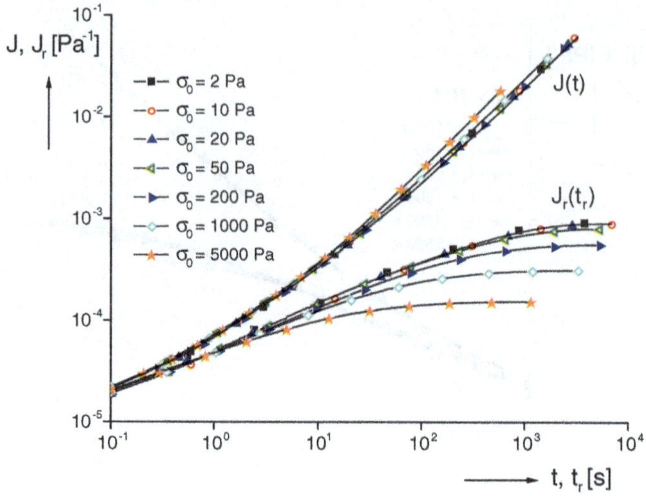

Fig. 11.10 Determination of the linear range of a creep recovery experiment at 180 °C for the polypropylene of Fig. 11.9 [21]. The physical quantities are the same as in Fig. 11.9

higher indicating a non-linear creep behavior with a lower viscosity typical of polymer melts. This finding indicates that the non-linearity is more strongly reflected by elastic than by viscous properties as the recovery curves show already a nonlinear behavior at a stress of 50 Pa. The same tendency is seen in Fig. 11.4 as the normal stress coefficient markedly decreases already at smaller shear rates than the viscosity.

As $J_r(t)$ approaches a constant value with time it follows from Eq. (5.27) that for long enough creep times the approximation

$$J(t) \approx t/\eta_0 \tag{11.12}$$

is valid, i.e., in a double-logarithmic plot the creep compliance as a function of t approaches a straight line with the slope 1 the intercept of which corresponds to the reciprocal viscosity. In Fig. 11.11, $J(t)$ at 5 and 10 Pa from Fig. 11.10 is shown once more to demonstrate that for the PP presented the slope of 1 is attained at long creep times. A more direct way to obtain η_0 from creep experiments is the plot of log $t/J(t)$ as a function of log t as shown in Fig. 11.11, too [22]. The plateau value corresponds to η_0.

Although Eq. (5.27) was derived for the linear range of viscoelasticity its validity has been shown empirically for nonlinear creep recovery experiments, too. In this case J_r and the viscosity become functions of t and the stress σ. For experiments performed long enough the steady-state compliance $J_e(\sigma)$ and steady-state viscosities $\eta(\sigma)$ can be determined by procedures corresponding to those in the linear regime.

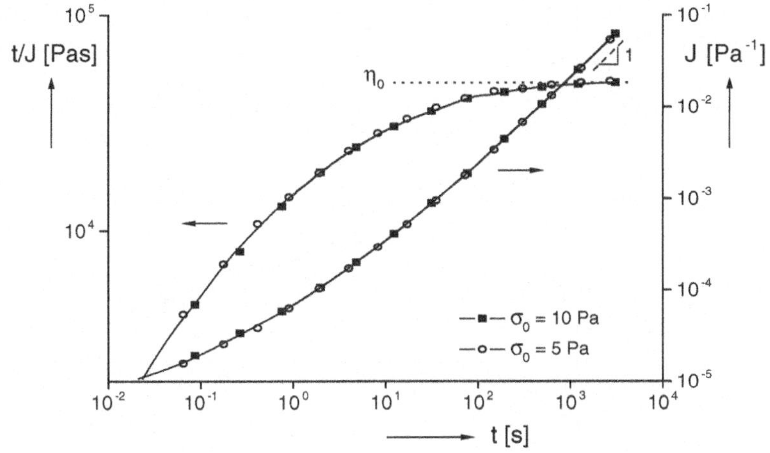

Fig. 11.11 Determination of the zero-shear viscosity from the creep compliance *J*(*t*) for the polypropylene of Fig. 11.9 [22]

11.4.1 Comparison of Creep Recovery with Dynamic-Mechanical Experiments

How accurately the correction procedure regarding the bearing influence works, is depicted by Fig. 11.12 which shows the loss compliance J'' and the storage compliance J' as functions of the angular frequency ω for a long-chain branched metallocene-catalyzed polyethylene [7]. The filled symbols were obtained from dynamic-mechanical tests down to the commonly applied frequency range of around $\omega = 10^{-2}\ \mathrm{s}^{-1}$. The open symbols were calculated from creep and creep recovery experiments making use of the relationships derived from the theory of linear viscoelasticity in Chap. 5 in the following way.

From the steady-state recoverable compliance, the retardation spectrum $f(\tau)$ was determined by Eq. (5.36) according to [23] and [24]. Using Eq. (5.99), J' was calculated and with Eq. (5.100) and η_0, obtained from the creep curve, J'' was assessed. The overlapping of the values from the different methods is excellent and gives confidence with respect to two aspects. The one is the hint to the reliability of the recovery measurements, the other that creep recovery experiments can be used to extend the range of dynamic-mechanical measurements to low frequencies difficult to attain with conventional rheometers. The results obtained are in perfect agreement with the theory of linear viscoelasticity regarding the frequency dependence of the real and imaginary part of the complex compliance.

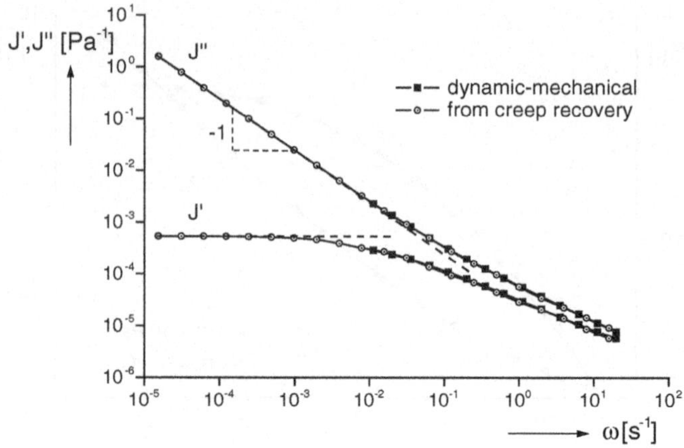

Fig. 11.12 Loss compliance J'' and storage compliance J' as functions of the angular frequency ω for a long-chain branched metallocene-catalyzed polyethylene as measured in a dynamic-mechanical test at 150 °C compared to the corresponding values calculated from a creep recovery experiment [7]

11.5 Capillary Rheometry

For viscosity measurements on polymer melts at higher shear rates, capillary rheometers are used. Common geometries for the capillaries are cylinders and slits. As discussed in detail in Sect. 15.1.1, the shear stress is calculated from the pressure acting on the melt and the shear rate from the throughput taking the geometry of the dic into account. Capillary rheometers are not well suited to measure the time dependence of the viscosity as with pressure-driven devices the output naturally cannot continuously be determined and for piston-driven instruments the correct pressure indication may be retarded by various experimental effects. In any case, the stationarity of the measurement has to be assured before reliable readings are made.

Elastic properties are reflected by the swell of the extrudate. An accurate determination of this quantity requires some experimental care. Strands of the extruded material are cut and their equilibrium diameters after a separate annealing procedure have to be measured to obtain data which are independent of the cooling conditions the samples experience after exiting the capillary. The extrudate swell is interesting in practice for an assessment of the geometry of extruded items. For fundamental research on elasticity, the extrudate swell is of minor importance as there is no theory available up to now which reliably correlates it with well-defined elastic quantities like the recoverable compliance, for example, and as the linear regime is generally difficult to attain in capillary rheometry.

11.5.1 Viscosity Functions

Capillary rheometry plays an important role for the determination of viscosities at higher shear rates relevant for processing. This is shown by Fig. 11.13, which displays a viscosity function of a commercial polystyrene measured over a range of shear rates covering seven decades. Various rheometers and the time-temperature shift were applied to obtain this curve. Conventionally, only positive shear rates are used in practice even in the case of capillary rheometers for which a physically correct description of the flow behavior leads to a negative sign of the shear rate (cf. Sect. 15.1.1).

From the constant zero-shear viscosity at small shear rates, the viscosity decreases by about four decades at $\dot{\gamma} = 10^4\ s^{-1}$. This immense amount of the so-called shear thinning is one key for the favorable processability of polymer melts as even for products with high molar masses necessary for good mechanical properties, but resulting in high zero-shear viscosities much lower viscosities and consequently smaller energy uptakes can be expected in the range of shear rates occurring in processing operations like extrusion or injection molding. As indicated in the figure, viscosity measurements with rotational rheometers have their upper limits at shear rates between 1 and 10 s^{-1} if performed in the stressing mode and may be extended up to some few hundred s^{-1} by dynamic-mechanical experiments making use of the *Cox-Merz relation* (cf. Eq. (11.6)). Processing relevant viscosity data are the domain of capillary rheometers. Furthermore, they play a dominant role for the determination of the so-called melt flow rate (MFR) which has got a strong position in quality control (cf. Sect. 17.1).

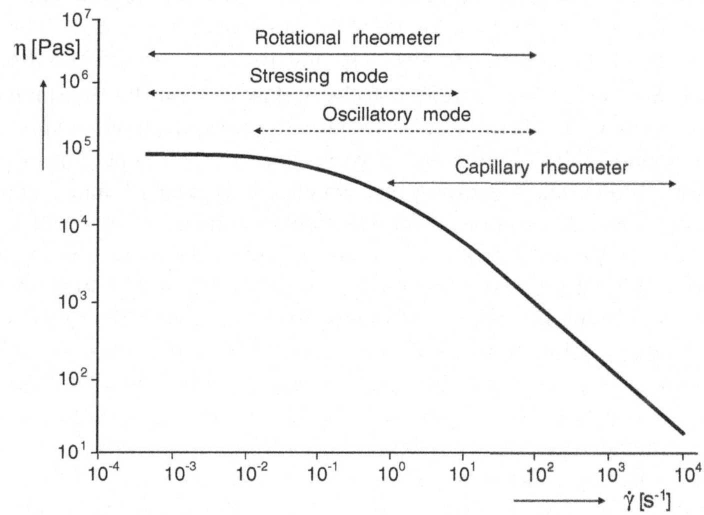

Fig. 11.13 Shear viscosity η as a function of the shear rate $\dot{\gamma}$ for a commercial polystyrene at a temperature of 190 °C measured by rotational and capillary rheometers

11.5.2 Numerical Descriptions of Viscosity Functions

Besides their constant viscosity at small shear rates, another feature of viscosity functions of polymer melts is their power-law behavior at high shear rates (cf. Fig. 11.13). This property is the base of the so-called *Ostwald-de Waele law*

$$\log \eta = \log C - k \log \dot{\gamma} \tag{11.13}$$

or

$$\sigma = \eta\dot{\gamma} = C\dot{\gamma}^{1-k} = C\dot{\gamma}^{n} \tag{11.14}$$

$n = 1 - k$ is called the power-law index with $-k$ representing the negative slope of the viscosity function as shown in Fig. 11.13. As the double-logarithmic slope of the viscosity function is found experimentally to lie between 0 and -1, i.e., $0 < k < 1$, it follows $0 < n < 1$. The Eqs. (11.13) and (11.14) are frequently used in practice for numerical calculations, but they have obviously to be restricted to rather high shear rates occurring in injection molding, for example.

For lower shear rates typical of extrusion, the *Carreau-Yasuda formula*

$$\eta = \frac{\eta_0}{\left(1 + (\lambda\dot{\gamma})^a\right)^{k/a}} \tag{11.15}$$

is often applied [25, 26]. It describes the viscosity function in a wider range of shear rates than the *Ostwald-de Waele law*. η_0 is the zero-shear viscosity, λ a characteristic relaxation time to be fitted. The fit parameter a describes the width of the transition from the Newtonian to the non-Newtonian regime and $-k$ is the linear slope of the viscosity function on the double-logarithmic scale. For $\lambda\dot{\gamma} \ll 1$, Eq. (11.15) gives the zero-shear viscosity and for $\lambda\dot{\gamma} \gg 1$ one gets Eq. (11.13) with $C = \eta_0\lambda^{-k}$. From Eq. (11.15) it follows that λ^{-1} can be interpreted as the shear rate at which the viscosity according to the *Ostwald- de Waele law* formally reaches the value of the Newtonian viscosity. For viscosity functions of polymers with uniform molar mass distributions which are typical of many commercial products, numerical descriptions can be obtained by the adaption of the parameter λ of the *Carreau-Yasuda formula* which are the better the more accurately η_0 and k are known. Difficulties arise if the molar mass distribution of a material contains components resulting in clearly distinguished relaxation times. In [27] it was shown that viscosity functions of such materials can be described by products of *Carreau-Yasuda formulas* of the type of Eq. (11.15) with different sets of parameters. Besides Eq. (11.15) other empirical formulas of a similar structure are sometimes applied which lead to approximations of comparable accuracy.

Obviously not applicable is the *Carreau-Yasuda formula* for the numerical descriptions of viscosity functions which do not exhibit a Newtonian region but an increase of the viscosity at small shear rates due to flow restrictions which can be interpreted by the existence of a yield stress.

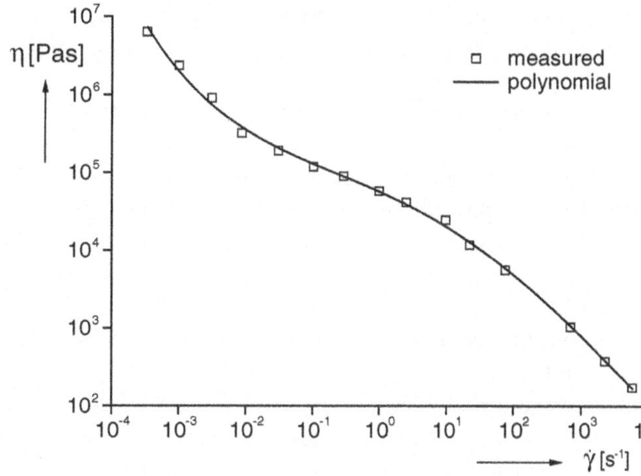

Fig. 11.14 Viscosity function of a styrene-acrylonitrile copolymer modified with 18 wt. % of an acrylic ester rubber (ASA) at $T = 190$ °C and its numerical description by a polynomial according to Eq. (11.17). (Reprinted from [28] with permission from John Wiley and Sons)

Materials of that kind are, for example, rubber-modified thermoplastics. A viscosity function of a styrene-acrylonitrile copolymer filled with 18 wt. % of an acrylic ester rubber is shown in the Fig. 11.14. The data over the wide range of shear rates and viscosities were obtained by various rheometers making use of the time-temperature superposition principle. At low shear rates the viscosity function possesses a slope of approximately -1 which indicates the existence of a yield stress. This interpretation can easily be verified by assuming that the acting stress σ is the sum of the constant yield stress σ_y and the variable stress σ_{fl} effecting the flow. If $\sigma_{fl} << \sigma_y$ one gets

$$\eta = \left(\sigma_y + \sigma_{fl}\right)/\dot{\gamma} \cong \sigma_y/\dot{\gamma} \tag{11.16}$$

which describes the slope of -1 in Fig. 11.14. At higher shear rates or shear stresses, respectively, the yield stress may be negligibly small compared to the total stress and the material shows the usual shear thinning advantageous for processing operations like extrusion and injection molding. Similar types of viscosity functions can be found for particle-filled materials which play an important role today.

Taking this general feature of viscosity functions into account, two inherent deficiencies of Eq. (11.15) become obvious. The one is its inaccuracy in the transient regime between the Newtonian and the power-law behavior and the other its inability to describe viscosity functions which miss a Newtonian region as in the case of heterogeneous polymer melts, for example.

For a comprehensive modeling of the flow of polymer melts sometimes a simple analytical description covering a wide range of the viscosity function is desirable. Polynomials are suitable pragmatic tools for that purpose, although they lack any physical background. In the case of viscosity functions, for a broad variety of commercial polymers the polynomial of the fourth order

$$\log \eta = A_0 + A_1 \log \dot{\gamma} + A_2 (\log \dot{\gamma})^2 + A_3 (\log \dot{\gamma})^3 + A_4 (\log \dot{\gamma})^4 \quad (11.17)$$

has proven its potential for a rather accurate numerical description [29]. Its limitation is that due to principal differences between the graphical representation of the polynomial (11.17) and curves of viscosity functions a good agreement can only be expected in a distinct region of shear rates and, therefore, extrapolations are not meaningful.

Regarding the limitations of numerical descriptions of viscosity functions a further remark has to be made. The *Carreau-Yasuda formula* is sometimes used in the literature to determine η_0 by fitting viscosity functions only known in a very restricted range of shear rates. Such a procedure usually leads to wrong values, particularly, if the Newtonian region is not approached closely enough by experimental data.

11.5.3 Entrance Pressure Loss

As shown in Sect. 16.3.1, the melt at the transition from the reservoir of a capillary rheometer to a die much smaller in diameter exhibits complex flow patterns which reduce the flow energy. Therefore, the pressure driving the extrusion of the melt through the die is lower than that in the reservoir. The so-called pressure loss has to be taken into account for the correct determination of the viscosity if for experimental reasons the pressure cannot be measured inside the die like for circular capillaries (cf. Sect. 15.1.1). On the other hand, the entrance pressure loss can become an essential part of the total extrusion pressure for short dies as in the case of the holes of pelletizing plates, for example. Therefore, its knowledge complements the rheological description of a polymer melt related to applications.

In Fig. 11.15, the entrance pressure loss of a polystyrene is presented as a function of the shear stress at the wall of the capillary. It was determined by so-called Bagley-plots (cf. Sect. 15.1.3). In the double-logarithmic presentation, the measured data for the four temperatures between 190 and 250 °C come to lie in good approximation on one straight line which follows the power law

$$p_e = c \sigma_w^m \quad (11.18)$$

with m and c being material-specific empirical parameters and σ_w the shear stress at the wall of the capillary. Equation (11.18) makes it possible to assess the entrance pressure losses for the polystyrene investigated at least in the temperature and stress ranges of Fig. 11.15. Similar relationships were found for other commercial polymers. The temperature insensitivity of p_e suggests that elastic effects may play a decisive role for the entrance pressure loss as they are not strongly

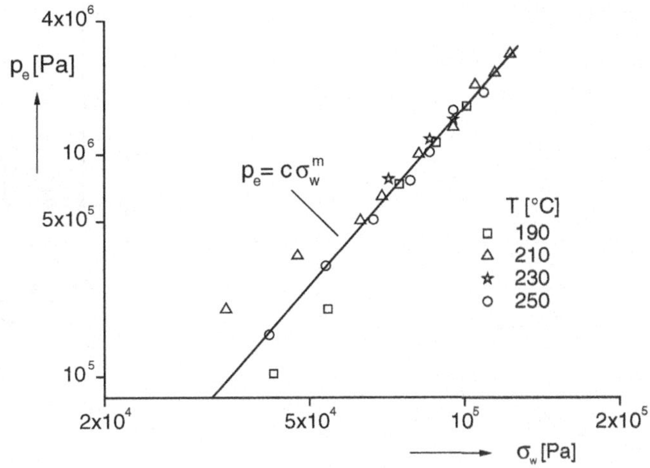

Fig. 11.15 Entrance pressure loss p_e as a function of the shear stress at the capillary wall σ_w in a double-logarithmic plot for a commercial polystyrene at various temperatures

dependent on temperature in contrast to viscosities (cf. Fig. 11.18). Such an interpretation gets some support from the measurements of the flow fields in the entrance region of a die in Sect. 16.3.1.2. They show a fast change of the elongational rates within short times for the entrance flow supporting the assumption of dominating elastic effects.

11.5.4 Extrudate Swell

The extrudate swell is schematicallyshown in Fig. 11.16. A strand exiting a die of the length L and the diameter d_0 attains the diameter d outside. The extrudate swell S is defined as

$$S = (d - d_0)/d_0 = d/d_0 - 1 \tag{11.19}$$

In the literature also the definition d/d_0 can frequently be found for the extrudate swell. In the following, preference is given to the definition of Eq. (11.19) as in the case of the absence of extrudate swell, S is zero, which is in accordance with a subject-specific logic.

As the velocity measurements by laser-Doppler velocimetry demonstrate, the entrance as well as the exit flow of polymer melts is rather complicated (cf. Chap. 16). Furthermore, the elastic deformation is composed of elongational components in the entrance region of the die and shear within the die. Therefore, the modeling of the extrudate swell is complex and has not been very straightforward up to now.

Moreover, determining the extrudate swell is not as easy as it may look at a first glimpse. One reason for it is that the extrudate may freeze before it reaches its

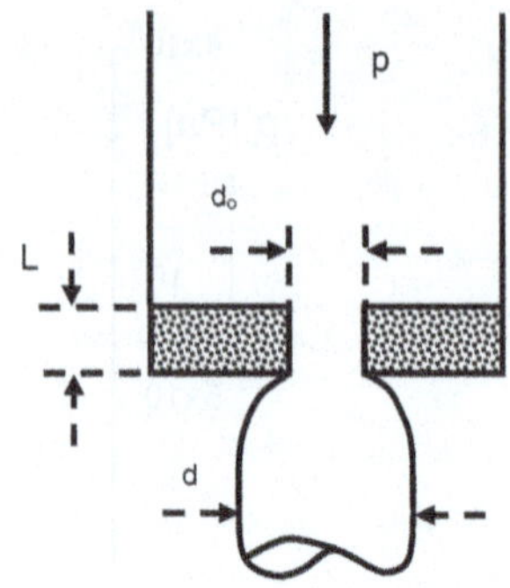

Fig. 11.16 Schematics of the extrudate swell. *p* is the driving pressure

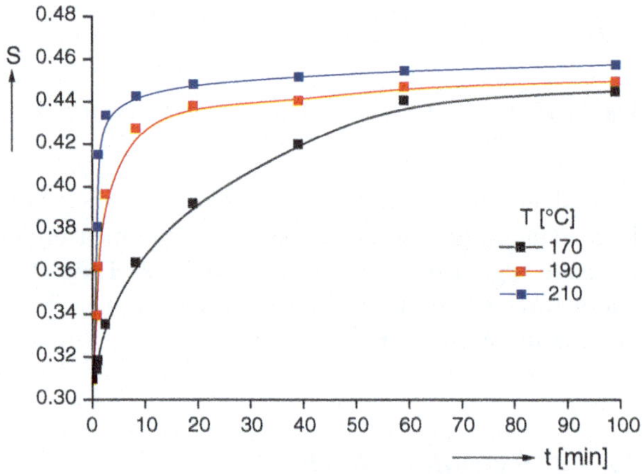

Fig. 11.17 Extrudate swell *S* as a function of the annealing time *t* for a commercial polystyrene at various annealing temperatures *T*. The extrusion pressure was $p = 6.8 \times 10^5$ Pa and the length to radius ratio of the capillary $L/R = 8$

maximum recovery. The other difficulty which is particularly evident in the case of samples of lower viscosities lies in the sagging of the exiting strand under its own weight. This effect can significantly be reduced by extruding the strand into a liquid of a specific weight similar to that of the melt (cf. Sect. 12.3.5). The equilibrium state can be approached by annealing the extrudates in a silicone oil bath above their glass or melting temperatures, respectively. This procedure needs some care, too, as the surface tension between the melt and the annealing liquid superimposes the recovery. An example of such an annealing procedure is given in Fig. 11.17 for a commercial polystyrene extruded through a circular capillary. At an annealing temperature of $T = 210$ °C the equilibrium value seems to be nearly reached on the linear timescale after 20 min, at 170 °C the extrudate swell even after 100 min is still lower by about 5 % than the values attained at the higher temperatures. This may be due to the fact that the stationary state has not been reached. Nevertheless, measurements of the extrudate swell have led to some qualitative insights into elastic properties and contribute to the characterization of polymeric materials.

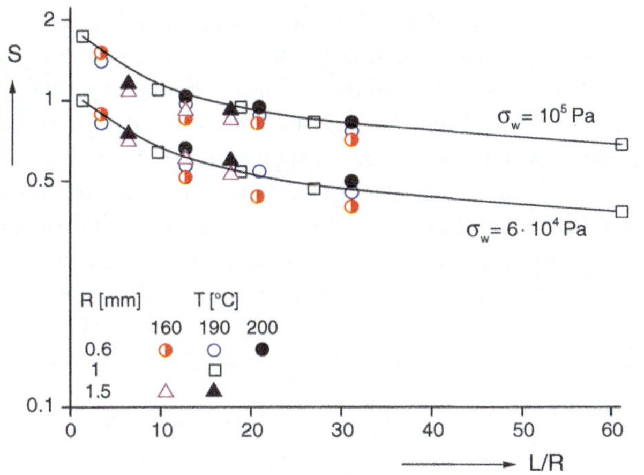

Fig. 11.18 Extrudate swell S of a commercial polystyrene after annealing as a function of the length to radius ratio L/R of the capillaries extruded at three temperatures T and two shear stresses at the wall σ_w

Figure 11.18 represents the extrudate swell after annealing as a function of the ratio of the length L of a capillary to its radius R for a commercial polystyrene at various conditions. The equilibrium data were taken from strands extruded at two wall shear stresses σ_w and three temperatures through dies with three different radii R. As it is clearly seen, the extrudate swell is a function of L/R and the shear stress as well. Interesting is the fact that within the accuracy of the experiments the extrudate swell does not depend on the radius of the capillary and the extrusion temperature, at least within the range of the parameters chosen. For the polystyrene of Fig. 11.18 the extrudate swell for the die of $L/R = 1$ is larger by a factor of two than that one for $L/R = 60$. The decrease of S is distinctly more pronounced at low L/R values than at higher ones. The extrudate swell significantly increases with stress. The two curves obtained run approximately in parallel in the semi-logarithmical plot indicating a constant factor between the values at the two stresses chosen. For the capillary of $L/R = 1$, i.e., of 1 mm radius and 1 mm in length, the shear deformation is small. Therefore, it can be concluded that the extrudate swell has its main origin in the recoverable part of the elongational flow exerted in the entrance region.

Due to the significant extensional flow in the entrance region, the higher extrudate swell at larger wall shear stresses is easily understood. As according to Fig. 11.15 the entrance pressure loss increases with the shear stress at the wall it can be assumed that the tensile stress becomes larger, too. From Fig. 12.18 it follows for an LDPE that the recoverable elongation increases with the tensile stress. This finding is qualitatively valid for other polymeric materials, too.

Taking into account that the elongational deformation in the entrance region is evidently independent of the length of the die, the extrudate swell decreasing with growing L/R can be explained in the following way. If R is assumed to be constant, two effects may come into play. The one is that the elastic deformation of the molecules generated at the die entrance recovers to some extent during the flow through the die. This recovery increases with time or with the length of the die, respectively, giving rise to a smaller extrudate swell. The other conclusion is that elastic deformations due to shear are small in comparison to those in elongation.

The temperature independency of the extrudate swell after annealing which follows from Fig. 11.18 is understood from the fact that the recoverable elongation does not depend on the temperature (cf. Sect. 12.5.4).

In summary it can be concluded, that for a given capillary geometry the extrudate swell of polystyrene after annealing is approximately independent of temperature, but a distinct function of shear stress. Similar dependencies are found for other polymeric materials as shown, for example, for polypropylenes in [30].

References

1. Meissner J (1972) J Appl Polym Sci 16:2877
2. Weissenberg K (1947) Nature 159:310
3. Meissner J (1975) Chem Pure Appl 42:553
4. Laun HM (1978) Rheol Acta 17:1
5. Meissner J, Garbella RW, Hostettler J (1989) J Rheol 33:843
6. Gleissle W (1980) In: Astarita G, Marruci G, Nicolais L (eds) Rheology II. Plenum Press, New York
7. Gabriel C (2001) Doctoral Thesis, University Erlangen-Nürnberg, Shaker Verlag, Aachen. ISBN 3-8265-8909-2
8. Cox WP, Merz EH (1958) J Polym Sci 28:619
9. Laun HM (1986) J Rheol 30:459
10. Van Gurp M, Palmen J (1998) Rheol Bull 67:5
11. Trinkle S, Walter P, Friedrich C (2002) Rheol Acta 41:103
12. Stadler FJ (2007) Doctoral Thesis, University Erlangen-Nürnberg, Sierke Verlag, Göttingen. ISBN 978-3-940333-24-7
13. Resch J (2010) Doctoral Thesis, University Erlangen-Nürnberg
14. Hyun K, Wilhelm M, Klein CO, Cho KS, Nam JG, Ahn KH, Lee SJ, Ewoldt RH, McKinley GH (2011) Prog Polym Sci 36:1697
15. Wilhelm M, Maring D, Spiess HW (1998) Rheol Acta 37:399
16. Wilhelm M (2002) Macromol Mater Eng 287:83
17. Gabriel C, Kaschta J (1998) Rheol Acta 37:358
18. Plazek DJ (1968) J Polym Sci A 6:621
19. Link G, Schwarzl FR (1985) Rheol Acta 24:211
20. Münstedt H (2014) J Rheol 58:565
21. Wolff F, Resch JA, Kaschta J, Münstedt H (2010) Rheol Acta 49:95
22. Wolff F (2008) Diploma Thesis, University Erlangen-Nürnberg
23. Kaschta J, Schwarzl FR (1994) Rheol Acta 33:517
24. Kaschta J, Schwarzl FR (1994) Rheol Acta 33:530
25. Carreau PJ (1972) Trans Soc Rheol 16:99
26. Yasuda K, Armstrong RC, Cohen RE (1981) Rheol Acta 20:163
27. Stadler FJ, Münstedt H (2008) J Nonnewton Fluid Mech 151:129
28. Münstedt H (1981) Polym Eng Sci 21:5
29. Münstedt H (1978) Kunststoffe 68:92
30. Resch JA (2010) Doctoral Thesis, University Erlangen-Nürnberg

Chapter 12
Extensional Rheology

For many processing operations of polymeric materials, elongational properties of their melts play an important role. Examples are fiber spinning, extrusion coating, thermoforming, film blowing, and foaming. With the exception of spinning which is dominated by uniaxial extension, the deformation kinematics of the other processes are generally much more complex making their quantitative description difficult. Other processes like extrusion or injection molding which are usually regarded to be determined by shear can contain some elongational components, too, which are not as obvious as for the operations mentioned above. In case of extrusion, any geometrical change of the flow channels leads to some kind of extension of the melt and in injection molding the molecules entering the tool are exposed to components of elongational flow.

12.1 Basics of Uniaxial Extension

As the extensional deformations are normally very large, the tensile stress, strain, and strain rates have to be defined differently from the case of small elongations discussed in Sect. 7.7. For large extensions, the strain and stress are related to the actual geometry taking the volume constancy into account. The tensile stress is given then by

$$\sigma = F/A \tag{12.1}$$

with F being the force acting on the actual cross-section A of the sample which changes according to

$$A = A_0 l_0 / l \tag{12.2}$$

A_0 and l_0 are the cross-section and the length, respectively, of the initial sample and l is the actual sample length.

H. Münstedt and F. R. Schwarzl, *Deformation and Flow of Polymeric Materials*,
DOI: 10.1007/978-3-642-55409-4_12, © Springer-Verlag Berlin Heidelberg 2014

In contrast to small deformations, the differential strain is defined by relating the change in length dl to the actual length l and not to the initial length l_0, i.e.,

$$\mathrm{d}\varepsilon_H = \mathrm{d}l/l$$

or

$$\varepsilon_H = \int_{l_0}^{l} \mathrm{d}\varepsilon = \int_{l_0}^{l} \mathrm{d}l/l = \ln l/l_0 = \ln \lambda \tag{12.3}$$

The quantity ε_H is called the Hencky strain or natural strain and $\lambda = l/l_0$ the stretching ratio (cf. Sect. 8.7.2). The Hencky strain, different from the so-called Cauchy strain or engineering strain $\varepsilon = (l - l_0)/l_0$, is preferably used in all the fields of polymer research and applications in which large elongational deformations are discussed. That is obviously the case when dealing with mechanical properties of rubbers and polymer melts. For small deformations, $A \approx A_0$ and $\varepsilon_H \approx \varepsilon$ and the definitions related to a constant geometry of a sample during deformation are valid as expected.

From Eq. (12.3) the Hencky elongation rate follows as

$$\dot{\varepsilon} = \frac{\mathrm{d}\varepsilon_H}{\mathrm{d}t} = \frac{\mathrm{d}(\ln l/l_0)}{\mathrm{d}t} = \frac{1}{l}\frac{\mathrm{d}l}{\mathrm{d}t} = \frac{v}{l} \tag{12.4}$$

with v being the stretching velocity. For the matter of convenience, the index "H" is omitted and $\dot{\varepsilon}$ stands for the Hencky elongational rate if not stated otherwise.

The elongational viscosity is defined in analogy to the shear viscosity as

$$\mu = \sigma/\dot{\varepsilon} \tag{12.5}$$

and the elongational compliance as

$$D = \varepsilon_H/\sigma \tag{12.6}$$

The time functions of these quantities depend on the experimental mode and, therefore, it has to be stated what the conditions of the experiments are, e.g., constant stress in a creep or constant elongational rate in a stressing test. The steady-state values are independent of the mode, of course.

The recoverable Hencky strain ε_r follows according to Eq. (12.3) as

$$\varepsilon_r = \int_{l_r}^{l} \mathrm{d}l/l = \ln l/l_r \tag{12.7}$$

with l being the elongated sample before recovery and l_r the length of the sample at any time t of the recovery process. l_r reaches an equilibrium value after complete retardation and the corresponding recoverable strain is denominated ε_{rs}.

The recoverable compliance in elongation D_r is defined then as

$$D_r = \varepsilon_r/\sigma \tag{12.8}$$

12.2 Multiaxial Extensions

In the three-dimensional case of pure elongation or contraction, the stress tensor defined in Sect. 7.1 consists of the three diagonal components σ_{11}, σ_{22}, and σ_{33}, only. They act as normal stresses on the three faces of a cubic volume element as shown in Fig. 12.1.

The stresses applied lead to the elongation rates $\dot{\varepsilon}_{11}$, $\dot{\varepsilon}_{22}$, and $\dot{\varepsilon}_{33}$ which are not independent of each other as the volume can be assumed to be constant during a deformation after which the edges l_0 of the cube in Fig. 12.1 may have attained the lengths l_1, l_2, and l_3, i.e.,

$$V(t) = l_1(t)\, l_2(t)\, l_3(t) = l_0^3 = V_0 \tag{12.9}$$

or

$$\ln(l_1(t)/l_0) + \ln(l_2(t)/l_0) + \ln(l_3(t)/l_0) = 0$$

and after differentiation with respect to time

$$\dot{\varepsilon}_{11} + \dot{\varepsilon}_{22} + \dot{\varepsilon}_{33} = 0 \tag{12.10}$$

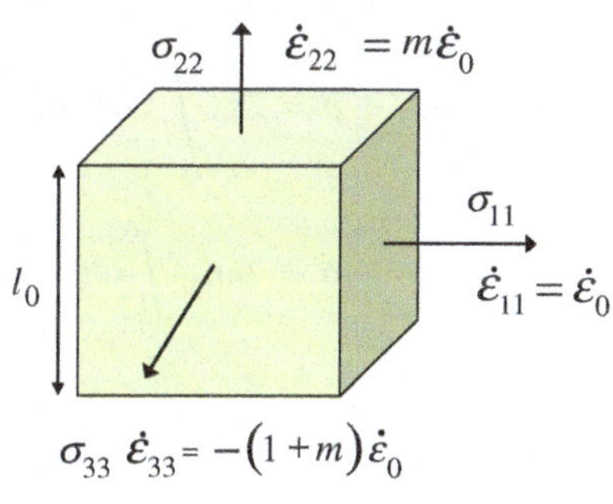

Fig. 12.1 Tensile stresses σ and elongation rates $\dot{\varepsilon}$ in multiaxial extension

Assuming that

$$\dot{\varepsilon}_{11} = \dot{\varepsilon}_0 \text{ and } \dot{\varepsilon}_{22} = m\dot{\varepsilon}_0 \tag{12.11}$$

with m being a time-independent constant, it follows from Eq. (12.10) that

$$\dot{\varepsilon}_{33} = -(1+m)\,\dot{\varepsilon}_0 \tag{12.12}$$

For a constant elongational rate $\dot{\varepsilon}_0$, the rate tensor $\underline{\underline{D}}$ in the principal coordinate system reads

$$\underline{\underline{D}} = \dot{\varepsilon}_0 \begin{pmatrix} 1 & 0 & 0 \\ 0 & m & 0 \\ 0 & 0 & -(1+m) \end{pmatrix} \tag{12.13}$$

The deformation rate induces a stress within the melt. Stress and strain rate are related to each other by the so-called constitutive equation. The search for a comprehensive constitutive equation covering the non-linear state of polymer melts is still an unsolved matter of research (cf. Chap. 10). In the linear range and at small deformations, however, the relationship between the stress tensor $\underline{\underline{S}}$ and the rate tensor $\underline{\underline{D}}$ for incompressible materials is given by (cf. Eq. 7.27)

$$\underline{\underline{S}}(t) + \underline{\underline{E}}p(t) = 2\int_{-\infty}^{t} G(t-t')\underline{\underline{D}}(t')\,dt' \tag{12.14}$$

with $G(t)$ being the time-dependent relaxation modulus in shear introduced in the theory of linear viscoelasticity and $\underline{\underline{E}}$ the unity tensor. $p(t)$ is an arbitrary hydrostatic pressure, which may always be added to the stress tensor of an incompressible material. This hydrostatic pressure may be used to satisfy the boundary conditions of the experiment. By introducing the rate tensor (12.13) into the constitutive Eq. (12.14) and using the condition, that $\dot{\varepsilon}_{ii}(t') = 0$ for $t' < 0$, one gets for the tensile stress components the equations

$$\begin{aligned} \sigma_{11} + p &= 2\dot{\varepsilon}_0 \int_0^t G(t')\mathrm{d}t' = 2\dot{\varepsilon}_0 \eta^0(t) \\ \sigma_{22} + p &= 2m\dot{\varepsilon}_0 \int_0^t G(t')\mathrm{d}t' = 2m\dot{\varepsilon}_0 \eta^0(t) \\ \sigma_{33} + p &= -2(1+m)\dot{\varepsilon}_0 \int_0^t G(t')\mathrm{d}t' = -2(1+m)\dot{\varepsilon}_0 \eta^0(t) \end{aligned} \tag{12.15}$$

Table 12.1 Specifications of the three extensional modes discussed

Mode	m	Stress conditions	Viscosities
Uniaxial	−0.5	$\sigma_{11} = \sigma$, $\sigma_{22} = \sigma_{33} = 0$	$\mu^0(\mathrm{t}) = \sigma/\dot{\varepsilon}_0 = 3\eta^0(\mathrm{t})$
Equibiaxial	1	$\sigma_{11} = \sigma_{22} = \sigma$, $\sigma_{33} = 0$	$\mu_e^0(\mathrm{t}) = \sigma/\dot{\varepsilon}_0 = 6\eta^0(\mathrm{t})$
Planar	0	$\sigma_{11} = \sigma$, $\sigma_{22} = \sigma/2$, $\sigma_{33} = 0$	$\mu_{1p}^0(\mathrm{t}) = \sigma_{11}/\dot{\varepsilon}_0 = 4\eta^0(\mathrm{t})$
			$\mu_{2p}^0(\mathrm{t}) = \sigma_{22}/\dot{\varepsilon}_0 = 2\eta^0(\mathrm{t})$

with

$$\int_0^t G(t')\,\mathrm{d}t' = \eta^0(t) \tag{12.16}$$

$\eta^0(t)$ is the *time-dependent linear shear viscosity* (cf. Eq. 11.4). Equations (12.15) are discussed in the following for the special cases of uniaxial, equibiaxial, and planar deformation. In the case of an uniaxial deformation in x_1-direction, $\dot{\varepsilon}_{11} = \dot{\varepsilon}_0$ and $\dot{\varepsilon}_{22} = \dot{\varepsilon}_{33}$ and according to Eqs. (12.11) and (12.12), m = −0.5. An equibiaxial extension is defined by equal elongation rates in two directions. Such a condition is fulfilled for m = 1 in Eq. (12.11) resulting in $\dot{\varepsilon}_{11} = \dot{\varepsilon}_{22} = \dot{\varepsilon}_0$ and $\dot{\varepsilon}_{33} = -2\dot{\varepsilon}_0$. A planar deformation is defined by a constant geometry in one direction, for example $\dot{\varepsilon}_{22} = 0$, i.e. m = 0, and from Eq. (12.12) $\dot{\varepsilon}_{11} = \dot{\varepsilon}_0$ and $\dot{\varepsilon}_{33} = -\dot{\varepsilon}_0$ follows. Any other type of deformation lying between these well--defined modes can be described by appropriate values of m between −0.5 and 1. The values of m for the extensions discussed are listed in Table 12.1

For the uniaxial deformation, i.e., m = −0.5, Eqs. (12.15) become

$$\begin{aligned}\sigma_{11} + p &= 2\dot{\varepsilon}_0\eta^0(t)\\ \sigma_{22} + p &= -\dot{\varepsilon}_0\eta^0(t)\\ \sigma_{33} + p &= -\dot{\varepsilon}_0\eta^0(t)\end{aligned} \tag{12.17}$$

From the conditions $\sigma_{22} = \sigma_{33} = 0$, it follows $p = -\dot{\varepsilon}_0\eta^0(t)$ and, therefore,

$$\sigma_{11} = 3\dot{\varepsilon}_0\eta^0(t) \tag{12.18}$$

$$\sigma_{11}/\dot{\varepsilon}_0 = \mu^0(t) = 3\eta^0(t) \tag{12.19}$$

In the linear range of deformation, the uniaxial time-dependent elongational viscosity $\mu^0(t)$ is three times the time-dependent shear viscosity. For the steady-state values this relationship was first found by Trouton and, therefore, is often named after him.

For the equibiaxial deformation, i.e., $m = 1$ and $\sigma_{33} = 0$, Eqs. (12.15) read

$$\begin{aligned} \sigma_{11} + p &= 2\dot{\varepsilon}_0 \eta^0(t) \\ \sigma_{22} + p &= 2\dot{\varepsilon}_0 \eta^0(t) \\ p &= -4\,\dot{\varepsilon}_0 \eta^0(t) \end{aligned} \tag{12.20}$$

Introducing the last of Eqs. (12.20) into the first two ones results in

$$\sigma_{11}/\dot{\varepsilon}_0 = \sigma_{22}/\dot{\varepsilon}_0 = \mu_e^0(t) = 6\eta^0(t), \tag{12.21}$$

i.e., the time-dependent linear equibiaxial elongational viscosity μ_e^0 is six times the time-dependent linear shear viscosity.

For the planar deformation, i.e., $m = 0$ and $\sigma_{33} = 0$, Eqs. (12.15) become

$$\begin{aligned} \sigma_{11} + p &= 2\dot{\varepsilon}_0 \eta^0(t) \\ \sigma_{22} + p &= 0 \\ p &= -2\,\dot{\varepsilon}_0 \eta^0(t) \end{aligned} \tag{12.22}$$

Introducing the last of Eqs. (12.22) into the first and second one gives the tensile stresses in the 1- and 2- direction, i.e., $\sigma_{11} = 2\sigma_{22}$ and the two planar viscosities follow as

$$\sigma_{11}/\dot{\varepsilon}_0 = \mu_{p1}^0(t) = 4\eta^0(t) \tag{12.23}$$

and

$$\sigma_{22}/\dot{\varepsilon}_0 = \mu_{p2}^0(t) = 2\eta^0(t) \tag{12.24}$$

which are four or two times, respectively, the time-dependent zero-shear viscosity.

The various viscosities are summarized in Table 12.1. They follow from the theory of linear viscoelasticity. That means they are only valid for very low elongational rates or small deformations and do not describe the often significant non-linear behavior of the elongational viscosity for polymer melts. The fulfillment of the linear relation is a good check, however, for the reliability and accuracy of the experimental method applied (cf. Sect. 12.5).

12.3 Experimental Devices for Uniaxial Extension

The determination of the stress–strain curves of materials in their solid state is a standard practice of their mechanical characterization. The tensile machines used for these purposes are not suitable for the measurement of the extensional behavior of polymer melts, however, for two main reasons.

One problem is fixing the samples to the measuring device as the melt flows out of the usual clamps. Another is the avoidance of sagging of the specimen under gravity. Furthermore, very sensitive force transducers are needed due to the low tensile stresses and small cross-sections occurring at the high deformations necessary for the characterization of polymer melts and last not least sophisticated control units are necessary to perform experiments at constant Hencky elongation rates (cf. Eq. (12.4)) or real constant stresses (cf. Eq. (12.1)).

For all these reasons, special devices for investigating elongational properties of polymer melts were designed. The experimental developments during the 1970s can be regarded as the base of the remarkable progress in better understanding the extensional rheology of polymer melts.

Some basic devices are shortly described in the following.

12.3.1 Elongational Rheometer After Meissner

Pioneering experimental work on the elongational behavior of polymer melts was performed by Meissner. The device he built up is presented in Fig. 12.2 [1, 2].

The preferably cylindrical polymer sample of the length L_0 is drawn by two pairs of toothed wheels Z_1 and Z_2 which counter rotate to each other with n_1 and n_2 revolutions per second, respectively. According to Eq. (12.4), the elongational rate follows as

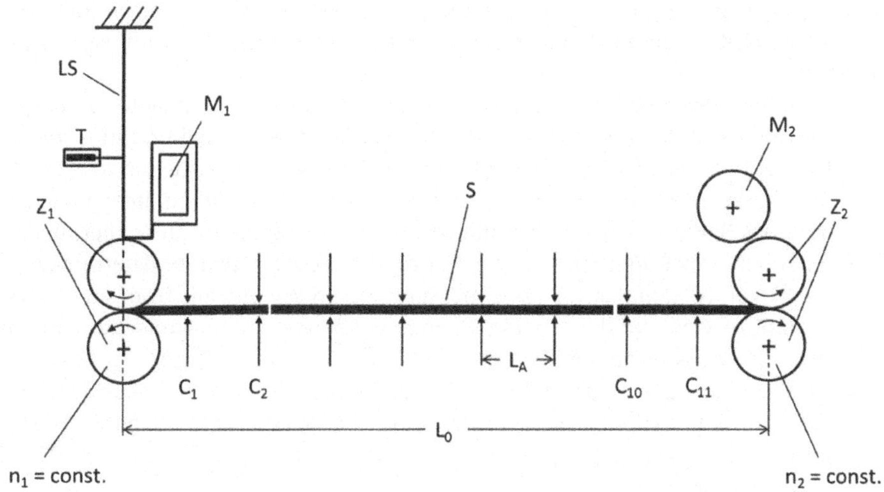

Fig. 12.2 Principle of the elongational rheometer developed by Meissner [1, 2]. *LS* leaf spring; *T* transducer; M_1, M_2 driving units; Z_1, Z_2 pairs of toothed wheels; *S* sample with length L_0; C_1 to C_{11} positions of 11 pairs of scissors, L_A distance between two scissors

$$\dot{\varepsilon} = \pi\, d\, (n_1 + n_2)/L_0 \tag{12.25}$$

with d being the diameter of the toothed wheels.

Due to the unchanged length of the sample during deformation, constant elongation rates can be achieved in a very simple way by keeping the revolutions of each pair of the wheels constant which is easily done by using commercially available driving motors M_1 and M_2. Z_1 and M_1 are suspended on an elastic leaf spring LF the deformation of which is measured by the transducer T and can be related to the tensile forces F by calibration. Assuming a uniform deformation of the sample and volume constancy, the actual cross-section $A(t)$ follows from Eq. (12.2) as

$$A(t) = A_0 \exp(-\varepsilon_H) = A_0 \exp(-\dot{\varepsilon}_0 t) \tag{12.26}$$

making use of the definition of the Hencky strain in Eq. (12.3). Relating the measured force to the actual cross-section, the elongational viscosities for given constant elongation rates can be determined according to Eq. (12.5).

At any state of the experiment, the stretched sample can be cut into pieces by scissors located at equal distances at 11 positions C_1 to C_{11} (cf. Fig. 12.2). The 10 strands serve two purposes. By weighing them, some information about the uniformity of the sample deformation can be obtained and from measuring their lengths after annealing, the recoverable portion of the total elongation can be determined according to Eq. (12.7).

The sample floats on a silicone oil bath which is heated to the measuring temperature. Together with a glass lid covering the oil bath it provides the required temperature constancy along the specimens up to 800 mm in length. In addition, the gravity of the sample is compensated by buoyancy and, therewith, sagging is avoided.

This unique machine, although large in size and not easy to handle, possesses great advantages. At various constant elongation rates, drawing ratios up to $\lambda = 400$ could be reached [3] which correspond to Hencky strains up to $\varepsilon_H = 6$. Accurate force measurements at such high elongations resulting in very small cross-sections and a control of the uniformity of elongation are the experimental base for reliable measurements. Furthermore, the recovery can be determined by averaging over 10 samples for each measurement. Very elegant from an experimental point of view are the rotating clamps which convey the stretched material away from the measuring area.

Clear disadvantages are the relatively large amounts of material necessary and the tedious work to extrude samples of uniform geometry over the lengths needed. Furthermore, the maximum measuring temperatures are limited to around 200 °C due to the lack of suitable silicone oils.

12.3.2 Rheometrics Extensional Rheometer

To overcome the difficulties mentioned above, Meissner designed a rheometer the principle of which is sketched in Fig. 12.3 [4]. It became the base of a commercial machine. The rotating clamp technique is preserved but a new type of clamp is used. The toothed wheels of the former design are replaced by metal conveyor belts. One clamp is mounted to a leaf spring the mechanical deformation of which is proportional to the force applied. It was shown that very low forces could be measured with high accuracy. Like for all the extensional apparatuses, the performance of the force transducer is one clue for reliable viscosity measurements at high elongations at which the sample cross-sections become very small. A detailed description of the device and results on different materials can be found in [4].

The essential improvements in comparison to the rheometer sketched in Fig. 12.2 are the following. The sample length is significantly reduced. Instead of cylindrical strands of several hundred millimeters, rectangular plates about 60 mm long, 10 mm wide, and 2 mm thick are used which can be prepared from compression-molded sheets. These dimensions enable investigations on samples which are available on the scale of grams, only.

The oil bath is avoided. The sample floats on a cushion of an inert gas which does not interact with the polymer material investigated. The gas stream supporting the molten sample is achieved by pressing the preheated gas through a flat metal frit.

Using inert gases as the supporting media and applying electrical heating, the temperatures are not restricted any longer to 200 °C, but can be chosen distinctly higher. Inert gases have the additional effect that they reduce the thermal degradation of the polymer samples. A good temperature constancy is reached by an electrical heating of the walls of the chamber in which the whole measuring device is located.

A window allows the visual observation of the sample deformation and its recording by a video system. Information of this kind is essential insofar, as the uniformity of sample deformation belongs to the most important preconditions for

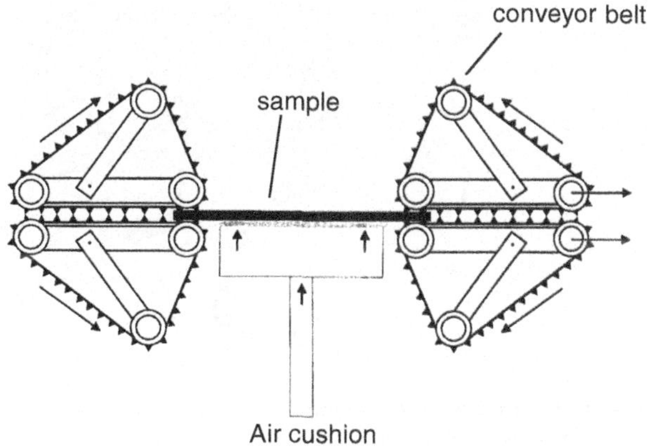

Fig. 12.3 Principle of the Rheometrics Extensional Rheometer (RME) [4]

reliable experimental results (cf. Sect. 12.7). On the other hand, slip of the melt at the rotating clamps can occur which may reduce the real elongation rate. Optical measurements using particle tracking are applied to determine the real deformation rates.

In spite of its sophisticated design and the broad range of experimental parameters, the RME has not become an instrument for the routine laboratory work.

12.3.3 Sentmanat Extensional Rheometer

The work horse in the field of extensional rheometry is based on a design developed by Sentmanat [5]. Its principle is represented in Fig. 12.4. A polymer sheet is clamped to two drums which counter rotate to each other driven by the shaft of the "master drum" the rotation of which is transferred to the "slave drum" by intermeshing gears as shown in Fig. 12.4. From the torque exerted on the "master drum," the tangential force acting on the sample can be determined. This device gets its popularity from the fact that it can be built into the housings of commercial rotational rheometers making use of their existing heating chamber, driving unit, and torque sensor.

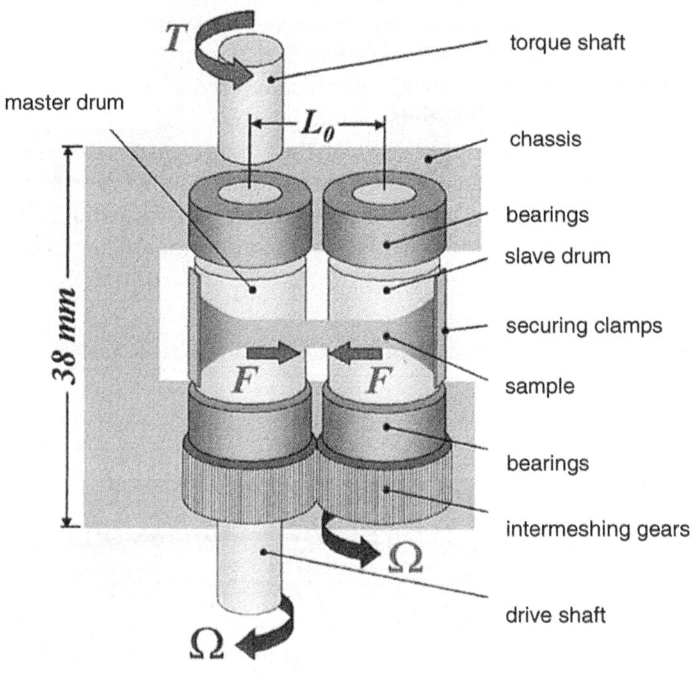

Fig. 12.4 Principle of the Sentmanat Extensional Rheometer (SER). L_0 unsupported length; Ω drive shaft rotation; T torque; F tangential force [5]

Several rheometer manufacturers offer this equipment commercially. It has proven its performance in the field of routine measurements, but its contribution to fundamental research has been very limited up to now for several reasons. The one is that sagging of the samples cannot be avoided even for the geometries chosen as a supporting medium is not applied. This effect is particularly pronounced in the case of samples with low viscosities. It reduces its versatility for measurements at high temperatures which, in principle, could easily be achieved by making use of the ovens of commercial rheometers. Another deficiency may arise from a too high friction of the bearings and the intermeshing gears which negatively affects the measurement of the torque acting on the sample. Furthermore, the uniformity of the sample deformation cannot be checked which is very important, however, for an assessment of the accuracy of the elongation rate and the tensile stress and, therewith, the elongational viscosity.

Measurements of the recovery of the sample are not possible due to the principal design of the device.

12.3.4 Filament Stretching Rheometer

Around the year 2000 several devices were presented in the literature which made use of the stretching of a sample between two coaxial cylindrical clamps by holding one clamp fixed and moving the other one apart. Most of these equipments were designed for measuring the elongational behavior of polymer solutions at room temperature. Some of them are described in a review article by Mc Kinley and Sridhar [6]. This principle is the base of a rheometer developed for polymer melts [7]. The cylindrical sample is stretched between two plates. The lower one is fixed and the upper one driven in a way that the radius in the middle of the sample decreases exponentially with time in order to keep up a constant Hencky strain rate. This assembly is placed in a temperature-controlled oven with glass covered slits allowing the measurement of the sample diameter by a laser device. The exerting force is registered by a load cell mounted to the lower plate. The upper plate and the laser are driven by the same motor. Using an appropriate gear, the laser is moved at half the velocity of the upper plate which results in a position just in the middle between the two plates. If the symmetry is preserved during the experiment, then the elongational viscosity at the given constant elongation rate can be determined. Typical sample geometries of initial diameters around 10 mm and lengths between 2.5 and 4.5 mm [7] lead to small aspect ratios which are favorable for high Hencky strains but are not free of problems regarding the fastening and the uniform deformation of the specimens.

This technique is not easy to handle and needs a lot of experimental skill. It is limited to samples of relatively high viscosities as sagging may come into play due to the absence of a supporting medium. Therefore, the filament stretching rheometer is not widely used for investigating elongational properties of polymer melts.

12.3.5 Tensile Rheometer After Münstedt

This type of rheometer goes back to the principle of tensile testing machines but avoids their disadvantages with respect to polymer melts which were shortly addressed in the introduction to this section. Cogswell [8] can be named to publish a first version using a load via a specially constructed mechanical cam to apply a constant actual stress on a sample suspended in a silicone oil bath to counterbalance the gravity of the sample by buoyancy. The one disadvantage of this creep rheometer was that a special cam had to be used for each initial sample length, the other that due to a certain friction of the air bearing supporting the rotation of the cam and the unavoidable moment of inertia, reliable experiments were only possible within a very limited range of stresses.

Based on the idea of a vertically suspended sample, Münstedt and co-workers developed a versatile instrument [9, 10] which has been used to characterize the elongational properties of a great number of polymer materials, particularly, in dependence on their molecular structure (see Sect. 13.3). For such kind of investigations the rheometer is particularly suited as comparably small amounts of samples are needed and, therefore, the elongational behavior of polymeric materials can be investigated which are available on a laboratory scale, only. As cylindrical samples are used, they can comfortably be prepared by extrusion through appropriate capillaries or the compression of a molten powder in preformed glass tubes which are removed after solidification. For typical sample diameters of around 5 mm and lengths between 10 and 25 mm, small amounts of material are sufficient to get comparative results on the elongational behavior of polymers. The problem of fixing the samples to the measuring device was solved by gluing them to metal clamps which are easy to handle. The gluing needs some know-how, but by adequate preparations using etching or plasma pretreatments a wide scope of polymers could be investigated.

The function of the rheometer becomes obvious from Fig. 12.5. The sample is vertically suspended in a silicone oil bath of a density adapted to that of the polymer melt. The one end is fixed to the force transducer and the other to the pulling device. Besides compensating the weight of the sample by buoyancy, the oil serves as a medium transferring the heat from the hollow wall of the glass vessel to its inner part which houses the sample during deformation. The heating is performed by an outside unit, pumping another temperature-controlled silicone oil through the hollow wall. An evacuated outer zone around the vessel provides a good thermal insulation without disturbing the visual observation of the sample. By such a device, a temperature gradient not larger than 1 °C over a sample length of 500 mm up to a measuring temperature of 200 °C could be obtained. Temperature constancy is an important condition for reliable elongational measurements as the viscosity of polymer melts is very temperature sensitive.

The vessel can easily be lowered and, thereby, opens up the possibility to freeze the samples at arbitrary stages of deformation and to remove them for further investigations like a check of the uniformity of sample elongation by manually

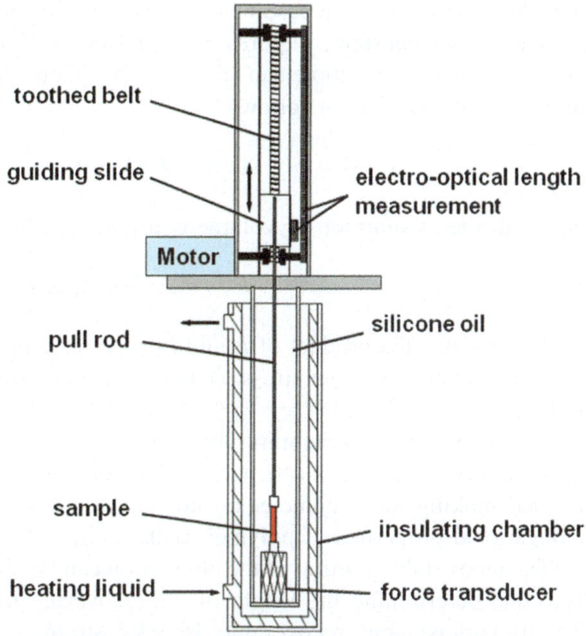

Fig. 12.5 Principle of the tensile rheometer developed by Münstedt (MTR)

measuring the diameters along the length or a screening of morphology changes in the case of polymer blends, for example.

Elongational rates between 0.001 and 5 s^{-1} can be applied. The maximum Hencky strain reaches a value of 4. The driving unit consists of a servomotor which moves the pulling rod by means of a toothed belt. The rod is guided by a sledge. The elongation of the sample is measured by a photoelectric transducer. The force resolution is around 0.2 mN. The drift of the force transducer is very small as it is directly located in the temperature-controlled oil bath.

Since the sample length does not remain constant as for the devices by Meissner, but becomes larger during stretching, the velocity v has to be enhanced proportional to the actual length according to Eq. (12.4) in order to apply a given constant Hencky strain rate, i.e.,

$$v = \dot{\varepsilon}_0 l \tag{12.27}$$

This relationship corresponds to an exponential increase of v with time as according to Eq. (12.3)

$$l = l_0 \exp\ \varepsilon_H = l_0\ \exp\ \dot{\varepsilon}_0\, \mathrm{t} \tag{12.28}$$

The proportionality between v and l is relatively easy to achieve by a corresponding electronic unit controlling the rotational speed of the servomotor.

The rheometer described so far has the unique feature that creep experiments can be performed which have the advantage of reaching a steady-state at smaller deformations than in the case of stressing experiments (cf. Sect. 12.5.1). Keeping the true stress constant, requires a special control of the force, however, as the cross-section of the sample can significantly change during elongation. In the case of a constant stress σ_0 it follows

$$\sigma = F/A = \sigma_0$$

and with the assumption of volume constancy during deformation

$$F = \sigma_0 A = \sigma_0 A_0 l_0 / l \tag{12.29}$$

That means, the product of F and l has to be kept constant in order to perform experiments at a constant stress. Details of the control loops applied can be found in [11].

By adequately programming the electronic control of the driving unit, relaxation and retardation experiments after various deformation histories can be performed making the rheometer a rather universal device for the investigation of elongational properties of polymer melts.

The recoverable portion of the elongation can be obtained by setting the force to zero and determining the length of the recovering sample as a function of time. The time-dependent recoverable Hencky strain ε_r follows then according to Eq. (12.7). However, this mode can be performed with a sufficient accuracy, only, if the recovery is not too fast. Easier to perform is the determination of the steady-state of the recovery.

As in all elongational experiments the uniformity of the sample deformation is the prerequisite for reliable experiments, an optical device was installed which makes it possible to get in situ images of the sample at different states of extension. Its principle features are described in [12]. The core is a high-resolution telecentric lens with the entrance pupil located at infinity. Its advantage is that in parallel light it delivers frames with very uniform brightness, and distortions, perspective errors, and edge position uncertainties are very small. This performance is a good base for the image acquisition by a video camera which is digitally synchronized with the rheometer driving unit. More detailed information on this technique can be found in [12, 13].

12.4 Experimental Devices for Multiaxial Extension

Biaxial and planar deformations play an important role in processing operations. In spite of this fact, the development of devices being able to measure rheological properties of polymer melts under such conditions has not attained a state comparable to that for the investigation of uniaxial flow. The reasons for this situation are the demanding experimental conditions which have to be fulfilled for getting reliable results. Furthermore, the range of elongational rates is rather limited.

12.4.1 Hachmann-Meissner Rheometer

The most advanced and versatile apparatus for measuring multiaxial properties is an apparatus described in detail in [14]. The basic elements of this rheometer are the rotating clamps already introduced in the presentation of the RME. Metal belts driven by a separate motor each and equipped with separate force transducers are used. The sample is located above a sintered bronze table and supported by a temperature-controlled inert gas flowing through it. The whole device is surrounded by a temperature-controlled housing. Eight clamps can symmetrically be arranged in a circle. In the case of an equibiaxial test, they rotate at the same speed and pull the different parts of the sample with an equal elongational rate as required per definition. For a planar extension, two clamps are placed opposite to each other. They do not rotate but measure the force necessary to keep the sample fixed. Rectangular to them, three parallel belts on each side deform the sample at the same rate. As two different stresses occur, the planar deformation is determined by two viscosities (cf. Eqs. 12.23 and 12.24).

The rheometer is extremely versatile as, in principle, uniaxial experiments can be performed by stretching samples between two clamps located opposite to each other.

A window allows the observation of the sample. For the different modes, the optical tracking of small particles put on the surface of the sample allows the control of the uniformity of the extension which is essential for reliable experiments.

Elongational rates between 0.003 and 0.1 s^{-1} are reported in the literature for the equibiaxial and the planar extension while total Hencky strains up to three and five, respectively, are documented [14].

Due to the complexity of the method, only one prototype does exist up to now. The experimental results it delivered so far are valuable contributions to the field of elongational rheology.

12.4.2 Lubricated Squeeze-Flow Rheometer

The physical principle of this type of multiaxial extensional rheometer is relatively simple. Two parallel plates with lubricated inner surfaces centered along one axis are moved toward each other and according to the geometry chosen a planar or an equibiaxial deformation is achieved. If the melt is squeezed between two plates and not restricted in its flow, the extension is equibiaxial; in the case of corresponding limiting walls it is planar. One prerequisite to avoid any shear deformation during squeezing is the careful lubrication of the walls in contact with the melt. The basics of a commercial biaxial rheometer are described in detail in [15], for example. The disk-shaped sample is located between two parallel plates. One of them is moved with respect to the other which is fixed squeezing the material out of the rim. The gap height can be changed in a controlled way by a suitable driving unit and the force is measured by a load cell. Relating the force to the cross-section of the sample, which remains constant during deformation, results in the stress. The viscosity is calculated then in the usual way.

The few experiments found in the literature were performed at constant elongational rates which can be obtained in analogy to Eq. (12.4). In contrast to the increase of the sample length during uniaxial elongation with the MTR, the distance between the plates becomes smaller during biaxial deformation. Therefore, the velocity has to be controlled in a way that it exponentially decreases with time. Of great importance for the accuracy of the measurements is once more the uniformity of the deformation which is strongly dependent on the perfection of the lubrication. It can be monitored indirectly, only. Comparing a circular pattern on the sample before and after the experiment is a way of checking the uniformity of the deformation.

12.5 Fundamental Features of Uniaxial Flow

12.5.1 Comparison of Basic Experiments at Constant Elongational Rate or Tensile Stress

For fundamental uniaxial experiments, a low density polyethylene named "IUPAC A" in the literature was used. Its density is 0.918 g/cm^3 at room temperature and its zero-shear viscosity $\eta_0 = 5.0 \times 10^4$ Pa at 150 °C. The thorough molecular and rheological characterizations of this material are documented in [16].

Extensional experiments on the LDPE IUPAC A at two different constant elongational rates are presented in Fig. 12.6 [3]. They were performed with the rheometer after Meissner described in Sect. 12.3.1. The initial length of the rod-like sample was 650 mm. It was elongated up to a stretching ratio of 400 which corresponds to a maximum Hencky strain of six. The uniformity of the sample deformation was determined by cutting the strand after stretching and weighing the various cuts obtained. At the elongational rate of $\dot{\varepsilon}_0 = 0.03\ \mathrm{s}^{-1}$ and the temperature of 150 °C, the diameters of 10 cuts along the stretched sample deviate from each other by maximal 2 % up to $\varepsilon_H = 5$ and by 5 % at $\varepsilon_H = 6$ (cf. [3]). These relatively small differences stand for an excellent uniformity of the samples whose data are presented in Fig. 12.6.

The curves in Fig. 12.6 demonstrate the typical tensile behavior of an LDPE melt in stressing experiments at the constant elongational rates of $\dot{\varepsilon}_0 = 0.03\ \mathrm{s}^{-1}$ and $\dot{\varepsilon}_0 = 0.1\ \mathrm{s}^{-1}$. After a steep increase of the stress as a function of strain at the beginning of the experiment, a region with a growing slope of the stress–strain curve follows for $1 < \varepsilon_H < 3$. Such a behavior is called strain hardening. For a polymer melt it was reported for the first time by Meissner [17]. For $\varepsilon_H > 4$, the stress remains constant within the accuracy of the experiments. The different symbols mark various samples and demonstrate the satisfactory reproducibility of

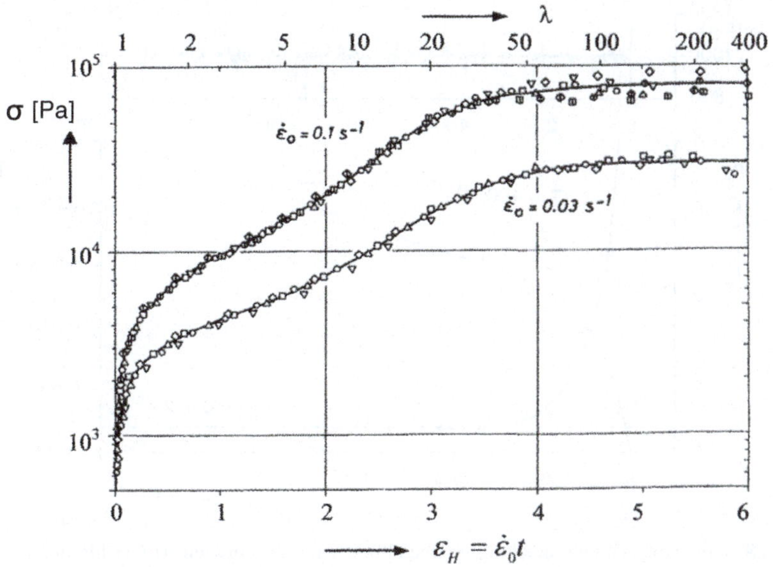

Fig. 12.6 Tensile stress σ as a function of the Hencky strain ε_H for an LDPE (IUPAC A) at two constant elongational rates $\dot{\varepsilon}_0$ and $T = 150$ °C. The different symbols mark experiments on different samples [3]. λ is the stretching ratio

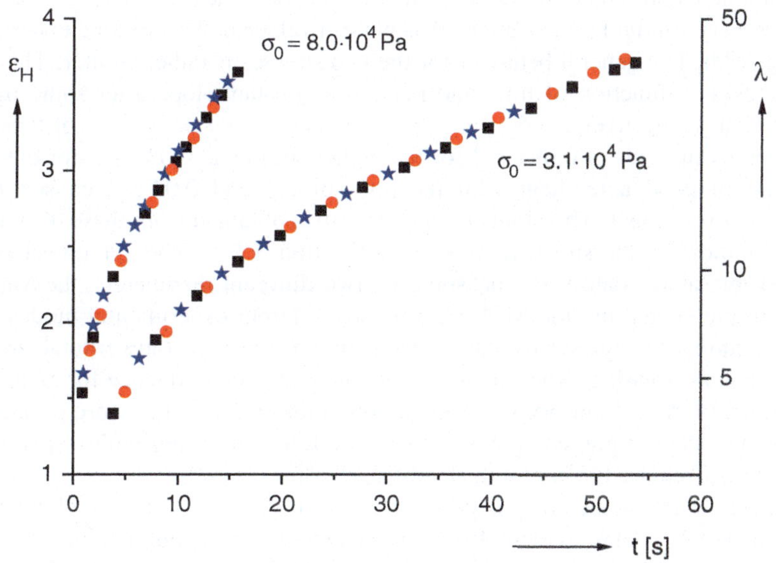

Fig. 12.7 Hencky strain ε_H as a function of time t for an LDPE (IUPAC A) at two constant stresses σ_0 and $T = 150$ °C. The different symbols represent measurements on different samples [3]. λ is the stretching ratio

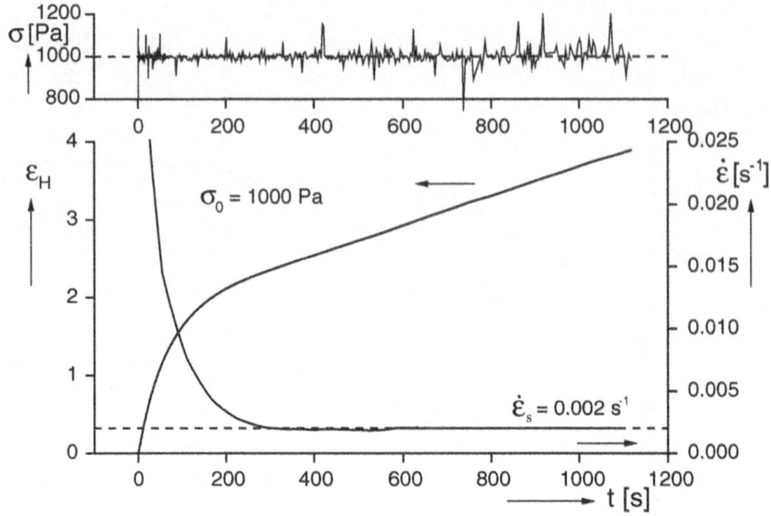

Fig. 12.8 Creep experiment on a long-chained branched polypropylene (LCB-PP) at T = 180 °C [11]. $\dot{\varepsilon}$ stands for the elongational rate obtained by the differentiation of the creep curve

the measurements, particularly, if one takes the large stretching ratio of 400 into account. At the higher elongational rate a larger stress level is reached.

The reliability of the values in the steady state gets some support from the measurements presented in Fig. 12.7. It exhibits creep measurements performed at stresses very similar to those attained as plateau values in the stressing experiments of Fig. 12.6. The general behavior for the two stresses is rather similar. The strain increases as a function of time and reaches a constant slope after some time or elongation, respectively, indicating the attainment of a steady-state of flow. The constant strain rates of 0.098 s^{-1} for the higher stress and 0.033 s^{-1} for the lower one are in good agreement with the rates of 0.1 and 0.03 s^{-1} chosen in the stressing experiment. The elongational viscosities obtained from these data are the same as those of the stressing experiments within ±3 %. The agreement is very remarkable as the data were measured by two different rheometers, the Meissner oil bath rheometer and the MTR, and in two different experimental modes.

The finding that the steady state of elongation is reached at lower total strains in creep than in stressing experiments has an interesting consequence for extensional experiments. As the uniformity of the deformation becomes the more problematic the thinner the sample, creep experiments are less demanding with respect to the total strains required to measure steady-state values.

The great advantage creep experiments offer if the steady state in elongation is investigated becomes evident from measurements on a long-chained branched polypropylene which are presented in the Figs. 12.8 and 12.9. In Fig. 12.8, a creep experiment is presented at a stress of 10^3 Pa the constancy of which is demonstrated in the upper part of the figure. The spikes are due to the electronic control unit and do not have an influence on the sample deformation. As can be seen from

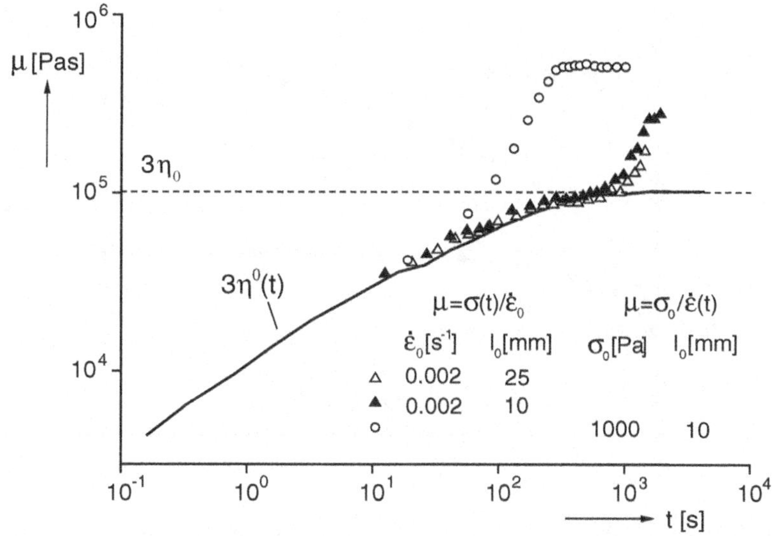

Fig. 12.9 $\mu(t) = \sigma(t)/\dot{\varepsilon}_0$ as a function of time t for a long-chain branched polypropylene at $T = 180$ °C and a constant elongation rate of $\dot{\varepsilon}_0 = 0.002$ s^{-1} [11] in comparison with $\mu(t) = \sigma_0/\dot{\varepsilon}(t)$ from the creep experiment in Fig. 12.8 (*open circles*). The creep test was performed with a constant stress of $\sigma_0 = 10^3$ Pa and attained a steady-state elongational rate of $\dot{\varepsilon}_0 = 0.002$ s^{-1}. The *full line* describes three times the time-dependent zero-shear viscosity. Two initial sample lengths l_0 were used for the stressing experiments

the elongational rate as a function of time, the steady-state of flow is reached at around 350 s corresponding to a total elongation of $\varepsilon_H = 2.5$. Over the duration of the experiment, i.e., up to a creep time of 1100 s or a total elongation of 3.8, there is no indication of a deviation from the constant elongational rate attained. The elongational rate as a function of time was calculated from the derivative of the spline used as a numerical description of the creep curve.

From the results of Fig. 12.8 $\sigma_0/\dot{\varepsilon}(t)$ is determined and plotted as a function of time in Fig. 12.9. This quantity reaches a pronounced stationary state from which a steady-state elongational viscosity of $\mu_s = 5 \times 10^5$ Pas follows.

As can be seen from Fig. 12.9, too, in the stressing experiment performed at a constant strain rate of $\dot{\varepsilon}_0 = 0.002$ s^{-1} up to the maximum total strain of $\varepsilon_H = 3.8$ a steady state of $\sigma(t)/\dot{\varepsilon}_0$ is not reached. The value of 3×10^5 Pas at this position is still distinctly lower than μ_s in the steady state of the creep experiment. Furthermore, the shapes of the two viscosity functions are totally different. From the two curves in Fig. 12.9 it becomes evident that the material property viscosity which should be independent of the experimental mode, of course, can be determined in the steady state of deformation, only. Therefore, to be precise the time-dependent ratios of stress and strain rate should be designated as *stress coefficients*. This distinction is rarely found in the literature and for the matter of convenience the expression "viscosity" is frequently used for the "stress coefficient", too.

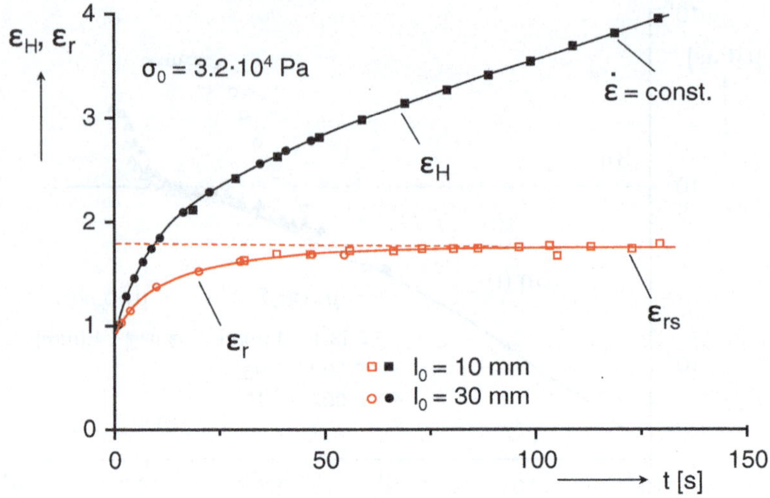

Fig. 12.10 Total strain ε_H and recoverable strain ε_r of an LDPE (IUPAC A) in creep recovery experiments performed at different creep times and $T = 125$ °C [18]. ε_{rs} is the steady-state recoverable strain, l_0 designates the initial sample length

Another quantity which can be determined from extensional experiments is the recoverable strain ε_r (cf. Eq. 12.7). In Fig. 12.10, this quantity measured in creep recovery experiments is shown as a function of time for the LDPE IUPAC A.

The creep experiments were performed with the MTR at $T = 125$ °C. Each plotted value of ε_r represents a sample elongated over different creep times after which the recovery to the stationary state was performed in the rheometer by setting the stress to zero. Compared with the measurement in Fig. 12.7 at nearly the same stress but at 150 °C, the constant strain rate is reached at distinctly longer creep times, the total elongations are very similar, however. The recoverable strain first increases but reaches its steady-state indicated by the plateau at a time which is somewhat higher than the corresponding time at which the creep curve attains its constant slope. The finding that the results on samples with two different initial lengths overlap each other is an experimental hint to the reliability of the experiments.

In Fig. 12.11, the recoverable strains attained in stressing experiments up to various elongations are plotted together with the corresponding curves for the stresses. Each circle of the curve of the recoverable strain stands for one sample and the different symbols of the stress–strain curve demonstrate the reproducibility of σ (ε_H) up to a Hencky strain of 6. Like in the creep experiments, a plateau ε_{rs} of the recoverable strain ε_r is reached within the accuracy of the experiments which goes along with that of the stress.

The creep and stressing experiments and the recoverable strains determined in these two different experimental modes, as well, convincingly demonstrate that a steady state of elongation exists for the LDPE investigated which is attained,

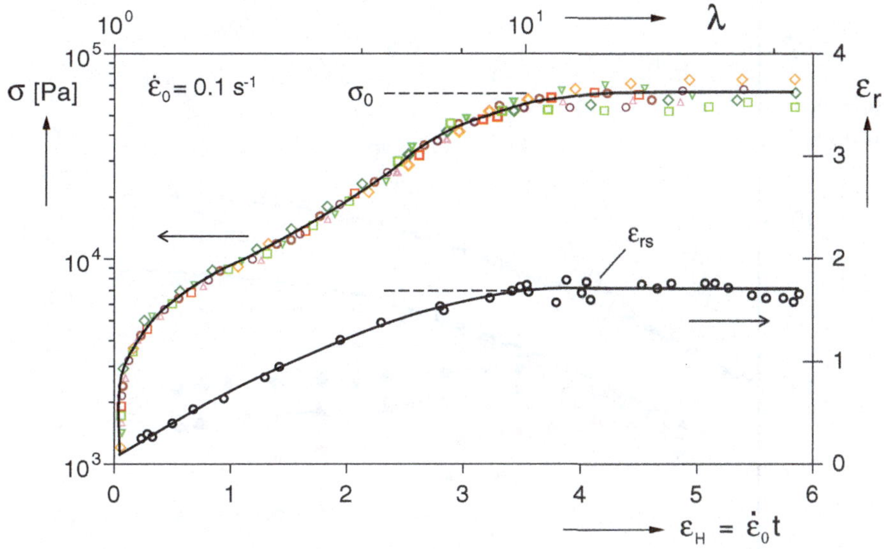

Fig. 12.11 Tensile stress σ and recoverable strain ε_r as functions of the total strain ε_H for stressing experiments at $\dot{\varepsilon}_0 = 0.1\ \mathrm{s}^{-1}$ and $T = 150\ °\mathrm{C}$ on the LDPE IUPAC A [18]

however, at Hencky strains which are rather high, especially in the case of tests at constant elongational rates. Thus, special care has to be taken with respect to a reliable performance of the measurements and, in particular, to the uniformity of sample deformation.

12.5.2 Dependence of the Elongational Viscosity on Elongational Rate

The "elongational viscosity" defined as $\mu(t) = \sigma(t)/\dot{\varepsilon}_0$ is represented in the upper part of Fig. 12.12 as a function of time in a double-logarithmic plot for constant elongational rates $\dot{\varepsilon}_0$ covering four decades from 0.001 to 10 s^{-1}.

The measurements of the stress σ as a function of the total elongation ε_H at $\dot{\varepsilon}_0 = 0.03\ \mathrm{s}^{-1}$ and $\dot{\varepsilon}_0 = 0.1\ \mathrm{s}^{-1}$ from Fig. 12.6 are now presented as $\mu(t)$ on a logarithmic time scale. The extended plateaus of Fig. 12.6 shrink to a small region due to the logarithmic time scale used. From such a presentation it may be difficult to decide whether a steady-state viscosity is reached, but taking Fig. 12.6 into account there is no doubt about its existence.

The presentation of Fig. 12.12 has the advantage, however, that the development of the elongational viscosity as a function of time becomes very obvious before it reaches its steady-state. At small elongational rates or short times, all the measurements in Fig. 12.12 come to lie on the curve $3\eta^0(t)$ indicating a linear behavior according to Eq. (12.19). This agreement is a hint to the reliability of the

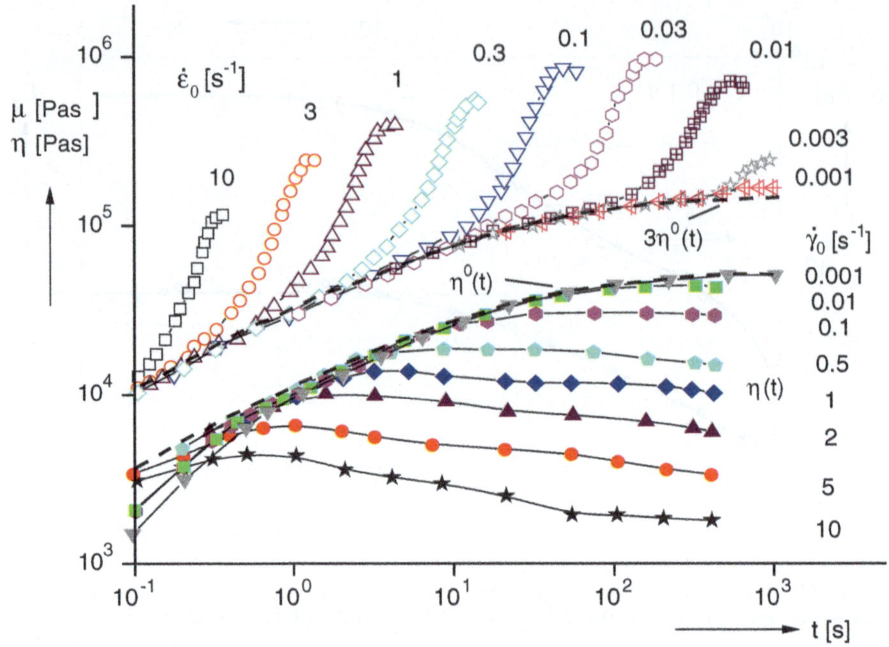

Fig. 12.12 Comparison of the time dependence of elongational viscosity μ and shear viscosity η for the LDPE IUPAC A at $T = 150$ °C and different elongational rates $\dot{\varepsilon}_0$ (*open symbols*) and shear rates $\dot{\gamma}_0$ (*closed symbols*) [19, 20]

extensional experiments. The linear range extends not further than $\varepsilon_H \approx 0.7$ indicating that for the LDPE investigated the linear regime is rather small in comparison to $\varepsilon_H \approx 4$ necessary to reach the steady-state.

The increase of the elongational viscosity as a function of time or strain, respectively, is a feature typical of the elongational flow of the LDPE investigated as becomes clear from the lower part of Fig. 12.12. It shows the shear viscosity as a function of time at various constant shear rates, taken from [19] and already presented in Fig. 11.3. After a weak maximum in the case of higher shear rates, the shear viscosities decrease as a function of time and approach constant values which, starting from the shear-rate independent viscosity $\eta^0(t)$ in the linear range, become smaller with increasing shear rate as it is well known from the shear thinning of viscosity functions of polymer melts (cf. Sect. 11.5.1).

From this comparison of the time dependencies of the shear and elongational viscosities, it becomes obvious that the elongational behavior in the non-linear range of deformation cannot generally be derived from shear data. Only in the linear range, the simple relationship $\mu^0(t) = 3\eta^0(t)$ holds for uniaxial flow.

For the description of the increase of the elongational viscosity, the so-called strain-hardening factor

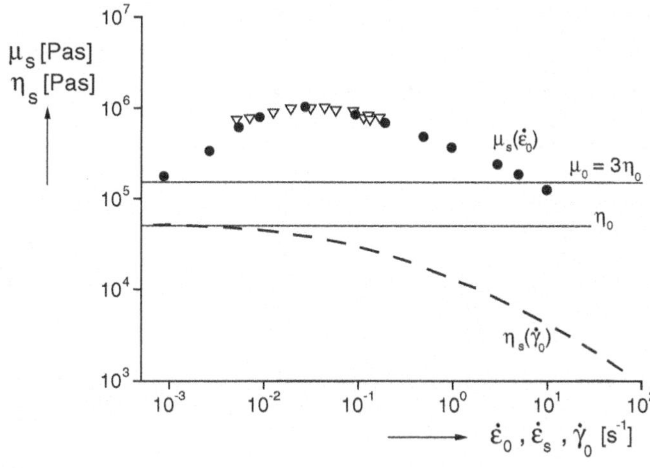

Fig. 12.13 Comparison of the steady-state elongational viscosity μ_s and shear viscosity η_s as functions of the corresponding rates [18] for the IUPAC A at 150 °C. The full circles represent data from stressing experiments, the open triangles those from creep tests

$$\chi\left((t, \dot{\varepsilon}_0\right) = \mu\left((t, \dot{\varepsilon}_0\right)/3\eta^0(\mathrm{t}) \tag{12.30}$$

is defined which is of some relevance for the assessment of the processing behavior of polymeric materials (cf. Sect. 17.5).

Besides the envelope of the elongational viscosity as a function of time which describes the linear viscoelastic limit of the deformation, the steady-state values can be regarded as distinguished data characterizing a material. For the LDPE IUPAC A presented in Fig. 12.12 the steady-state elongational viscosity μ_s is shown as a function of elongational rate in Fig. 12.13. The data obtained from stressing and creep experiments are in good agreement. Starting from $3\eta_0$ at small elongational rates, μ_s runs through a pronounced maximum which is seven times higher than the linear value. In addition, the shear viscosity as a function of shear rate is plotted in Fig. 12.13 showing the typical shear thinning behavior. The direct comparison demonstrates how different the two viscosity functions are. In elongation, the LDPE investigated exhibits strain-hardening (thickening) starting from the linear regime which is followed by strain thinning similar to that in shear.

12.5.3 Dependence of the Recoverable Tensile Compliance on Stress

Analogous to shear, the elasticity in elongation is described by a material function, the recoverable elongational compliance D_r defined by Eq. (12.8). As demonstrated in Fig. 12.10 and Fig. 12.11, the recoverable strain and following from that

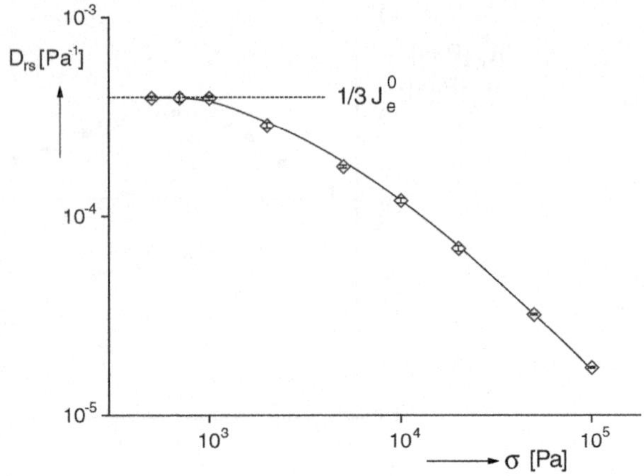

Fig. 12.14 Steady-state recoverable elongational compliance D_{rs} of an LDPE as a function of tensile stress σ obtained in creep experiments [21]. J_e^0 is the linear steady-state recoverable compliance in shear. The error bars within the symbols indicate the reproducibility of the experiments

the recoverable compliance is a function of time or elongation, respectively. In order to investigate the elasticity in the steady state of extension, creep experiments were performed as stationarity is reached at smaller total strains than in measurements at constant strain rates.

Comprehensive recovery experiments using the MTR rheometer were carried out on an LDPE with the same density but a higher molar mass than the LDPE IUPAC A [21]. To obtain time-independent values which are the base for correlations with molecular parameters it has to be made sure that steady states in the preceding creep as well in the following recovery experiment have been reached similar to creep experiments in shear (cf. Sect. 11.4).

In Fig. 12.14 the recoverable elongational compliance in the steady state D_{rs} is shown as a function of the stress previously applied. It starts from a plateau at small stresses indicating the linear regime. Its value is one-third of the linear steady-state recoverable shear compliance predicted from the theory of linear viscoelasticity. At higher stresses, D_{rs} decreases, i.e., values far above the linear range as in the case of the elongational viscosity of strain-hardening polymers are not found. The attainment of stationary states in elongation at the high stresses of Fig. 12.14 is demonstrated in [22].

12.5.4 Temperature Dependence

The temperature dependence of the extensional behavior of the LDPE IUPAC A was investigated by performing creep experiments at different temperatures. The creep compliance D defined by Eq. (12.6) is presented in the left part of Fig. 12.15 as a function of the time in a logarithmic scale at temperatures between 120 and 180 °C and a tensile stress of $\sigma_0 = 3.1 \times 10^4$ Pa. As the right part of Fig. 12.15 demonstrates, a master curve can be obtained by shifting the compliances with respect to the time axis.

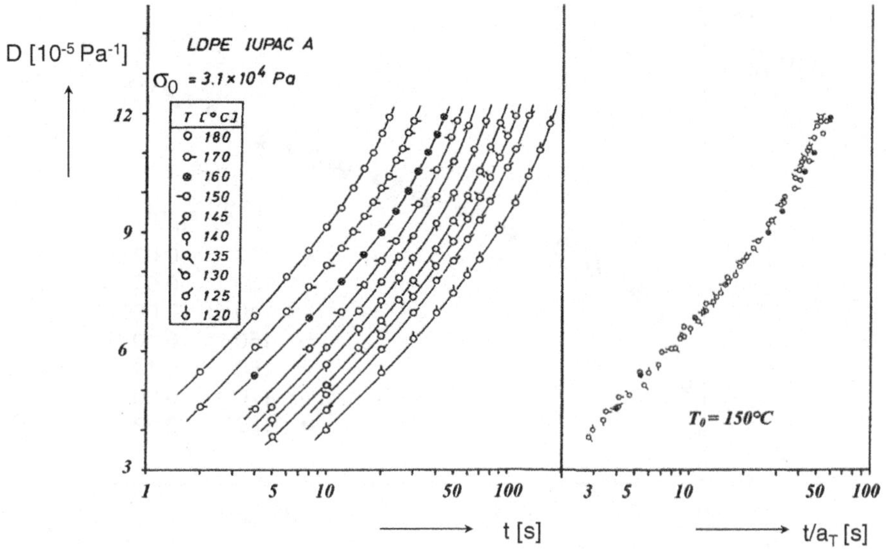

Fig. 12.15 Temperature dependence of the tensile creep compliance D as a function of the time t and its representation by a master curve at the reference temperature $T_0 = 150$ °C [20]. a_T is the shift factor

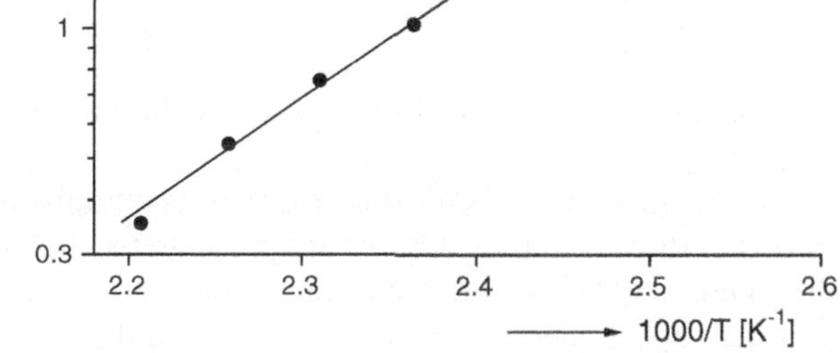

Fig. 12.16 Shift factor a_T as a function of the reciprocal absolute temperature 1/T for IUPAC A [18]. The reference temperature is $T_0 = 150$ °C. The symbols mark the results obtained from the creep experiments in Fig. 12.15, the *full line* represents a_T from shear experiments

In Fig. 12.16 the resulting shift factors a_T are plotted semi-logarithmically as a function of the reciprocal absolute temperature 1/T. They come to lie on the full line representing the factors obtained from the shift of viscosity functions in shear measured for the same LDPE (cf. Sect. 6.5). The agreement indicates that the time–temperature superposition is the same in shear and elongation. Similar results were found for a polystyrene [23]. This result supports the assumption that the temperature dependence of the molecular processes underlying the shear and elongational viscosity is the same.

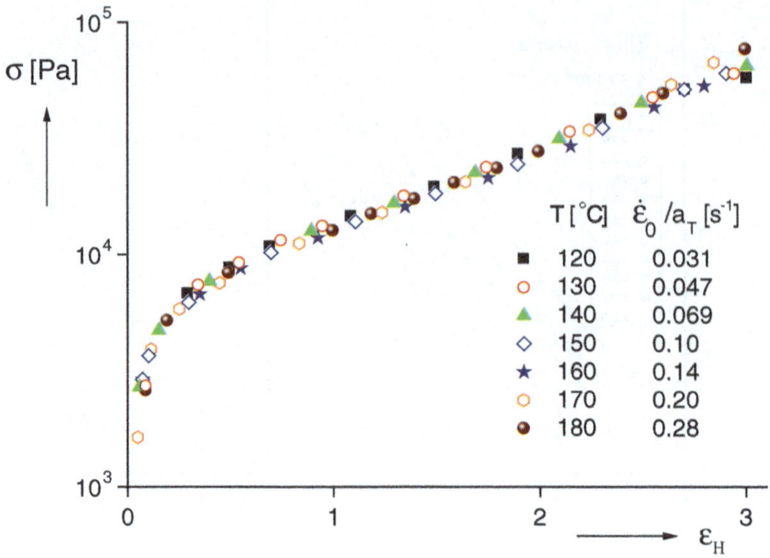

Fig. 12.17 Stress-strain curves for the LDPE IUPAC A at various temperatures and elongational rates chosen according to the corresponding shift factors. The elongational rate at the reference temperature $T_0 = 150$ °C is $\dot{\varepsilon}_0 = 0.1\ \text{s}^{-1}$ [20]. The shift factors are taken from Fig. 12.16

In constant elongational rate experiments, the validity of the time–temperature superposition principle can be checked in an indirect way, only. It implicates that the stress–strain curve measured at a temperature T_0 and a constant elongational rate $\dot{\varepsilon}_0(T_0)$ should remain unchanged at another temperature T if $\dot{\varepsilon}_0(T)$ is chosen according to

$$\dot{\varepsilon}_0(T) = \dot{\varepsilon}_0(T_0)/a_T \tag{12.31}$$

From Fig. 12.17 it is obvious that all the data measured come to lie on one curve.

In Fig. 12.18 the steady-state recoverable strains ε_{rs} measured in creep experiments at three different stresses are presented as functions of the temperature. For the LDPE IUPAC A it is found that ε_{rs} does not depend on temperature within the accuracy of the measurements. According to these results the elastic properties in elongation may be regarded to be temperature independent in good approximation.

These results have consequences interesting for the application of data of elongational flow. Using the shift factors from shear experiments which are easier to determine than in elongation, rheological properties determined at one temperature in extension, only, can be modeled for other temperatures. For example, extensional data comfortably measurable at temperatures and strain rates lower than those occurring in processing can be extrapolated to higher temperatures and elongational rates approaching conditions common in practice.

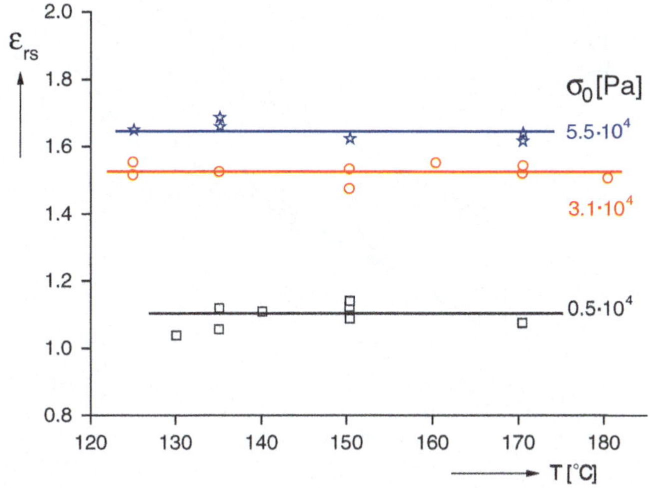

Fig. 12.18 Steady-state recoverable strain ε_{rs} of LDPE IUPAC A at various constant tensile stresses σ_0 as a function of the temperature T [20]

12.6 Comparison of Uniaxial and Multiaxial Extensions

Multiaxial extensional experiments are less frequently found in the literature than uniaxial ones due to the demanding experimental conditions and the small number of reliable instruments available. Moreover, the range of total elongation and elongational rates which can experimentally be achieved is not so broad compared to uniaxial elongation.

Extensional data are generally still far away from being quantitatively used for modeling processing operations but, particularly, the effect of strain hardening is exploited to qualitatively predict some processing aspects (cf. Sect. 17.5). As most of the processing operations involve biaxial or planar extensional components it may be helpful to get at least some knowledge about the comparability of the strain-hardening effects in the different stretching modes.

In [24] elongational viscosities obtained for various polymeric materials in uniaxial and equibiaxial extension are compared. For the uniaxial experiments, the MTR rheometer described in Sect. 12.3.5 was used, the equibiaxial experiments were performed by the lubricated squeeze-flow rheometer (cf. Sect. 12.4.2). Figure 12.19 shows the results on the LDPE of a higher molar mass than the LDPE IUPAC A which was already used before for the investigations of the recoverable compliance (cf. Fig. 12.14). Whereas the uniaxial experiments were performed up to a Hencky strain of nearly 4, in the equibiaxial tests only elongations up to 1.5 were possible.

This strain level is marked by the dotted line added to the uniaxial elongation curves as base for a comparison. The numbers on the curves give the elongational rates at which the experiments were carried out. For a better distinction, the equibiaxial data are shifted by a factor of 10 along the viscosity axis. Also plotted in the

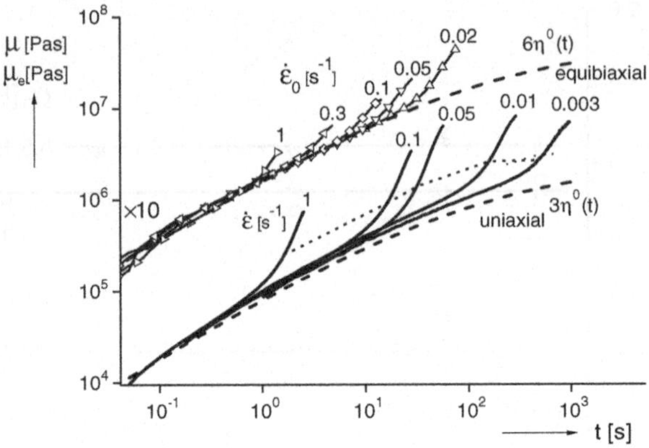

Fig. 12.19 Comparison of the uniaxial and equibiaxial elongational viscosities μ and μ_e as functions of the time t for an LDPE [24]. The *dotted line* in the diagram of the uniaxial elongational viscosity marks the total elongation ε_H corresponding to that achieved in the equibiaxial experiments. The curve of the equibiaxial viscosity is shifted by a factor of 10 along the viscosity axis for the matter of clarity

figure are the curves which relate shear and elongational data in the linear range of deformation to each other (cf. Eqs. 12.19 and 12.21). The satisfying agreement between the envelope of the extensional viscosities and the corresponding linear curves derived from shear supports the confidence in the methods applied.

Strain hardening is seen for both deformation modes. But while strain hardening increases with decreasing rate for the equibiaxial extension, remarkable differences in strain hardening for the strain rates applied cannot be seen in the uniaxial case at the relatively low total strain of 1.5 which is still far away from the elongations at which plateaus may be attained.

For a long-chain branched polypropylene, strain hardening is found in both modes, too [24]. Compared to the relatively low total strain, a significant rate dependence of the strain hardening cannot be stated, however. This finding is similar to that on the LDPE presented in Fig. 12.19.

With the Hachmann-Meissner rheometer introduced in Sect. 12.4.1, equibiaxial measurements up to a total strain of about 3 were performed. An LDPE and an HDPE were investigated and compared with results from uniaxial flow [14]. Qualitatively it can be stated that the strain hardening found for the uniaxial elongation occurs in the equibiaxial mode, too, but less pronounced.

This rather universal rheometer makes it possible to apply a planar deformation. In this case two viscosities can be defined (cf. Eqs. 12.23 and 12.24). From the data in [14] a comparison between the uniaxial and the two planar viscosities is possible. For the LDPE and the HDPE investigated, it can be concluded that the strain hardening in the direction of planar deformation is very similar to that obtained in uniaxial extension. For the other direction, a viscosity independent of the elongational rate was found for the LDPE and values slightly decreasing with $\dot{\varepsilon}_0$ for the HDPE up to total elongations of 5.

From the few reliable data on multiaxial extension available in the literature it can be concluded that if strain hardening is existent in uniaxial extension, the probability is very high to find it in equibiaxial and planar deformations, too, which are the dominating extensional modes in processing.

12.7 Role of the Uniformity of Sample Deformation for the Accuracy of Extensional Experiments

For all the extensional devices presented in Sect. 12.3, a uniform sample deformation is one of the experimental preconditions to attain a good accuracy of the measurements. In the case of the MTR, this fact becomes evident from the basic equations underlying the control of the driving unit (cf. Sect. 12.3.5), but in a corresponding way it is valid for the other devices, too. The uniformity of a sample during extension is naturally limited by the experimental condition that its ends have to be fastened in some way or the other to a pulling device. The influence of these end effects obviously is the larger the shorter the sample. Other reasons which can lead to a lack of uniformity are deficiencies of the initial sample geometry, temperature gradients, sagging effects, or the formation of cracks. The latter becomes the more probable the thinner the sample, i.e., the larger the elongation. Therefore, an in situ visualization of the sample during the experiment is a desirable experimental tool but it is available for some few devices, only.

The role of sample uniformity comes particularly into play when the existence of an overshoot or a maximum of the elongational viscosity is discussed. This topic is less relevant for an assessment of processing than for proving the validity of theoretical models. Whereas from the pioneering measurements displayed in Fig. 12.6 a distinct steady state of the elongational viscosity is obvious, in some papers maxima are presented for polystyrene [25] and polyethylene [4, 10]. Using the filament stretching rheometer described in Sect. 12.3.4, even an overshoot of the elongational viscosity is reported in [26] for a low density polyethylene, i.e., after running through a maximum the viscosity approaches a steady state at very high Hencky strains of around 5. Reports on the sample uniformity at such large elongations are not given, however.

A "pom-pom" model modified for randomly long-chain branched polyethylenes [27] was shown to match the elongational viscosities measured for several low density polyethylenes as a function of time. It predicts plateaus at high total elongations which were not reached in the experiments, however. Other models support the appearance of an overshoot [28] or are even able to fit a maximum or a steady state [29]. For all the experiments described by models, nothing is said about the reliability of the experiments and, particularly, about the uniformity of the sample during extension.

The visual in situ observation of a sample during elongation as presented in [12] is an experimental tool to get a direct insight into the uniformity during deformation. In the left part of Fig. 12.20, the elongational viscosity is represented as a function of time and in the right part as a function of the Hencky strain for a

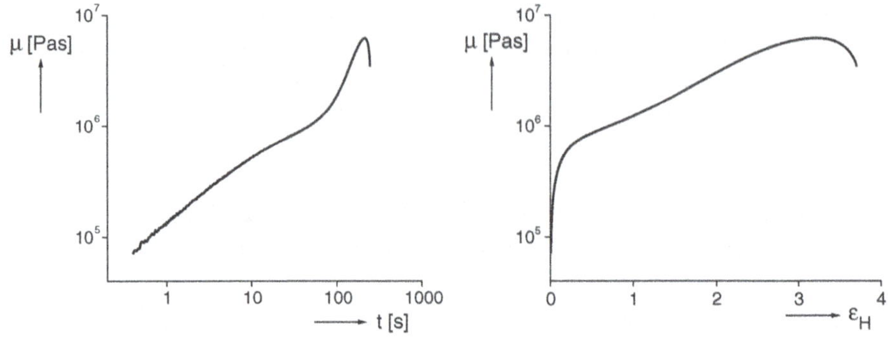

Fig. 12.20 Elongational viscosity μ as a function of time (*left*) and of the total Hencky strain ε_H (*right*) for an LDPE at $\dot{\varepsilon}_0 = 0.015\ \mathrm{s}^{-1}$ and a temperature of 140 °C. (Reprinted from [12] with permission from Elsevier)

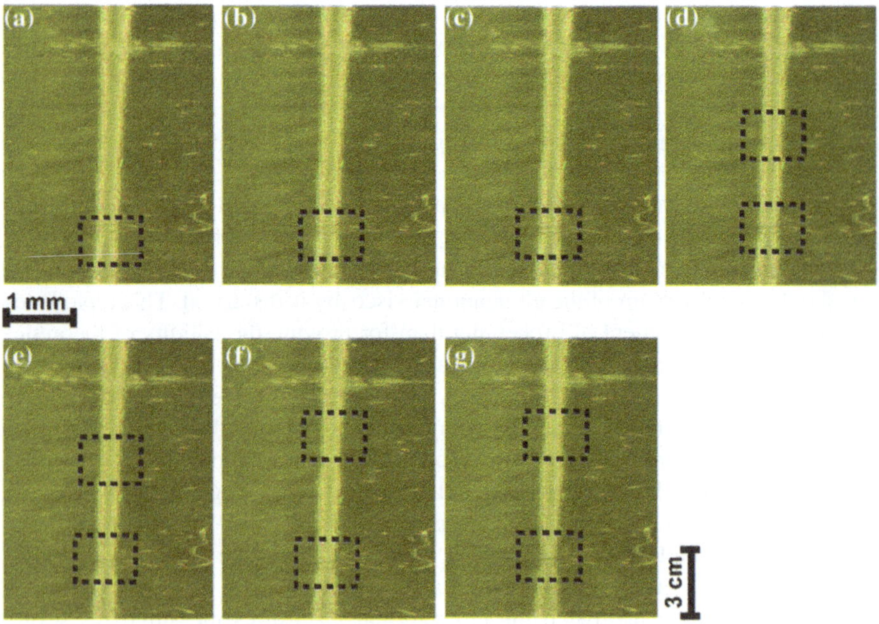

Fig. 12.21 Geometry of the LDPE sample elongated at $\dot{\varepsilon}_0 = 0.015\ \mathrm{s}^{-1}$ and $T = 140$ °C at different states of deformation [12]: **a** $\varepsilon_H = 3.55$, **b** $\varepsilon_H = 3.56$, **c** $\varepsilon_H = 3.58$, **d** $\varepsilon_H = 3.59$, **e** $\varepsilon_H = 3.60$, **f** $\varepsilon_H = 3.62$, **g** $\varepsilon_H = 3.63$. The images are not on scale. (Reprinted from [12] with permission from Elsevier)

commercial LDPE. A maximum in viscosity is shown as frequently reported in the literature.

The pictures of the sample at different states of elongation are displayed in Fig. 12.21. They clearly exhibit necks marked by the dotted squares occurring around a total elongation at which the elongational viscosity starts to decrease.

At somewhat higher extensions, a second neck can be observed which forms within a rather narrow increase of ε_H. Necking lowers the tensile stress and, consequently, the elongational viscosity. According to the in situ imaging presented, the maximum of the elongational viscosity in the case of the polyethylene investigated can be related to necking. For other materials possibly other geometrical non-uniformities may come into play which form during deformation and lead to the occurrence of similar maxima. More detailed discussions on the appearance of a maximum in elongational viscosity can be found in [12].

These results underline once more the requirement of careful observations of the sample uniformity during an extensional experiment if the reliability of data particularly at high deformations has to be proven. From this point of view, the measurements presented in Fig. 12.6 on the LDPE IUPAC A gain a particular importance as concomitant investigations demonstrated the uniformity of the sample deformation which support the reliability of the experimental findings (cf. Sect. 12.5). The results obtained clearly point to a significant strain hardening of the material which changes over to a steady state of the elongational viscosity at high elongations. Other findings which describe a maximum of the elongational viscosity of long-chain branched polyethylenes or even an overshoot may be an artifact caused by the formation of necks within a sample.

References

1. Meissner J (1969) Rheol Acta 8:78
2. Meissner J (1971) Rheol Acta 10:230
3. Laun HM, Münstedt H (1976) Rheol Acta 15:517
4. Meissner J, Hostettler J (1994) Rheol Acta 33:1
5. Sentmanat ML (2004) Rheol Acta 6:657
6. McKinley GH, Sridhar T (2002) Annual Rev Fluid Mech 34:375
7. Bach A, Rasmussen HK, Hassager O (2003) J Rheol 47:429
8. Cogswell FN (1968) Plast Polym 36:109
9. Münstedt H (1979) J Rheol 23:421
10. Münstedt H, Kurzbeck S, Egersdörfer L (1998) Rheol Acta 37:21
11. Kurzbeck S (1999) Doctoral Thesis, University Erlangen-Nürnberg
12. Burghelea TI, Stary Z, Münstedt H (2011) J Nonnewton Fluid Mech 166:1198
13. Burghelea TI, Stary Z, Münstedt H (2009) J Rheol 53:1363
14. Hachmann P, Meissner J (2003) J Rheol 47:989
15. Takahashi M, Isaki T, Matsuda T (1993) J Rheol 37:827
16. Meissner J (1975) Chem Pure Appl 42:553
17. Meissner J (1972) Rheol Acta 10:230
18. Laun HM, Münstedt H (1978) Rheol Acta 17:415
19. Meissner J (1972) J Appl Polym Sci 16:2877
20. Münstedt H, Laun HM (1979) Rheol Acta 18:492
21. Wolff F, Resch JA, Kaschta J, Münstedt H (2010) Rheol Acta 49:95
22. Resch J (2010) Doctoral Thesis, University Erlangen-Nürnberg
23. Münstedt H (1975) Rheol Acta 14:1077
24. Stadler FJ, Nishioka A, Stange J, Koyama K, Münstedt H (2007) Rheol Acta 46:1003
25. Hepperle J, Münstedt H (2005) Rheol Acta 45:717

26. Rasmussen HK, Nielsen JK, Bach A, Hassager O (2005) J Rheol 49:369
27. Read DJ, Auhl D, Das C, den Doelder J, Kapnistos M, Vittorias I, McLeish TCB (2011) Science 333:1871
28. Wagner MH, Raible T, Meissner J (1979) Rheol Acta 18:427
29. Wagner MH, Rolon-Garrido VH (2008) J Rheol 52:1049

Chapter 13
Rheological Properties and Molecular Structure

As shown before, rheological properties are dependent on temperature and mechanical quantities like stress or deformation rate, respectively. Some relations describing these dependencies were presented in the preceding two chapters. They are very useful for an assessment of the processing behavior. Moreover, the rheological behavior of polymer melts is significantly governed by the structure of the molecules which for a given chemical composition is described by the molar mass distribution and the architecture of the polymer chain. The different moments of the molar mass distribution are defined in Sect. 2.2. In common use are the first moment or number average M_n, the second moment or weight average M_w, and the third moment or centrifuge average M_z of the molar mass. The polydispersity factors M_w/M_n and M_z/M_w are related to the width of the distribution curve.

Characterizing the branching structure is difficult insofar as the branches are distinguished by their chemical composition, their length, concentration, and distribution. In the case of a polymerization with a chosen comonomer the first three characteristic features may be known, but the fourth one is difficult to determine. Even if only one monomer is used as for polyethylenes, for example, the branching analysis may become difficult as the widely used nuclear magnetic resonance techniques are not able to discriminate between branches with more than six carbon atoms. By gel permeation chromatography coupled with multi-angle laser light scattering (GPC/MALLS) the radius of gyration can be measured. From a comparison with the coil dimensions of the corresponding linear species qualitative conclusions with respect to the existence of branches can be drawn, but an analysis regarding their length and distribution is not possible (cf. Sect. 2.7).

A differentiation between short-chain and long-chain branches is rather arbitrary, but it is not unreasonable to designate those branches as long-chain ones which are able to form entanglements. In practice, for polymerising short-chain branched polyethylenes, α-olefins up to octenes are used. For the latter the length of the branch comprises six carbon atoms. Typical long-chain branched polyolefins are polyethylenes made under high pressures and temperatures (LDPE). Their structure is very complicated as branches of various lengths are generated during polymerization which can be branched, too, giving rise to a tree-like architecture.

H. Münstedt and F. R. Schwarzl, *Deformation and Flow of Polymeric Materials*, DOI: 10.1007/978-3-642-55409-4_13, © Springer-Verlag Berlin Heidelberg 2014

Modern metallocene catalysts make it possible to attach few but long chains to the backbone. Their structure is difficult to determine, however.

Relationships between molecular parameters and rheological properties are interesting from two points of view. On one hand, they can contribute to tailor materials with a special rheological behavior and, consequently, to predetermine some aspects of processing. On the other, rheological measurements are able to support the molecular analysis of unknown species.

In general, relations between the viscoelastic behavior of polymers and their molecular structure are very complex. Therefore, the discussion is divided into that of viscous, elastic, and elongational properties in dependence on molar mass, polydispersity, and branching in each case.

13.1 Viscous Properties of Polymer Melts

Due to the significant non-linearity of the viscosity relevant for processing and the importance of the linear quantities for analytical purposes, linear and non-linear properties are discussed separately.

13.1.1 Viscosity Functions in Dependence on Molecular Structure

In Fig. 13.1 the shear viscosity η as a function of the shear rate $\dot{\gamma}$ of three polystyrenes with different molar masses is presented. Polystyrene is a very suitable material to investigate the effects of molar masses and their distributions as the molecules are linear. From the inset of Fig. 13.1, it can be seen that the molar mass distributions of the three samples are very similar, but their weight average molar masses M_w are clearly different. Their viscosity functions are typical of polymer melts. Starting from the constant zero-shear viscosity η_0, the viscosity decreases by around two decades at shear rates of several hundred s^{-1} which are of relevance for processing. This shear thinning effect observable for various types of polymer melts is one of their advantages in processing. It is understood by the concept of entangled macromolecules, which loose the more of their entanglements the larger the shear rates or shear stresses are. The zero-shear viscosity becomes distinctly higher with growing molar mass. However, the differences of the viscosities get smaller with growing shear rate.

The influence of the molar mass distribution on the viscosity function is shown in Fig. 13.2 for two polystyrenes with the same weight average molar mass of $M_w = 340$ kg/mol, but distinctly different molar mass distributions (cf. inset). As expected, the zero-shear viscosities are the same. The viscosity of PS 2 with the broader distribution starts to decrease already at shear rates lower than those of the more narrowly distributed PS 4. The curves of the two samples approach each

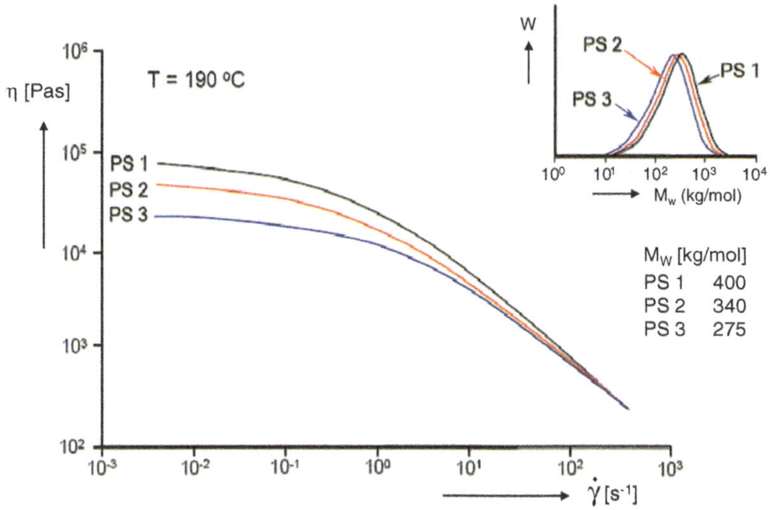

Fig. 13.1 Viscosity η as a function of the shear rate $\dot{\gamma}$ for three polystyrenes with different weight average molar masses M_w, but very similar molar mass distributions [1]

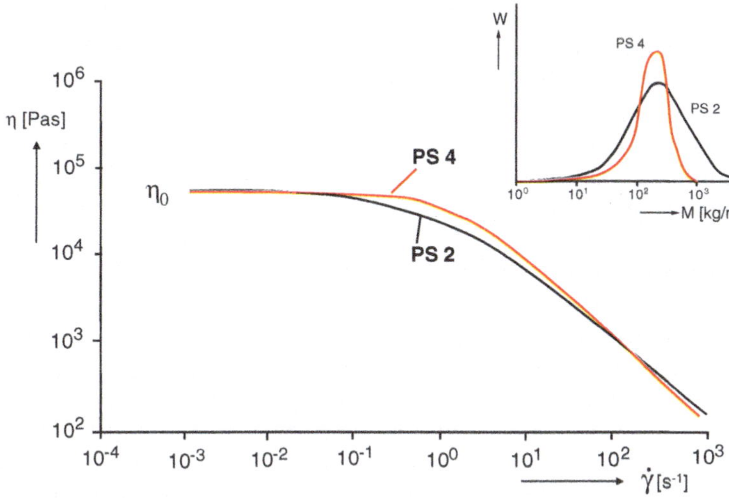

Fig. 13.2 Viscosity η as a function of the shear rate $\dot{\gamma}$ for two polystyrenes with equal weight average molar masses M_w, but different molar mass distributions [1]

other at higher shear rates and even intersect in the shear rate range between $10^2\ s^{-1}$ and $10^3\ s^{-1}$. This effect of a broader molar mass distribution on shear thinning has been observed for many classes of polymers and is one of their common features. The viscosities of the two polystyrenes were measured over nearly six decades of shear rates by using rotational rheometry for the lower shear rates and capillary rheometry for the higher ones.

Table 13.1 Molecular data and zero-shear viscosity η_0 of the three polyethylenes presented in the Fig. 13.3 [2]

	M_n (kg/mol)	M_w (kg/mol)	M_z (kg/mol)	M_w/M_n	M_z/M_w	η_0[a] (kPas)	ρ[b] (g/cm^3)
mHDPE 1	46	98	170	2.1	1.7	6	0.935
mLLDPE1	55	114	188	2.1	1.7	14	0.900
LCB-mLLDPE 1	48	99	170	2.1	1.7	42	0.902

[a] measured at 150 °C
[b] measured at 23 °C

Besides the molar mass distribution, long-chain branches have a significant effect on viscosity functions, too. This finding is not new, but in order to separate these two influences, samples with well-known molecular structures have to be investigated. Polyethylene exists in a wide range of molecular variations and, therefore, this type of polymer was chosen to demonstrate the effect of long-chain branching on the shape of the viscosity functions.

Table 13.1 gives the molecular data of three types of polyethylenes with different branching structures, but very similar molar mass distributions according to the values of M_w/M_n and M_z/M_w. The relatively small polydispersity factor M_w/M_n of 2.1 was obtained by using metallocene catalysts as indicated in the nomenclature by "m." The mHDPE 1 contains a tiny amount of butene, the mLLDPE 1 is a copolymer with a rather high butene content. LCB-mLLDPE is a long-chain branched metallocene-catalyzed copolymer with octene as comonomer. More detailed information about these three samples can be found in [2].

The viscosity functions of these three samples normalized by the zero-shear viscosity η_0 are shown in Fig. 13.3. The data were obtained from dynamic-mechanical experiments. Applying the Cox-Merz relation (cf. Sect. 11.3.1) the conclusions can be expected to be valid for the viscosity η as a function of the shear rate, too. The normalization with the zero-shear viscosities η_0 given in Table 13.1 makes the assessment of the shear thinning more distinct, as it brings the regions of constant viscosities to the same level. It is evident from Fig. 13.3 that the decrease of the viscosity of the long-chain branched LCB-mLLDPE 1 starts at a distinctly lower reduced angular frequency $\eta_0\omega$ than that of the two linear products. In addition, it can be stated that the concentration of comonomers, which is significantly different between mHDPE 1 and mLLDPE 1 does not seem to influence the shape of the viscosity function as the corresponding curves come to lie on each other.

This result has a direct consequence for processing. The disadvantages of metallocene-catalyzed polyethylenes in extrusion caused by their only small shear thinning can be overcome by introducing long-chain branches. The amount of LCB necessary to create such an effect does not seem to be high as the radius of gyration as a function of the molar mass only weakly deviates from the curve of the linear polyethylenes (see Fig. 2.18 and [2]).

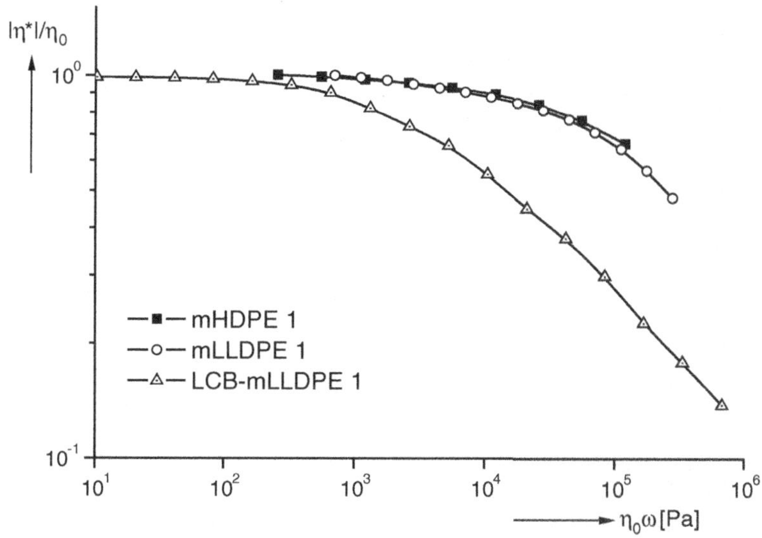

Fig. 13.3 Normalized viscosity functions of the three narrowly distributed polyethylenes with different branching structures listed in Table 13.1 [2]. $|\eta^*|$ is the magnitude of the complex viscosity, ω the angular frequency. The zero-shear viscosities η_0 are given in Table 13.1. The measuring temperature was 150 °C

Besides on the typical molecular parameters discussed, the shape of the viscosity function of a polymer depends on its chemical composition. This interesting finding is demonstrated by the Figs. 13.4 and 13.5 which were taken from [2]. In Fig. 13.4, the molar mass distributions of three metallocene-catalyzed polyolefins (two polyethylenes and one polypropylene) and one polystyrene are compared. All the four samples have a linear structure. As they possess different weight average molar masses, the distribution curves are shifted to superimpose their maxima. From Fig. 13.4 follows that their distributions are very similar. This fact allows a qualitative assessment of the influence of the chemical structure on the shear thinning which is shown in Fig. 13.5. The viscosity functions are plotted in their normalized form in order to facilitate the comparison of their shapes. The curves of mHDPE 1 and mLLDPE 1 which only differ by the amount of the comonomer butene incorporated are not distinguishable from each other as was already shown in Fig. 13.3. The polypropylene exhibits a distinctly stronger shear thinning than the two polyethylenes, but that of the polystyrene is still more pronounced.

The following conclusions on the viscosity functions can be drawn:

- The comparison of the linear polyethylene, polypropylene, and polystyrene indicates that for comparable molar mass distributions the shear thinning may be the stronger the bulkier the monomer unit.
- For a given type of polymer the non-linearity of the shear viscosity as a function of shear rate or shear stress, respectively, is the more pronounced the broader the molar mass distribution and/or the higher the "efficiency" of long-chain branches.

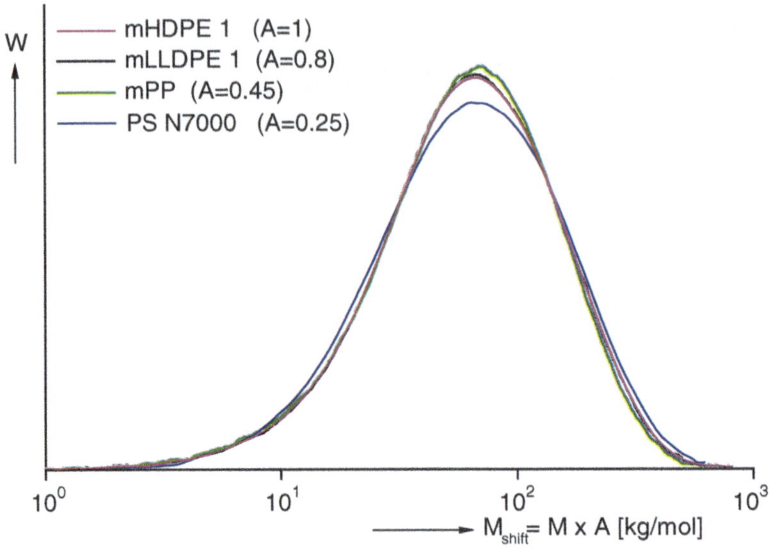

Fig. 13.4 Molar mass distributions of polymers with similar polydispersities, but different chemical structures [2]. The curves are shifted along the axis of the molar mass by the factors A

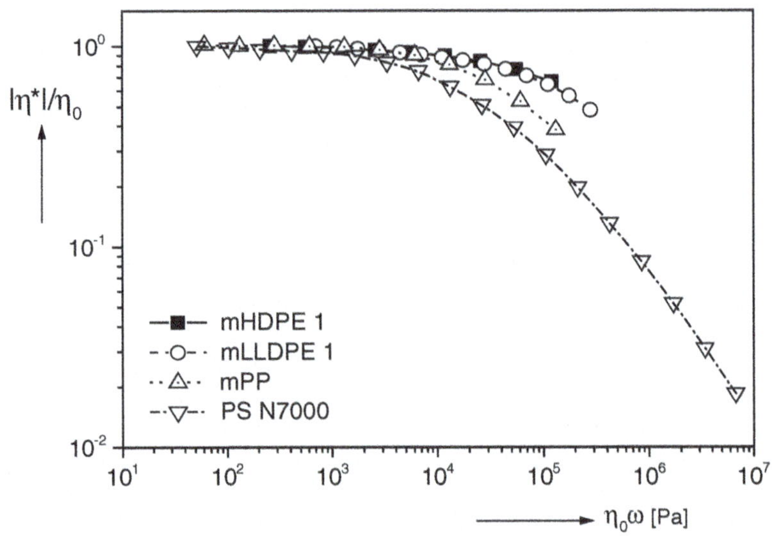

Fig. 13.5 Normalized viscosity functions of the polymers presented in Fig. 13.4 [2]. The data for PS N 7,000 are taken from [3]

- Therefore, from viscosity functions even qualitative conclusions with respect to molecular parameters can only be drawn for linear samples.
- Although only qualitative relationships between viscosity functions and molecular parameters are known, viscosity functions play an important role, however, for the assessment of the flow behavior regarding processing.

13.1.2 Zero-Shear Viscosities

As it is well known, simple power-laws between η_0 and the absolute value of M_w can be found for a variety of linear polymers, e.g., [4]. Nearly any textbook on rheology contains corresponding data which underline that

$$\eta_0 = K_1 M_w \ \text{for} \ M_w < M_c \tag{13.1}$$

and

$$\eta_0 = K_2 M_w^{\alpha} \ \text{for} \ M_w > M_c \tag{13.2}$$

K_1 and K_2 are parameters depending on the type of polymer and the temperature. The unit of K_1 follows as Pas kg^{-1} mol, that one of K_2 is a fractional rational depending on the exponent α which varies in the literature, but can be assumed to lie between 3.4 and 3.6. M_c is a critical molar mass which is approximately two to three times the molar mass M_e between entanglements. This quantity is significantly different for various polymers, e.g., [5]. Up to now there is no theory which provides the exact value of the exponents of Eq. (13.2).

Different opinions exist in the literature about the question whether the power-law relation is independent of the molar mass distribution. For blends of polystyrenes [6] and blends of polyisobutylenes [7], it was shown that $\eta_0(M_w)$ can be described by Eq. (13.2), i.e., the molar mass distribution which significantly changes within a series of blends does not influence the simple relationship between the zero-shear viscosity and the weight average molar mass. For linear polyethylenes this topic was addressed in [8]. The results are given in Fig. 13.6, which presents η_0 as a function of M_w for 24 samples covering the range of nearly four decades of molar masses up to 10^3 kg/mol. The molar masses and the polydispersity factors of the samples can be found in Table 13.2. Down to molar masses of 20 kg/mol, M_w was determined by gel permeation chromatography coupled with multi-angle laser light scattering (GPC-MALLS) (cf. Sect. 3.4), the data of the products with very low molar masses (samples W1–W7) were calculated from the known chemical structure of these ethylene waxes and the molar masses in between were obtained using the universal calibration method (cf. Sect. 3.4). For $M_w > M_c =$ 3 kg/mol, the data measured follow the full line the parameters of which are given in Fig. 13.6. As the polydispersity factors range from 1.8 to 16 there is some experimental evidence that $\eta_0(M_w)$ can be regarded as independent of the molar mass distribution. It should be mentioned that the samples A7 and A8 are bimodally distributed and C2, C4, and A5 contain a high molar mass shoulder.

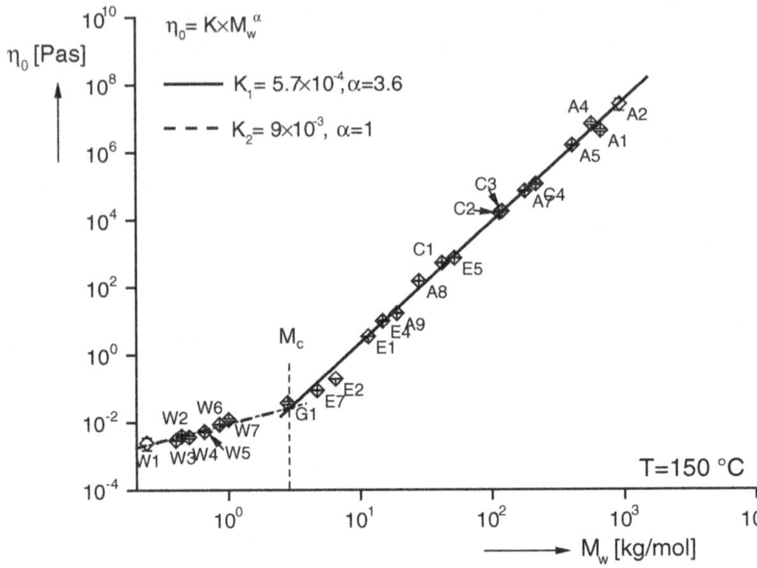

Fig. 13.6 Zero-shear viscosity η_0 at $T = 150$ °C as a function of the weight average molar mass M_w for linear polyethylenes of different molar mass distributions. The molecular data of the samples are given in Table 13.2 [8]. M_c follows as about 3 kg/mol. For the units of K_1 and K_2 see text

For $M_w \leq 3$ kg /mol a linear relationship between the zero-shear viscosity and the weight average molar mass is found in agreement with results from the literature for other polymers. If one agrees that M_c is two to three times the entanglement molar mass M_e, this quantity follows to lie between 1 and 1.5 kg/mol in agreement with the range of 0.9 and 1.8 kg/mol found in the literature for polyethylene, e.g., [5].

There are few papers which state zero-shear viscosities above the power-law of Fig. 13.6 for linear polyethylenes, but these laboratory samples possess an extremely high molar mass tail resulting in M_w/M_n values around 10 and M_z/M_w ratios of about 20, e.g., [9]. These products were shown to follow the relation

$$\eta_0 \cong 0.51\ \mathrm{KM}_w^{\alpha} \left(\frac{M_z}{M_w}\right)^{\beta} \tag{13.3}$$

with $\alpha = 3.4$ and $\beta = 0.8$. Equation (13.3) is an approximation of a rather complicated expression based on a double-reptation approach [10].

As it is obvious from Table 13.2, the M_z/M_w values of the listed samples lie between 2 and 3. Then the differences between Eqs. (13.3) and (13.2) are in the range of the given uncertainties of the measurements, particularly with respect to the determination of M_w and M_z. A relation similar to Eq. (13.3) was published in [11] together with the assessment that the maximum deviation of η_0 is about 20 % for M_w/M_z ratios between 2 and 5 if compared to the power-law of Eq. (13.2).

Table 13.2 Weight average molar mass and polydispersity factors for the polyethylenes presented in Fig. 13.6

Sample	M_w (kg/mol)	M_w/M_n	M_z/M_w
W1	0.24	1.0	–
W2	0.44	1.1	–
W3	0.4	1.1	–
W4	0.5	1.1	–
W5	0.655	1.1	–
W6	0.85	1.1	–
W7	1	1.1	–
G1	2.8	2.9	–
E7	4.7	2.5	2.0
E2	6.5	2.1	1.5
E1	11.5	1.8	1.8
E4	14.8	2.5	2.1
A9	19	6.5	3.8
A8[1]	28	7.8	3.5
C1[2]	42	3	2.0
E5	52	2	1.8
C2[3]	114	16	2.3
C3	120	2	1.6
A7[1]	178	4	3.5
C4[4]	224	3	1.7
A5[4]	403	2.6	2.3
A1	665	3.5	1.8
A4	564	4.3	2.1
A2	923	3.8	1.9

[1] bimodal [2] low molar mass shoulder [3] low- and high molar mass shoulder [4] high molar mass shoulder

Even in the case that Eq. (13.3) gives an appropriate description of the influence of the molar mass distribution on $\eta_0(M_w)$ as shown in [9] for extremely broad distributions, it has to be stated that there are only very few commercial linear polyethylenes with $M_z/M_w > 3$. Similar arguments hold for other linear polymeric materials like polypropylene, polystyrene, and polyamide. Therefore, the widely used power-law (13.2) can be assumed to be a very reasonable relation for linear polymers to assess η_0 from M_w and vice versa.

Furthermore, in [12] it is reported for ethylene copolymers that various comonomers from butene (4 carbon atoms) to hexacosene (26 carbon atoms) with contents between 2.2 and 6.8 mol % do not influence the relationship $\eta_0(M_w)$ found for linear polyethylenes at 150 °C. Such a finding is confirmed in [13] for ethylene copolymers based on butene, hexene, and octene between 0.7 and 8 mol % with M_w/M_n not exceeding 5 and M_z/M_w distinctly below this value. This result is somewhat surprising insofar as the activation energy increases with the comonomer content (cf. Sect. 14.2), i.e., the agreement found at 150 °C may become worse at other temperatures. For example, the largest activation energy of $E_a = 34$ kJ/mol determined in [13] for an ethylene copolymer with 8 mol % butene leads to an expected difference of about 20 % for η_0 at 190 °C compared to

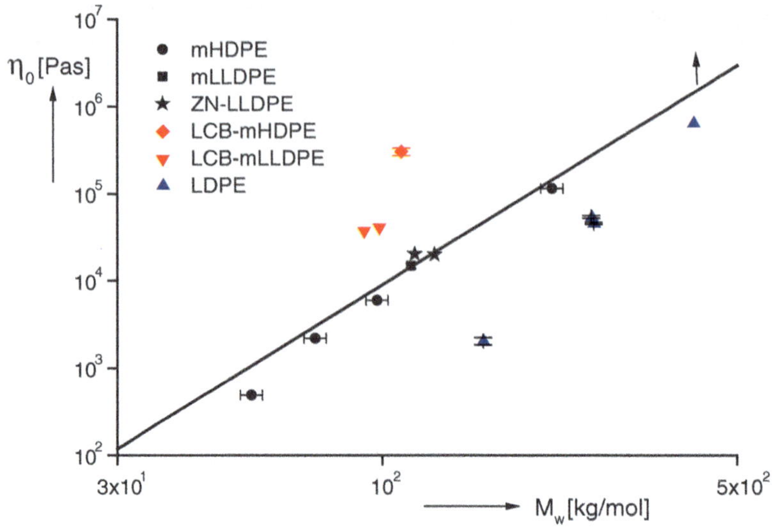

Fig. 13.7 Zero-shear viscosity η_0 at 150 °C as a function of the weight average molar mass M_w for linear and long- chain branched polyethylenes [2]. The full line represents the power-law for linear polyethylene. The arrow indicates that the steady state for the viscosity was not reached

a linear homopolymer with the same zero-shear viscosity at 150 °C, but an activation energy of 27 kJ/mol. Such a deviation is hard to be seen, however, in a double-logarithmic plot like that of Fig. 13.6.

The existence of a power law between η_0 and M_w for linear polyethylenes only weakly dependent on molar mass distribution and usual comonomer composition is of importance insofar, as long-chain branches effect significant deviations. Examples become obvious from Fig. 13.7 which presents the results on a variety of commercial products with different branching structures. Their generic designations are listed in the figure. The linear ethylene homopolymers (mHDPE) described by the circles follow the relationship of Fig. 13.6 for the linear laboratory products (full line). The values of the three copolymers, two of them polymerized with Ziegler-Natta catalysts (ZN-LLDPE) (stars) and one by metallocenes (mLLDPE) (square) are in good agreement with those of the corresponding linear homopolymers. The results on the long-chain branched samples significantly deviate from the curve for the linear products. But there exists a striking difference between them. The LDPE, for which a tree-like branching structure is assumed due to their radical polymerization, lie below the reference line, the metallocene products (LCB-mHDPE and LCB-mLLDPE) with postulated star-like structures above it.

Theories related to such a behavior are not very straightforward for the reason of the complicated and variable molecular architecture of long-chain branches. Nevertheless, for star-like molecules three papers give a rather similar expression for the dependence of η_0 on the ratio of the molar mass of the branched arm M_a to the entanglement molar mass M_e [14–16]

$$\eta_0 \approx (M_a/M_e)^{\alpha}\exp(\nu M_a/M_e) \tag{13.4}$$

They only differ with respect to the values of α and ν. α is in the order of 1, ν around 0.5. In the case of symmetric star polymers with the functionality f, the molar mass of the arm follows from the total molar mass M as

$$M_a = M/f \tag{13.5}$$

Although the influence of long-chain branches on the zero-shear viscosity is not understood quantitatively up to now, measurements of this rheological quantity and its discussion as a function of the absolute weight average molar mass can be used to get a qualitative insight into the branching architecture of polymers. If η_0 (M_w) lies distinctly above the reference line for linear materials, a star-like structure is probable. Values significantly below indicate a more complicated structure which may be similar to that of LDPE which contains branching points on other branches. Such an analysis has to be done with care, however, as Fig. 13.7 demonstrates. The LDPE with the highest molar mass possesses a zero-shear viscosity, which is very close to that of the linear reference with a corresponding molar mass. Further results on the zero-shear viscosities of LDPE and their relationship to the molecular structure are given in [17].

From the foregoing results and discussions the following conclusions can be drawn with respect to relations between the zero-shear viscosity and the molecular structure of polymers:

- The zero-shear viscosity as a function of the absolute weight average molar mass can be regarded to be approximately independent of the molar mass distribution for usual linear polymers and follows a power law.
- The high sensitivity of the zero-shear viscosity to the weight average molar mass makes the power law suitable for investigating the quantitative change of the molar mass after thermal or mechanical treatments in the case of linear polymers.
- Long-chain branched materials show distinct deviations from the power law.
- The deviations from the reference line for linear products can be used for a classification of long-chain branches into categories with mainly star-like or tree-like structures.

13.1.3 Analysis of Long-Chain Branched Polypropylenes Using $\eta_o(M_w)$

Despite the limitations discussed above, η_0 (M_w) can contribute to the characterization of long-chain branched materials of an unknown structure. This power is demonstrated on electron beam irradiated polypropylenes. The preparation of a variety of different samples starting from an isotactic polypropylene homopolymer is described in detail in [18]. The samples and some of their molecular properties are listed in Table 13.3. The doses of the electron irradiations are given in the

Table 13.3 Molecular data of electron beam irradiated polypropylenes. The number behind PP gives the irradiation dose in kGy

	d (kGy)	M_w (kg/mol)	M_w/M_n	Branching degree
PP-0	0	669	4.2	–
PP-1	1	604	3.8	Not detectable
PP-2	2	565	3.6	Not detectable
PP-5	5	473	3.8	Weak
PP-10	10	444	3.6	Weak
PP-20	20	384	3.5	Weak
PP-60	60	285	3.4	Medium
PP-100	100	279	3.4	Strong
PP-150	150	268	3.5	Strong

second column of this table. They range from 1 to 150 kGy and are indicated by the numbers of the denominations of the samples. As expected, M_w goes down with increasing dose. Interesting is the finding that the polydispersity factor does not change much. The molar mass distribution becomes slightly narrower up to doses of 10 kGy, but then remains nearly constant up to the highest applied dose of 150 kGy.

In Fig. 13.8 investigations of the degree of long-chain branching of the sample by GPC-MALLS are presented. As can be seen from the radius of gyration as a function of the absolute molar mass, for 5 kGy a significant indication of branching is detectable which becomes more distinct at higher doses. The slope of 0.625 for the non-irradiated linear polypropylene is higher than 0.5 indicating the non-theta conditions of the used solvent trichlorobenzene at 150 °C (cf. Eq. (3.38)). The slope of the experimentally found straight lines in the double-logarithmic plot of the irradiated samples decreases with the dose and reaches a value of 0.42 at 150 kG.

Figure 13.9 displays the zero-shear viscosities of the irradiated samples in comparison to various commercial isotactic polypropylenes. For the commercial linear PP, the η_0 data plotted as a function of the absolute weight average molar mass determined by GPC-MALLS can approximately be connected by a straight line in a double-logarithmic plot indicating the validity of the power-law relationship Eq. (13.2) as known from the literature. For the exponent $\alpha = 3.5$ and for the prefactor $K_2 = 1.25 \times 10^{-4}$ at 180 °C are found. The data points of the samples irradiated with doses up to 60 kGy lie distinctly above the reference line of the linear products, the samples PP-100 and PP-150 just below. As the polydispersity factor remains nearly the same, the viscosity increase can be related to the existence of long-chain branches without any doubt. In comparison to the polyethylenes of different branching structure shown in Fig. 13.7, it can be concluded that the samples irradiated with doses up to 60 kGy exhibit a star-like architecture which gradually seems to change into a more tree-like one at higher doses. Such a picture is not unreasonable, of course, as one can imagine that at higher doses functionalized sites occur at branches previously formed giving rise to branches on branches.

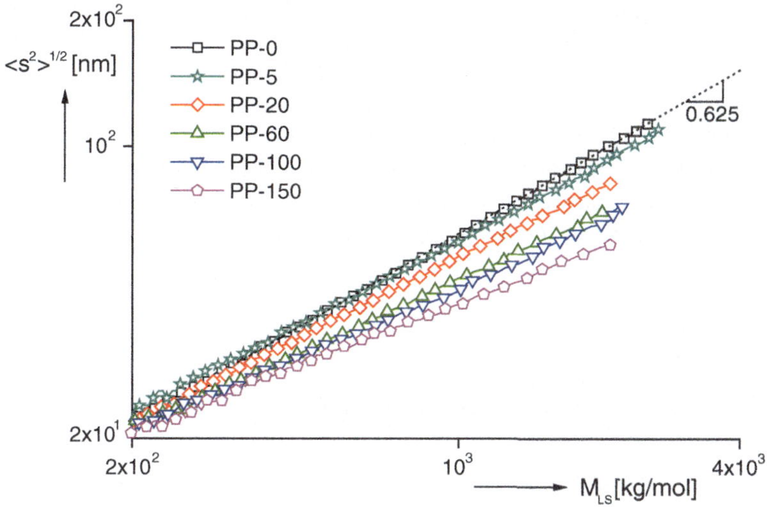

Fig. 13.8 Root of the mean-square radius of gyration $< s^2 >^{1/2}$ as a function of the molar mass M_{LS} determined by light scattering for the differently irradiated polypropylenes. The solvent was trichlorobenzene (TCB), the measuring temperature 150 °C. (Reprinted from [18] with permission from American Chemical Society)

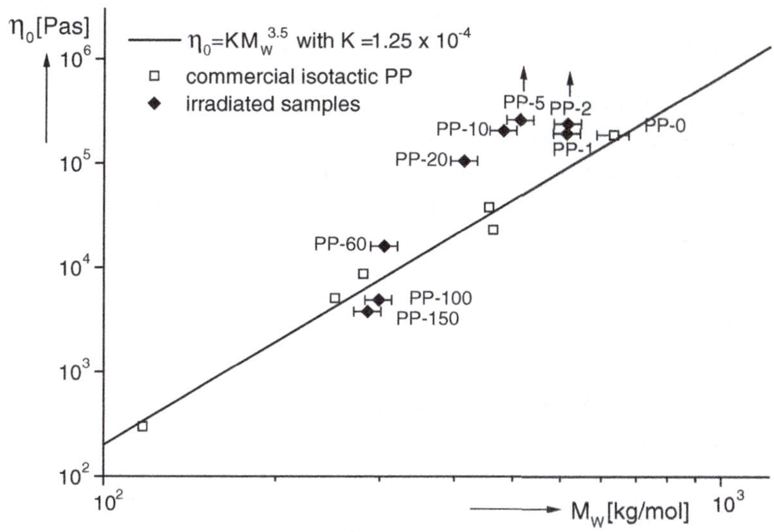

Fig. 13.9 Zero-shear viscosity η_0 at $T = 180$ °C as a function of the absolute weight average molar mass M_w for polypropylenes irradiated with different doses. The numbers give the doses in kGy. The arrows indicate that the zero-shear viscosity was not reached experimentally. (Reprinted from [18] with permission from American Chemical Society)

Impressive is the high sensitivity of the rheological characterization in comparison with that of GPC-MALLS for small degrees of long-chain branching. Whereas the radii of gyration of PP-1 and PP-2 cannot be distinguished from that of the non-irradiated PP and, therefore, are not plotted in Fig. 13.8, $\eta_0(M_w)$ for these samples already differs from the corresponding linear reference by a factor of around two. Although very sensitive, the rheological method applied—like all the others—is not able, however, to give a quantitative insight into the branching structure with respect to the length, concentration, and distribution of branches.

13.2 Elastic Properties

13.2.1 Quantities Reflecting Elasticity

Whereas the viscosity has widely been used for the characterization of the molecular structure of polymers and an assessment of their processability as was shown in previous sections, elastic effects play a comparably minor role in the literature of today, although they are able to very sensitively reflect some molecular features. Furthermore, the elasticity may affect properties of manufactured items like the geometry of extruded parts or internal stresses within injection molded parts. A very obvious appearance of elasticity is the so-called extrudate swell, which can be observed while extruding a melt through a die (cf. Sect. 11.5.4). Other phenomena reflecting the elasticity of polymer melts are the occurrence of normal stress differences during shearing (cf. Sect. 11.1) and the existence of distinct storage moduli and storage compliances in dynamic-mechanical experiments (cf. Sect. 11.3). Recovery experiments after deformation in shear and elongation can be used to get an insight into elasticity, too, (cf. Sect. 11.4).

These effects are the base of experimental devices for the determination of elasticity the principles of which are described in Chap. 15 of this book. Using the various methods and quantities, one has to be aware of their advantages and disadvantages. These are compiled in Table 13.4.

13.2.2 Extrudate Swell in Dependence on Molecular Structure

Measurements of the extrudate swell in dependence on the molar mass and the molar mass distribution were performed on various polystyrenes. The extrudate swell as a function of shear stress for the three polystyrenes of very similar molar mass distributions, but distinctly different M_w (cf. Fig. 13.1) is shown in Fig. 13.10. The data were obtained by extrusion through a capillary with a length to radius ratio of $L/R = 16$ at different shear stresses σ_w at the wall. It is clearly seen that the extrudate swell defined as

Table 13.4 Methods for the measurement of elastic quantities and their special features

Method	Measured quantity	Advantage	Disadvantage
Capillary rheometry	Extrudate swell	• Rheometers available	• Annealing necessary for equilibrium values • Sagging has to be avoided
		• Only measurement of diameter necessary	• Flow not well defined • Linear regime difficult to attain
Rotational rheometry			
Dynamic-mechanical analysis	Storage modulus and elastic compliance	• Sophisticated frequency analysis available • Direct measurement of elastic compliance	• Frequency range and time window limited • Terminal regime for some products difficult to attain
Stressing experiment	Normal stress differences		• No direct measurement of elastic compliance • Method difficult to handle
Creep recovery in shear	Elastic shear compliance	• Linear range attainable • Large time window accessible • Direct measurement of elastic compliance	• Drift correction necessary
Extensional rheometry			
Creep recovery in elongation	Elastic elongational compliance	• Straightforward method • Visual observation of samples possible	• Equipment not commercial • Measurement of time dependence not very accurate • Linear range difficult to attain

$$S = (d - d_0)/d_0 = d/d_0 - 1 \tag{11.19}$$

for samples with similar molar mass distributions are not or at least very weakly dependent on M_w. Figure 13.10 shows the interesting experimental finding that the extrudate swell S as a function of the wall shear stress σ_w follows a straight line in the double-logarithmic plot. According to the results of Fig. 11.18, it can be assumed to be independent of the temperature at which it was measured. That means, $S(\sigma_w)$ can be described by the power law

$$S = a\sigma_w^s \tag{13.6}$$

with a and s being material-specific parameters if the geometry of the capillary is kept constant. A physical background of this empirical relation has to be left open. What can be said from the measurements of the velocity fields in the die entrance

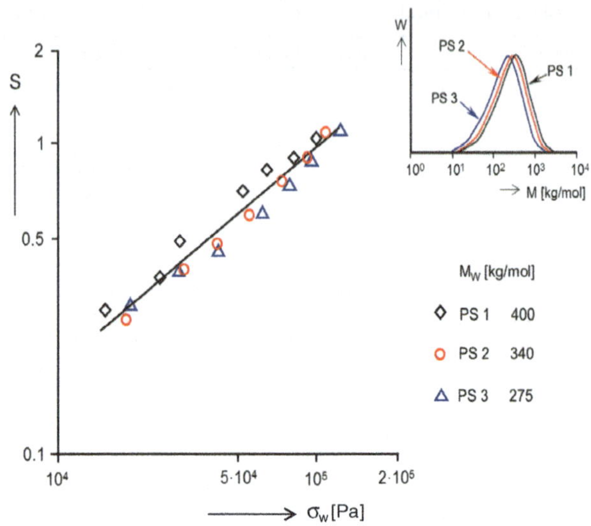

Fig. 13.10 Extrudate swell S as a function of the wall shear stress σ_w for three polystyrenes of different molar masses but very similar molar mass distributions. The extrusion temperature was 190 °C, the length to radius ratio of the capillary $L/R = 16$. The samples were annealed to their equilibrium after extrusion

region by laser-Doppler velocimetry, however, is that the elongational rates and therewith the recoverable strains become higher with larger output rates which are connected with an increase of the wall shear stresses σ_w, of course, (cf. Sect. 16). Therefore, Eq. (13.6) has to be regarded as an approximation which can be used for some kind of cautious interpolation, if the order of magnitude of swell data at stresses different from the measured ones is requested. But the dependence on the length to diameter ratio L/R of the die has to be taken into account (cf. Fig. 11.18).

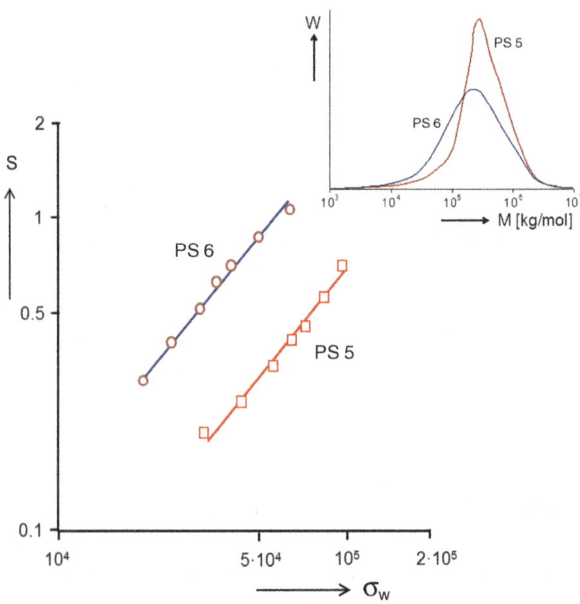

Fig. 13.11 Extrudate swell S as a function of wall shear stress σ_w of two polystyrenes with similar M_w, but different molar mass distributions. The length to radius ratio was $L/R = 20$ and the extrusion temperature $T = 170$ °C. The samples were annealed to their equilibrium after extrusion

Clear differences of the extrudate swell are found in Fig. 13.11 for two polystyrenes of similar weight average molar masses, but distinctly different molar mass distributions. PS 6 with the broader molar mass distribution exhibits a significantly larger extrudate swell than the distinctly more narrowly distributed PS 5. This result indicates that the polydispersity has a strong effect on the elasticity of a polymer melt.

All the measurements of the extrudate swell of the polystyrenes following the power law of Eq. (13.6) were performed in the non-linear range of deformation which is the domain of processing. For a more quantitative insight into the effect of the molecular structure on the elasticity of polymer melts, other quantities which allow a description by the linear theory of viscoelasticity have to be discussed.

13.2.3 Steady-State Linear Recoverable Compliance

The steady-state linear recoverable compliance J_e^0 defined by Eq. (5.36) is a very suitable quantity to characterize the elasticity of a polymer as it can directly be related to the retardation spectrum (cf. Sect. 5.5) which is a fingerprint of molecular motions. J_e^0 can be determined by different methods (cf. Chap. 11), but the most direct one is the creep-recovery experiment as described in detail in Sect. 11.4. It is preferentially used as the experimental base for the relationships discussed in the following.

As J_e^0 is a steady-state linear quantity, it is independent of time and stress per definition. From Fig. 13.12, it follows for a commercial isotactic polypropylene that the recoverable compliance J_r as a function of time depends on temperature, but that it becomes a temperature-independent quantity in its steady state at least in the temperature range between 180 and 220 °C [19]. Similar results were obtained for three other linear polypropylenes of various molar masses and some linear polyethylenes [19]. The temperature independence of J_e^0 is not a general feature of all kind of polymers, however, as it is discussed in Sect. 14.2.3.

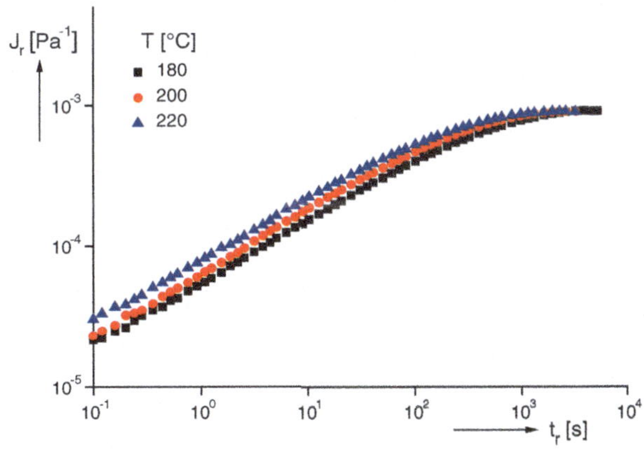

Fig. 13.12 Creep-recovery compliance J_r as a function of the recovery time t_r for a commercial isotactic polypropylene at three measuring temperatures [19]

13.2.3.1 Dependence on Molecular Structure

Separating the influence of various molecular parameters on elastic properties is demanding even in the case of linear materials as the molar mass distribution has a significant effect according to the results of the extrudate swell shown in Fig. 13.11. The molar mass distribution is sometimes not easy to determine in all its details due to high molar mass components which may be difficult to detect. Anionic polystyrenes are distinguished by a very narrow molar mass distribution and, therefore, they are very suitable to investigate the influence of the molar mass on the recoverable compliance.

Figure 13.13 shows J_e^0 for various anionic polystyrenes with molar masses M_w between 200 and 800 kg/mol [3]. In this range, J_e^0 is constant, indicating that the elasticity is independent of the molar mass. This result is in agreement with the textbook knowledge that J_e^0 above a critical value of about five times the molar mass between entanglements M_e becomes independent of M_w (e.g., [20]). For smaller molar masses J_e^0 linearly increases with the molar mass according to the Rouse-Bueche theory of unentangled polymers [21, 22]. Compared to $M_e \approx 16$ kg/mol for polystyrene, the molar masses of the samples of Fig. 13.13 are distinctly higher than the critical value of about 5 M_e. Similar results on polystyrene are reported in [23] and on polyisobutylene in [24]. These correlations explain the finding that elastic properties of an engineering polymer are not dependent on M_w as long as their polydispersities are not much different (cf. Fig. 13.10). Molar masses of engineering polymers are generally large in comparison to M_e.

Elastic effects are strongly dependent on the molar mass distribution, however. The significant effect on J_e^0 which the addition of a high molar mass component to a matrix of a lower molar masses can have, for example, is depicted in Fig. 13.14.

Some few percents of an anionic PS of the molar mass 700 kg/mol within another anionic polystyrene of $M_w = 40$ kg/mol result in an increase of J_e^0 by

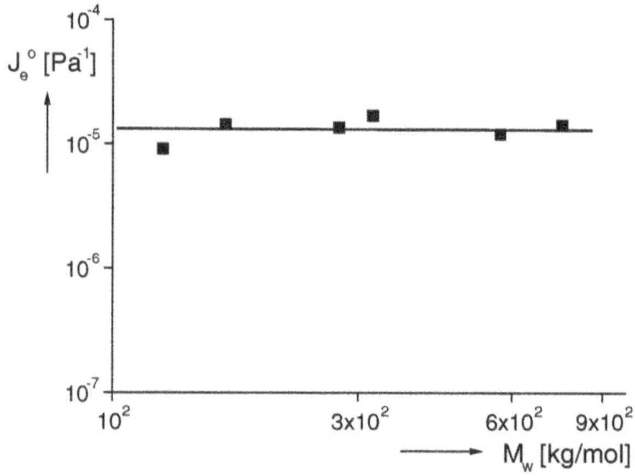

Fig. 13.13 Steady-state linear recoverable compliance J_e^0 as a function of M_w for anionic polystyrenes of various molar masses [3]

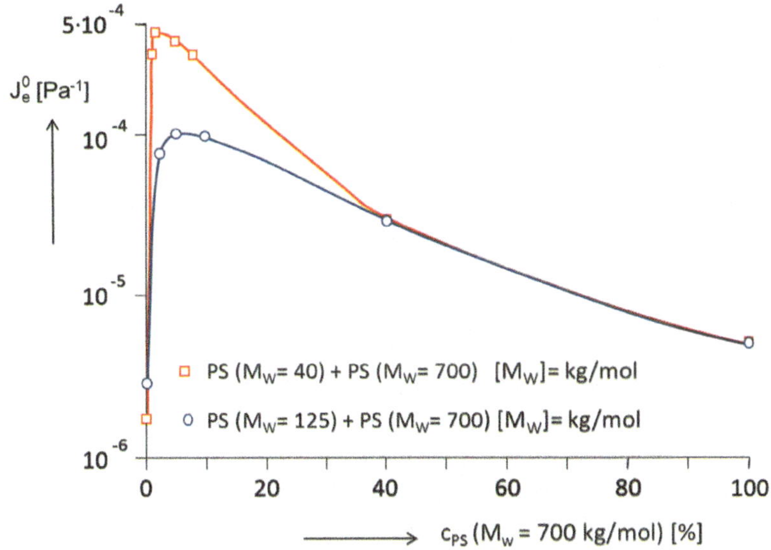

Fig. 13.14 Steady-state linear recoverable compliance J_e^0 of polystyrene blends as a function of the mass percentage c_{PS} of the high molar mass component *(*M_w = 700 kg/mol*)*. The lower molar masses are M_w = 40 kg/mol *or* M_w = 125 kg/mol*,* respectively, [25]

about two decades. J_e^0 runs through a maximum at around 1 wt. % of the larger molar mass component and approaches a recoverable compliance J_e^0, which is higher than that of the lower molar mass component, but lies significantly below the maximum. The enhanced J_e^0 for the anionic sample of larger molar mass can be explained by its somewhat broader molar mass distribution. A similar but not so pronounced effect is obtained, if an anionic PS of M_w = 125 kg/mol is used as the matrix (cf. Fig. 13.14). These results show the significant effect a high molar mass component can have on the elasticity in the linear range of deformation. It may be mentioned that 1 wt. % of the high molar mass component with M_w = 700 kg/mol in a matrix with M_w = 40 kg/mol or M_w = 125 kg/mol is very hard to detect by gel permeation chromatography. A similar effect of increasing the elasticity of a polymer significantly by adding a small amount of a high molar mass component to a matrix of lower molar mass was reported for blends of polyisobutylenes [7].

The results on the polystyrenes shown in Fig 13.14 indicate, however, that the effect of increasing J_e^0 of a matrix by the addition of a high molar mass component may be the less distinct the closer the molar mass of the species added is to that of the matrix. Therefore, the elasticity of a material with an already broad molar mass distribution is enhanced to a much smaller extent than that of a narrowly distributed one.

The relationships obtained for polystyrenes are found for polypropylenes of various molar masses and polydispersities, too, which show a linear molecular structure due to the polymerization process. In Table 13.5, the weight average

Table 13.5 Molecular data and J_e^0 of various polypropylenes [19]

	M_w (kg/mol)	M_w/M_n	J_e^0 [a] (10^{-4} Pa^{-1})
PP 1	265	2.5	1.0 ± 0.1
PP 2	244	3.5	2.7 ± 0.1
PP 3	263	6.4	9.3 ± 0.1
PP 4	268	7.7	12.1 ± 0.1
PP 5	525	6.0	9.1 ± 0.2
PP 6	738	6.0	11.5 ± 1.0

[a] measured at T = 180 °C

molar masses M_w and the polydispersity factors M_w/M_n are listed together with the linear steady-state elastic compliances J_e^0 for six polypropylenes [19]. PP 1–PP 4 possess very similar molar masses, but distinctly different polydispersity factors ranging from 2.5 to 7.7. J_e^0 of these samples clearly increases with M_w/M_n. PP 3, PP 5, and PP 6 have polydispersity factors around 6, but various molar masses. For these samples, the values for J_e^0 can be regarded as similar, if the uncertainties of the measurements of the compliances are taken into account. The differences of M_w by a factor of about three are not reflected in the compliances of these samples as it should be expected according to $M_e \approx 7$ kg/mol for polypropylene [5].

As the molar mass distribution very significantly influences elastic properties of a polymer melt, it has to be well known, if the influence of long-chain branches is to be reliably assessed. For such an investigation, the linear mLLDPE 1 and the long-chain branched LCB-mLLDPE 1 of Table 13.1 are well suited, whose molar mass distributions shown in Fig. 13.15 are indistinguishable. A distinct high molar mass component not detected by GPC is improbable to exist due to the polymerization kinetics.

The steady-state recoverable compliances of the two LLDPE presented in Fig. 13.16 show remarkable differences. This quantity is higher by a decade for the long-chain branched sample LCB-mLLDPE 1 in comparison to the linear mLLDPE 1 demonstrating the strong influence long-chain branches can have on

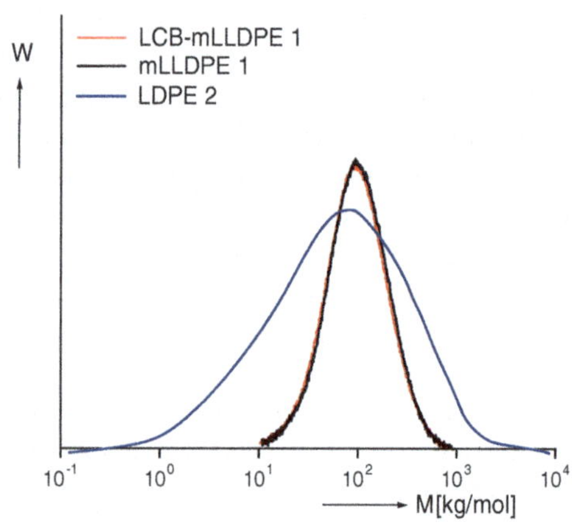

Fig. 13.15 Molar mass distributions of the linear mLLDPE 1 and the long-chain branched LCB-mLLDPE 1 of Table 13.1 in comparison with a classical low density polyethylene (LDPE 2). The samples were dissolved in trichlorobenzene (TCB) at 135 °C [26]

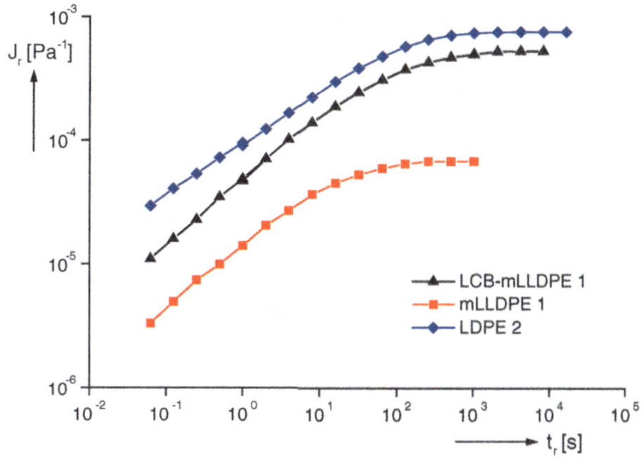

Fig. 13.16 Recoverable compliances J_r of mLLDPE 1, LCB-mLLDPE 1, and LDPE 2 as a function of the recovery time t_r at $T = 150$ °C [26]

elastic properties of polymer melts [27]. Furthermore, it is obvious from Fig. 13.16 that the time-dependent recoverable compliance J_r *(t)* of the LCB-mLLDPE 1 reaches its steady state after significantly longer times than the mLLDPE 1. This behavior points to the longer retardation times due to the long-chain branches.

Although the effect of long-chain branches on elastic properties is obviously very pronounced, it is not straightforward to employ this quantity for a branching analysis. For such purposes, the knowledge of the molar mass distribution is one prerequisite. But even if that is known, only qualitative conclusions can be drawn due to the fact that a quantitative correlation between the polydispersity factor and a well-defined elastic quantity like J_e^0 does not exist, which could be used for taking the contribution of the molar mass distribution into account. One reason for this deficiency is that the interactions between molecules of different lengths are too complicated to be related to one number. Moreover, the polydispersity factor describing the width of a distribution curve is not bijective by its nature and can stand for various distributions.

From the comprehensive investigations of the elasticity of polymer melts in [2] it is evident that the architecture of long-chain branches affects the elasticity measured by the linear recoverable compliance in a complicated way. This is demonstrated by a comparison of the LCB-mLLDPE 1 with the highly branched LDPE 2 of a broad molar mass distribution which is shown in Fig. 13.15. The branching structure of these two samples is totally different, as can be concluded from Fig. 13.17, which shows the expectation value of the squared radius of gyration as a function of the molar mass for these two materials in comparison to the relationship for linear polyethylenes as obtained by GPC coupled with MALLS (cf. Sect. 3.4). The LDPE 2 is significantly long-chain branched even at relatively low molar masses, whereas for the LCB-mLLDPE 1 a weak indication of branching can be detected only above a molar mass of 500 kg/mol. This finding indicates that the metallocene-catalyzed sample may consist of a blend of linear and branched

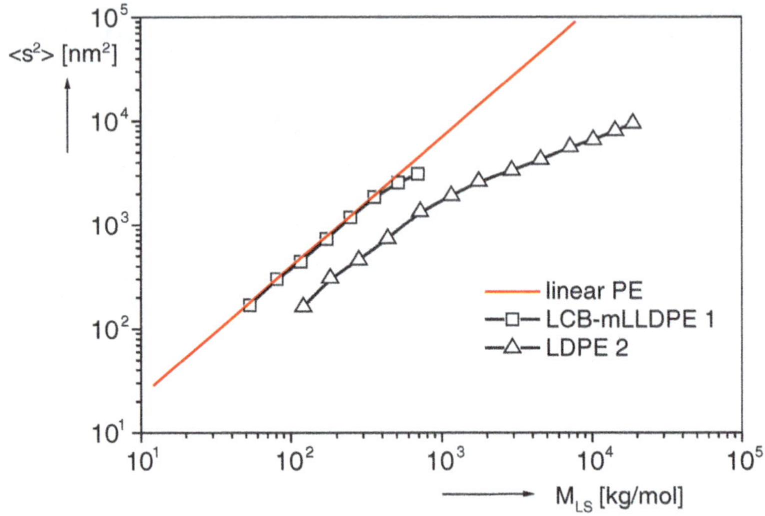

Fig. 13.17 Expectation value of the squared radius of gyration $<s^2>$ as a function of the weight average molar mass M_{LS} determined by light scattering for the two long-chain branched polyethylenes LCB-mLLDPE 1 and LDPE 2 in comparison to linear polyethylenes. The solvent is trichlorobenzene (TCB) at $T = 135$ °C [2]

molecules with the long-chain branches mainly restricted to the molecules of high molar mass.

Furthermore, according to the polymerization processes different branching architectures have to be assumed. The structure is tree-like, i.e., branches on branches, in the case of LDPE and star-like for the metallocene-catalyzed product. As follows from Fig. 13.16, J_r(t) of the LDPE 2 is only slightly higher than that of the LCB-mLLDPE with long-chain branches, which are only weakly reflected by the radius of gyration. This may be due to their small concentration.

A hint to an explanation of the high elasticity of the LCB-mLLDPE carrying few branches, only, may be obtained from Fig. 13.14, which demonstrates the immense influence some few, but long molecules can have on the recoverable compliance of linear polystyrenes. Taking these results into account, it may not seem surprising any more that a similar increase of elasticity can be affected by a small amount of very long branches existing in the LCB-mLLDPE.

13.2.4 Recoverable Compliances at Higher Stresses

By a variation of the stresses applied in the creep-recovery experiment, investigations of the non-linear range of deformation can be performed. In Fig. 13.18, the stress dependence of the steady-state recoverable compliance J_e is shown for some

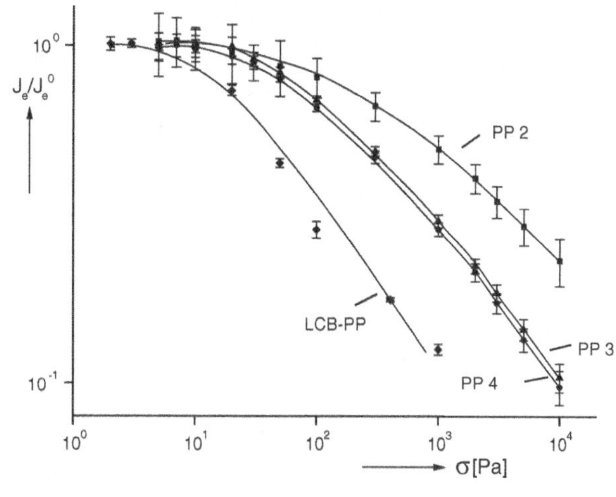

Fig. 13.18 Steady-state recoverable compliances J_e normalized by the steady-state linear recoverable compliance J_e^0 as a function of the shear stress σ for polypropylenes with different polydispersities (cf. Table 13.5) in comparison to a long-chain branched polypropylene (LCB-PP) at $T = 180$ °C [19]

of the linear polypropylenes from Table 13.5 and the long-chain branched polypropylene LCB-PP. The linear polypropylenes possess different polydispersity factors (cf. Table 13.5). For an easier discussion, the steady-state recoverable compliances J_e obtained at different stresses are normalized by the corresponding steady-state linear recoverable compliances J_e^0 which are listed in Table 13.5, too. The normalized compliances decrease as a function of the shear stress from their constant linear value. It is obvious that this non-linearity starts at the lower stresses the broader the molar mass distribution characterized by the polydispersity factor increasing from PP 2 over PP 3 to PP 4.

The influence of long-chain branching on the stress dependence of the recoverable compliance becomes obvious from the curve of the LCB-PP. Its linear range is fairly small and the decrease of the steady-state compliance J_e with stress is distinctly more pronounced than that of the linear sample PP 4 with the highest polydispersity factor. The results that a broader molar mass distribution or long-chain branching shorten the extension of the linear range of the recoverable compliance is in a formal agreement with the findings on the transition of the viscosity from the linear to the non-linear regime (cf. Figs. 13.2 and 13.3).

13.2.5 Conclusions on the Relationships Between Elasticity in Shear and Molecular Structure

The following conclusions can be drawn from the results of the dependencies of elastic properties of polymer melts on the molecular structure:

- An influence of the weight average molar mass M_w on the melt elasticity has not been found for various polymers of distinctly higher M_w than the molar mass between entanglements M_e.

- The molar mass distribution has a significant effect on the melt elasticity of linear polymers. The broader the distribution, the larger are the extrudate swell, or the recoverable compliance, for example.
- Small amounts of high molar mass components can lead to a large increase of the linear steady-state recoverable compliance.
- Long-chain branching contributes to an enhancement of elasticity.
- Long-chain branching or broadening of the molar mass distribution increase the non-linearity of the recoverable compliance.

13.3 Elongational Properties of Polymer Melts

In Sect. 12.5 it was demonstrated that rheological properties in elongation generally cannot be derived from those in shear. Only in the linear range of deformation a simple relationship exists between the viscosities in shear and uniaxial elongation, i.e., the elongational viscosity is three times the shear viscosity (cf. Eq. 12.19). In Fig. 12.12 the effect that long-chain branching leads to a significant increase of the elongational viscosity with time or strain, respectively, the so-called strain hardening, is presented. This typical feature is discussed in the following in dependence on the molecular structure of the polymeric materials.

13.3.1 Influence of Molar Mass and Molar Mass Distribution

Three polypropylenes with different molar masses and polydispersity factors were chosen to investigate the influence of these parameters on the elongational viscosity of this linear polymer [28]. The molecular data of the polypropylenes are listed in Table 13.6. The weight average molar masses differ by a factor of about two and the polydispersity factors indicate broad molar mass distributions. The elongational viscosities of these samples as a function of time at various elongation rates are shown in Fig. 13.19. Up to the total maximum elongation of $\varepsilon_{H\max} = 3$ and for strain rates $\dot{\varepsilon}_0$ covering a factor of around 300, strain hardening does not occur. All curves come to lie on the full lines which represent $3\eta^0(t)$ measured in shear. From these results it can be concluded that for the linear polypropylenes neither the molar mass nor the polydispersity give rise to strain hardening at least in the ranges chosen.

Table 13.6 Molar masses and polydispersity factors of three linear polypropylenes [28]

	M_w (kg/mol)	M_w/M_n
PP-A	669	4.2
PP-B	452	5.8
PP-C	345	3.3

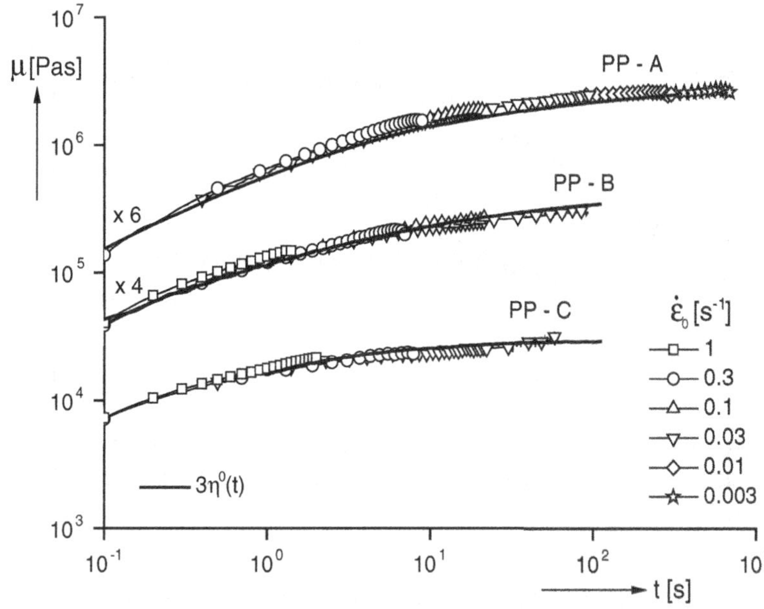

Fig. 13.19 Elongational viscosity μ as a function of time t for three linear polypropylenes with different molar masses and polydispersity factors at various constant elongation rates $\dot{\varepsilon}_0$. The full lines correspond to three times the linear shear viscosities measured. The factors on the curves indicate the shift along the viscosity axis with respect to PP-C which was applied for a clearer presentation [28]

That this conclusion cannot be generalized for linear polymers, however, becomes evident from results on polystyrenes which possess a linear molecular structure according to their polymerization process, too. Figure 13.20 shows the elongational viscosities of the two anionic polystyrenes PS II and PS III whose molar mass distributions are displayed in the inset. The narrowly distributed PS III does not exhibit any strain hardening in the range of the elongational rates applied. The addition of a small high molar mass component to a very narrow distribution (cf. the tiny high molar mass peak of the distribution curve of PS II in the inset of Fig. 13.2) leads to a very pronounced strain hardening which diminishes at very small elongational rates [29].

The strain-hardening effect due to a high molar mass component observed for anionic polystyrenes can be found principally but less pronounced for broadly distributed polystyrenes, too, as the Figs. 13.21 and 13.22 demonstrate. The radically polymerized PS-r-95 (M_n = 95 kg/mol) was mixed with small amounts of anionic PS with the different molar masses of 600 kg/mol (PS-a-600) and 1.800 kg/mol (PS-a-1800), respectively. PS-r-95 itself does exhibit a small strain hardening effect at the two highest strain rates of 1.0 and 0.3 s^{-1} (cf. Fig. 13.21). At the lower strain rates of 0.1 and 0.06 s^{-1} the elongational curves are in good agreement with $3\eta^0(t)$. The addition of 1 wt. % of the PS-a-600 does not show any

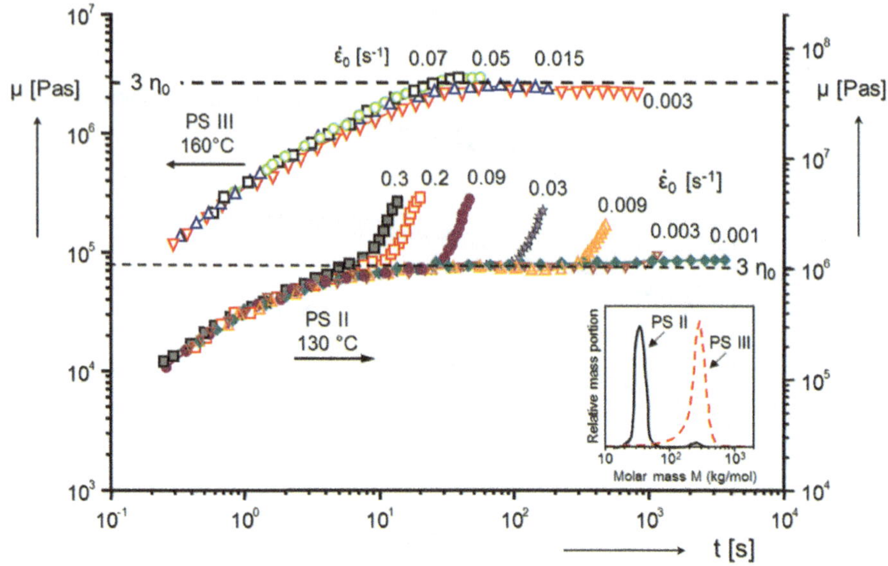

Fig. 13.20 Elongational viscosity μ as a function of time t at various constant elongational rates $\dot{\varepsilon}_0$ for two anionic polystyrenes with different molar mass distributions. (Reprinted from [29] with permission from The Society of Rheology)

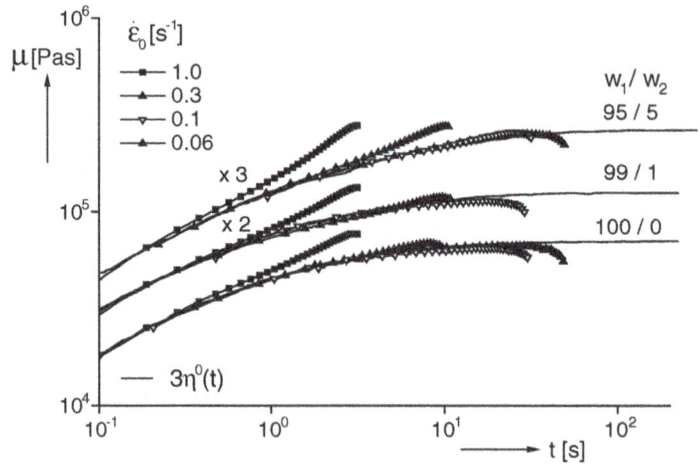

Fig. 13.21 Elongational viscosities of blends of the radically polymerized polystyrene PS-r-95 of $M_n = 95$ kg/mol and $M_w/M_n = 1.6$ with $w_2 = 1$ wt. % and 5 wt. % of the anionic polystyrene PS-a-600 of a molar mass of 600 kg/mol measured at $T = 169$ °C. The curves are shifted along the viscosity axis by the factors indicated [30]. The weak maxima at small elongational rates are due to the beginning of the nonuniformity of sample deformation

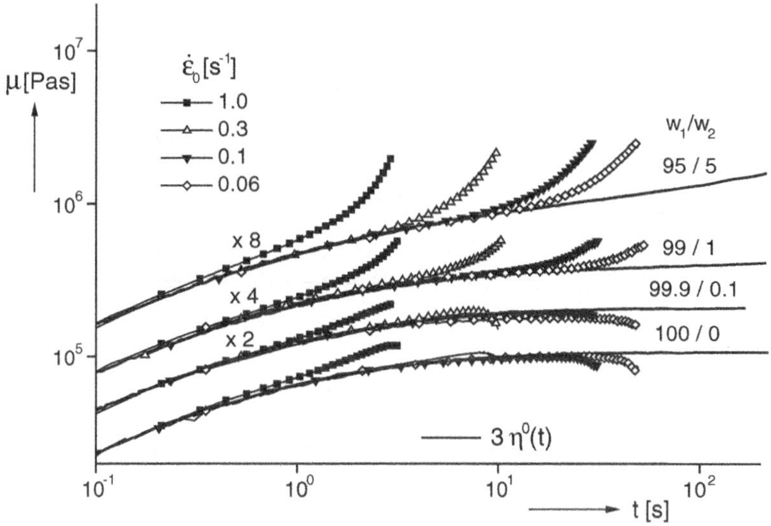

Fig. 13.22 Elongational viscosities of blends of the radically polymerized polystyrene PS-r-95 of $M_n = 95$ kg/mol and $M_w/M_n = 1.6$ with 0.1, 1 and 5 wt. % (w_2) of the anionic polystyrene PS-a-1800 of a molar mass of 1800 kg/mol measured at $T = 169$ °C. The curves are shifted by the factors indicated [30]. The weak maxima at small elongational rates are due to the beginning of the nonuniformity of sample deformation

effect, 5 wt. % of the anionic polystyrene enhance the strain hardening, however. The effect of adding the PS-a-1800 is more pronounced (cf. Fig. 13.22). 1 wt. % leads to a significantly higher strain hardening than for the matrix PS-r-95 which becomes still more pronounced for 5 wt. %. These results shine some light on the effect of high molar mass components on strain hardening. Their efficiency is the more pronounced the larger the ratios of the molar masses. This finding indicates the molecular route one can go if strain hardening for a linear polymer is intended to be generated.

13.3.2 Influence of Long-Chain Branches

As follows from the polystyrene blends in the Figs. 13.20 to 13.22 and the LDPE in Fig. 12.12, strain hardening can be related to two molecular parameters, high molar mass components or long-chain branches. In order to separate these two effects, elongational experiments were performed on the two metallocene polyethylenes mLLDPE 1 and LCB-mLLDPE 1 whose molar mass distributions cannot be distinguished from each other as shown in Fig. 13.15. As Fig. 13.23 demonstrates, mLLDPE 1 which should not contain any long-chain branches according to the GPC/MALLS analysis and the way of polymerization does not show any strain hardening. The slight decrease of the curves is related to a nonuniformity of sample

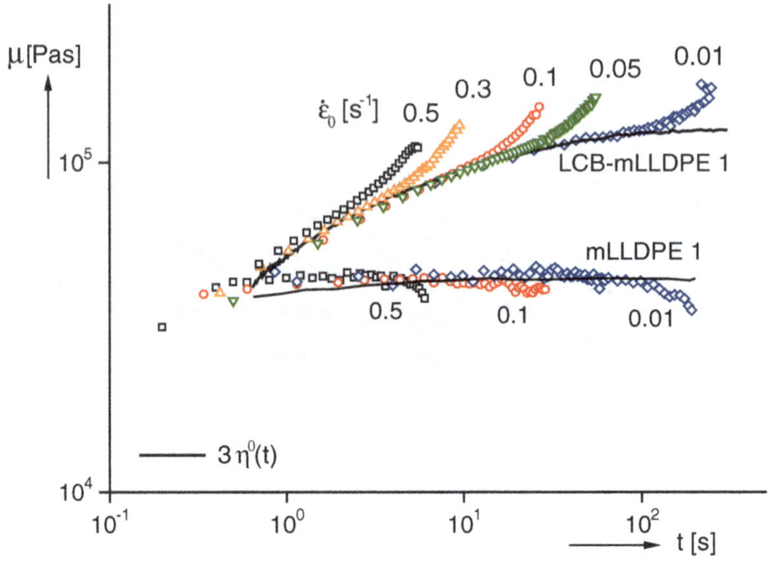

Fig. 13.23 Influence of long-chain branching on the elongational viscosity μ of a metallocene-catalyzed linear (mLLDPE 1) and a long-chain branched polyethylene (LCB-mLLDPE) with very similar molar mass distributions at $T = 150$ °C and a maximum Hencky strain $\varepsilon_{H\max} = 2.7$. η^0(t) describes the linear time-dependent shear viscosity. (Reprinted from [31] with permission from The Society of Rheology)

deformation. The viscosities measured at different elongation rates come to lie on one curve within the accuracy of the measurements. The elongational viscosity of the long-chain branched LCB-mLLDPE clearly increases as a function of time at all the elongational rates covering two decades. This result is a clear indication that long-chain branching induces strain hardening [31].

Another experimental hint to the strain-hardening effect of long-chain branches are measurements on two polypropylenes [32]. The linear polypropylene (PP) and the long-chain branched polypropylene (LCB-PP) are commercial products. The elongational viscosities of these samples are shown in Fig. 13.24. They were measured over a wide range of elongational rates $\dot{\varepsilon}_0$ covering three decades. The linear PP does not exhibit any indication of strain hardening similar to the linear polypropylenes shown in Fig. 13.19. Its molar mass of $M_w = 409$ kg/mol and its polydispersity factor of $M_w/M_n = 4.1$ lie within the range of the other linear PP investigated (cf. Table 13.6). The increase of the elongational viscosity of the long-chain branched sample is very pronounced. At the maximum elongation of $\varepsilon_{H\max} = 2.8$ no indication of a bending off of the curves can be seen. Experimentally very convincing is the finding that the elongational viscosities of the linear PP fall together with $3\eta^0(t)$ and that the envelope of the curves of the branched sample matches the relationship following from the theory of linear viscoelasticity, too.

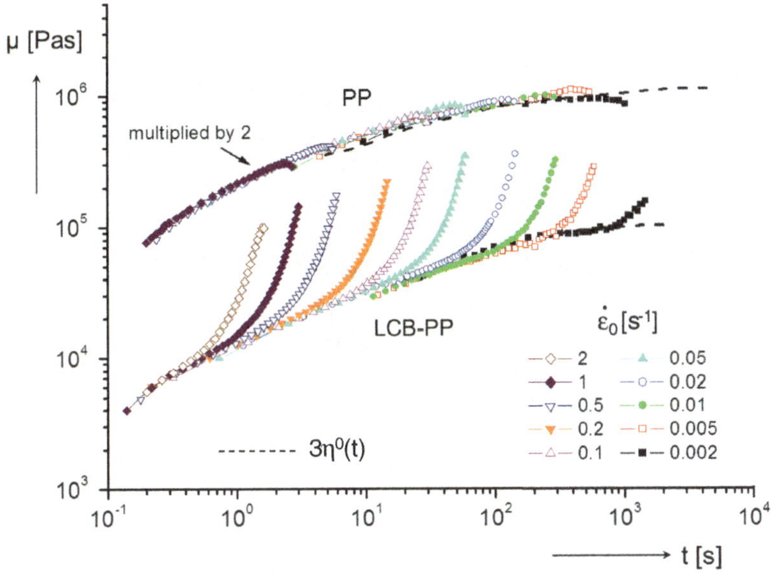

Fig. 13.24 Elongational viscosities μ of a linear polypropylene (PP) and a long-chain branched polypropylene (LCB-PP) at various elongational rates $\dot{\varepsilon}_0$ and $T = 180$ °C. (Reprinted from [32] with permission from The Society of Rheology)

Although the elongational behavior of polymer melts and, particularly, their special feature of strain hardening is affected by high molar mass components and long-chain branches as well, it can be very helpful for the analysis of branching in some cases. This power is demonstrated by Fig. 13.25, which shows the elongational viscosities of polypropylenes electron irradiated with small doses [18]. Their molecular data are given in Table 13.3. For the linear non-irradiated sample PP-0 which is identical with PP-A presented in Fig. 13.19, strain hardening cannot be observed at all the elongational rates applied. For PP-1 which was irradiated with the lowest dose of 1 kGy a faint but significant viscosity increase can already be detected at low elongational rates. This strain-hardening effect becomes more pronounced by increasing the dose up to 10 kGy. According to Table 13.3, the polydispersity factor slightly decreases at low irradiation doses and remains nearly unaffected by higher doses. Therefore, a broadening of the molar mass distribution can be excluded as the source of the strain hardening observed and it can be taken as an indication of long-chain branching. This result demonstrates that strain hardening very sensitively reacts on long-chain branching as for PP-1 and PP-2 branching could not be found by GPC/MALLS. This method reveals first indications of branching at doses of 5 and 10 kGy (cf. Table 13.3).

In Fig. 13.26 the strain-hardening coefficient χ defined by Eq. (12.30) is plotted as a function of the constant elongational rate. It was determined at a Hencky strain of $\varepsilon_H = 2.7$. The strain-hardening coefficient increases with the irradiation

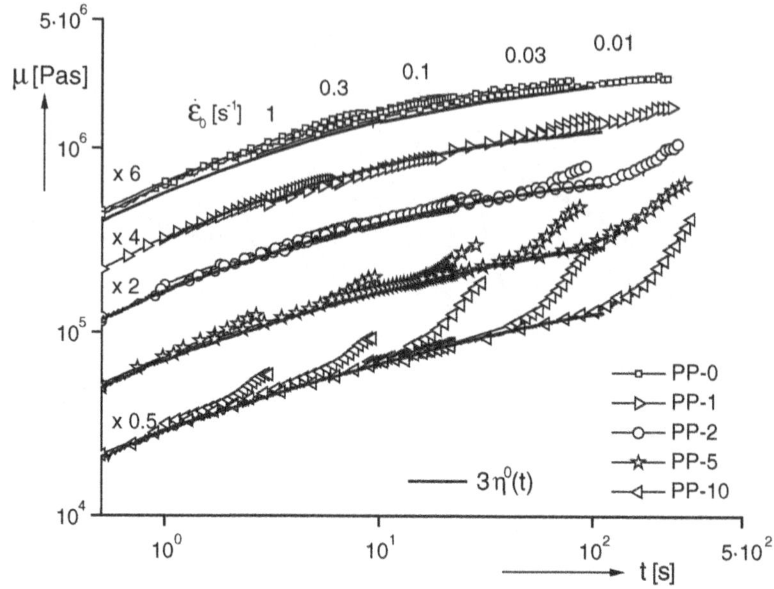

Fig. 13.25 Elongational viscosities μ of polypropylenes electron beam irradiated with different doses which are indicated in kGy by the digits of the sample denominations at different constant elongational rates $\dot{\varepsilon}_0$ and $T = 180$ °C. For a better distinction the curves are shifted along the viscosity axis by the numbers indicated. (Reprinted from [18] with permission of American Chemical Society)

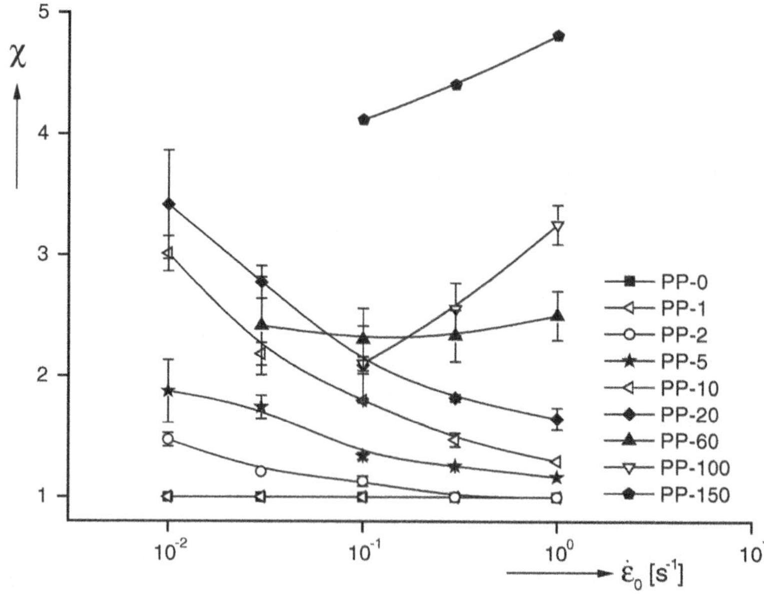

Fig. 13.26 Strain-hardening coefficient χ as a function of the elongational rate $\dot{\varepsilon}_0$ for the polypropylenes electron beam irradiated with different doses. The strain-hardening coefficient was determined at a total Hencky strain of $\varepsilon_H = 2.7$ and $T = 180$ °C [28]

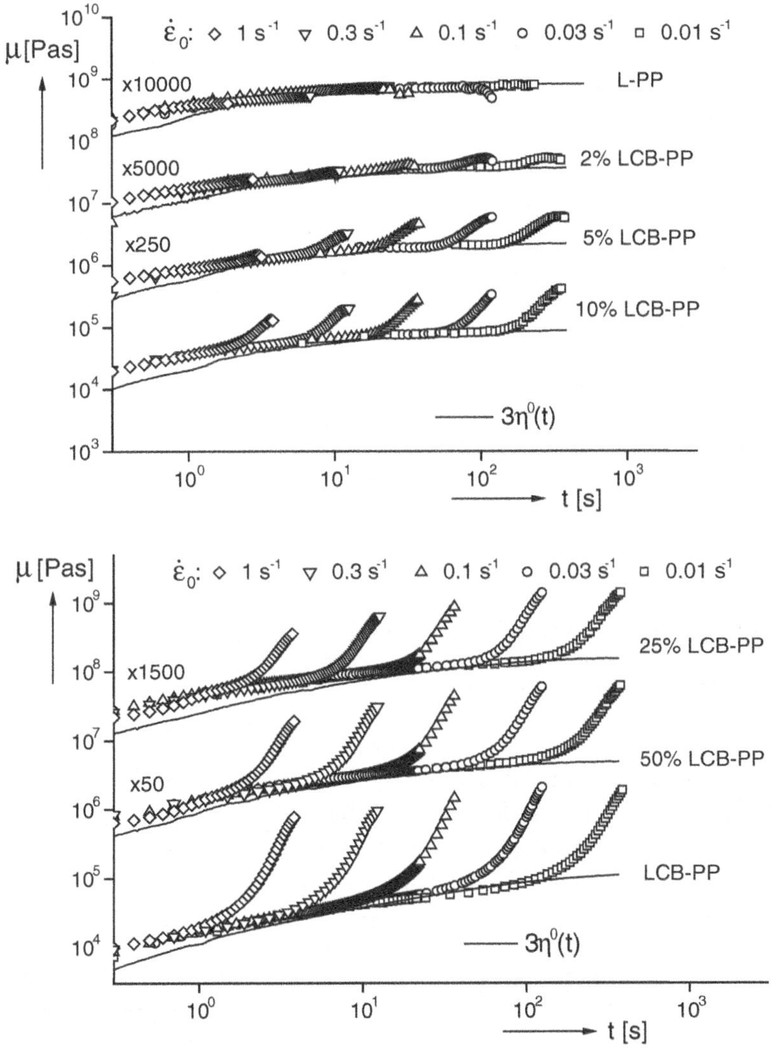

Fig. 13.27 Elongational viscosities of blends of a linear and a long-chain branched polypropylene at $T = 180$ °C. Above: lower weight percentages, below: higher weight percentages. The curves were shifted according to the factors indicated. (Reprinted from [33] with the permission from The Society of Rheology)

dose. Up to doses of 20 kGy, χ decreases with growing elongational rate, for 100 and 150 kGy a distinct increase is found.

Definite conclusions on the architecture of the branched molecules cannot be drawn, however, from the elongational viscosity alone. This insight is demonstrated by the upper and lower graph of Fig. 13.27, which exhibit the elongational viscosities of blends of a linear and a long-chain branched polypropylene similar to

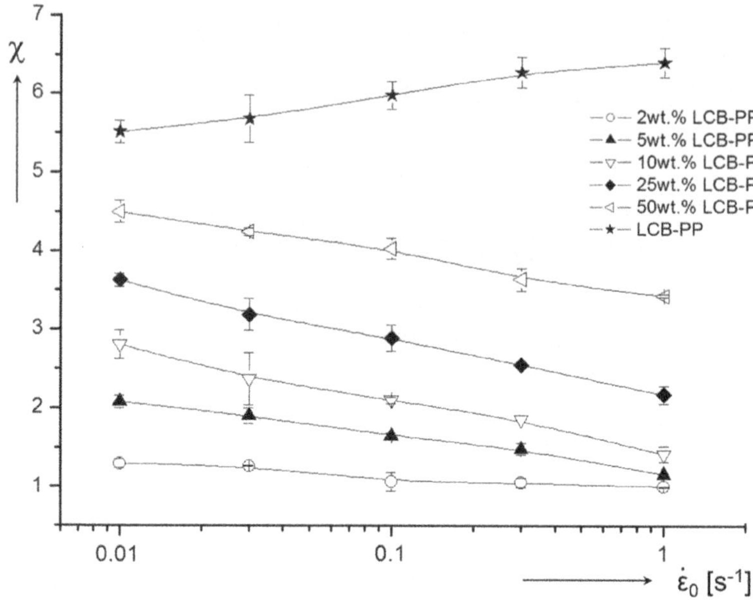

Fig. 13.28 Strain hardening coefficient χ at a Hencky strain of $\varepsilon_H = 2.7$ and $T = 180$ °C as a function of the elongational rate $\dot{\varepsilon}_0$ for the blends of Fig. 13.27. (Reprinted from [33] with the permission from The Society of Rheology) [33]

that shown in Fig. 13.24 [33]. Whereas the viscosity function of the linear sample is in good agreement with $3\eta^0(t)$, already 2 wt. % of the branched material lead to an indication of strain hardening which becomes more pronounced with growing amount of the long-chain branched species. The samples were elongated up to a maximum Hencky strain of $\varepsilon_H = 2.7$. The strain-hardening coefficient decreases with increasing elongational rate up to the highest content of 50 wt. % of the branched material as depicted by Fig. 13.28 similar to the samples irradiated with low and medium doses in Fig. 13.26. For the branched blend component, a slight increase of the strain-hardening coefficient with elongational rate occurs within the range of strain rates applied.

This result demonstrates that the strain hardening obtained by electron beam irradiation can qualitatively be mimicked by blending a long-chain branched polypropylene to a linear one. This result poses the question whether the irradiated sample is only partly branched, i.e., whether linear molecules exist besides some branched ones. A strong argument for branching structures of the irradiated samples different from those of the blends can be derived, however, from Fig. 13.29 which shows $\eta_0(M_w)$ for the blends. All data points are distinctly located below the straight line for linear polypropylenes. They differ significantly from the irradiated samples for which $\eta_0(M_w)$ comes to lie above the reference line in a wide range of doses (cf. Fig. 13.9).

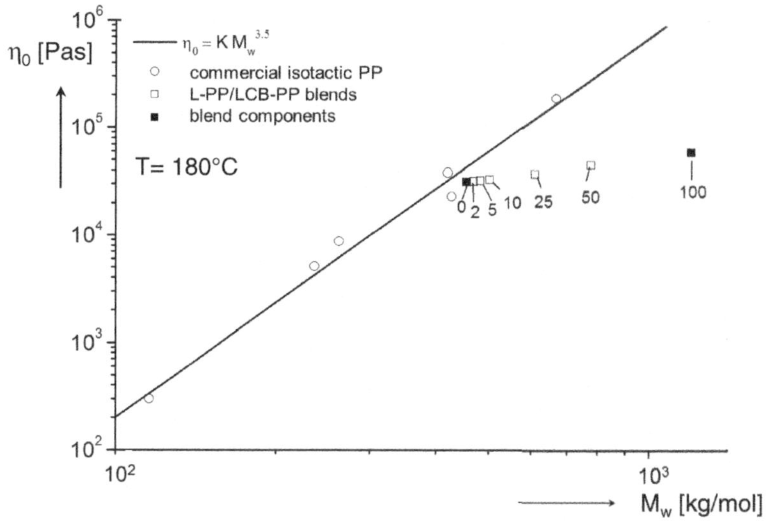

Fig. 13.29 $\eta_0(M_w)$ for the blends of a linear and a long-chain branched polypropylene at $T = 180$ °C. The numbers below the symbols correspond to the weight percentage of the long-chain branched component. (Reprinted from [33] with the permission from The Society of Rheology)

13.3.3 Conclusions on the Relationships Between Elongational Viscosity and Molecular Structure

The following conclusions can be drawn from the results presented:

- For polymers with linear chains, e.g., polypropylenes and anionic polystyrenes, the elongational viscosity as a function of time is three times the time-dependent zero-shear viscosity within a wide range of elongational rates. The elongational viscosity in this regime is only determined by the influence of the molecular structure on the time-dependent zero-shear viscosity and no strain hardening occurs.
- High molar mass components can lead to a significant increase of the elongational viscosity above the linear regime, the so-called strain hardening.
- Strain hardening is found for long-chain branched polymers, too.
- The elongational viscosity very sensitively reacts on long-chain branches.
- From the strain hardening alone conclusions with respect to the branching architecture cannot be drawn, however.

References

1. Münstedt H (2011) Soft Matter 7:2273
2. Gabriel C (2001) Doctoral Thesis, University Erlangen-Nürnberg, ISBN 3-8265-8909-2, Shaker Verlag Aachen
3. Wolf M (1992) Doctoral Thesis, University Erlangen-Nürnberg
4. Berry GC, Fox TG (1968) Adv Polym Sci 5:261
5. Fetters LJ, Lohse DJ, Colby RH (2006) Mark JE (ed) Physical properties of polymers handbook, Springer, Berlin
6. Matsuda T, Kitagawa K, Inoue T (1970) Macromolecules 3:116
7. Pechhold WV, Soden W, Stoll B (1981) Makromol Chem 182:573
8. Stadler F, Piel C, Kaschta J, Rulhoff W, Kaminsky W, Münstedt H (2006) Rheol Acta 45:755
9. Vega JF, Otegui J, Ramos J, Martinez-Salazar J (2012) Rheol Acta 51:81
10. Nobile MR, Cocchini F (2000) Rheol Acta 39:152
11. Den Doelder J (2006) Rheol Acta 46:195
12. Stadler FJ, Münstedt H (2008) J Rheol 52:697
13. Kessner U (2010) Doctoral Thesis, University Erlangen-Nürnberg, ISBN 13:978-3-86844-300-4, Sierke Verlag Göttingen
14. Pearson DS, Helfland E (1984) Macromolecules 17:888
15. Ball RC, McLeish TCB (1989) Macromolecules 22:1911
16. McLeish TCB, Milner ST (1999) Adv Polym Sci 143:195
17. Gabriel C, Lilge D (2006) Rheol Acta 45:995
18. Auhl D, Stange J, Münstedt H, Krause B, Voigt D, Lederer A, Lappan U, Lunkwitz K (2004) Macromolecules 37:9465
19. Resch JA (2010) Doctoral Thesis, University Erlangen-Nürnberg
20. Dealy JM, Larson RG (2006) Structure and Rheology of Molten Polymers, Carl Hanser Verlag, Munich
21. Ferry JD (1980) Viscoelastic properties of polymers. Wiley, New York
22. Bueche F (1952) J Chem Phys 20:1959
23. Onogi S, Masuda T, Kitagawa K (1970) Macromolecules 3:109
24. Zosel A (1971) Kolloid-Z Z Polym 246:657
25. Münstedt H (1986) Fließverhalten von Stoffen und Stoffgemischen, Kulicke, WM edn. Hüthig & Wepf Basel, Heidelberg
26. Gabriel C, Münstedt H (1999) Rheol Acta 38:393
27. Gabriel C, Münstedt H (2002) Rheol Acta 41:323
28. Auhl D (2006)Doctoral Thesis. University Erlangen-Nürnberg
29. Münstedt H (1980) J Rheol 24:847
30. Hepperle J (2003) Doctoral Thesis. University Erlangen-Nürnberg, Shaker Aachen
31. Gabriel C, Münstedt H (2003) J Rheol 47:619
32. Kurzbeck S, Oster F, Münstedt H, Nguyen TQ, Gensler R (1999) J Rheol 42:359
33. Stange J, Uhl C, Münstedt H (2005) J Rheol 49:1059

Chapter 14
Thermorheological Behavior of Various Polymer Melts

As discussed in Chap. 6, a great number of polymer melts follow the time-temperature superposition principle. The temperature dependence of the shift factors is different, however. An Arrhenius equation with specific activation energies was found for the polyethylenes investigated. These results open up the question on relationships between activation energies and the molecular structure of semi-crystalline polymers. The shift factors for the linear polystyrene in Sect. 6.4 follow a WLF-equation with two invariants as known for amorphous polymers. For this kind of polymeric material it is of practical and fundamental interest in which way the coefficients of the WLF-equation depend on the type of polymer and its molecular structure.

14.1 Amorphous Polymers

In Fig. 14.1 the shift factors of a variety of polystyrenes with different molecular structures are presented. They were obtained from dynamic-mechanical measurements in the flow region exhibiting a thermorheologically simple behavior for all the samples investigated [1]. The shift factors a_T of the polystyrenes listed in the Fig. 14.1 are indistinguishable from each other and come to lie on a curve typical of a WLF-equation as described by Eq. (6.26). This relation corresponds to Eq. (6.7) with the coefficients c_1 and c_2 valid in the glass transition.

"r" and "a" describe radically and anionically polymerized samples, respectively, the numbers denote the molar masses M_n in kg/mol. The standard deviations of a_T lie within the accuracy of the experiments, although M_n of the samples distinctly differ and the molar mass distributions vary over a wide range. These results mean that the coefficients a_T and following from them the invariants of the WLF-equation are independent of the molar mass and its distribution. The mean values of the invariants d_1d_2 and T_∞ (cf. Eqs. (6.20) and (6.21)) of the blends and their components are listed in Table 14.1 together with their standard deviations. Similar results can be found in the literature [2, 3].

H. Münstedt and F. R. Schwarzl, *Deformation and Flow of Polymeric Materials*,
DOI: 10.1007/978-3-642-55409-4_14, © Springer-Verlag Berlin Heidelberg 2014

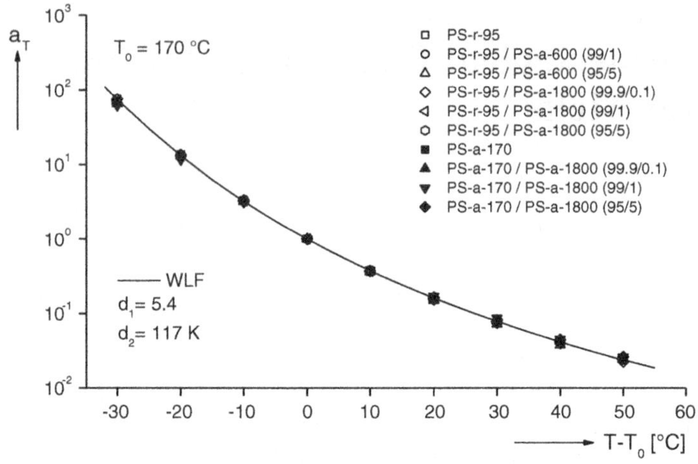

Fig. 14.1 Shift factor a_T as a function of $T-T_0$ for various anionic (*PS-a*) and radical (*PS-r*) polystyrenes and their blends [1]. The numbers within the designations of the products denote M_n in kg/mol, the numbers in brackets give the blend ratios. The reference temperature is $T_0 = 170$ °C. d_1 and d_2 are the coefficients of the corresponding WLF-equation (cf. Eq. 6.26)

Table 14.1 The glass temperature T_g measured by differential scanning calorimetry (DSC) and mean values of the invariants d_1d_2 and $T_\infty$of polystyrenes and polycarbonates with various molecular structures [1] and polymethylmethacrylate from the literature [4]. T_g represents the value extrapolated to the heating rate zero for PS and PC, for PMMA it was determined at a heating rate of 10 K/min. "st" denotes the syndiotacticity of the PMMA

	T_g [°C]	$d_1d_2[K]$	$T_\infty[°C]$
Polystyrene blends	99.6 ± 1.7	631 ± 23	53 ± 2
Branched polystyrenes	98.3 ± 1.1	648 ± 27	51 ± 2
Linear and branched polycarbonates	145.2 ± 1.2	401 ± 22	110 ± 2
PMMA (81 % st) [4]	130	1648 ± 28	15 ± 1
PMMA (~59 % st) [4]	117	1536	13.3
PMMA (~55 % st) [4]	105	1211	47.0
PMMA [5]	–	1270	30

In Fig. 14.2 the shift factors of nine polystyrenes with different branching architectures are plotted. Their structure follows from the nomenclature given in the caption to the figure.

Again no differences can be seen within the accuracy of the plot. The invariants given in the third and fourth columns of Table 14.1, show small standard deviations which indicate that there is no influence of the various branching topographies on the shift factors as a function of the temperature. A comparison of the invariants of the blends and the long-chain branched samples does not point to any significant difference, either. The similarity of the WLF-equations valid for the glass transition and in the molten state of amorphous polymers suggests to relate d_1 to the free

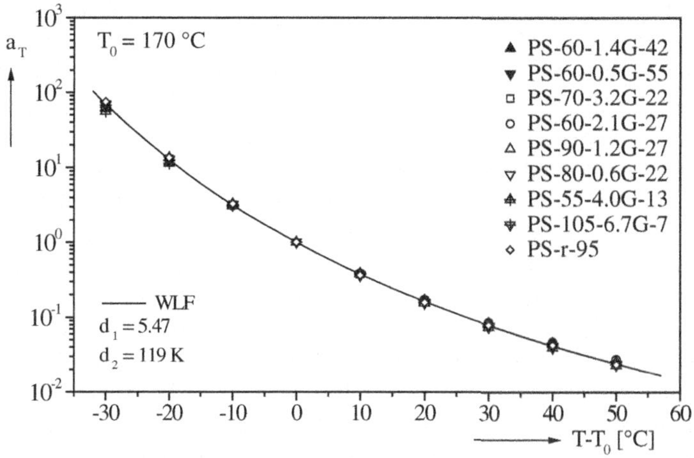

Fig. 14.2 Shift factor a_T as a function of $T-T_0$ for polystyrenes with different branching architectures. The samples are designated as PS-x-pG-y with x denoting M_n of the backbone in kg/mol, p the average number of grafted branches G per molecule and y the number average molar mass M_n of the branches in kg/mol. d_1 and d_2 are the coefficients of the WLF-equation. The reference temperature is $T_0 = 170$ °C [1]

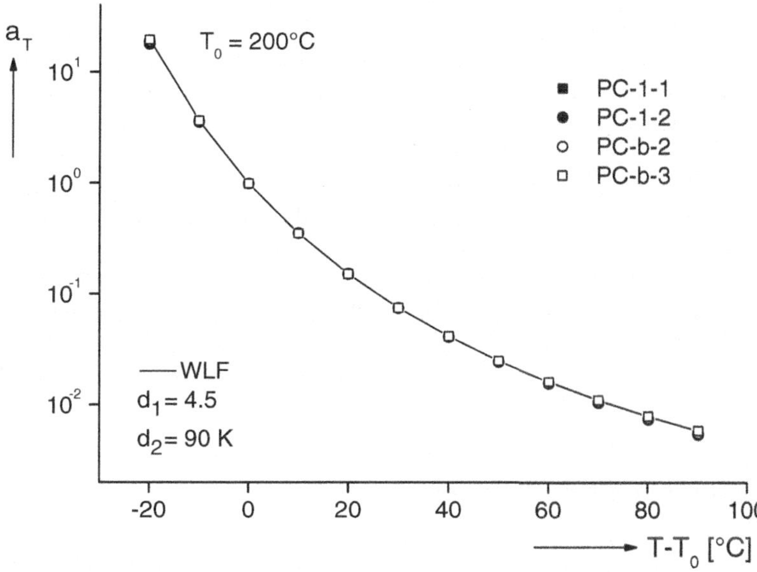

Fig. 14.3 Shift factor a_T as a function of the temperature difference $T-T_0$ for four different polycarbonates. "b" denotes branching. d_1 and d_2 are the coefficients of the WLF-equation. The reference temperature is $T_0 = 200$ °C [1]

volume and $d_1 \cdot d_2$ to the expansion coefficient according to Eq. (6.18) for c_1 and Eq. (6.20) for $c_1 \cdot c_2$ [6]. Therefore, one can assume from these experimental findings that neither the thermal expansion coefficient nor the size of the descriptive, but physically not well defined free volume are affected by the length and the distribution of the macromolecules attached to the backbone, at least in the case of the relatively small numbers of branches of the polystyrenes investigated.

In Fig. 14.3, the shift factors of two linear polycarbonates (PC) and two branched polycarbonates (PC-b) are plotted semilogarithmically as a function of T-T_0. They were obtained from temperature-frequency shifts of G' and G'' which do not show any indication of a thermorheological complexity [1]. PC-1-2 has a slightly larger M_n than PC-1-1, $M_w/M_n = 1.9$ is the same for both samples. PC-b-3 is distinguished from PC-b-2 by a higher degree of branching and a somewhat larger M_n. The polydispersity indices are $M_w/M_n = 2.3$ for PC-b-2 and $M_w/M_n = 2.7$ for PC-b-3. All the measured points can be connected by one curve the shape of which is typical of the WLF-equation. Like for the polystyrenes in the Figs. 14.1 and 14.2, the molecular structure of the polycarbonates has no influence on the shift factor as a function of the temperature. The mean invariants $d_1 d_2$ and T_∞ are listed in the Table 14.1. They are similar to those found in the literature [7] on linear polycarbonates. The standard deviations of the invariants as determined from measurements on four different samples are of the order of those of the polystyrenes. The values are significantly different, however, indicating other properties of the free volume.

The Table 14.1 contains the data of another commercially important amorphous polymer, namely polymethylmethacrylate [4]. The invariants for the PMMA of 81 % syndiotacticity are the averages of seven samples covering a range of M_n between 41 and 340 kg/mol. The polydispersity factor is about 1.1 for all the samples. Similar to polystyrene and polycarbonate, the invariants are not dependent on the molar mass as the standard deviation is small. The invariants are markedly different from those of the polycarbonate and the polystyrene, however. This finding is not surprising at all if one takes the various chemical structures of the three materials into account.

The results on the PMMA of different syndiotacticities in Table 14.1 demonstrate that the syndiotacticity has a remarkable influence on the invariants and following from that on the parameters of the free volume. A systematic effect of the syndiotacticity cannot be derived, however, from the limited number of data available.

14.2 Semicrystalline Polymers

As shown in Fig. 6.17, the activation energies of the linear polyethylene (HDPE) and the branched polyethylene (LDPE) are different, i.e., contrary to the polystyrene and the polycarbonate discussed in the previous section, long-chain branching has an influence on the temperature dependence of the shift factor of the melt. In addition, another distinction comes into play as not for all polyethylenes a thermorheological simplicity is found. This feature becomes obvious from

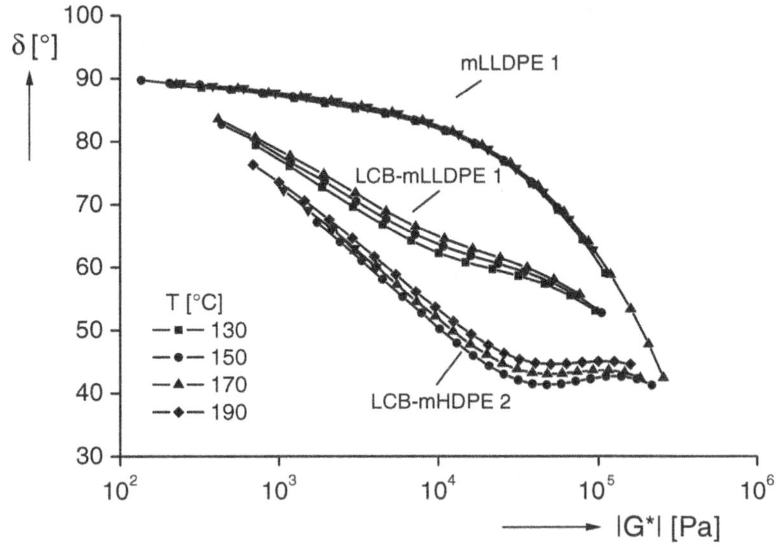

Fig. 14.4 Phase angle δ as a function of the magnitude of the complex modulus $|G^*|$ for the linear polyethylene mLLDPE 1 and the two long-chain branched polyethylenes LCB-mLLDPE 1 and LCB-mHDPE 2 at various temperatures. *m* indicates the synthesis with metallocene catalysts [8]

Fig. 14.4, which shows the phase angle δ as a function of the magnitude of the complex modulus $|G^*|$. As discussed in Sect. 11.3.2, all the data measured at different temperatures should come to lie on one curve in the case of thermo-rheological simplicity. Deviations from such a common curve are an indication of thermorheological complexity. Obviously, the metallocene-catalyzed linear low density polyethylene mLLDPE 1 behaves thermorheologically simple. For the other two samples, temperature-independent master curves are not obtained, i.e., their behavior is thermorheologically complex. The two long-chain branched polyethylenes were synthesized by metallocene catalysts which are assumed to generate a mixture of linear and star-like branched molecules. This example shows that investigations on the thermorheological behavior of a polymer may be used to answer the question whether a material is purely linear or contains long-chain branches.

14.2.1 Thermorheological Simplicity and Its Application to the Analysis of Short-Chain Branching

Linear homopolymers like HDPE or PP behave thermorheologically simple, i.e., the data can be shifted over a wide time or frequency scale to obtain master curves. An answer to the question whether the activation energy depends on the molar

Table 14.2 Activation energies E_A of linear ethylene and propylene homopolymers with various molar masses M_w and polydispersity factors M_w/M_n. The data for the mHDPE are taken from [9], those for the isotactic PP from [10]. T_m is the melting temperature

	T_m [°C]	M_w [kg/mol]	M_w/M_n	E_A [kJ/mol]
mHDPE 1	135	205	2.6	27
mHDPE 2	134	45	2.7	27
PP 1	165	219	4.9	40
PP 2	165	325	5.9	40
PP 3	165	525	6.0	42

mass or the molar mass distribution is given by the data in Table 14.2. For the two HDPE of significantly different molar masses, but very similar polydispersity factors the same activation energies are found. For the three polypropylenes which differ in molar mass and polydispersity the same activation energies E_A follow within the accuracy of the measurements, but they are distinctly higher than those of the polyethylenes. Taking these data and others from the literature into account, it can be concluded that molar mass and molar mass distribution do not have an influence on the activation energy of linear polyethylenes and polypropylenes.

A thermorheologically simple behavior can be found for short-chain branched polyethylenes, too, as demonstrated by the plot of δ ($|G^*|$) in Fig. 14.5 for ethylene copolymers with various contents of the comonomer hexene. The concentrations range from 1.2 mol % for L6-1 to 3.0 mol % for L6-4 [11]. The inset with a higher resolution of the phase angle close to 90° demonstrates the excellent superposition of the curves measured at temperatures between 130 and 210 °C even in a range, in which the requirements regarding the accuracy of a measurement are very challenging. A thermorheologically simple behavior was found for butene and octene copolymers with comonomer contents up to 7.6 mol %, too [11].

Although the presentation of δ ($|G^*|$) is very sensitive and versatile to discriminate between thermorheologically simple or complex materials, the activation energy cannot be determined from it, however. For that purpose, a rheological quantity has to be plotted as a function of time or frequency at various temperatures in order to get the shift factor as a function of temperature from the master curve. In principle, all the functions can be used the time-temperature behavior of which is governed by relaxation or retardation spectra, respectively. Frequently, G' and G'' are measured as functions of the angular frequency at different temperatures and shifted onto a master curve.

An interesting question is in which way the short-chain branches of ethylene/olefin copolymers affect the activation energy. Due to the finding that the activation energy is not dependent on the molar mass and its distribution (cf. Table 14.2), in principle clear correlations between activation energies and comonomers should be possible to get. A difficulty, however, is the detailed knowledge of the real amount of comonomers built into a chain and their

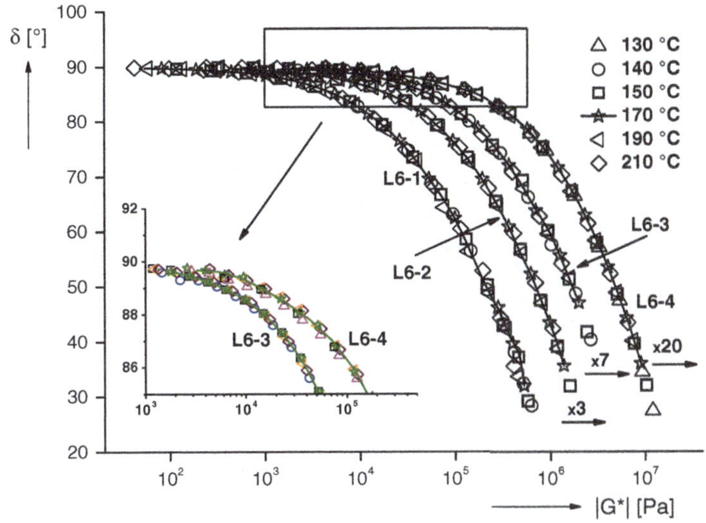

Fig. 14.5 Phase angle δ as a function of the magnitude of the complex modulus $|G^*|$ at different temperatures for ethylene/hexene copolymers with various hexene contents [11]. L6-1: 1.2 mol %, L6-2: 2.6 mol %, L6-3: 2.6 mol %, L6-4: 3.0 mol %. For a better distinction the curves are shifted along the $|G^*|$- axis by the factors indicated

Table 14.3 Activation energies E_A of ethylene copolymers with various comonomers. n_c is the comonomer content in mol-%, s_c denotes the weight percentage of the short-chain branches [12]

Name	Comonomer	n_c [mol-%]	s_c [wt%]	E_A [kJ/mol]
mHDPE 1	–	0	–	27
mHDPE 2	–	0	–	27
L4	Butene	6.8	6.4	32
L6	Hexene	5.9	10.6	33
L8	Octene	1.0	2.9	28
F18F	Octadecene	3.0	21.8	35
F18G	Octadecene	4.4	26.0	39
F26C	Hexacosene	2.3	21.6	37
F26F	Hexacosene	2.7	26.5	39

distribution along the backbone, as in many cases the polymerization is not easy to perform in a definite way. The amounts of the comonomers n_c listed in mol-% in the Table 14.3 for various copolymers were obtained from melt-state or solution NMR and in some cases from the fast Fourier transform infrared spectroscopy (FT-IR). The side-chain content s_c of the short-chain branches was calculated according to

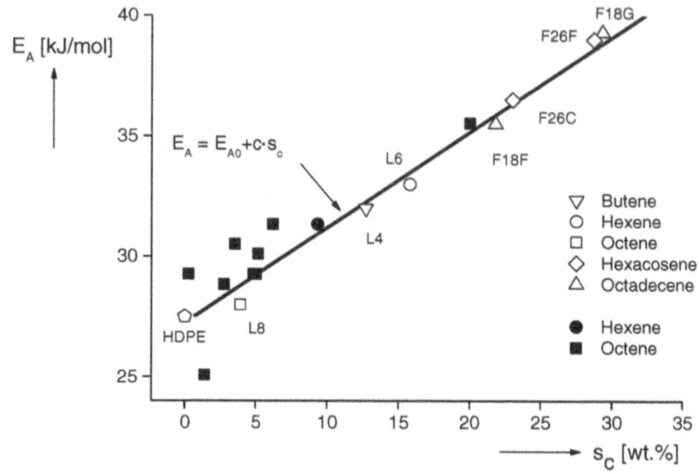

Fig. 14.6 Activation energy of flow E_A as a function of the weight percentage s_c of various short-chain branches in ethylene copolymers. The filled symbols were taken from [14]. E_{A0} describes the activation energy without any short-chain branches, i.e., that of an HDPE. c is the slope of the straight line. (Reprinted from [9] with permission from John Wiley and Sons)

$$s_c = \frac{n_c(l-2)}{n_c(l-2)+2} \tag{14.1}$$

with l being the number of carbon atoms of the comonomer and n_c its molar ratio. L4, L6, and L8 in Table 14.3 are commercially available butene, hexene, and octene copolymers. The copolymers with octadecene and hexacosene are laboratory products. Detailed information on their polymerization processes and some of their rheological properties can be found in [9] and [12].

From the data given in the Table 14.3 the tendency can be seen that the activation energy E_A increases with the molar content n_c for octadecene and hexacosene. From results in [13] on various butene, hexene and octene copolymers a trend of a slight increase of E_A with the number of carbon atoms of the comonomer may be deduced, but the experiments are not elaborate enough to draw founded conclusions. A clearer tendency occurs from Fig. 14.6 which presents the activation energies E_A of the samples listed in Table 14.3 as a function of the side-chain content s_c of the short-chain branches generated by the comonomers built into the ethylene backbone. One gets a linear relationship between the activation energy and the side-chain content in weight percent. The dependency presented Fig. 14.6 is physically reasonable insofar as the extrapolation of the straight line to $s_c = 0$, i.e., to linear chains, gives an activation energy typical of a HDPE.

Values from the literature for ethylene copolymers with hexene and octene as comonomers presented in Fig. 14.6 as filled symbols [14] do not contradict the straight line found, particularly, if one takes the uncertainties of a reliable determination of the comonomer content into account.

Intuitively, the increase of the activation energy with the content of short-chain branches following from Fig. 14.6 is plausible. Due to the fact that for all the copolymers investigated the temperature of the measurement lies distinctly above the glass transition temperature, the overcoming of potential barriers within the chain can be regarded as the decisive mechanism for a local displacement of molecule segments (cf. Sect. 6.5). The energy necessary for this process will be the higher the more side groups along a chain hinder the motions of particular segments. Additionally, the bulkiness of a side group will affect the mobility of the chain segment to which it is coupled. Therefore, the weight percentage of the short-chain branches may be a not too bad parameter for a relationship between activation energies of ethylene copolymers and their comonomer contents.

14.2.2 Thermorheological Complexity and Its Potential for the Analysis of Branching

In the case of thermorheological complexity the time-temperature shift factor is not the same for all the relaxation times, and, therefore, it becomes dependent on the value of the physical quantity, at which it is taken. As was explained before, thermorheological simplicity implies that the time-temperature shift is equal for all the functions related to a definite relaxation or retardation spectrum or, vice versa, it should not matter which material function is chosen for the determination of the shift factor. Such a behavior is not valid anymore for thermorheologically complex materials as the Fig. 14.7 shows for an LDPE [15]. While from the shift of $\delta(\omega)$ a value of 66 kJ/mol follows for E_A which is independent of δ within the accuracy of the measurements, the activation energies determined from the frequency-temperature shifts at various values of the moduli are somewhat smaller and decrease with increasing modulus. Within the uncertainties of the evaluation, the curves related to the moduli do not differ significantly from each other.

These results can formally be explained by assuming that the moduli G' and G'' change with temperature by a temperature-dependent factor $b_T\,(T)$ i.e.,

$$G'(\omega, T) = b_T(T) \cdot G'(a_T\omega, T_0) \qquad G''(\omega, T) = b_T(T) \cdot G''(a_T\omega, T_0) \tag{14.2}$$

which corresponds to a vertical shift in a double-logarithmic plot. Therefore, $b_T(T)$ is called the vertical shift factor.

As for $\tan\delta = G''/G'$ the shift factor b_T is eliminated, i.e.,

$$\tan\delta(\omega, T) = \tan\delta(a_T\omega, T_0) \tag{14.3}$$

it follows that δ at a given angular frequency is independent of a vertical temperature shift and, therefore, a constant activation energy is found, if all relaxation times change with the same temperature-dependent factor a_T. That means in the case of the LDPE the nonexistence of master curves for G' and G'' as functions of

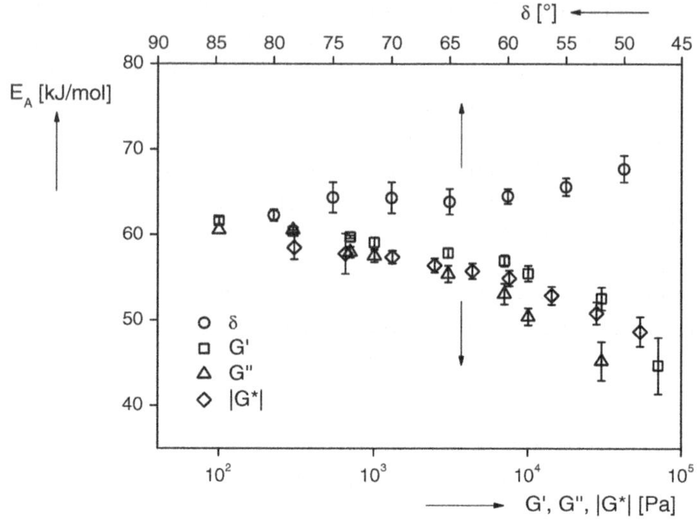

Fig. 14.7 Activation energy E_A for LDPE 2 determined from the frequency-temperature shifts at different values of the material functions $\delta(\omega)$, $G'(\omega)$, $G''(\omega)$ and $|G^*|\ (\omega)$. (Reprinted from [15] by permission from The Society of Rheology)

ω, but their existence for δ can be explained by a shift factor in vertical direction which is temperature dependent.

A constant activation energy from the frequency-temperature dependence of the phase angle δ is not found for all long-chain branched polyethylenes, however, as Fig. 14.8 demonstrates [16]. $E_A\ (\delta)$ for the LCB-mLLDPE 1 and LCB-mLLDPE 2 synthesized with metallocene catalysts show activation energies which do not change much down to phase angles of about 60° and are close to those of the LDPE. For smaller angles, they distinctly decrease with declining phase angle and seem to approach the constant activation energy of the linear sample. Obviously, these metallocene long-chain branched polyethylenes reveal a thermorheological complexity which is different from that of the LDPE. As a vertical shift is eliminated if δ is plotted as a function of ω at different temperatures, a non-constant activation energy points to different shift factors for the relaxation times of the material. In the case of the two LCB-mLLDPE of Fig. 14.8 higher activation energies at larger phase angles corresponding to longer relaxation times, and lower activation energies approaching that of linear species at smaller phase angles are found. This result can be interpreted assuming the existence of linear and branched molecules within the LCB-mLLDPE. It is obvious that the branched molecules possess the longer relaxation times which come into play at smaller frequencies or higher phase angles, respectively, if compared to the linear molecules.

Such an interpretation is supported by the closed symbols in Fig. 14.8, which present results on a blend of 25 wt. % of the linear mLLDPE L6-1 and 75 wt. % of the long-chain branched LDPE 2 with totally different maximal relaxation times of

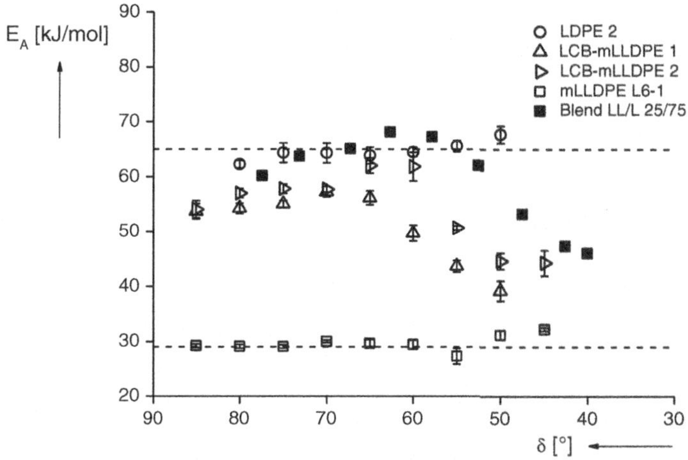

Fig. 14.8 Activation energy E_A as a function of the phase angle δ for the LDPE of Fig. 14.7, the two metallocene long-chain branched LCB-mLLDPE 1 and LCB-mLLDPE 2, the linear copolymer mLLDPE L6-1 and its blend with 75 wt. % of the LDPE 2 (blend LL/L 25/75). (Reprinted from [16] by permission from Elsevier)

$\tau_{max} = \eta_0 J_e^0 = 35$ s at 150 °C for the LDPE 2 and $\tau_{max} = 0.5$ s for the mLLDPE L6-1 [11]. At higher phase angles, i.e., lower angular frequencies, the longer relaxation times of the LDPE component come into play and, therefore, it is not surprising that its thermorheological properties dominate and the activation energies of the blend are very close to those of the LDPE component. At lower phase angles, corresponding to higher frequencies, the molecular motions within the LDPE cannot follow the oscillatory deformation and the linear component with its distinctly shorter maximal relaxation times determines the rheological behavior.

Taking these results into account, the activation energies of the LCB-mLLDPE presented in Fig. 14.8 could be interpreted in the way that the composition of the samples is not homogeneous, but consists of a blend of linear and long-chain branched molecules. This example demonstrates, in which way thermorheological measurements could contribute to an analysis of unknown branched polyethylenes.

In [11] it was shown that activation energies independent of the phase angle can be obtained for various LDPE. However, at a given density their values depend on the molar mass. For example, within a series of products of the same density of 0.919 g/cm^3 at ambient temperatures made by the same tubular process, E_A spreads from 58 kJ/mol for $M_w = 150$ kg/mol over 64 kJ/mol for $M_w = 250$ kg/mol to 72 kJ/mol for $M_w = 370$ kg/mol. This finding may be explained by the assumption that branches become longer with increasing molar mass and give rise to an entanglement network which makes the thermally induced motion of molecule segments more difficult.

The different activation energies of LDPE and the fact that long-chain branched polyethylenes polymerized by metallocene catalysts do not show constant activation energies, even if the phase angle is used for the shift, demonstrate that the effect of the highly variable architectures of branched molecules on the thermal activation is not easy to understand quantitatively. Nevertheless, from thermorheological properties qualitative conclusions can be drawn which support the analysis of branched polymers.

14.2.3 Interpretation and Determination of the Vertical Shift Factor

For accurate time-temperature shifts in the glass transition a vertical shift factor corresponding to the density correction $\rho_0 T_0/\rho T$ is usually applied (cf. Sect. 6.4). For polyethylene melts this factor is not sufficient for an explanation of the experimental findings, as follows from the interpretation of the vertical shift factor according to [17] and its determination by measurements.

In the terminal range

$$G'(\omega, T) = \eta_0(T)^2 \cdot J_e^0(T) \cdot \omega^2 \tag{5.106}$$

and

$$G''(\omega, T) = \eta_0(T) \cdot \omega \tag{5.105}$$

are valid (cf. Sect. 5.8). Taking the shift factors according to Eq. (14.2) into account, Eq. (5.106) reads

$$\eta_0(T)^2 \cdot J_e^0(T) \cdot \omega^2 = b_T a_T^2 \eta_0(T_0)^2 J_e^0(T_0) \omega^2 \tag{14.4}$$

or

$$\frac{\eta_0(T)^{\,2}}{\eta_0(T_0)^2} \frac{J_e^0(T)}{J_e^0(T_0)} = b_T a_T^2 \tag{14.5}$$

From the loss modulus in Eq. (5.105) it follows correspondingly

$$\eta_0(T)\omega = b_T a_T \eta_0(T_0)\omega \tag{14.6}$$

or

$$\frac{\eta_0(T)}{\eta_0(T_0)} = b_T a_T \tag{14.7}$$

Inserting Eq. (14.7) into Eq. (14.5) results in

Table 14.4 Linear steady-state recoverable compliances J_e^0 in $[10^{-5}\ Pa^{-1}]$ for polyethylenes of different molecular structures [11]

T [°C]	130	140	150	170	190	210
mLLDPE L6-1	n.d.	3.8 ± 0.4	3.7 ± 0.3	3.8 ± 0.2	3.9 ± 0.3	3.9 ± 0.8
LDPE 2	81.3 ± 0.4	n.d.	73.5 ± 0.6	66.6 ± 0.9	59.9 ± 0.8	55.0 ± 0.5
LCB-mLLDPE 1	31.8 ± 0.7	n.d.	28.4 ± 0.3	26.0 ± 0.4	24.6 ± 0.8	n.d.
LCB-mLLDPE 2	61.3 ± 1.1	n.d.	54.6 ± 1.3	47.6 ± 1.0	43.4 ± 1.8	n.d.

n.d.: not determined

$$b_T = \frac{J_e^0(T_0)}{J_e^0(T)} \tag{14.8}$$

As a consequence, for thermorheologically simple materials, which in a first approximation do not need a vertical shift to obtain master curves, i.e., $b_T = 1$, J_e^0 should be independent of temperature.

As is discussed in Sect. 11.4, J_e^0 can be determined by creep recovery experiments. The data obtained for the linear mLLDPE L6-1, the LDPE 2, the LCB-mLLDPE 1 and the LCB-mLLDPE 2 are listed in the Table 14.4. J_e^0 for the linear sample is found to be constant within the accuracy of the measurement, i.e., the density correction according to Rouse (cf. Eq. (6.16)) is not reflected. This result is reasonable due to the weak temperature dependence of $T_0\rho_0/T\rho$ for polyethylenes (cf. Fig. 14.9). For $T = T_0 \pm 40$ °C, $T_0\rho_0/T\rho$ changes by ±7 % according to Fig. 14.9, for example. This correction is distinctly smaller than the uncertainties of J_e^0 given in Table 14.4.

In the Fig. 14.9 $J_e^0(T)$ related to the values at the reference temperature $T_0 = 170$ °C obtained from the Table 14.4 are plotted linearly as a function of the reciprocal absolute temperature. All of them are distinctly stronger functions of $1/T$ than $\rho_0 T_0/\rho T$.

In case of the LDPE 2 with $b_T \neq 1$, master curves for G', G'' or $|G^*|$ have to be expected if these quantities are corrected by b_T determined from Table 14.4. Figure 14.10 does show, indeed, that the resulting constant activation energies do not differ from each other within the accuracy of the measurements and are the same as the value from the shift of $\delta(\omega)$. These findings demonstrate that for LDPE a method-invariant activation energy E_A can be found by using the vertical shift factor b_T determined from the temperature dependence of the linear steady-state compliances. Consequently, in the case of LDPE from the shift of $\delta\ (a_T\omega, T)$ the constant activation energy can be obtained without knowing b_T and, therefore, the time-temperature behavior of the phase angle is very suitable to determine the material-specific activation energy.

As Fig. 14.8 demonstrates, however, constant activation energies cannot be obtained for the two long-chain branched polyethylenes made by metallocene catalysts even if a vertical shift factor is eliminated using the phase angle for the frequency-temperature shift. Such a behavior can formally be interpreted by

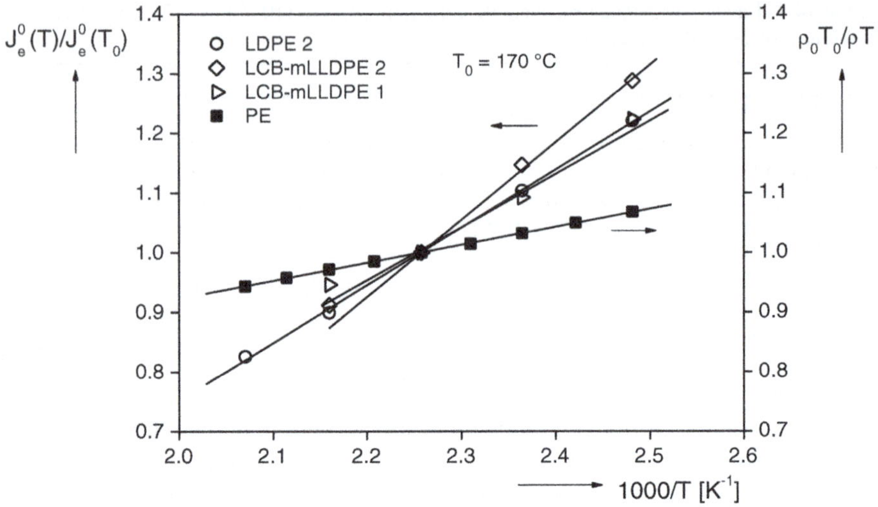

Fig. 14.9 $J_e^0(T)/J_e^0(T_0)$ as functions of the reciprocal absolute temperature 1000/T for various long-chain branched polyethylenes (*open symbols*) in comparison to $\rho_0 T_0/\rho T$ for polyethylenes (*filled squares*) [11]

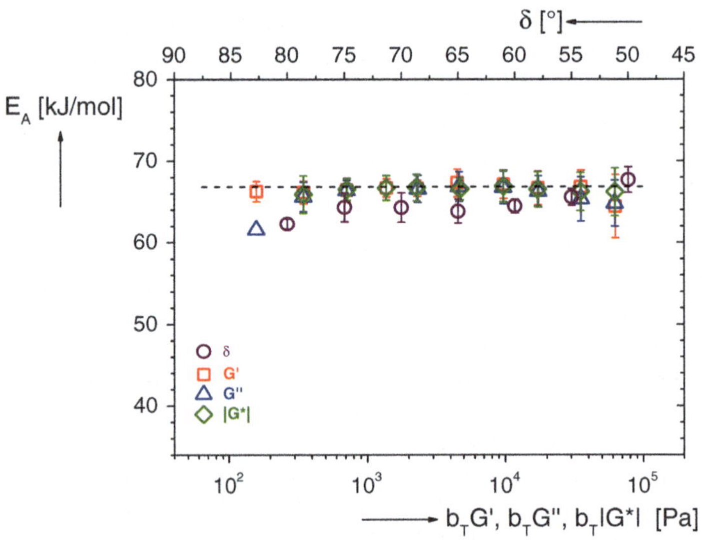

Fig. 14.10 Activation energies E_A of LDPE 2 obtained from the frequency-temperature shifts of the phase angle δ and the moduli G', G'' and $|G^*|$ corrected by the vertical shift factor b_T defined by Eq. (14.8) [11]

assuming different temperature dependencies of the relaxation times of various parts of the spectrum. Due to the dependence of the activation energies of polyethylene molecules on the kind and degree of branching, this interpretation is conclusive in the case of samples consisting of various species, as the results on the blend in Fig. 14.8 demonstrate.

Considering the results presented, for long-chain branched polyethylenes two types of thermorheological complexity could be distinguished. In the case of several LDPE investigated, master curves of rheological quantities as a function of angular frequency were obtained by introducing a vertical shift factor b_T. Using the phase angle for shifting, eliminates b_T and directly provides a method-invariant activation energy. But there are other long-chain branched polyethylenes which cannot be shifted to a master curve in this way. For such materials, activation energies may be determined which are dependent on the rheological quantity chosen. Such findings point to a heterogeneous composition of the sample. For example, blends of linear and long-chain branched polyethylenes do show such a behavior, which was found for metallocene catalyzed long-chain branched polyethylene, too (cf. Fig. 14.8).

From the results on LDPE it could be concluded that its molecular structure is rather homogeneous.

References

1. Hepperle J (2002) Doctoral thesis, University Erlangen-Nürnberg, Shaker Verlag, Aachen, ISBN 3-8322-1228-0
2. Onogi S, Masuda T, Kitagawa K (1970) Macromolecules 3:109
3. Montfort JP, Marin G, Monge P (1984) Macromolecules 17:1551
4. Fuchs K, Friedrich C, Weese J (1996) Macromolecules 29:5893
5. Masuda T, Kitagawa K, Onogi S (1970) Polym J 1:418
6. Ferry JD (1980) Viscoelastic properties of polymers. John Wiley, New York
7. Lomellini P (1992) Macomol Chem 193:69
8. Gabriel C (2001) Doctoral thesis, University Erlangen-Nürnberg, Shaker Verlag, Aachen, ISBN 3-8265-8909-2
9. Stadler FJ, Gabriel C, Münstedt H (2007) Macromol Chem Phys 208:2449
10. Wolff F (2008) Diploma thesis, University Erlangen-Nürnberg
11. Kessner U (2010) Doctoral thesis, University Erlangen-Nürnberg, Sierke Verlag, Göttingen, ISBN 13:978-3-86844-300-4
12. Stadler FJ (2007) Doctoral thesis, University Erlangen-Nürnberg, Sierke Verlag, Göttingen, ISBN 978-3-940333-24-7
13. Kessner U, Kaschta J, Stadler FJ, Le Duff CS, Drooghaag X, Münstedt H (2010) Macromolecules 43:7341
14. Vega JF, Santamaria A, Munoz-EscalonaA, Lafuente P (1998) Macromolecules 31:3639
15. Kessner U, Kaschta J, Münstedt H (2009) J Rheol 53:1001
16. Kessner U, Münstedt H (2010) Polymer 51:507
17. Verser DW, Maxwell B (1970) Polym Engng Sci 10:122

Chapter 15
Rheometry

Due to the importance of rheological measurements for applications and basic research, various rheometers have been in use to determine viscous and elastic properties of polymer melts. They cover a wide range of shear rates or shear stresses, respectively, and are based on different geometries and principles. During the early stages of rheology, viscosity measurements were performed with capillary rheometers, only. Instruments of such kind have preferably been used for the rheological characterization of polymer solutions discussed in Sect. 3.2 and are still the basic equipment for the determination of viscosity functions of polymer melts in the range of high-shear rates relevant for processing. Rotational rheometers are widespread tools, today, for basic research and quality control on a variety of polymeric systems. In order to understand the principles of the measuring techniques and to assess their performances, the equations underlying the data evaluation are given in the following.

15.1 Capillary Rheometry

Figure 15.1 illustrates the principle of a capillary rheometer. From a barrel the melt is pushed through a capillary with the radius R and the length L. The quantities directly measured are the pressure p acting on the melt and the passing volume per time $\dot{V}$. This so-called throughput can either be determined from the given velocity of a moving piston and the barrel cross section or in case of a pressure-driven rheometer by weighing the strands extruded during a certain time interval and calculating their volume at the measuring temperature using the corresponding density. From the pressure the shear stress is calculated and from the throughput the shear rate. For the characterization of the flow behavior of the melt the viscosity is considered either as a function of shear rate or shear stress. In addition to the viscous properties, some insight into the elasticity can be obtained by determining the swell of the extruded strand. This quantity has been discussed in some detail in Sect. 13.2.2. Another interesting effect occurring at high shear rates is the distortion of the surface of the extruded strands which is commonly

H. Münstedt and F. R. Schwarzl, *Deformation and Flow of Polymeric Materials*,
DOI: 10.1007/978-3-642-55409-4_15,

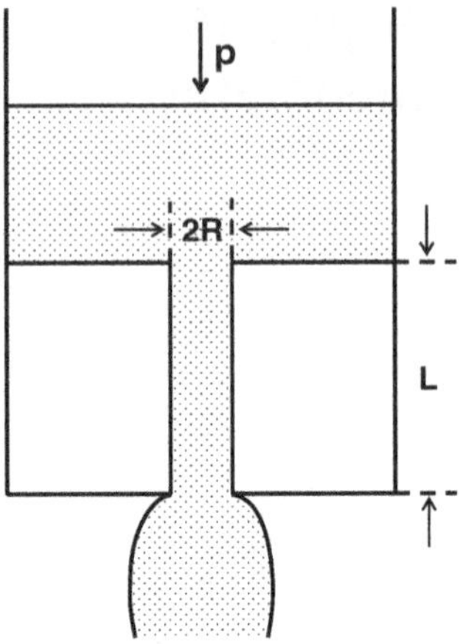

Fig. 15.1 Principle of the capillary rheometer. L is the length and R the radius of the capillary. p denotes the pressure acting on the melt. The swell of the extrudate is due to the elasticity of the melt

called melt fracture. Capillary rheometry is a very useful method to investigate this phenomenon interesting for processing. It is discussed in Sect. 16.3.4 in connection with quantitative investigations of the flow fields.

Due to the pronounced geometrical differences between the barrel and the capillary, for viscoelastic materials complex flow patterns in the entrance region can occur which extend into the capillary. These effects normally can only be captured indirectly as at the high temperatures and pressures necessary for the investigation of polymer melts steel is commonly used as the wall material which obviously makes accurate measurements of local velocity fields impossible. Generally, a capillary rheometer is a "black box" with respect to the detailed flow behavior of polymer melts (cf. Sect. 15.1.3 on the Bagley-correction). Chapter 16 on the laser-Doppler velocimetry applied to the capillary flow of polymer melts demonstrates which interesting insights can be obtained by this technique. Thus, it is shown, for example, that a flow pattern independent of the axial position within a die can only be achieved if the capillary is long enough. The equations for this type of flow are derived in the following under the assumption of a uniform flow field, i.e., L >> R.

15.1.1 Calculation of Shear Stress and Shear Rate

A cylindrical coordinate system with the coordinates r, φ, z is chosen whereby the z-axis represents the symmetry axis of the capillary. The position $z = 0$ designates the entrance of the capillary and $z =$ L its exit. It is assumed that the z-coordinate describes the direction of flow. Then r is the direction of the velocity gradient and φ the neutral direction. The velocity components for a fully developed flow are

$$v_r = 0 \quad v_\varphi = 0 \quad v_z = f(r) \tag{15.1}$$

Three additional assumptions are made:

1. The melt adheres to the capillary wall, i.e., $v(r = R) = 0$.
2. The flow is isothermal.
3. The viscosity is independent of pressure.

Whereas the assumption 1 is fulfilled for a great number of polymer melts in good approximation, 2 and 3 are the better matched the smaller the shear rates and the lower the extrusion pressures, respectively.

From inserting the velocities (15.1) into Eq. (8.41a) the components of the deformation rate tensor follow as

$$d_{rr} = d_{r\varphi} = d_{\varphi\varphi} = d_{\varphi z} = d_{zz} = 0 \quad 2d_{rz} = df(r)/dr = \dot{\gamma} \tag{15.2}$$

Due to $d_{r\varphi} = 0$ and $d_{\varphi z} = 0$, the corresponding components of the stress tensor $\sigma_{r\varphi}$ and $\sigma_{\varphi z}$ are zero and as the stresses do not depend on φ for symmetry reasons, the equation of motion for the z-direction which is relevant for the capillary flow follows from Eq. (7.6) for a creeping flow, i.e., a flow with Reynolds numbers much smaller than 1, and negligible volume forces as

$$\frac{\partial \sigma_{rz}}{\partial r} + \frac{\partial \sigma_{zz}}{\partial z} + \frac{1}{r}\sigma_{rz} = 0 \tag{15.3}$$

or

$$\frac{1}{r}\frac{\partial}{\partial r}(r\sigma_{rz}) = -\frac{\partial \sigma_{zz}}{\partial z} \tag{15.4}$$

Under the assumption that σ_{rz} is a unique function of d_{rz}, only, the left side of this equation does not depend on z, consequently σ_{zz} in its most general form can be written as

$$\sigma_{zz} = Az + B \tag{15.5}$$

and with the boundary conditions $\sigma_{zz} = -p_1$ at $z = 0$ and $\sigma_{zz} = -p_2$ at $z = L$ one gets

$$A = \frac{p_1 - p_2}{L} = \frac{\Delta p}{L} \tag{15.6}$$

which is a positive quantity as per definition $p_1 > p_2 . p_1$ is called the entrance pressure and p_2 the exit pressure which can be neglected in many cases. Introducing σ_{zz} from Eq. (15.5) into Eq. (15.4) for the shear stress σ_{rz}, leads by integration to

$$\sigma_{rz} = -\frac{1}{2}Ar + \frac{b}{r} \tag{15.7}$$

As σ_{rz} has to be finite at any position within the die, i.e., for $r = 0$, too, b is zero and the shear stress in dependence on the radius follows as

$$\sigma_{rz} = -\frac{\Delta pr}{2L} \tag{15.8}$$

It attains the maximum negative value at the wall ($r = R$), i.e.,

$$\sigma_{rz}(r = R) = \sigma_w = -\frac{\Delta pR}{2L} \tag{15.9}$$

Equation (15.8) for the shear stress which was derived from the balance of forces in combination with the additional assumption $\sigma_{rz} = f(d_{rz})$ can be obtained by a simple consideration based on the force balance on the wall of a cylindrical element of the melt with the radius r and the length L. From the balance of the driving force

$$F_d = \pi r^2 \Delta p \tag{15.10}$$

and the force acting on the curved surface of the cylinder

$$F_s = 2\pi r L \sigma_{rz} \tag{15.11}$$

it follows straightforward Eq. (15.8) above

$$\sigma_{rz} = -\frac{\Delta pr}{2L} \tag{15.8}$$

The shear rate necessary for the determination of the viscosity is calculated from the measurable volume flow rate $\dot{V}$ in the following way.

The volume flow rate $d\dot{V}$ through a differential cross section element $2\pi rdr$ of the cylindrical capillary is given by

$$d\dot{V} = 2\pi r\, v(r)dr$$

with $v(r)$ being the flow velocity in z-direction. $\dot{V}$ then follows as

$$\dot{V} = 2\pi \int_0^R rv(r)dr \tag{15.12}$$

and by partial integration one gets

$$\dot{V} = \pi r^2 v(r)|_0^R - \pi \int_0^R r^2 (dv/dr) dr \qquad (15.13)$$

With the boundary condition $v(r = R) = 0$ and the relations $dv/dr = \dot{\gamma} = \sigma_{rz}/\eta$, this equation becomes

$$\dot{V} = -\pi \int_0^R r^2 \dot{\gamma}\, dr = -\pi \int_0^R r^2 \sigma_{rz}/\eta\, dr \qquad (15.14)$$

This integral can be evaluated in closed form for some special cases, only. For a Newtonian fluid the viscosity is constant, i.e., $\eta = \eta_0$, and with Eq. (15.8) for the shear stress σ_{rz} it follows

$$\dot{V} = \frac{\pi \Delta p}{2L\eta_0} \int_0^R r^3 dr = \frac{\pi R^4 \Delta p}{8L\eta_0} \qquad (15.15)$$

which is the so-called Hagen-Poiseuille law (cf. Eq. 3.27)

For non-Newtonian fluids the viscosity in general is a function of the shear rate or the shear stress, respectively, and, therefore, not constant over the capillary diameter. But Eq. (15.8) describing the distribution of the shear stress across the capillary and Eq. (15.9) for the stress at the wall are independent of the material as they are derived without making use of the rheological law of a fluid. For Newtonian fluids with a linear distribution of the shear rates across the radius of the capillary (cf. Fig. 15.2) the shear rate D at the wall follows from Eq. (15.15) as

$$D = -\frac{4\dot{V}}{\pi R^3} \qquad (15.16)$$

which can be regarded as a reduced volume rate.

In the case of a non-Newtonian fluid Eq. (15.16) is often used, too. D is then called the *apparent shear rate at the wall* due to the fact that it represents an approximation as the velocity distribution is not linear anymore.

However, the *true shear rate* $\dot{\gamma}_w$ *at the wall* can be determined from D in a straightforward way as derived in the following.

Substituting in Eq. (15.14) the integration variable r by σ_{rz}, i.e., $r = -\sigma_{rz} 2L/\Delta p$ results in

$$\dot{V} = \pi \left(\frac{2L}{\Delta p}\right)^3 \int_0^{\sigma_w} \sigma_{rz}^3/\eta\, d\sigma_{rz} \qquad (15.17)$$

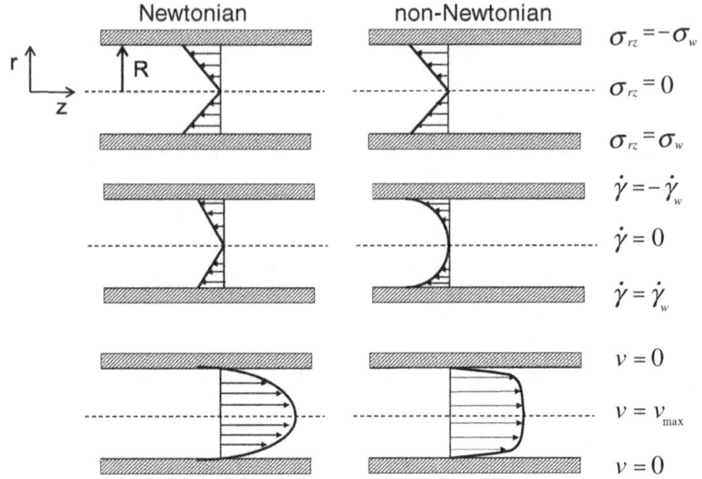

Fig. 15.2 Schematics of shear stress, shear rate, and flow velocity of a Newtonian in comparison to a non-Newtonian fluid. z marks the flow direction and $\sigma_{rz}$is the shear stress in cylinder coordinates. σ_w is the shear stress and $\dot{\gamma}_w$the shear rate at the wall

and with Eqs. (15.9) and (15.16) in

$$D\sigma_w^3 = 4\int_0^{\sigma_w} \sigma_{rz}^3/\eta \, d\sigma_{rz} \tag{15.18}$$

By differentiation of Eq. (15.18) with respect to σ_w and with $\sigma_w/\eta = \dot{\gamma}_w$ one gets

$$\frac{dD}{d\sigma_w}\sigma_w^3 + 3D\sigma_w^2 = 4\sigma_w^3/\eta = 4\sigma_w^2\dot{\gamma}_w \tag{15.19}$$

or

$$\dot{\gamma}_w = \frac{1}{4}\frac{dD}{d\sigma_w}\sigma_w + \frac{3}{4}D \tag{15.20}$$

Replacing the linear differentials by the logarithmic ones which may be convenient in case of the widely used double-logarithmic plots it follows the so-called *Weissenberg-Rabinowitsch relation* [1]

$$\dot{\gamma}_w = \frac{1}{4}\frac{d\,\log(-D)}{d\,\log(-\sigma_w)}D + \frac{3}{4}D = \frac{3+\tilde{n}}{4}D \tag{15.21}$$

with

$$\tilde{n} = \frac{d\,\log(-D)}{d\,\log(-\sigma_w)} \qquad (15.22)$$

The minus sign has to be introduced as D and σ_w are both negative according to their derivation. In practice, for the shear rates and the shear stresses as well their absolute values are used. This convention is applied in all the parts of this book where material functions depending on the shear rate or the shear stress are discussed.

15.1.2 Velocity Profiles

From the definition of the viscosity it follows for the velocity distribution in a circular capillary

$$dv/dr = \dot{\gamma} = \sigma_{rz}/\eta = -\frac{\Delta pr}{2L\eta} \qquad (15.23)$$

or

$$v = \int dv = -\frac{\Delta p}{2L}\int r/\eta dr \qquad (15.24)$$

This integral can be evaluated in closed form for special viscosity functions, only. For example, in the case of a constant Newtonian viscosity η_0 Eq. (15.24) reads

$$v = -\frac{\Delta p}{2L\eta_0}\int rdr = -\frac{\Delta p}{4L\eta_0}r^2 + C$$

and with the boundary condition $v = 0$ at $r = R$ one gets

$$v = \frac{\Delta p}{4L\eta_0}\left(R^2 - r^2\right) \qquad (15.25)$$

The velocity profile of a Newtonian fluid according to Eq. (15.25) is plotted in Fig. 15.2 together with the profiles for the shear stress and the shear rate. These functions are compared with the profiles of a non-Newtonian shear thinning fluid. Obviously, the local distribution of the shear stress has to be independent of rheological properties. The shear stress is zero in the middle of the capillary and increases linearly up to its maximal absolute value at the wall. The region around the maximum of the parabolic flow profile for the Newtonian fluid flattens as according to Eq. (15.25) the non-Newtonian viscosity smaller than η_0 enhances v. The slope near the wall becomes distinctly steeper due to $v = 0$ at the wall. The shear rate being the gradient of the velocity follows as a simple consequence. Near the center of the capillary it increases more weakly than for Newtonian fluids, but more pronounced by approaching the wall.

15.1.3 Viscosity Functions and Bagley-Correction

As a first result from capillary rheometry the extrusion pressure as a function of the apparent shear rate is commonly presented. In Fig. 15.3 an example of measurements on a commercial polystyrene is shown with capillaries of a radius of $R = 0.6$ mm and different lengths described by the length to radius ratio *L/R*. First qualitative considerations regarding the flow behavior of a material can already be deduced from such a presentation.

To determine the viscosity functions which have a great importance in practical rheology a special evaluation has to be applied which is explained by making use of the data in Fig. 15.3. At seven different apparent shear rates marked by the broken lines the corresponding pressures are presented as functions of *L/R* in Fig. 15.4. This so-called *Bagley-plot* [2] reveals two features. The points obtained at a distinct apparent shear rate D can be connected by a straight line which is expected according to Eqs. (15.15) and (15.16) for a viscosity independent of the pressure applied. The slope corresponds to twice the shear stress at the wall (cf. Eq. 15.26). Positive curvatures of $p(L/R)$ indicate a viscosity increasing with pressure.

The second feature which has some interesting consequences is the finding that the pressure at the capillary entrance ($L = 0$) is not zero, but positive and becomes larger with increasing shear rates. This so-called entrance pressure loss has its origin in particular flow patterns. Extensional flow occurring due to the pronounced differences between the cross sections of the barrel and the capillary is one reason, the other can be seen in the formation of vortices in the entrance region observed for some polymers. These flow fields were quantitatively investigated by laser-Doppler velocimetry and are discussed in Chap. 16.

For an accurate determination of the viscosity the driving pressure in the capillary Δp has to be determined as sketched in the Fig. 15.5. Neglecting the

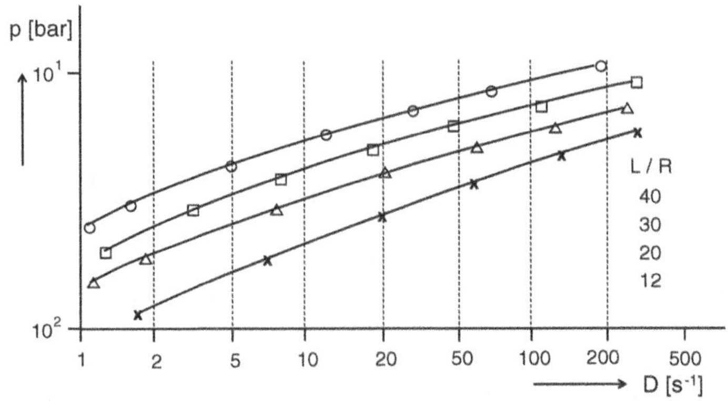

Fig. 15.3 Double-logarithmic plot of the extrusion pressure p as a function of the apparent shear rate D for a commercial polystyrene at $T = 190$ °C and different length to radius ratios *L/R* of the die ($R = 0.6$ mm)

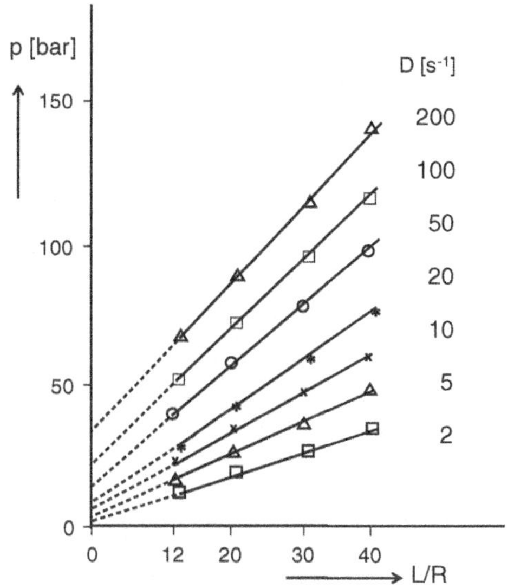

Fig. 15.4 Presentation of the extrusion pressure p as a function of the length to radius ratio *L/R* of the capillaries for various apparent shear rates *D* (*Bagley-plot*). The experimental data were taken from Fig. 15.3

small exit pressure, the true shear stress at the wall σ_w then follows from the slope $\tan\alpha$ as

$$\sigma_w = \frac{p - p_e}{2L/R} = \frac{1}{2}\tan\alpha \tag{15.26}$$

The entrance pressure loss p_e is of importance insofar as in the case of short dies it can represent a remarkable portion of the total pressure. The practical relevance of this fact is discussed in more detail in Sect. 11.5.3 on shear rheology. From Fig. 15.5 it becomes obvious that the Bagley-plot sometimes is formally interpreted by assuming a virtual capillary which is longer than the real one.

For practical purposes the *apparent viscosity* η_a calculated as the quotient of the true shear stress obtained by means of the *Bagley-plot* and the apparent shear rate at the wall is often discussed as the latter one is directly correlated with the volume rate $\dot{V}$ (cf. Eq. 15.16). Making use of the numerical description of the apparent viscosity $\log\eta_a$ as a function of $\log D$ by a polynomial as presented in Sect. 11.5.2, the true shear rate $\dot{\gamma}_w$ at the wall can be determined by calculating the factor $\tilde{n}$ for the *Weissenberg-Rabinowitsch correction* from its definition in Eq. (15.22) according to

$$\frac{1}{\tilde{n}} = \frac{d\log\sigma_w}{d\log D} = \frac{d\log\eta_a}{d\log D} + 1 \tag{15.27}$$

As the true viscosity calculated from measured quantities at the wall is a material function and, therefore, does not depend on the location at which it was determined the index "w" is usually omitted.

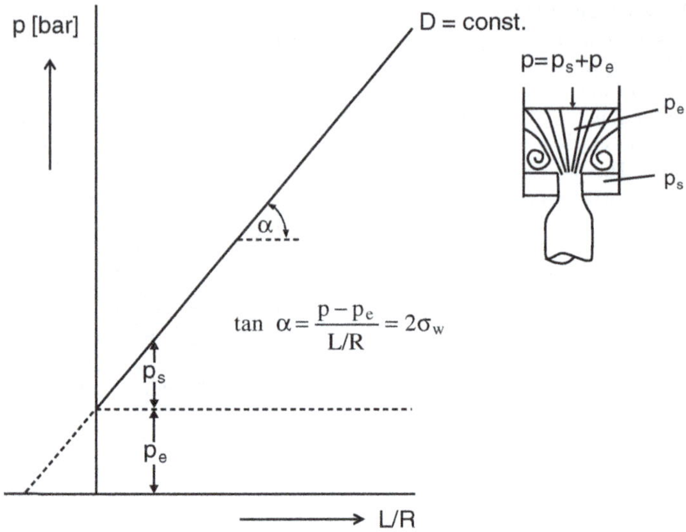

Fig. 15.5 Interpretation of the *Bagley-plot*. p_e is the entrance pressure loss and p_s the pressure loss due to the flow through the capillary

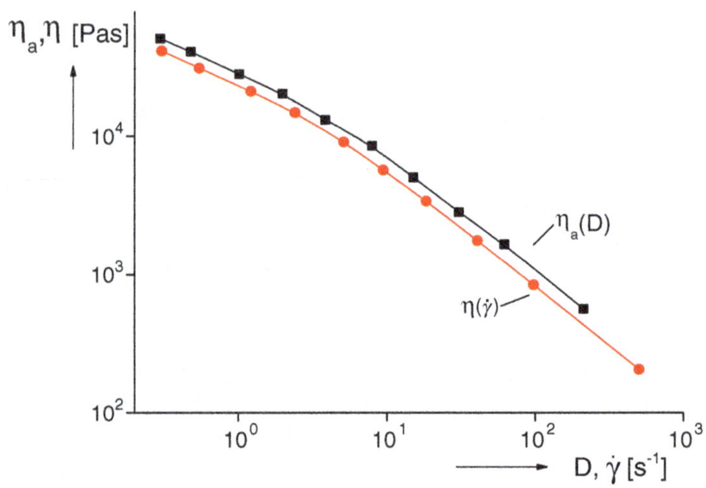

Fig. 15.6 Apparent viscosity η_a as a function of the apparent shear rate D in comparison with the true viscosity η as a function of the true shear rate $\dot{\gamma}$ for a standard polystyrene with $M_w = 400$ kg/mol at 190 °C. The curves were obtained by a time-temperature shift. (Reprinted from [3] with permission from Carl Hanser Verlag, München)

In Fig. 15.6 the apparent viscosity η_a as a function of the apparent shear rate D is compared with the true viscosity η as a function of the true shear rate $\dot{\gamma}$ for a standard polystyrene with $M_w = 400$ kg/mol at a temperature of 190 °C.

The differences between the apparent and the true viscosities taken at the same values of the corresponding shear rates are small and, therefore, are often neglected for applications. According to Eqs. (15.27) and (15.21) the true and the apparent shear rates are the same in the Newtonian region. They become the more different the more pronounced the shear thinning is, i.e., the more the viscosity function approaches the slope of -1.

15.2 Slit Rheometry

The elaborate method of determining the viscosity function by measurements with circular capillaries of different *L*/*R* ratios and the use of the *Bagley-plot* could be avoided if the pressure gradient were obtained directly from transducer readings. In the case of a circular capillary, it is not possible for geometrical reasons to place suitable gages within the wall. The slit geometry, however, allows the flush-mounted installation of transducers as sketched in the Fig. 15.7. The first who published this idea and measurements with a self-constructed slit rheometer was Janeschitz-Kriegl [4, 5, 6]. Later on this method became commercialized.

The positions of the pressure gages have to be chosen far enough from the inlet of the die to be able to investigate the locally constant flow. As it is shown in Sect. 16.3.1.2 the effect of the entry flow can still be seen within the slit at some distance from the entrance.

For the calculations of the shear stress σ_w and the shear rate D at the wall, a slit element of the length ΔL between the two pressure gages is regarded. A Cartesian coordinate system as sketched in Fig. 15.8 is introduced. According to the convention widely found in the literature, for the discussions of the flow in slits the x-coordinate is chosen in the direction of flow, the y-coordinate in that of the velocity gradient, and z describes the neutral direction.

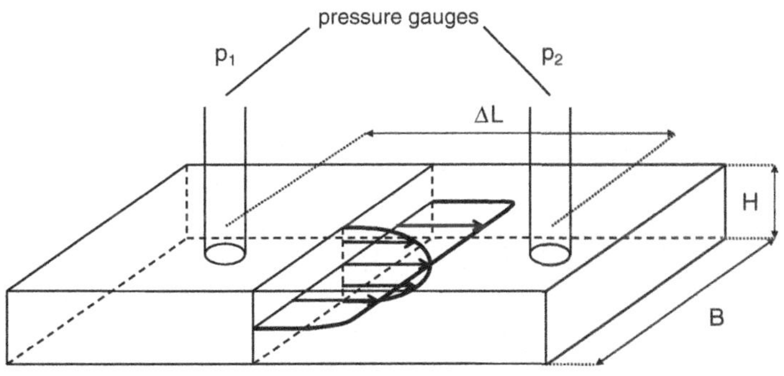

Fig. 15.7 Flow profiles and positions of the pressure transducers p_1 and p_2 in a slit die. B is the width and H the height of the slit

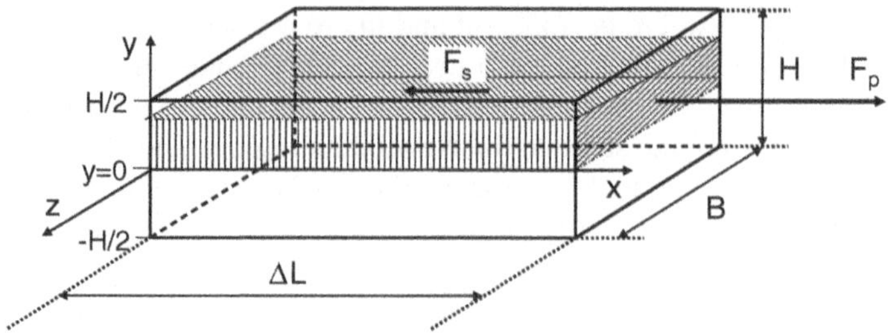

Fig. 15.8 Coordinate system used for the slit die and schematics of the force balance for the marked volume element. H and B are the height and the width of the slit, ΔL is the distance between the pressure transducers. F_s is the tangential force, F_p the force exerted by the pressure

From a simple balance of the forces acting on the volume element sketched in Fig. 15.8 for a slit with the height H much smaller than the width B, the shear stress σ_{xy} follows as

$$\sigma_{xy} = -\frac{\Delta p y}{\Delta L} \tag{15.28}$$

Δp is the pressure difference measured by the pressure gages mounted at the distance ΔL from each other. The shear stress σ_w at the upper wall, i.e., for $y = H/2$, then reads

$$\sigma_w = -\frac{\Delta p H}{2\Delta L} \tag{15.29}$$

that one at the lower wall, i.e., $y = -H/2$, is positive.

By considerations similar to those for the circular capillary the velocity distribution within the slit for a Newtonian liquid with the viscosity η_0 follows as

$$v = \frac{\Delta p}{2\Delta L \eta_0}\left(\frac{H^2}{4} - y^2\right) \tag{15.30}$$

the volume rate $\dot{V}$ as

$$\dot{V} = \frac{BH^3 \Delta p}{12\Delta L \eta_0} \tag{15.31}$$

and the apparent shear rate D at the lower die wall as

$$D = \frac{6\dot{V}}{BH^3} \tag{15.32}$$

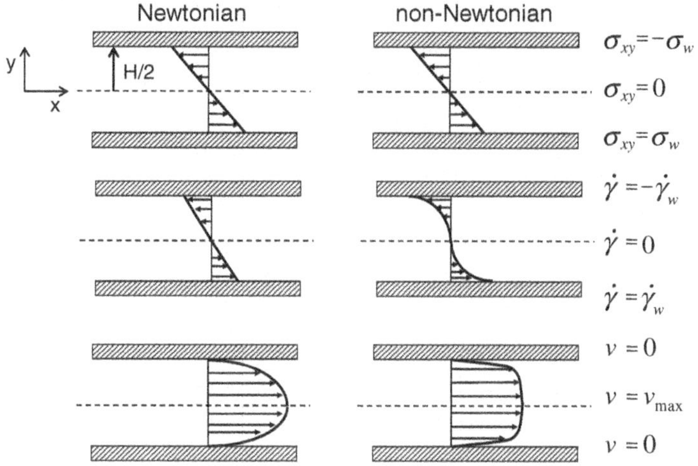

Fig. 15.9 Shear stress σ_{xy}, shear rate $\dot{\gamma}$, and velocity field v in a slit die for a Newtonian and a non-Newtonian fluid

The true shear rate at the wall $\dot{\gamma}_w$ can be calculated by a relation similar to that of *Weissenberg and Rabinowitsch* for the circular capillary, viz.,

$$\dot{\gamma}_w = D\left(\frac{2}{3} + \frac{1}{3}\frac{d \log D}{d \log \sigma_w}\right) = \frac{2+\tilde{n}}{3} D \text{ with } \tilde{n} = \frac{d \log D}{d \log \sigma_w} \tag{15.33}$$

The shear stress σ_{xy} (Eq. 15.28), the flow velocity v (Eq. 15.30) and the shear rate $\dot{\gamma} = dv/dx$ are schematically presented in the Fig. 15.9 for a Newtonian and a non-Newtonian liquid. In both cases the shear stresses obviously are the same as they are independent of material properties. The velocity fields differ insofar as the parabola for the Newtonian fluid flattens in case of the non-Newtonian liquid and consequently the linear increase of the velocity gradient over the slit height typical of a Newtonian fluid changes remarkably (cf. lower parts of Fig. 15.9). The velocity fields sketched in Fig. 15.9 for a non-Newtonian liquid are verified in Fig. 16.13 by investigations with the laser-Doppler velocimetry.

By measuring the pressure p_0 in the barrel of a slit rheometer the entrance pressure loss p_e can be determined according to Fig. 15.10. Assuming a linear pressure distribution within the slit, the pressures at the slit entrance ($x = 0$) and exit ($x = L$) can be calculated from the pressure readings p_1 at the positions $x = L_1$ and p_2 at $x = L_2$. The pressure at the exit is generally small for polymer melts and, therefore, not much discussed in the literature. The entrance pressure loss p_e which can be of some relevance for processing (cf. Sect. 11.5.3) follows as

$$p_e = p_0 - p(x=0) \tag{15.34}$$

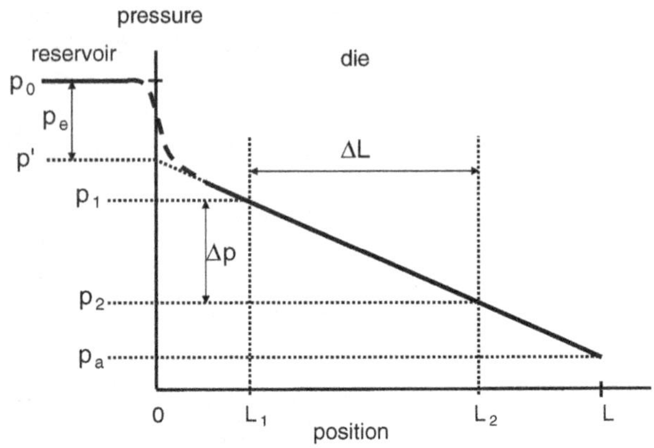

Fig. 15.10 Schematics of the pressure distribution in a slit die. p_e: entrance pressure loss; p_a: exit pressure loss; p_0, p_1, p_2: pressure readings in the barrel and at the positions $x = L_1$ and $x = L_2$ within the slit, respectively

The fact that in spite of their obvious methodical elegance slits are not as widely used as circular capillaries for the determination of viscosity functions goes back to very simple technical reasons. Pressure transducers are to handle with great care as their membranes are prone to mechanical damage, particularly, due to fouling by solidified melt residues. Furthermore, precise slits of small heights are not easy to manufacture and cleaning them may be hard. In all the cases for which plane walls are required as for optical investigations on flowing polymer melts, for example, the slit is the geometry of choice, however, (cf. Sect. 16.2.2).

15.3 Rotational Rheometry

One great advantage of rotational rheometers is the small amount of material needed for a comprehensive rheological characterization. Moreover, they are distinguished by a wide variety of experimental modes which can be applied to the samples. These are described in Chap. 11. The principle of rotational rheometers is that the material is sheared between two walls, one of them fixed and the other rotating. Widely used today in the field of melt rheology are plate-plate and cone-and-plate geometries. Measuring quantities are the angular velocity and the torque. In the following it is presented how the shear rate and the shear stress and, consequently, material functions characterizing a polymer melt can be derived from them.

15.3.1 Plate-Plate Rheometer

Figure 15.11 shows the principle of a plate-plate assembly. Between the two parallel plates with the radius R the sample of the height H is placed. The upper plate may be driven with a constant angular velocity Ω and the resulting torque M measured at the

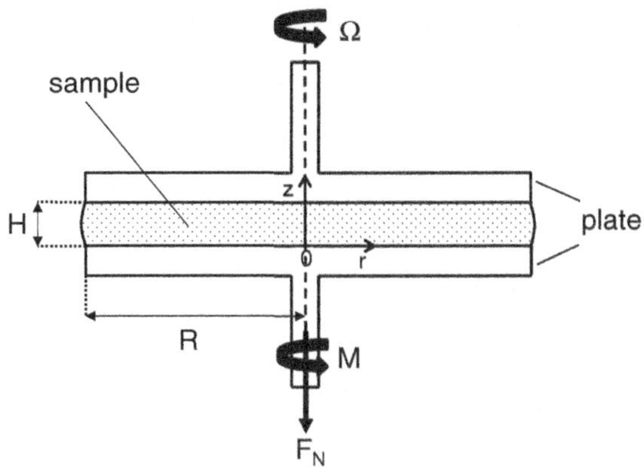

Fig. 15.11 Schematics of a plate-plate assembly. *H* sample height, *R* sample radius, *Ω* angular velocity, *M* torque, F_N normal force

lower plate. In addition, the normal force F_N can be registered. For a quantitative description of the flow and stress fields a cylindrical coordinate system is introduced with z pointing into the direction of the rotational axis of the system and its origin located at the upper side of the lower plate (cf. Fig. 15.11). The azimuth angle is denominated as φ. The φ-direction is the direction of flow (1-direction), the z-direction that of the velocity gradient (2-direction) and r designates the neutral direction (3-direction). It is assumed that in the stationary state all fluid particles at a distinct z-coordinate perform a rotation around the driving axis with the angular velocity $\omega(z)$. The velocities in the three directions then follow as:

$$v_r = 0 \quad v_z = 0 \quad v_\varphi = r\omega(z) \tag{15.35}$$

and the components of the deformation rate tensor from Eq. (8.41a) as

$$d_{rr} = d_{zz} = d_{\varphi\varphi} = d_{r\varphi} = d_{rz} = 0 \tag{15.36}$$

$$2d_{\varphi z} = r\frac{\partial \omega}{\partial z} = \dot{\gamma} \tag{15.37}$$

As the shear stresses $\sigma_{r\vartheta}$ and σ_{rz} are zero and due to symmetry reasons the stress field does not depend on the φ-coordinate, the equations of motion for a creeping flow without volume forces then follow from Eq. (7.6) as

$$\frac{\partial \sigma_{rr}}{\partial r} + \frac{\sigma_{rr} - \sigma_{\varphi\varphi}}{r} = 0 \tag{15.38}$$

$$\frac{\partial \sigma_{\varphi z}}{\partial z} = 0 \tag{15.39}$$

$$\frac{\partial \sigma_{zz}}{\partial z} = 0 \tag{15.40}$$

$\sigma_{\varphi z}$ depends on $d_{\varphi z}$ only and, therefore, from the Eqs. (15.39) and (15.37) follows that

$$\frac{\partial^2 \omega}{\partial z^2} = 0$$

Taking the boundary conditions $\omega(z = 0) = 0$ and $\omega(z = H) = \Omega$ into account the solution of this differential equation is

$$\omega(z) = \frac{\Omega}{H} z \tag{15.41}$$

and with (15.37)

$$\dot{\gamma} = \frac{\Omega}{H} r \tag{15.42}$$

The shear rate is a linear function of the radius r. It becomes zero at the rotation axis and attains

$$\dot{\gamma}_R = \frac{\Omega R}{H} \tag{15.43}$$

at the rim.

Whereas in the case of the capillary and slit rheometers the stress distributions are given for a distinct geometry, for the plate-plate rheometer it is the shear rate which according to Eq. (15.42) is the preset quantity. The rheological law then determines the stress distribution which is reflected by the integral of the torque M acting on the plate, i.e.,

$$M = \int_0^{2\pi} \int_0^R r \sigma_{\varphi z} r dr d\varphi = 2\pi \int_0^R r^2 \sigma_{\varphi z} dr \tag{15.44}$$

With $\sigma_{\varphi z} \equiv \sigma$ and the new integration variable $\dot{\gamma} = \Omega r / H$ the integral (15.44) becomes

$$M = 2\pi (H/\Omega)^3 \int_0^{\dot{\gamma}_R} \dot{\gamma}^2 \sigma(\dot{\gamma}) d\dot{\gamma} \tag{15.45}$$

For a further evaluation one needs the rheological law, i.e., a relation between σ and $\dot{\gamma}$. In the simplest case of a Newtonian liquid $\sigma = \eta_0 \dot{\gamma}$ is valid and the torque follows as

$$M = \frac{\pi}{2}\frac{R^4}{H}\eta_0\Omega = \frac{\pi}{2}R^3\eta_0\dot{\gamma}_R = \frac{\pi}{2}R^3\sigma_w \tag{15.46}$$

with σ_w being the shear stress at the rim. From this equation it is obvious that the radial dimension of a sample has to be determined very carefully to get reliable values for the viscosity as inaccuracies propagate with the fourth power of R.

For a non-Newtonian fluid the exact stresses can be determined experimentally by a procedure similar to that one applied in case of the shear rates for the capillary rheometer (cf. Sect. 15.1.1). For this purpose, an apparent shear stress S at the rim is defined according to Eq. (15.46) by

$$S = \frac{2M}{\pi R^3} \tag{15.47}$$

assuming a Newtonian viscosity. With M from Eq. (15.45) and $\dot{\gamma}_R$ from Eq. (15.43) the following relationship between the apparent and the true stress is obtained

$$S = \frac{4}{\dot{\gamma}_R^3}\int_0^{\dot{\gamma}_R} \dot{\gamma}^2\sigma(\dot{\gamma})d\dot{\gamma} \tag{15.48}$$

Differentiation of the product $S\,\dot{\gamma}_R^3$ with respect to $\dot{\gamma}_R$ leads to the equation which relates the true stress at the rim $\sigma(\dot{\gamma}_R)$ to the apparent stress S, viz.,

$$\sigma(\dot{\gamma}_R) = S\left[\frac{3}{4} + \frac{1}{4}\frac{d\log S}{d\log\dot{\gamma}_R}\right] \tag{15.49}$$

This equation is of a similar structure as that of Weissenberg and Rabinowitsch (cf. Eq. 15.21). As the differential quotient in Eq. (15.49) comes to lie between 0 and 1 for polymer melts it follows

$$\frac{3}{4}S \leq \sigma_R < S \tag{15.50}$$

i.e., real and apparent shear stresses at the wall are not extremely different from each other.

Shear rate and shear stress distributions in a plate-plate assembly are sketched in the Fig. 15.12. The shear rate increases linearly from zero to its maximal value at the rim. It is determined by the rotational velocity and the plate geometry and, therewith, independent of material properties. Evidently, for a Newtonian liquid the shear stress rises linearly from zero to its value S at the wall. The true stress of a non-Newtonian liquid becomes smaller than its Newtonian counterpart by approaching the rim.

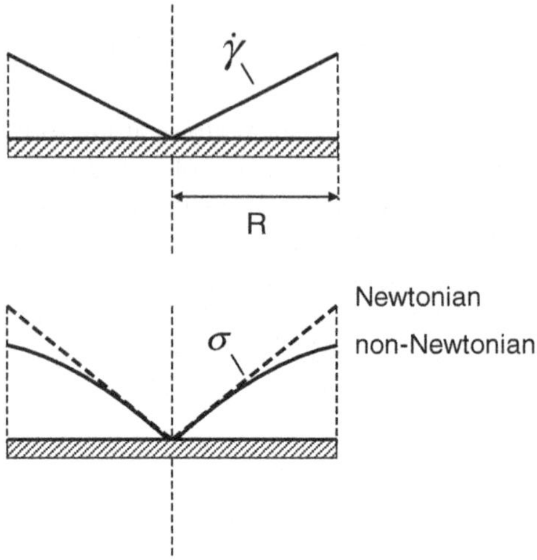

Fig. 15.12 Schematics of the shear rate $\dot{\gamma}$ and the shear stress σ in a plate-plate assembly for a Newtonian (*broken line*) and a non-Newtonian (*full line*) fluid

From measurements of the normal force acting on the rheometer plates some information about the normal stress difference can be obtained as shown in the following. Differentiating the identity

$$\sigma_{zz} = \sigma_{rr} + (\sigma_{zz} - \sigma_{rr})$$

with respect to r, then multiplying the result with r and using Eq. (15.38) one gets

$$r\frac{\partial \sigma_{zz}}{\partial r} = \left(\sigma_{\varphi\varphi} - \sigma_{rr}\right) + r\frac{\partial}{\partial r}(\sigma_{zz} - \sigma_{rr})$$

or by introducing the notations σ_{11} for the stress component in the flow direction, σ_{22} for that in the direction of the velocity gradient and σ_{33} for the neutral direction it follows

$$r\frac{\partial \sigma_{zz}}{\partial r} = (\sigma_{11} - \sigma_{22}) + (\sigma_{22} - \sigma_{33}) + r\frac{\partial}{\partial r}(\sigma_{22} - \sigma_{33}) \tag{15.51}$$

With

$$\sigma_{11} - \sigma_{22} = n_1(\dot{\gamma})\dot{\gamma}^2$$

being the first normal stress difference (cf. Eq. 11.3),

$$\sigma_{22} - \sigma_{33} = n_2(\dot{\gamma})\dot{\gamma}^2$$

the second normal stress difference and replacing the differentiation after r by that after $\dot{\gamma}$ according to

$$r\frac{\partial}{\partial r} = \dot{\gamma}\frac{\partial}{\partial \dot{\gamma}}$$

one gets from Eq. (15.51)

$$\frac{d\sigma_{zz}}{d\dot{\gamma}} = n_1(\dot{\gamma})\dot{\gamma} + n_2(\dot{\gamma})\dot{\gamma} + \frac{d}{d\dot{\gamma}}\left[n_2(\dot{\gamma})\dot{\gamma}^2\right] \tag{15.52}$$

$n_1(\dot{\gamma})$ and $n_2(\dot{\gamma})$ are the corresponding normal stress coefficients.

This equation can be integrated for constant normal stress coefficients $n_1(\dot{\gamma}) = n_{1,0}$ and $n_2(\dot{\gamma}) = n_{2,0}$ which are valid in good approximation at small shear rates. Substituting $\dot{\gamma}$ by r and setting $\sigma_{zz}(r = R) = \sigma_{zz}(R)$ one obtains

$$\sigma_{zz}(r) = \frac{1}{2}\left(n_{1,0} + 3n_{2,0}\right)\frac{\Omega^2}{H^2}\left(r^2 - R^2\right) + \sigma_{zz}(R) \tag{15.53}$$

Assuming that the radial stress at the rim is zero, i.e.,

$$\sigma_{rr}(R) = 0$$

from the second normal stress difference at the rim

$$\sigma_{22} - \sigma_{33} = n_{2,0}\dot{\gamma}_R^2 = \sigma_{zz} - \sigma_{rr}$$

it follows

$$\sigma_{zz}(R) = n_{2,0}\dot{\gamma}_R^2$$

and, therefore,

$$\sigma_{zz}(r) = \frac{1}{2}\left(n_{1,0} + 3n_{2,0}\right)\frac{\Omega^2}{H^2}\left(r^2 - R^2\right) + n_{2,0}\frac{\Omega^2}{H^2}R^2 \tag{15.54}$$

For the normal force F_N acting on the lower plate one gets with Eq. (15.54)

$$F_N = 2\pi\int_0^R \sigma_{zz}(r)\, r dr = -\frac{\pi}{4}\left(n_{2,0} - n_{1,0}\right)\frac{R^4\Omega^2}{H^2} \tag{15.55}$$

The normal force increases quadratically with the angular velocity Ω whereas the torque increases linearly with Ω (cf. Eq. 15.46). Again the dominant role of the sample diameter for the accuracy of the measurements becomes obvious due to its influence by the fourth power. As Eq. (15.55) shows, some information on the normal stress coefficients can be obtained from a measurement of the normal force. The results are of a limited value insofar, as the sum of the first and the second normal stress coefficients can be determined from Eq. (15.55) in the linear range, only. Changing the experimental parameters R, H or Ω, in principle offers a possibility to calculate the two normal stress coefficients as long as the experiments are performed in the linear range. But the measurements are generally not accurate enough to be of much practical importance. A better method for normal stress measurements is the cone-and-plate rheometry which is discussed in the next section.

15.3.2 Cone-and-Plate Rheometer

A schematic of the cone-and-plate assembly is presented in the Fig. 15.13. A cone with small angles β up to 5° which just touches the fixed plate by its somewhat flattened tip is driven with the angular velocity Ω. The sample which approximately forms a spherical sector is sheared between cone and plate. For an evaluation of the rheological data it is assumed that the melt adheres to the surfaces of both of them. The torque and the normal force generated at the fixed plate are measured. Technical details can be found, for example, in [7].

For a quantitative description of the flow kinetics a polar coordinate system is chosen, the origin of which comes to lie at the cone tip. The polar angle ϑ is counted from the rotation axis and becomes 90° at the plate. The azimuth angle is denoted by φ, and r designates the distance of a volume element from the origin. Its projection on the plate is given by $\rho = v\cos(90° - \vartheta) = v\sin\vartheta$. It is assumed that in the stationary state the surface of a cone formed by the material rotates with the angular velocity $\omega(\vartheta)$ around the symmetry axis which is Ω at $\vartheta = 0$ and becomes 0 on the plate at $\vartheta = 90°$. The velocities follow as:

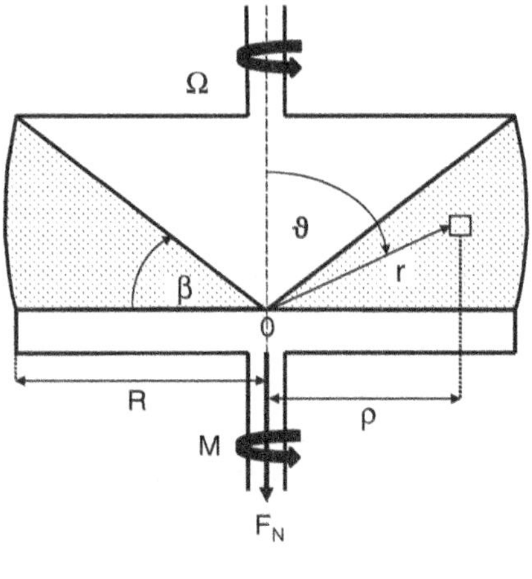

Fig. 15.13 Schematics of a cone-and-plate assembly. β cone angle, ϑ polar angle, R plate radius, r radial position of a volume element, ρ projection of its position on the lower plate, Ω angular velocity, M torque, F_N normal force

$$v_r = 0 \quad v_\vartheta = 0 \quad v_\varphi = \rho\omega(\vartheta) = r\omega(\vartheta)\sin\vartheta \tag{15.56}$$

According to Eq. (8.41b) the components of the strain rate tensor in polar coordinates are

$$d_{rr} = d_{\varphi\varphi} = d_{\vartheta\vartheta} = d_{r\vartheta} = d_{r\varphi} = 0 \tag{15.57}$$

$$2d_{\vartheta\varphi} = \frac{1}{r}\frac{\partial v_\varphi}{\partial\vartheta} - \frac{\cot\vartheta}{r}v_\varphi = \sin\vartheta\frac{d\omega(\vartheta)}{d\vartheta} = \dot{\gamma}(\vartheta) \tag{15.58}$$

These relations correspond to a viscometric flow shearing the surfaces of the cones (ϑ-areas) in φ-direction which is the direction of flow (1-direction). The ϑ-direction is that of the velocity gradient (2-direction) and the r-direction is the neutral direction (3-direction).

The equations of motion for creeping flow without volume forces follow from Eq. (7.8) under the assumptions that for symmetrical reasons the components of the stress tensor are independent of φ and the two shear stresses $\sigma_{r\varphi}$ and $\sigma_{r\vartheta}$ are zero as

$$\frac{\partial\sigma_{rr}}{\partial r} + \frac{2\sigma_{rr} - \sigma_{\vartheta\vartheta} - \sigma_{\varphi\varphi}}{r} = 0 \tag{15.59}$$

$$\frac{\partial\sigma_{\vartheta\vartheta}}{\partial\vartheta} + \left(\sigma_{\vartheta\vartheta} - \sigma_{\varphi\varphi}\right)\cot\vartheta = 0 \tag{15.60}$$

$$\frac{\partial\sigma_{\vartheta\varphi}}{\partial\vartheta} + 2\sigma_{\vartheta\varphi}\cot\vartheta = 0 \tag{15.61}$$

The shear stress $\sigma \equiv \sigma_{\vartheta\varphi}$ follows as;

$$\sigma = C/\sin^2\vartheta$$

which can directly be verified by insertion into Eq. (15.61). If the shear stress at the lower plate ($\vartheta = \pi/2$) is called σ_w one gets

$$\sigma = \sigma_w/\sin^2\vartheta \tag{15.62}$$

and for the torque acting on the lower plate

$$M = -2\pi\int_0^R r\sigma_w r dr = -\frac{2\pi}{3}\sigma_w R^3 \tag{15.63}$$

The minus sign has to be chosen as the gradient $\partial\omega/\partial\vartheta$ and according to Eq. (15.58) the shear rate and following from that the shear stress is negative.

According to Eq. (15.62) the shear stress is minimal at the plate ($\vartheta = \pi/2$) and reaches its maximum at the cone, i.e., $\vartheta = \pi/2 - \beta$. For small cone angles β the differences of the stresses at cone and plate become negligible as

$$\sigma(\pi/2 - \beta) = \sigma_w/\sin^2(\pi/2 - \beta) = \sigma_w/\cos^2\beta \tag{15.64}$$

With a cone angle of $\beta = 5°$ it follows $\sigma = 1.008\sigma_w$. As for such a marginal change the viscosity does not vary much, the shear rate $\dot{\gamma}$ can be regarded as constant, i.e., from Eq. (15.58) follows with $\sin\vartheta = \sin(\pi/2 - \beta) \cong 1$ in good approximation

$$\dot{\gamma} \cong \frac{d\omega}{d\vartheta} = const.$$

and with the boundary conditions $\omega(\vartheta = \pi/2) = 0$ and $\omega(\vartheta = \pi/2 - \beta) = \Omega$ one gets

$$\omega(\vartheta) = \frac{\Omega}{\beta}(\pi/2 - \vartheta) \tag{15.65}$$

and

$$\dot{\gamma} \cong \frac{d\omega}{d\vartheta} = -\frac{\Omega}{\beta} \tag{15.66}$$

This direct relation between the shear rate and the angular velocity together with the shear stress according to Eq. (15.63) leads to a simple formula which allows the determination of the viscosity from the directly measured quantities torque and angular velocity by

$$\eta(\dot{\gamma}) \cong \frac{3\beta}{2\pi R^3}\frac{M}{\Omega} \tag{15.67}$$

From Eqs. (15.66) and (15.67), data of the viscosity function can be obtained directly without any corrections. Similar to capillary rheometry, only positive values of the shear rate and the shear stress are conventionally used (cf. Sect. 15.1.1). The range of the shear rate is limited, however, as at values relevant for processing a fracture of the melt at the rim may occur which leads to inaccurate measurements, e.g., [8, 9]. Therefore, the cone-and-plate rheometer is a preferable instrument for the determination of viscosities at small shear rates used for the characterization of polymers (cf. Chap. 13).

The analysis of the normal stress distribution and following from it the method to determine normal stresses with the cone-and-plate rheometer start from Eq. (15.59) and (15.60) by identifying the φ-coordinate with the 1-direction, the ϑ-coordinate with the 2-direction, and the r-coordinate with the 3-direction. One gets

$$r\frac{\partial\sigma_{33}}{\partial r}+(2\sigma_{33}-\sigma_{11}-\sigma_{22})=r\frac{\partial\sigma_{33}}{\partial r}-(\sigma_{11}-\sigma_{22})-2(\sigma_{22}-\sigma_{33})=0 \quad (15.68)$$

$$\frac{\partial\sigma_{22}}{\partial\vartheta}+(\sigma_{22}-\sigma_{11})\cot\vartheta=0 \quad (15.69)$$

The following considerations based on the special conditions valid within the slit of a cone-and-plate rheometer with a small enough cone angle lead to interesting conclusions regarding the determination of the normal stress differences. According to Eq. (15.69), at the plate ($\vartheta = 90°$) it follows $\partial\sigma_{22}/\partial\vartheta = 0$, i.e., σ_{22} depends there on r, only. As in good approximation $\dot{\gamma}$ is constant within the gap, the normal stress differences $(\sigma_{11}-\sigma_{22})$ and $(\sigma_{22}-\sigma_{33})$, which depend on $\dot{\gamma}^2$, are constant there, too, and their derivatives with respect to r are zero. One consequence is that $\partial\sigma_{22}/\partial r = \partial\sigma_{33}/\partial r$ and according to Eq. (15.68)

$$-\frac{\partial\sigma_{22}}{\partial r}=-[(\sigma_{11}-\sigma_{22})+2(\sigma_{22}-\sigma_{33})]\frac{1}{r} \quad (15.70)$$

As the normal stress differences are independent of r, from the integration of Eq. (15.70) follows with the boundary condition $\sigma(r=R)=\sigma(R)$

$$-\sigma_{22}(r)=-[(\sigma_{11}-\sigma_{22})+2(\sigma_{22}-\sigma_{33})]\ln r/R-\sigma_{22}(R) \quad (15.71)$$

At the rim, the stress component in r-direction has to be zero, i.e., $\sigma_{33}(R)=0$ and

$$\sigma_{22}(R)=(\sigma_{22}(R)-\sigma_{33}(R))=n_2(\dot{\gamma})\dot{\gamma}^2 \quad (15.72)$$

The normal force F_N acting on the plate follows from Eq. (15.71) as

$$\begin{aligned}F_N&=-2\pi\int_0^R\sigma_{22}(r)rdr\\&=-2\pi\sigma_{22}(R)\int_0^R rdr-2\pi[(\sigma_{11}-\sigma_{22})+2(\sigma_{22}-\sigma_{33})]\int_0^R r\ln r/Rdr\end{aligned} \quad (15.73)$$

and with partial integration of the second integral

$$F_N=-\pi\sigma_{22}(R)R^2+\frac{\pi}{2}[(\sigma_{11}-\sigma_{22})+2(\sigma_{22}-\sigma_{33})]R^2 \quad (15.74)$$

As the term in the squared bracket is the same for all r it can be taken at $r=R$ and with the consequence $\sigma_{33}(R)=0$ it follows

$$F_N = \frac{\pi R^2}{2}(\sigma_{11} - \sigma_{22}) \tag{15.75}$$

This result means that in the case of small cone angles β ($\cos\beta \cong 1$) the first normal stress difference can directly be determined from a measurement of the force acting on the plate of a cone-and-plate rheometer. By changing the angular velocity the first normal stress difference can be obtained as a function of the shear rate.

In [10] an interesting way is proposed how to measure the first and the second normal stress difference from cone-and-plate experiments. The basic idea is to partition the plate with the radius R into two coaxial parts by a circular slit. The normal force F_N is monitored then on the inner plate with the radius R_i. The stress acting on the inner plate has to be integrated up to R_i instead of R in case of the non-separated one. Equation (15.73) becomes with the boundary conditions already introduced before

$$\begin{aligned} F_{Ni} &= -2\pi \int_0^{R_i} \sigma_{22}(r) r dr \\ &= \frac{\pi R_i^2}{2}(\sigma_{11} - \sigma_{22}) + \pi R_i^2[(\sigma_{11} - \sigma_{22}) + 2(\sigma_{22} - \sigma_{33})] \ln R/R_i \end{aligned} \tag{15.76}$$

This equation contains the first and the second normal stress difference as the two unknown quantities. Changing R or R_i results in the corresponding normal forces. From the at least two conditional equations both normal stress differences can be determined as a function of shear rate. In practice, the inner plate connected with the sensors is changed. This obvious concept requires a lot of effort, however, and is not much in use, therefore. As reported in [9] the results obtained for commercial polystyrene were found to be slightly dependent on the cone angle.

To reduce the experimental time for determining the shear stress and the two normal stress differences in parallel, in [11] a partitioned cone-and-plate assembly is introduced which avoids the time-consuming change of the inner plate. The torque and the normal forces are measured at the inner plate and the outer ring. The viscosity and the first normal stress difference can be obtained directly from the readings of the sensors connected with the inner plate. Based on Eq. (15.76) a direct relation for the second normal stress difference can be derived [11]. It was found, however, that this elegant experimental method gives less stable results for the second normal stress differences than the one with exchangeable inner plates as the signals of the outer plate are more prone to occurring edge fractures and flow irregularities [11].

These experimental difficulties are the reason why there are not many reliable measurements of the second normal stress difference in the literature. Consent exists, nowadays, that it is negative and dependent on the shear rate applied. A value of around 10 % of the first normal stress difference may be used as an approximation.

The great advantage of the cone-and-plate rheometer is the relatively good spatial constancy of the shear rate within the gap at correspondingly chosen cone angles.

Therefore, this instrument is very suitable for direct measurements of viscosity function and the shear rate dependence of the first normal stress difference in parallel. It is very versatile as by a special control of the drive unit a variety of modes can be applied. Stressing and creep experiments are possible and dynamic-mechanical tests are very common, today.

A limitation of the cone-and-plate rheometer is the fact that at longer experimental times and higher shear rates secondary flow phenomena can occur which disturb the linear flow profiles as was explicitly shown for polymer solutions in [8] and for polymer melts in [9], for example. Additionally, a fracturing of the melt at the rim has been observed by several authors the probability and intensity of which become the higher the larger the shear rate. Such irregularities affect the measurements of shear and normal stresses even in the case of a partitioned plate assembly [9]. Therefore, in the case of polymer melts the cone-and-plate rheometer is a suitable instrument for investigations of rheological properties at shear rates not too far away from the Newtonian region. An extension to higher shear rates interesting for applications can be attained from dynamic-mechanical measurements when use of the Cox-Merz relation can be made (cf. Sect. 11.3.1). But for this mode, too, an undisturbed flow profile within the gap is the base of reliable experiments.

15.3.2.1 Alternative Derivation of the Equations for Shear Rate and Shear Stress

A simple derivation of the equations for shear rate and shear stress of a cone-and-plate geometry is given in the following. The shear rate $\dot{\gamma}$ defined as the gradient of the velocity v_φ in φ-direction with respect to the ϑ-direction can approximately be written as (cf. Fig. 15.13)

$$\dot{\gamma} \approx -\frac{dv_\varphi}{dr \sin(\pi/2-\vartheta)} = -\frac{dv_\varphi}{rd \cos\vartheta} = \frac{dv_\varphi}{r \sin\vartheta d\vartheta} \tag{15.77}$$

The velocity v_φ follows as

$$v_\varphi = \omega(\vartheta)\rho = \omega(\vartheta) r \sin\vartheta \tag{15.78}$$

For the small cone angles β chosen, it follows $\sin\vartheta \approx 1$ and then the shear rate $\dot{\gamma}$ becomes

$$\dot{\gamma} \approx \frac{d\omega(\vartheta)}{d\vartheta} \tag{15.79}$$

As the plate is fixed, i.e., $\omega = 0$ at $\vartheta = \pi/2$, and the cone ($\vartheta = \pi/2-\beta$) rotates with the angular velocity Ω one gets for the angular velocity of a sample adhering to the walls

$$\omega(\vartheta) = \frac{\Omega}{\beta}(\pi/2 - \vartheta) \tag{15.65}$$

and

$$\dot{\gamma} \approx -\frac{\Omega}{\beta} \tag{15.66}$$

For a shear rate which is independent of the position within the slit, the shear stress σ is constant, too, and the torque exerted on the rotational axis becomes

$$M = -2\pi \int_0^R r\sigma r dr = -\frac{2\pi}{3}\sigma R^3 \tag{15.63}$$

From the measured rotational velocity Ω and torque M of a cone-and-plate geometry with the cone angle β and the plate radius R the shear viscosity can be determined. In practice, the minus signs for the shear rate and the shear stress following from the exact physical derivation are neglected.

The equations for the shear rate and the shear stress were derived without using polar coordinates and the somewhat complicated corresponding components of the stress and deformation rate tensors as in Sect. 15.3.2. Such a simplified consideration cannot be applied, however, for the evaluation of the normal stress differences from cone-and-plate measurements.

References

1. Weissenberg K (1929) In: Rabinowitsch B (ed) Z Phys Chemie A 145:1
2. Bagley EB (1957) J Appl Phys 18:624
3. Münstedt H (1978) Kunststoffe 78:92
4. Eswaran R, Janeschitz-Kriegl H, Schijf J (1963) Rheol Acta 3:83
5. Wales JLS, den Otter JL, Janeschitz-Kriegl H (1965) Rheol Acta 4:146
6. Janeschitz-Kriegl H (1965) In: Lee EH, Copley AL (eds) Proceedings of the 4th international congress rheology part I, 143, Interscience, New York
7. Meissner J (1972) J Appl Pol Sci 16:2877
8. Giesekus H (1967) Rheol Acta 6:339
9. Schweizer T (2002) Rheol Acta 41:337
10. Meissner J, Garbella RW, Hostettler J (1989) J Rheol 33:843
11. Schweizer T, Hostettler J, Mettler F (2008) Rheol Acta 47:943

Chapter 16
Measurements of Flow Fields of Polymer Melts by Laser-Doppler Velocimetry

16.1 Motivation

The formulas for the evaluation of material functions from rheological experiments as presented in the foregoing sections are based on certain assumptions concerning the underlying flow fields. Adhesion to the wall is the fundamental condition for all the derivations. But not all polymer melts stick to the wall and there are cases for the processing of polymer melts in which slippage on the walls of a tool is deliberately generated (cf. Sect. 17.3). Therefore, it is of fundamental interest to know the conditions under which the flow of a polymer melt takes place.

Besides that, there are phenomena particularly connected with capillary flow which can only be understood by looking at the flow fields in detail. Examples are the extrudate swell and the various appearances of melt fracture. In general, the flow in a capillary rheometer, which is frequently applied to determine viscosity functions and which is the base of the widely used melt indexer, has to be regarded as very complex as it is composed of an entry flow region and the shear flow in the capillary.

Some attempts were undertaken to get an insight into the flow pattern by observing differently dyed polymer layers during or after extrusion. A big disadvantage of this method is that the single fluid elements cannot be distinguished from each other and, therefore, a quantitative evaluation of the velocities is difficult.

Another visualization method is the streak photography. It is based on tracer particles added to the fluid the motion of which is followed up by photography or more recently by particle image velocimetry. From these methods, in principle, particle velocities can be determined, but the results from the literature show that the accuracy is not high enough for getting reliable quantitative results for geometrically complex flows.

A method to measure velocity fields with a much higher accuracy is the laser-Doppler velocimetry (LDV). Developed for the determination of velocities in gases, it was used in the 1960's for measurements in water [1]. Later, first LDV measurements were performed on polymer solutions, e.g., [2, 3, 4, 5].

H. Münstedt and F. R. Schwarzl, *Deformation and Flow of Polymeric Materials*,

DOI: 10.1007/978-3-642-55409-4_16,

LDV-measurements on polymer melts are very scarce in the literature. First results were reported 1980 on a low density polyethylene [6], and later ones on low and high density polyethylenes [7], on a high density polyethylene [8] and a polybutadiene [9]. Particularly for polymer melts, the accuracy of the measurements was not good enough to get a quantitative picture of the velocity distributions in the secondary flow regions which are of special interest, however, with respect to the features of the entry flow and their dependence on the molecular structure of the polymers. One reason for this is the high temperature sensitivity of the viscosity of polymer melts, and following from that, unsymmetrical flow fields due to a nonuniformity of the temperature along the ducts occurred. The other reason are vibrations from external sources which superimpose the motion of the scattering particles in the melt.

Further developments in applying the LDV-technique [10, 11] made it possible to investigate flow phenomena in polymer melts with high accuracy. Therefore, laser-Doppler velocimetry can now be regarded as a reliable and interesting experimental method to investigate the flow behavior of polymer melts and to add new information to the knowledge already obtained with classical rheological methods as discussed before.

16.2 Measuring Principle and Setup of a Slit Die

16.2.1 Measuring Principle of Laser-Doppler Velocimetry

The requirements for LDV measurements on polymer melts are very special from two points of view. Firstly, the velocities are small in comparison to those in gases and liquids and, therefore, very prone to disturbances from outside. Secondly, the temperature distribution has to be very uniform, as the viscosity of polymer melts and following from it the velocity distributions are very temperature sensitive. This fact represents a great experimental challenge, particularly, in the case of large and complicated flow geometries.

The physical base of the laser-Doppler velocimetry is the optical Doppler effect. If a light source of a frequency f_S moves with the velocity v relative to an observer, then a frequency f_D is measured which for $v \ll c$ reads

$$f_D = f_S/(1 - v/c) \tag{16.1}$$

with c being the velocity of light and v the velocity component in the direction given by the connecting lines between the light source and the observer. The fundamentals of LDV can be found, for example, in [12]. They are not simple enough to become a part of this book. But a few features are outlined here which are necessary for an understanding of some of the results. In practice the so-called two beam method is applied which is sketched in Fig. 16.1. The laser-Doppler velocimeter consists of two main components, the optics and the burst spectrum

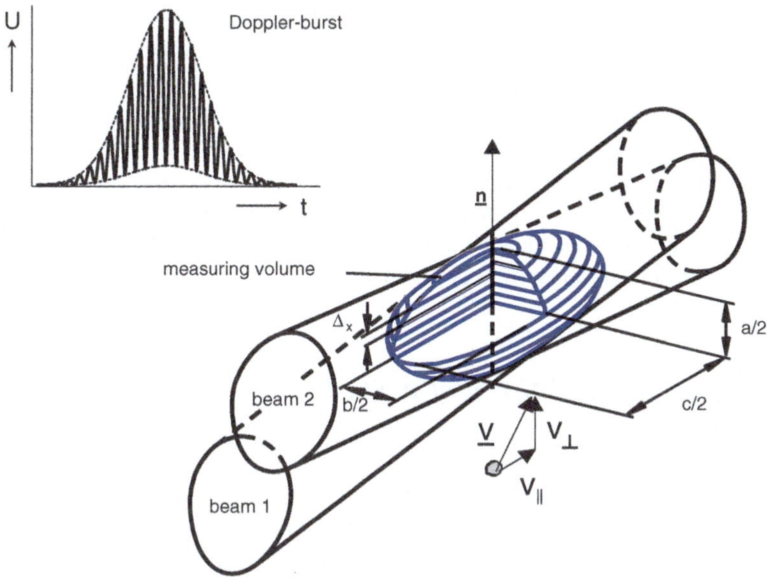

Fig. 16.1 Interference pattern and Doppler burst [11]

analyzer for the evaluation of the laser signals. The incident laser beam is optically split up into two beams which are guided by a glass fibre to the measuring probe and focused on the sample volume to be investigated. At their intersection, the interference of the two beams forms a fringe pattern with bright and dark planes separated by the distance Δx from each other. The ellipsoidal measuring volume has typical dimensions of $a \approx 30$ μm and $c \approx 300$ μm which determine the spatial resolution of the method (cf. Fig. 16.1). The light which is backscattered from a particle traversing the measuring volume contains the velocity information, i.e., the Doppler frequency f_D. The scattered signal is detected by the measuring probe and conducted through glass fibers to a photomultiplier which transforms it into an electrical signal as shown in Fig. 16.1. This so-called Doppler burst contains the Doppler frequency f_D which is related to the velocity component $v_\perp$ of the particle vertical to the main axis of the measuring volume by

$$f_D = v_\perp / \Delta x = v_\perp 2\sin(\Phi_{air}/2)/\lambda_{0air} \tag{16.2}$$

Φ_{air} is the angle between the two beams crossing each other and λ_{0air} the wave length of the laser light measured in air.

The Doppler frequency f_D can be obtained by a Fourier analysis of the photomultiplier signal a typical example of which is shown in Fig. 16.1. This so-called back scattering device needs only one window whereas two windows are needed if the forward scattering is exploited.

The temporal resolution of the LDV-system used was 2 ms, the velocity range covered 50 μm/s to 50 m/s. The parallel velocity component $v_{||}$ can be obtained by an additional measurement after having rotated the optical plane around the optical axis by 90° or making use of the second wavelength of a two color laser. This method is very elegant, but needs separate optics. The modern glass fiber technique for guiding the light beams allows to mount the optics to a three-dimensional translation stage which offers the possibility to position the measuring volume with a spatial resolution better than 10 μm. More detailed information on the technique underlying the results presented is given in [11].

16.2.2 Experimental Setup of a Slit Die

In principle, the LDV-device described can be used to measure velocity profiles of polymer melts in various geometries. The only prerequisite is a window to let the laser light pass. Therefore, in addition to laboratory experiments, LDV has the potential to investigate the flow of polymer melts in processing equipments. Results of flow profiles of a polypropylene melt extruded through a flat die of 300 mm in width are reported in [13] and shortly addressed in Sect. 17.4. For fundamental investigations, particularly on the influence of the geometry of a tool or the molecular structure of the polymeric material on the flow behavior, laboratory-scale equipments are used. Rectangular ducts and slit dies are preferred as they are simpler to handle with respect to the optical techniques than circular geometries.

An example of a die which was particularly designed for LDV-experiments is represented in Fig. 16.2. The flow channel consists of a reservoir of a squared cross section of 14 × 14 mm and a slit capillary of 50 mm in length, 1.0 mm in height, and 14 mm in width. The entrance angle from the reservoir to the slit is 180°, the contraction ratio 14:1. On one side the channel is covered by a glass window. The Cartesian coordinate system is chosen as shown in Fig. 16.2. The slit dimensions and the entrance geometry can be varied rather easily by changing the insets forming the die. The flow channel is equipped with four pressure transducers. One is located at $x = -35$ mm, i.e., in the reservoir, the others are mounted along the slit at $x = 5$ mm, $x = 21$ mm, and $x = 37$ mm.

The challenge of measuring flow profiles of polymer melts accurately sets some standards for the flow channel with respect to a uniform temperature distribution. In order to get a symmetrical velocity field the temperature has to be constant independent of the local position. This goal is not easy to achieve in a device for which obviously the window must not be covered by heating elements which normally are not transparent for light, of course. For the experimental device shown in Fig. 16.2, a local temperature constancy of 0.2 °C could be reached

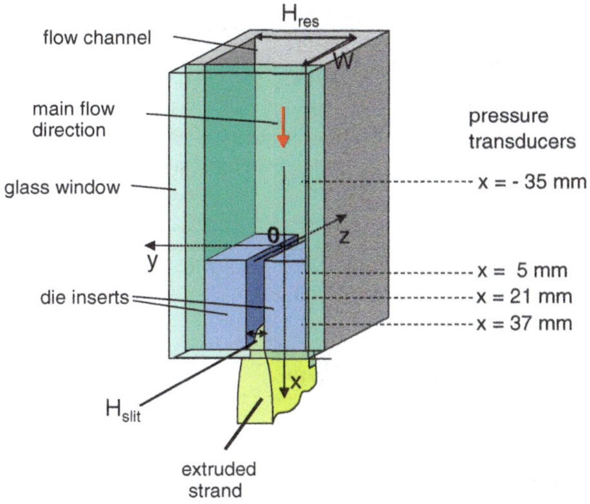

Fig. 16.2 Flow channel. $W = H_{res} = 14.0$ mm. H_{slit} is variable. The contraction ratio is defined as $CR = H_{res}/H_{slit}$

within the channel by using 12 separately controlled heat cartridges. The channel can be connected to machines which are able to supply a continuous and steady flow of melt without generating disturbing vibrations of the channel. A laboratory extruder in combination with a melt pump was found to fulfil these requirements.

16.3 Flow Fields in Different Sections of the Slit Die

The flow fields in different parts of the channel presented in the Fig. 16.2 are of interest from the processing and the fundamental point of view. It is well known that polymer melts can show complex patterns when flowing through a narrowing duct as it is the case at the entrance of a capillary, for example. These phenomena are dependent on external parameters like throughput, temperature, and geometry as well as on the molecular structure of the material.

Of great importance for an understanding of the processing of polymers is the knowledge about the flow behavior of their melts in circular and rectangular ducts. Slippage or adhesion on the wall of a die, for example, determine some important features of the processing performance of polymer melts. Investigations on the velocity fields at the die exit are very helpful in understanding some aspects of melt fracture. Furthermore, they give some insight into the kinematics of the extrudate swell.

Table 16.1 Characteristic data of the polyolefins investigated

	LDPE 2	LDPE 3	HDPE	LCB-PP	L-PP
ρ (25 °C)[g/cm^3]	0.918	0.918	0.953	0.909	0.908
M_w [kg/mol]	250	410	300	587	245
M_w/M_n	17	24	15	10	5
η_0 [kPas]	26[a]	800[b]	>660[c]	34[c]	24[c]
Strain hardening	++	++	+	++	−
Elastic properties	++	++	++	++	+

[a] $T = 170$ °C, [b] $T = 150$ °C, [c] $T = 180$ °C
++very significant, +weak, −absent

16.3.1 Entrance Flow

It has been known for a long time that contraction flows of polymer melts and polymer solutions can be accompanied by flow irregularities although the Reynolds numbers are far below the critical value for turbulence. These irregularities which are called *secondary flow* appear as vortices in the corners of the contraction. An example for a polymer melt, in this case a low density polyethylene, has been published in [14]. The results were obtained by streak photography which gives a good overall picture of flow phenomena, but is not a quantitative method for determining flow profiles. They describe the somewhat surprising feature that the pronounced secondary flow found for a long-chain branched polyethylene cannot be observed for a linear polyethylene.

This puzzling result was tried to be explained in [15] by the assumption that the larger elastic properties of the long-chain branched material is the source of the secondary flow. From this interpretation, the notation "elastic turbulence "may have been derived which is sometimes used in the literature instead of secondary flow. It should be taken into consideration, however, that special linear polyethylenes can also exhibit an amount of elasticity similar to that of long-chain branched species (cf. Table 16.1). Later on, it was postulated, that the differences in strain hardening are responsible for various secondary flow patterns as long-chain branched polyethylenes show strain hardening, but linear polyethylenes not [16]. Therefore, the molecular origin of secondary flow has to be left open.

The secondary flow phenomena have initiated some efforts to model them e.g., [17, 18]. In many cases, these attempts have been of limited value, only, as accurate quantitative measurements of the velocity fields and, therefore, a comparison of theory and experiment were not available at that time.

Besides a general fundamental interest in the phenomena of secondary flow there are some aspects regarding the processing of polymer melts. If the flow pattern is stable and the velocities within the vortices are low then the residence times for a polymer molecule can be very long giving rise to thermal degradation which may go thus far that gels or carbon particles originate. In processing, these degradation products are called "fish eyes" or "black specs", respectively.

16.3.1.1 Flow Profiles in the Reservoir

In Fig. 16.3 the velocity profiles v_x (y) in the reservoir at different positions x upstream the slit are plotted for LDPE 2, some properties of which are given in Table 16.1. The measurements were performed at an apparent shear rate of $D = 122\ \mathrm{s}^{-1}$ and a temperature of $T = 150$ °C. All the profiles show an excellent symmetry which demonstrates the high reliability of the measurements and a sufficient uniformity of the temperature distribution across the channel. The shapes of the profiles change completely within a distance of about 10 mm upstream the slit entry. They indicate a pronounced convergent flow. At $x = -$ 20 mm, i.e., 20 mm above the entrance plane in the reservoir, the velocity profile $v_x(y)$ is fully developed and no longer influenced by entrance effects as it does not differ from the profiles at $x = -30$ and -40 mm (For the matter of clarity the corresponding curves are not explicitly shown). From Fig. 16.3 it is obvious that there exist strong elongations of the melt near the center of the reservoir. 1 mm in front of the slit, i.e., at $x = -1$ mm, the velocity on the channel axis, i.e., $y = 0$ mm and $z = 0$ mm, is about eight times that in the fully developed region, whereas 2 mm away from the center line ($|y| = 2$ mm) the velocities become comparably small.

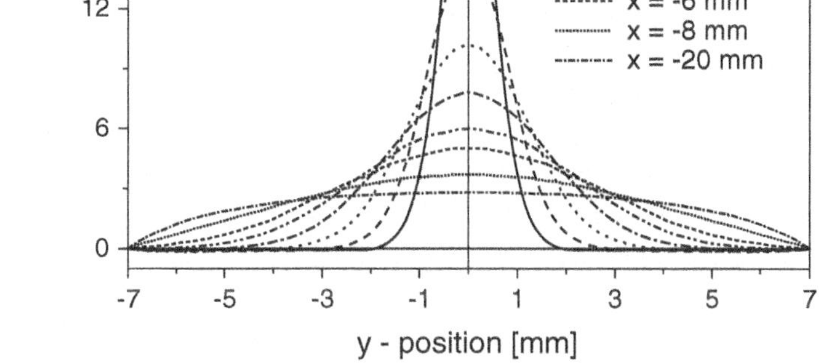

Fig. 16.3 Velocity components v_x (x,y) over the width of the reservoir at different distances x from the entrance plane for LDPE 2. (Reprinted from [10] by permission from The Society of Rheology)

16.3.1.2 Flow Along the Channel Axis

The flow along the axis of the channel which is purely elongational and changes significantly close to the entrance region deserves special attention as it offers interesting insights into features related to rheological measurements with the capillary rheometer, the basics of which are discussed in Chap. 15. Assumptions are made the validity of which, strictly taken, have to be proven for each material separately. In Fig. 16.4, the velocity component v_x (x) along the axis $(y = z = 0)$ of the flow channel consisting of the reservoir and the slit is shown for the LDPE 2.

Four different flow regions can be distinguished:

1. The fully developed flow upstream the slit entry $(x < -10 \text{ mm})$
2. The converging flow $(-10 \leq x \leq 0 \text{ mm})$
3. The developing flow within the slit $(0 \leq x \leq 15 \text{ mm})$
4. The fully developed flow within the slit $(x > 15 \text{ mm})$

Within the fully developed flow regime in the reservoir, a constant velocity of $v = 3$ mm/s is found for an apparent shear rate in the slit of $D = 122 \text{ s}^{-1}$. By approaching the slit entry plane $x = 0$, the velocity v_x increases rapidly. In the slit $(0 < x < 0.5 \text{ mm})$ the velocity profile shows an overshoot. The origin of this phenomenon which was observed, for example, in [6] too, was an open question for many years, particularly, as an overshoot was not found for LLDPE. From LDV-measurements conducted with very high accuracy it was concluded in [19] that the three-dimensional flow of the LDPE in the entrance region can be regarded as the reason for the overshoot confirming the modelling predictions in [18].

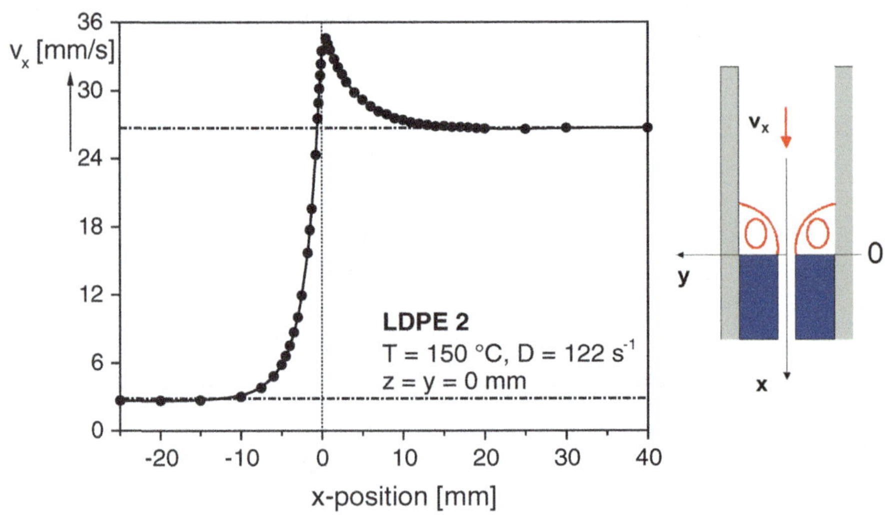

Fig. 16.4 Velocity profile v_x (x) along the center line of the channel, i.e., $z = y = 0$ mm, for LDPE 2 [11]

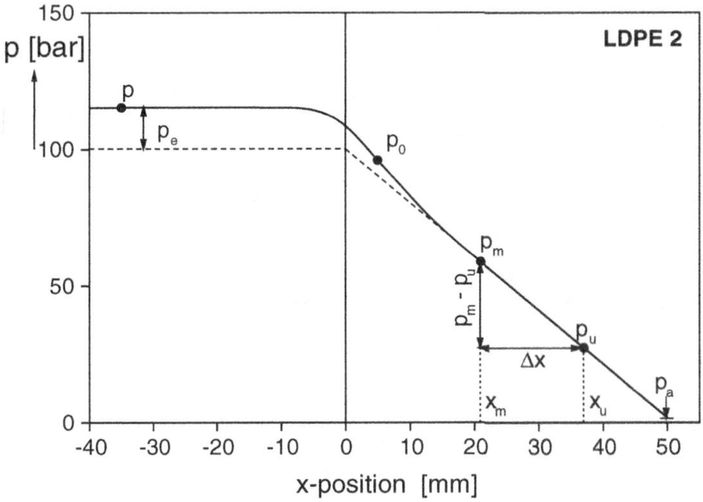

Fig. 16.5 Pressure profile measured in the flow channel at the three positions p_o, p_m, p_u in the slit and p in the reservoir for LDPE 2. p_e is the entrance pressure loss and p_a the exit pressure loss. The temperature was $T = 150°$ and the apparent shear rate $D = 122\ s^{-1}$ [20]

Besides the overshoot, Fig. 16.4 points to a behavior of the LDPE melt which has some relevance for viscosity measurements using capillary rheometers. Within the slit, the shear flow is fully developed not before $x \approx 15$ mm, i.e., a length of 15 slit heights of 1 mm is needed to achieve the constant velocity of $v_x = 26$ mm/s under the conditions given. This feature has to be considered, if measurements of the shear viscosity are carried out using a slit viscometer. A pure shear flow does not exist along a rather considerable distance from the slit entrance within the capillary for the LDPE 2. That means, in this range the basic assumptions for the determination of the shear viscosity are not fulfilled. Therefore, the lengths of slit dies chosen for rheometers have to take the properties of the material investigated into account and the pressure gauges have to be positioned correspondingly.

This insight is underlined by the readings of the four pressure gauges mounted to the rear of the flow channel (cf. Fig. 16.2) which are presented in Fig. 16.5. It is clearly seen that the reading of the pressure gauge at $x = 5$ mm is higher than the linear extrapolation according to the other two gauges located in the region of the fully developed pure shear flow. Thus, the pressure gradient has to be determined from the transducers at $x = 21$ mm and $x = 37$ mm. The die entry pressure loss p_e (cf. Sect. 15.1.3) is obtained by means of the pressure difference between the reading of the transducer in the reservoir and the pressure extrapolated to the die entry plane at $x = 0$. The exit pressure loss p_a is found to be very small in comparison to the entrance pressure loss p_e for the long-chain branched polyethylene investigated.

From the pronounced changes of v_x with x around the contraction as shown in Fig. 16.4, a large elongational deformation becomes obvious. The elongational

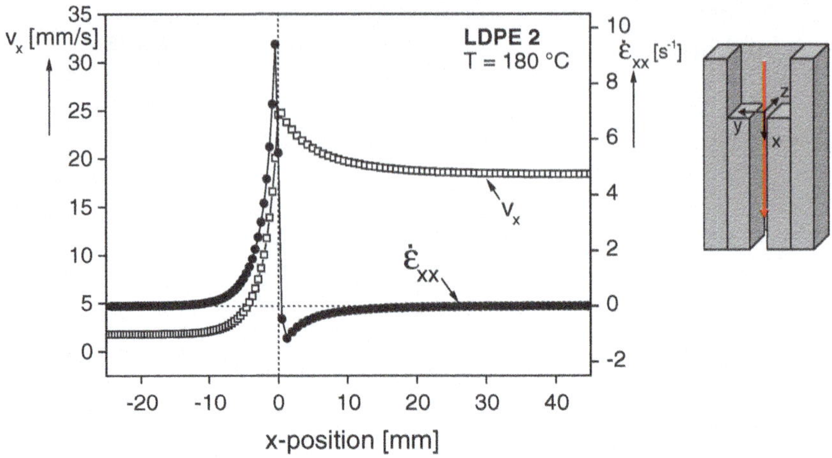

Fig. 16.6 Velocity $v_x(x)$ and elongational rate $\dot{\varepsilon}_{xx}$ along the channel axis for LDPE 2 at $T = 180$ °C and an apparent shear rate $D = 100\ \mathrm{s}^{-1}$

rate $\dot{\varepsilon}_{\mathrm{xx}} = \partial v_x/\partial x$ along the channel axis is plotted in Fig. 16.6 in addition to $v_x(x)$. In the regions of the fully developed flow upstream and downstream of the contraction the elongation rates $\dot{\varepsilon}_{\mathrm{xx}}$ are zero, of course. In the converging flow region, $\dot{\varepsilon}_{\mathrm{xx}}$ strongly increases, reaches its maximum shortly before the slit entrance, and then decreases almost spontaneously to negative values which are caused by the velocity overshoot shortly behind the slit entrance. Though the corresponding

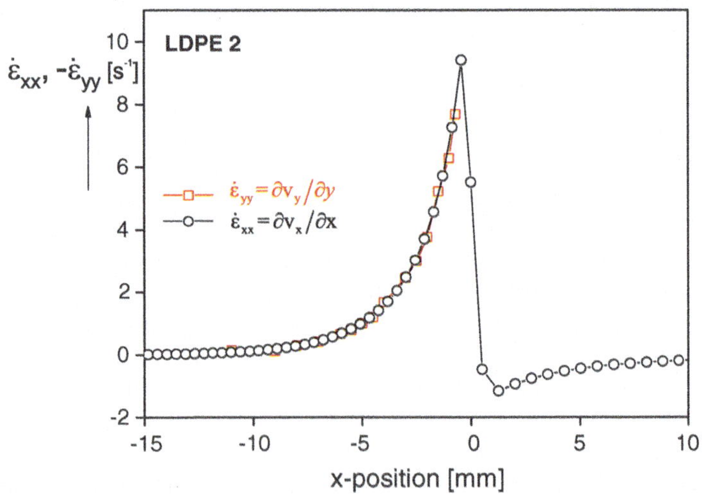

Fig. 16.7 Elongational rate $\dot{\varepsilon}_{xx}$ in flow direction and $\dot{\varepsilon}_{yy}$ perpendicular to it measured along the center line for LDPE at $T = 180$ °C and the apparent shear rate $D = 100\ \mathrm{s}^{-1}$ (Reprinted from [10] by permission from The Society of Rheology)

apparent shear rate of around 100 s^{-1} is at the lower limit of those occurring in processing operations, the maximum elongational rate of 9 s^{-1} is already rather high in comparison to values which can be achieved by laboratory elongational rheometers (cf. Sect. 12.3).

An important aspect of the elongational deformation along the center line of the channel becomes obvious if the two components $\dot{\varepsilon}_{xx}$ and $\dot{\varepsilon}_{yy}$ of the strain rate tensor are compared (cf. Fig. 16.7). The experiment on the LDPE shows that $\dot{\varepsilon}_{xx}(x)$ is equal to $-\dot{\varepsilon}_{yy}(x)$. As the velocity components v_y cannot be measured in the slit due to geometrical restrictions, this result was experimentally only verified within the reservoir. For the incompressible polyethylene melt the equality of $\dot{\varepsilon}_{xx}$ and $-\dot{\varepsilon}_{yy}$ means that the elongation is planar for $z = 0$ (cf. Sect. 12.2). It has to be considered, however, that this relation is only valid within the accuracy of the measurement as it was shown by the more sophisticated investigations in [19] that the flow of the LDPE is not truly two-dimensional even in the central plane.

16.3.1.3 Flow Beyond the Channel Axis

As Fig. 16.3 shows, v_x decreases from the center to the wall and seems to become zero at some distance from the barrel axis. If the velocities v_x in the marked area of the left part of Fig. 16.8 are plotted at a higher resolution, however, it is obvious that negative v_x components exist which indicate a mass flow opposite to the main stream (right part of Fig. 16.8). Their values are small, but due to the excellent performance of the LDV-device used they can be measured with a rather high accuracy. The extension of the negative velocities along the y-direction is the larger the smaller the distance from the entry plane. The velocities at the walls of

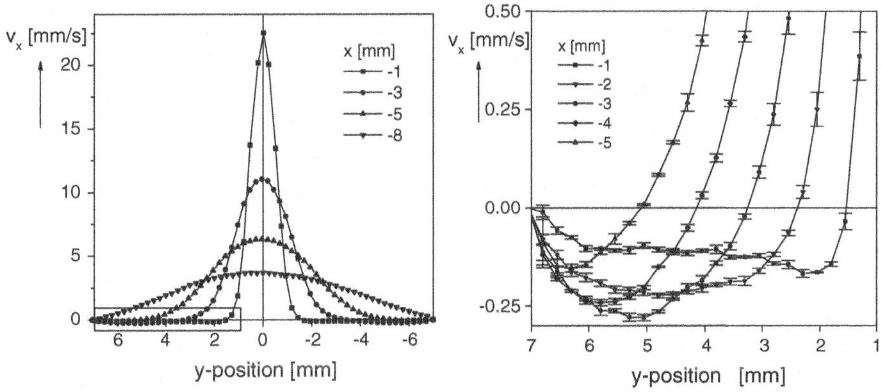

Fig. 16.8 Velocity components v_x over the width of the reservoir y at $z = 0$ and different positions x from the entrance plane for LDPE 2. The temperature was $T = 180$ °C and the mass throughput $\dot{m} = 0.15$ g/s. The right figure presents the section marked in the *left* one at higher resolution. The *bars* indicate the repeatability of the measurements. (Reprinted from [10] by permission from The Society of Rheology)

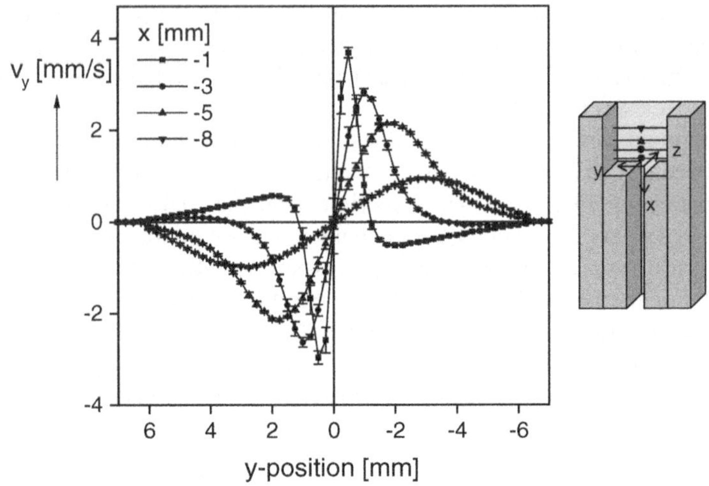

Fig. 16.9 Velocity components v_y over the reservoir width at $z = 0$ and different distances x from the entrance plane for LDPE 2 at $T = 180$ °C and the apparent shear rate $D = 100\ s^{-1}$. The *error bars* indicate the reproducibility of the measurements [22]

the reservoir ($|y| = 7$ mm) diminish as it should be for a melt sticking to the wall. The distribution of the velocity component v_y along the y-axis which is represented in Fig. 16.9 shows a point-symmetric feature. Whereas 8 mm away from the entry plane the velocity distribution reflects the main flows into the directions of the slit inlet, distinct components opposite to the main flow can be observed by approaching the entry plane. These velocity components indicate that in distinct regions the melt flows towards the wall and not to the center of the channel. Velocity vectors resulting from the separate measurements of v_x and v_y in the Figs. 16.8 and 16.9 are plotted in Fig. 16.10 and give a quantitative picture of the vortices already qualitatively described in [14]. As the flow fields in the channel were found to be highly symmetric with respect to the center line of the reservoir only one half is plotted and discussed in the following. The velocities within the vortex presented in Fig. 16.10 lie distinctly below 1 mm/s. In comparison to those, the velocities in the region of the main flow towards the slit inlet are much higher. In the literature the main flow is often designated as primary flow and the vortices are called secondary flow.

For a more detailed investigation of the influence of different experimental conditions and molecular parameters on the secondary flow phenomena, a quantification of the vortices is introduced which is described in detail in [20]. Based on a velocity balance, the full line in Fig. 16.10 was determined separating the vortex from the main flow. Using this procedure, the vortex area A can be quantified.

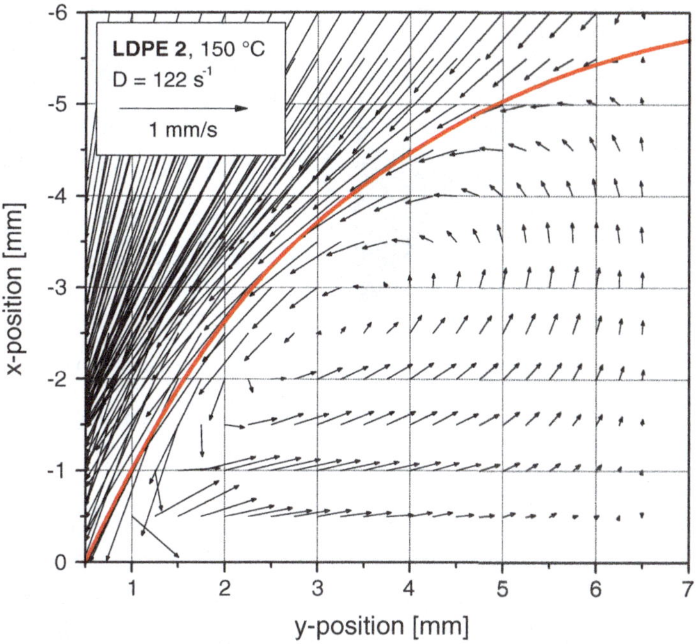

Fig. 16.10 Vortices in the entrance region of the slit with a height of 1 mm for LDPE 2 at $T = 150$ °C and an apparent shear rate of $D = 122\ s^{-1}$. (Reprinted from [10] by permission of The Society of Rheology)

16.3.1.4 Secondary Flow of Linear and Long-Chain Branched Polyolefins

In [14] and [21] visually observed vortices were reported for a long-chain branched, but not for a linear polyethylene. A similar behavior was found in [22] for a linear and a long-chain branched polyethylene (cf. Fig. 16.11) and a linear and long-chain branched polypropylene (cf. Fig. 16.12) which were quantitatively investigated using the LDV-technique. The branching structure—although important for some properties of a polymeric material—is surely not the quantity from which a correlation with secondary flow phenomena can be expected. Rather, the rheological behavior should be considered as it provides direct insights into the flow of polymer melts.

Some characteristic data of the LDPE, the HDPE and the two PP are given in Table 16.1. The two long-chain branched polymers LDPE 3 and LCB-PP show a pronounced strain hardening at elongational rates larger than 0.05 s^{-1} whereas the strain hardening of the HDPE is small in this range, but becomes higher at lower rates [22]. The elasticity measured by the extrudate swell is significantly larger for the LCB-PP than for the linear PP. The extrudate swell data for the three polyethylenes are very similar, as shown in [22]. This finding is surprising and points to

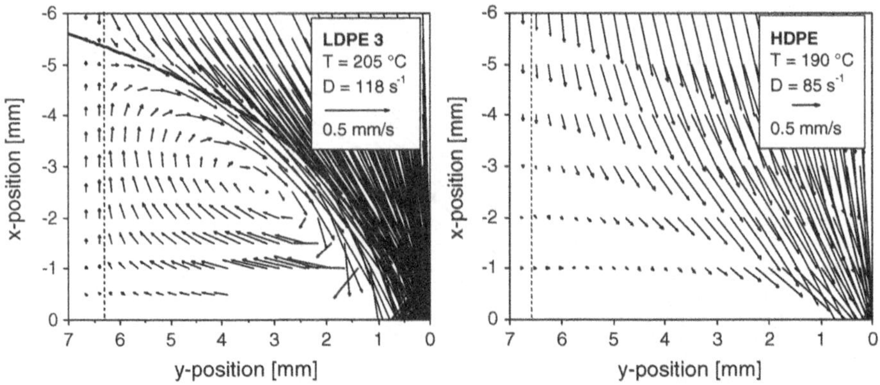

Fig. 16.11 Entrance flow pattern of a long-chain branched and a linear polyethylene. Some characteristic data of the materials are given in Table 16.1 [22]

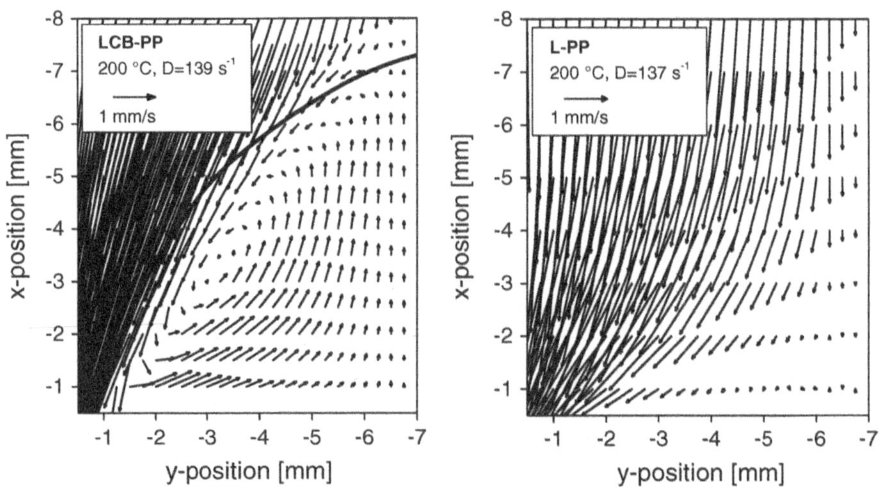

Fig. 16.12 Entrance flow pattern of a long-chain branched and a linear polypropylene. Some characteristic data of these materials are given in Table 16.1 [22]

a high molar mass component of the HDPE which could not be detected by gel permeation chromatography, however. Whereas from the results on the polypropylenes one could suppose that the elasticity plays a decisive role for the formation of vortices, this heuristic deduction does not hold in case of the polyethylenes and, therefore, is not of a general validity. The findings are inconsistent with assertions from the literature that elasticity has to be regarded as the origin of secondary flow resulting in the annotation "elastic turbulence" for the occurrence of vortices.

The correlation between vortices and strain hardening as proposed in the literature (e.g., [23]) does not hold at a first glance, either, since the HDPE which does not exhibit vortices shows at least an indication of strain hardening. It is very small in the range of the higher elongational rates, however, which dominate in the entrance region (cf. Fig. 16.6). Empirical correlations discussed in [24] give at least a hint that strain hardening and secondary flow may be related to each other.

From the investigations of the flow patterns in the entrance region it becomes obvious that the significant elongation of the melt and the vortices may contribute to the measured entrance pressure loss. As for the linear polypropylene and polyethylene, which do not exhibit any secondary flow, entrance pressure losses are found it may be concluded that the elongational deformation is their main source.

16.3.2 Flow Inside the Slit Capillary

In Fig. 15.9 qualitative pictures of the shear rate and the velocity profile of a non-Newtonian liquid in a slit are shown. With the LDV setup presented, it is possible to quantitatively measure the velocity across a slit capillary and to derive the shear rate from it. In this way the validity of assumptions underlying the use of a slit rheometer to determine viscosity functions can be checked. An example of such a flow profile measured for LDPE 2 is shown in Fig. 16.13. For an analytical presentation of the experimental data two formulas were used. The polynomial with the coefficients a_i

$$v_x(y) = \sum_{i=1}^{m} a_i \left(1 - (2y/H)^2\right)^i \tag{16.3}$$

provides a more formal description without a physical background, whereby the relationship

$$v_x(y) = v_0 \left(1 - |2y/H|^{(n+1)/n}\right) \tag{16.4}$$

is based on the power law (cf. Eq. 11.14). v_0 is the velocity in the center of the slit, H the slit height and the parameter n specifies the power-law index. A detailed derivation of Eq. (16.4) can be found in [20]. According to the presupposition made it is obvious that with Eq. (16.4) good results can only be expected if the experiment is conducted in the range of the viscosity functions following a power-law behavior, i.e., it is not as generally applicable as the polynomial.

From Fig. 16.13 it can be concluded that for $m = 4$ the measured velocity data are described with good accuracy. The fitted curve resulting from Eq. 16.4 with $n = 0.36$ can hardly be distinguished from the polynomial. The fit parameter n is in good agreement with n obtained for LDPE 2 directly from its viscosity function (cf. Fig. 11.4).

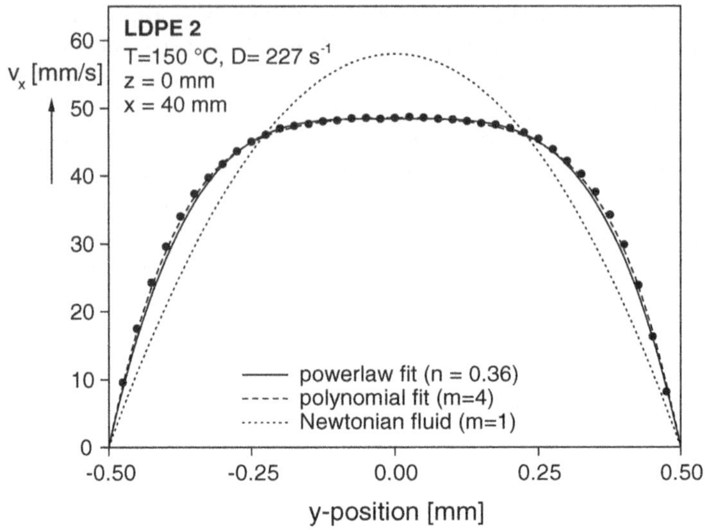

Fig. 16.13 Velocity profile of LDPE 2 measured across the slit in the non-Newtonian regime compared to that calculated in the Newtonian region. LDPE 2 is very similar to the LDPE IUPAC A discussed before [11]

It is interesting to note that the two different fit functions were obtained by setting the slit height H to 1.0 mm. A velocity of $v_x = 0$ at the walls is obtained from both of the fits, i.e., the melt of LDPE 2 sticks to the wall as assumed for the determination of viscosity functions by measurements of the pressure gradient and the throughput in a capillary rheometer (cf. Sect. 15.1.1).

For comparison, the velocity profile of a Newtonian liquid of the same output rate is plotted as a dotted line. It becomes obvious to what extent the shear thinning of the polyethylene for this output rate flattens the parabolic flow profile.

In Fig. 16.14 the shear rate $\dot{\gamma}(\mathrm{y}) = \mathrm{d}v_x/\mathrm{dy}$ obtained as the derivative of the velocity profile is plotted as a function of the position y across the slit height. The magnitude of the shear rate in the Newtonian range achieves its maximum at the walls of the capillary and decreases linearly to zero in the center of the slit at $y = 0$. In the non-Newtonian regime the curve of the shear rate exhibits an S-shape which was qualitatively derived according to the considerations presented in Fig. 15.9. The shear stress shown by the dotted line in Fig. 16.14 was obtained from the pressure gradient (cf. Fig. 16.5).

Using the shear rates obtained from the difference quotients of the point-wise measured velocities and the shear stresses at various y-positions, the viscosity as a function of the shear rate can be presented. The values are plotted as open triangles in Fig. 16.15. They show a big scatter at low shear rates due to the inaccuracy of the measurement of small velocities. The full line represents the viscosity according to the description of the LDV data by the power law. The filled circles stem from measurements of the volume output rate applying the usual evaluation

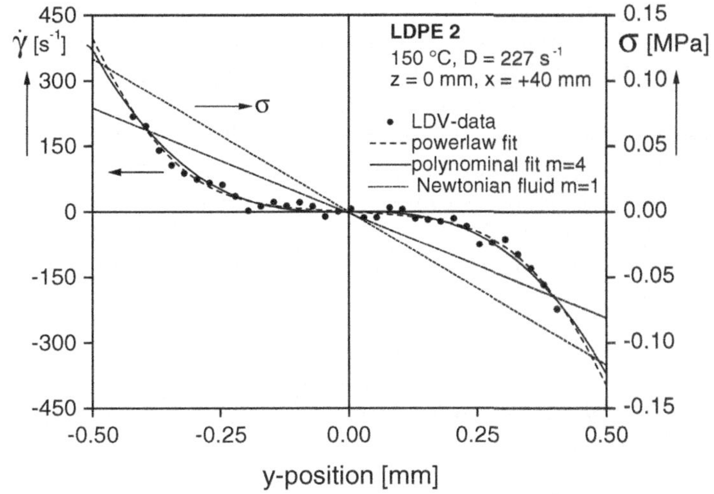

Fig. 16.14 Shear rate across the slit as obtained by numerical differentiations of the curves in Fig. 16.13 and the shear stress determined from the pressure gradient according to Fig. 16.5 [11]

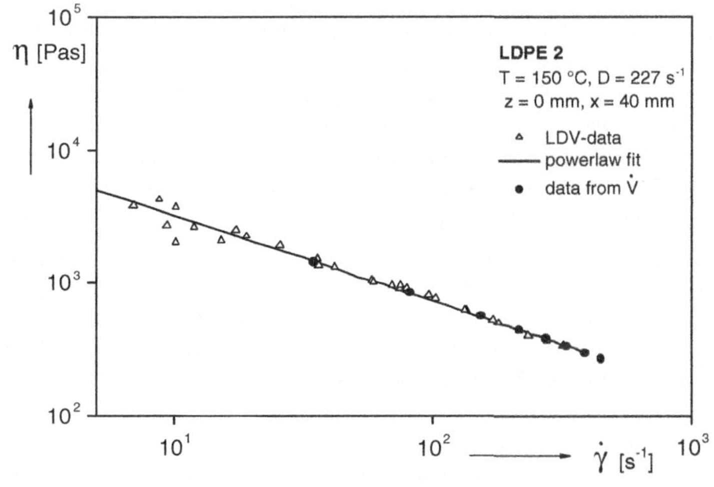

Fig. 16.15 Comparison of viscosity data obtained by different methods [11]

procedure described in Sect. 15.2 for the determination of the viscosity function from output rate and pressure gradient in a slit die.

As can be seen from Fig. 16.15, the agreement between data from LDV and capillary rheometry is excellent in the range of shear rates which is reasonably accessible by measuring the output rate. At smaller shear rates which are the domain of rotational rheometry the data from the LDV measurements show a large scatter due to the inherent limited accuracy of the technique. The drawn curve which goes back to a numerical description of the measured velocity profiles by a power law has a physical meaning for higher shear rates, only, as the power law is principally not valid for small shear rates (cf. Sect. 11.5.2).

From Fig. 16.15 it becomes obvious how narrow the measuring regime of LDV is if this method would be taken into consideration to determine viscosity functions. Moreover, measurements with LDV are much more complicated than those using conventional rheometers. Laser-Doppler velocimetry demonstrates its potential if velocity profiles are required to assess the flow behavior of polymer melts within ducts.

16.3.3 Stick-Slip Phenomena Investigated by Laser-Doppler Velocimetry

Determining viscosity functions from capillary flow is based on the assumption that the melt sticks to the wall. This condition is not always fulfilled as qualitative experiments on PVC [25] and HDPE [26], for example, show. Laser-Doppler velocimetry can offer a quantitative answer to the question which flow conditions one has to deal with.

The occurrence of stick and slip plays a decisive role in the case of melt fracture and, particularly, for the occurrence of the so-called "shark skin" (cf. Sects. 16.3.4.1 and 17.3) which can be a limiting factor for the efficiency of extrusion processes.

Another phenomenon which occurs in capillary flow and has some practical importance for processing operations like extrusion is the so-called "spurt effect", i.e., a spontaneous fast flow under certain conditions. It is found for linear polymers as high density polyethylenes, but has never been observed for long-chain branched polyethylenes. There are some hints that slip effects are responsible for this behavior which is occasionally exploited for the processing of high density polyethylenes as it decreases the pressure necessary for extrusion.

An example of the peculiarities of the flow of a HDPE, some characteristic molecular and rheological data of which are given in Table 16.1, is presented in Fig. 16.16.

Three regions for the extrusion pressure as a function of the apparent shear rate can be distinguished. In region I, one gets pressure readings constant with time at the shear rates applied which result in a smooth flow curve. Region II is defined by the range of shear rates in which pressure oscillations occur, and in region III the pressure p as a function of time is constant again but the slope of p (D) is smaller than in the first part of region I. It is interesting to note that the pressure at the beginning of region III starts at a value which is lower than that at the end of region I which means that the melt flows more easily at least in some parts of region III. Therefore, this behavior is often called superflow.

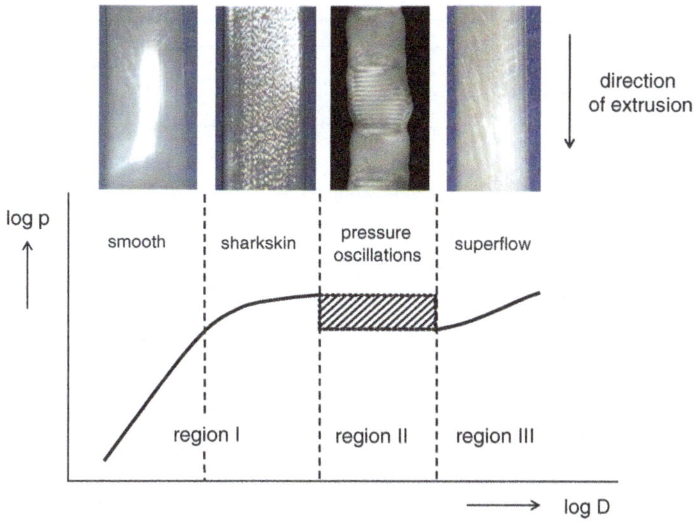

Fig. 16.16 Schematic presentation of the extrusion pressure p as a function of the apparent shear rate D in a double-logarithmic scale during the extrusion of a linear polyethylene in a capillary rheometer. The photographs display the various features of melt fracture which can be observed

Related to the flow characteristics are various features of melt fracture as shown in the pictures above the curves in Fig. 16.16. Up to a distinct pressure or shear rate, respectively, the extruded strand is smooth, followed by a surface roughness which is called "shark skin". Within the range of pressure oscillations smooth and rippled sections follow each other in a periodical manner. At the beginning of region III the strand is smooth again, but it gets heavily distorted at still higher shear rates (not shown here). This feature is typical of gross melt fracture found in a variety of geometries for all polymers at high enough pressures or shear stresses, respectively.

LDV makes it possible to measure the flow fields in the different regions offering some insights into the origin of the phenomena observed.

16.3.3.1 Velocity Profiles in Region I

Figure 16.17 shows the velocities across the slit at different apparent shear rates D for the HDPE in the regime of stable flow and Fig. 16.18 those of LDPE 2 for comparison. The curves of the LDPE can numerically well be described by assuming adhesion of the melt to the wall at all the apparent shear rates measured. The most remarkable differences of the profiles for the HDPE are that the velocities close to the wall do not become zero, i.e., that a measurable slippage exists, even in the range of a smooth flow curve. If a slip velocity v_s at the wall is

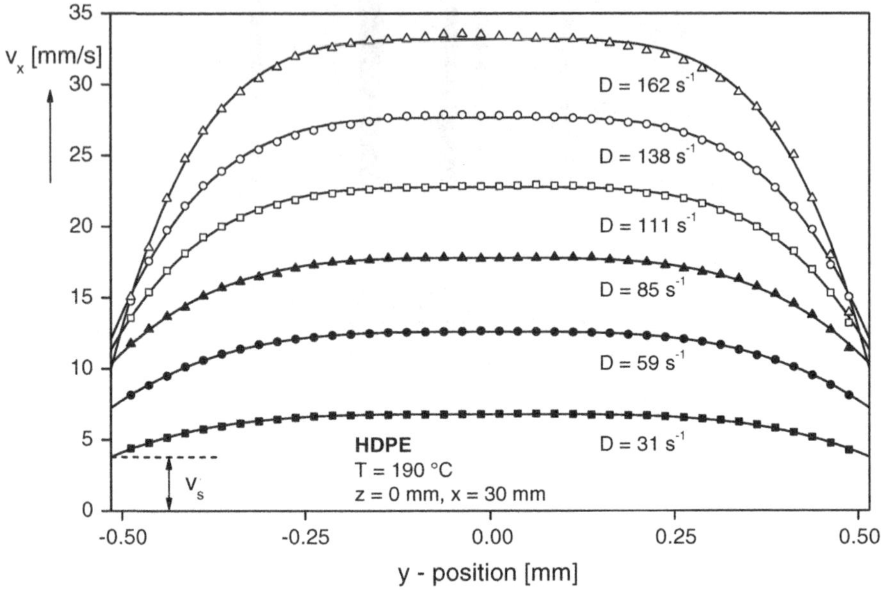

Fig. 16.17 Velocity profiles of the HDPE across the slit 30 mm downstream the slit entrance at different apparent shear rates *D*. v_s is the slip velocity. The full lines are numerical fits using Eq. 16.3 with a slip velocity v_s added. (Reprinted from [33] by permission from The Society of Rheology)

added to Eqs. 16.3 or 16.4, the experimental data can numerically be described with good accuracy as Fig. 16.17 demonstrates.

It is obvious that v_s depends on the shear rate or the shear stress at the wall σ_w, respectively. The values of v_s taken from Fig. 16.17 are plotted double-logarithmically as a function of σ_w in Fig. 16.19. log v_s linearly increases with log σ_w first and then bends off. The linear region in Fig. 16.19 can be described by

$$v_s = a\,\sigma_w{}^m \tag{16.5}$$

with $a = 743$ and $m = 2.31$, if v_s is measured in mm/s and σ_w in MPa.

A power law for the stress dependence of the slip velocity of a high density polyethylene was reported in [27] and [28] using Mooney's method of measurements with different capillary radii [29]. The values found for *a* and m are different, however. But neither the wall materials nor the polymer samples used can be assumed to be comparable.

The finding that a significant amount of slip at the wall is detectable for the HDPE investigated even for smooth pressure readings has an important consequence with regard to the determination of viscosity functions using a capillary rheometer. In such a case the total volume output rate $\dot{V}$ is composed of a fraction $\dot{V}_s$ related to slip and $\dot{V}_v$ following from viscous flow, i.e.,

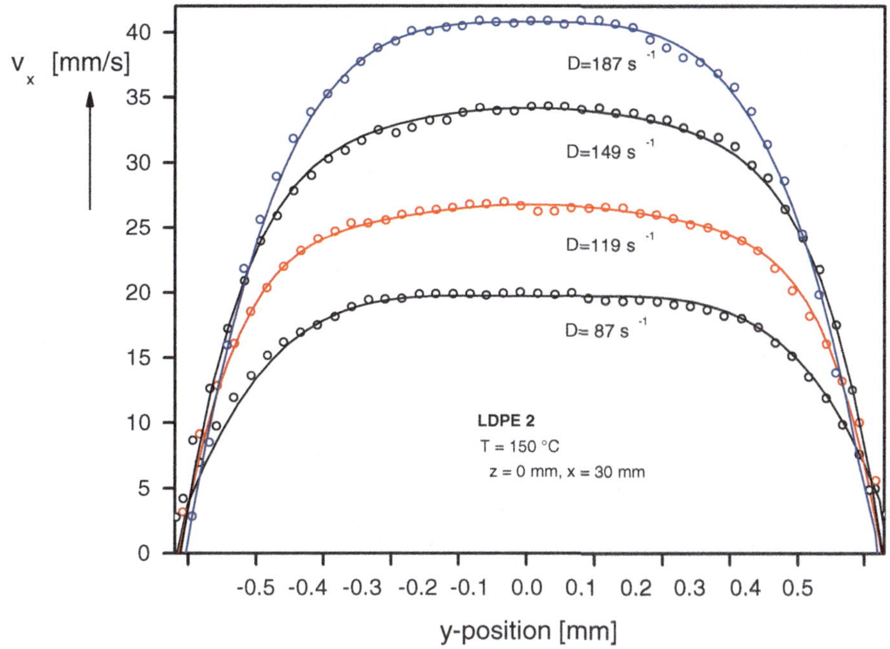

Fig. 16.18 Velocity profiles of LDPE 2 across the slit 30 mm downstream the slit entrance at different apparent shear rates D. The *full lines* are the numerical descriptions according to Eq. (16.3)

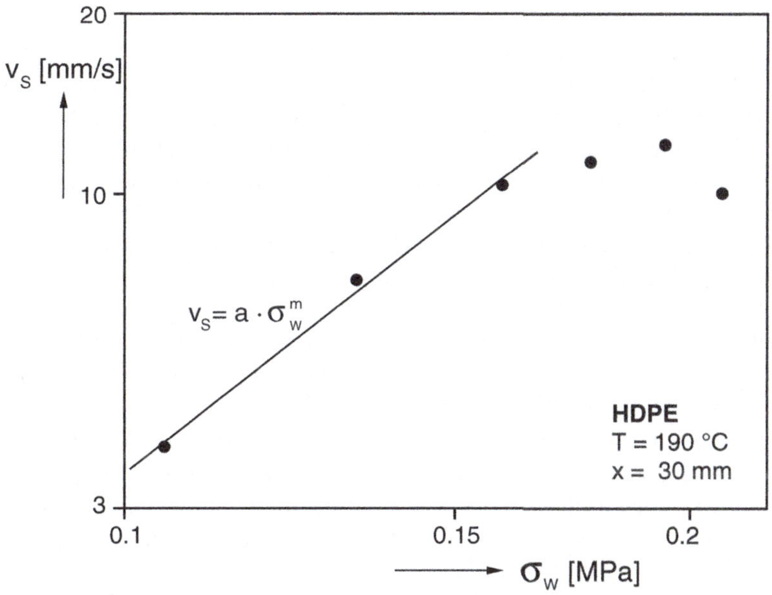

Fig. 16.19 Double-logarithmic plot of the slip velocity v_s as a function of the wall shear stress σ_w obtained from the flow profiles of Fig. 16.17. (Reprinted from [33] by permission from The Society of Rheology)

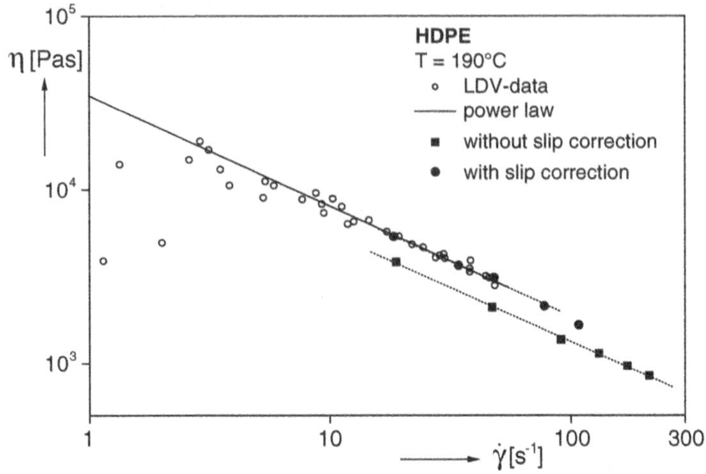

Fig. 16.20 Viscosity function of the HDPE determined without and with slip corrections. (Reprinted from [33] by permission from The Society of Rheology)

$$\dot{V} = \dot{V}_s + \dot{V}_v \tag{16.6}$$

with

$$\dot{V}_s = WHv_s \tag{16.7}$$

W is the width and H the height of the slit.

From $\dot{V}_v$ and the corresponding pressure the viscosity function can be calculated which is a material property and, therefore, has to be independent of the geometry of the measuring device. This becomes obvious from Fig. 16.20 which displays viscosity functions for the HDPE determined in different ways. The shear stress at the wall was calculated from the pressure gradient measured. The open circles in Fig. 16.20 were obtained by using the shear rates from a graphical differentiation of the velocity curves in Fig. 16.17. The large scatter at small shear rates goes back to the fact that the numerical determination of the shear rates can only be carried out with some kind of uncertainty due to an inherent inaccuracy of the measurements. The full line follows from the description of the measured velocity data using a power law. The filled squares represent viscosity data based on the shear rates determined from the total volume rate $\dot{V}$ applying the Weißenberg-Rabinowitsch correction. Their values are obviously too low compared with those from the velocity profiles. If the viscous part $\dot{V}_v$ of the total volume rate $\dot{V}$ is taken, i.e., the flow due to slip is subtracted, then the filled circles are obtained. They are in good agreement with the open circles.

This finding demonstrates on the one hand, that the slip of the HDPE observed in region I is not an artefact of the LDV measurement, on the other, that one has to be sure that the melt adheres to the wall of a capillary if the usual procedure to determine viscosity functions with capillary rheometers is applied. In the case of slip, the viscosity function defined as a material property of a material can only be obtained if the slip velocities are known whose determination requires some experimental effort, however. For example, in the case of circular dies the method after Mooney needs measurements with three capillaries of different radii *R*, but constant length to radius ratios [34]. For rheological methods based on parallel plates and cone-and-plate geometries comprehensive and time-consuming experimental work is necessary to determine the slip velocity [35].

Interesting is the fact that in region I "shark skin" above a certain shear rate is observed, although neither oscillations of the pressure nor of the flow velocity can be found. This topic is addressed in more detail in Sect. 16.3.4 on the flow behavior at the exit of the die.

16.3.3.2 Velocity and Pressure Profiles in Region II

Region II is characterized by an oscillating pressure. The transition from the steady to the unsteady behavior is rather sharp as Fig. 16.21 demonstrates. The small

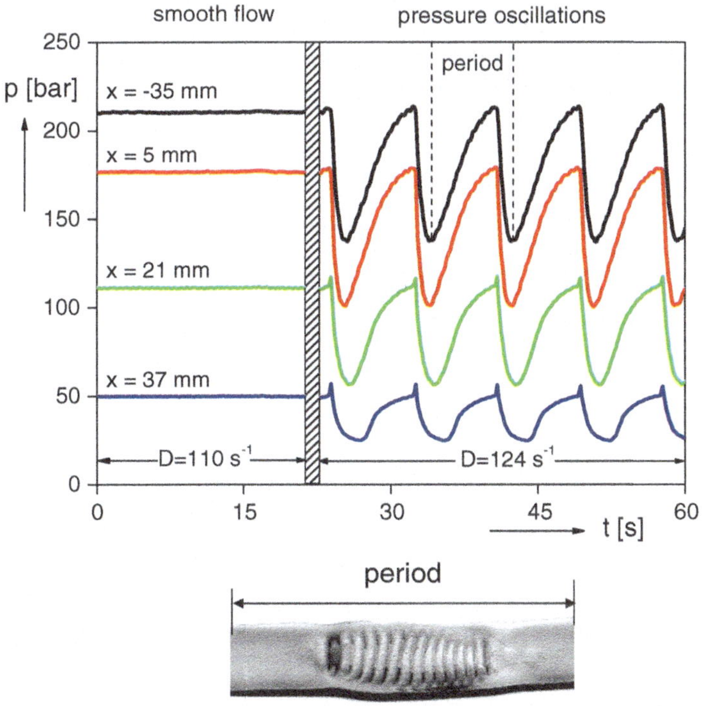

Fig. 16.21 Pressure oscillations of the HDPE as a function of time at different positions x within the channel (one in the reservoir three in the slit) at $T = 170$ °C. The hatched area marks the time necessary to change the revolution of the extruder screw. *D* designates the reduced output rate according to Eq. (15.32) (Reprinted from [33] by permission from The Society of Rheology)

increase of the apparent shear rate from $D = 110\ s^{-1}$ to $D = 124\ s^{-1}$ changes the pressure readings in the channel significantly. Whereas for the smaller shear rate the pressure is constant as a function of time at all the four positions of the pressure transducers, oscillations occur at the higher shear rate which show an excellent reproducibility with time. The critical shear stress for the onset of the pressure oscillations is found to be 0.19 MPa which is in good agreement with the literature [28]. The frequencies of the oscillations at the various positions are the same, but their shapes differ. The pressure amplitudes 35 mm upstream the entrance plane ($x = -35$ mm) and 5 mm downstream ($x = 5$ mm) look very similar, but further down the slit the pressure minimum becomes broader. At the positions $x = 21$ mm and $x = 37$ mm in the slit, a very distinct spike occurs.

As the photograph of a strand extruded in region II demonstrates, the feature of the occurring melt fracture can be related to the period of the pressure change (cf. Fig. 16.21). Around the pressure minima the surface of the strand is smooth, but it becomes wavy with increasing pressure. This pattern very regularly follows the pressure oscillations.

In parallel to the pressure fluctuations, significant changes of the velocity within the slit are observed as Fig. 16.22 shows. Pressure and velocity are plotted as a function of $f{\cdot}t$ with f being the frequency of the oscillations. It is obvious from Fig. 16.22 that velocity and pressure oscillate with the same frequency, but the shapes of their curves are very different. Two features of the velocity curve are evident. First, the velocity does not become zero at any time and second, the velocity at the maximum is eight times higher than the minimum value. Starting

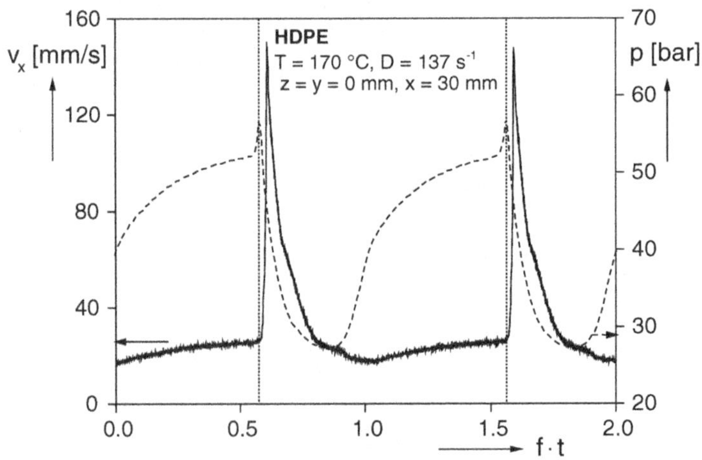

Fig. 16.22 Velocity v_x on the slit axis and pressure p at the wall for the HDPE at 170 °C as a function of the product of frequency f and time t. The pressure was measured at $x = 37$ mm, the velocity at $x = 30$ mm. D is the reduced output rate according to Eq. 15.32. (Reprinted from [33] by permission from The Society of Rheology)

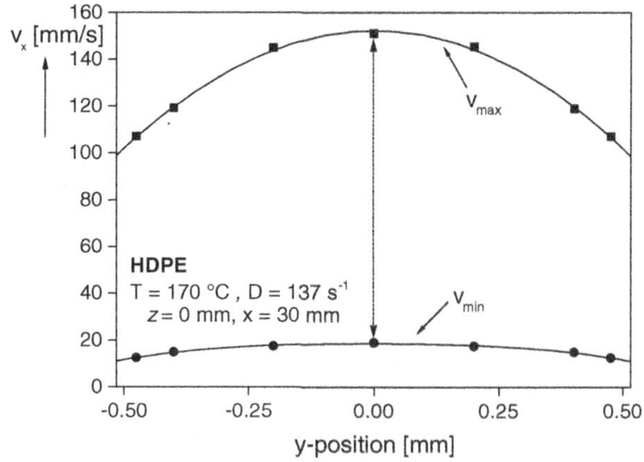

Fig. 16.23 Minimum and maximum velocities v_x of the HDPE across the slit height at $x = 30$ mm. D is the reduced output rate according to Eq. 15.32. The *lines* are drawn to guide the eye. (Reprinted from [33] by permission from The Society of Rheology)

from the minimum, the velocity increases slowly, but then jumps up steeply and comes down again forming a small shoulder before reaching the minimum again.

The minimum and maximum velocities of the cycles do not vary much across the slit height as Fig. 16.23 demonstrates. Near the wall, significant velocities are still measured, giving evidence of a pronounced slip at the wall. Although the minimum slip velocities close to the wall are around ten times smaller than the maxima values, they still exist showing that slip never diminishes during a pressure and velocity cycle in case of the HDPE investigated.

16.3.3.3 Velocity and Pressure Profiles in Region III

As shown in Fig. 16.24, for HDPE a smooth flow can be obtained again by further increasing the output rate. Changing the reduced output rate from $D = 163\ \mathrm{s}^{-1}$ to $D = 176\ \mathrm{s}^{-1}$ results in a disappearance of the pressure oscillations. The constant pressure readings do not exhibit any indication of an unsteady behavior over a long period of time. The hatched area in Fig. 16.24 marks the time for setting the extruder to higher revolutions. Due to the way of the melt from the extruder head to the channel it takes about 15 s before the pressure corresponding to the changed flow conditions is obtained.

The velocity profiles in the superflow region at different positions in the slit are shown in Fig. 16.25. The shape of the velocity profiles indicates a very pronounced slip velocity at the wall which is approximately independent of the position along the axis of the slit. The velocities in the center of the slit are not much higher than at the walls indicating a kind of plug flow in the superflow region.

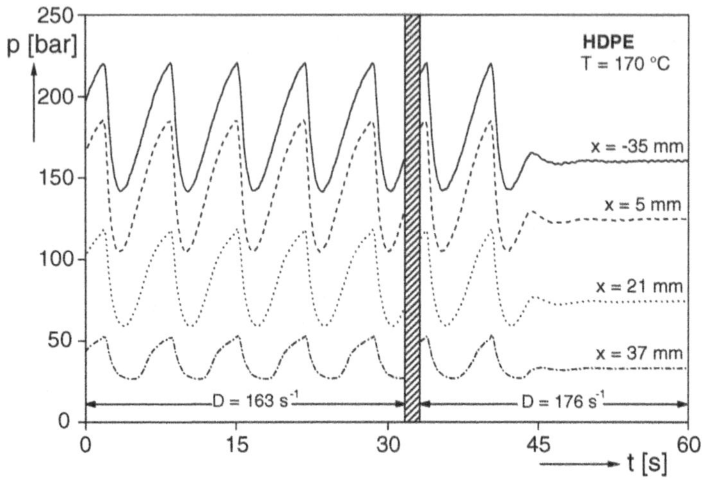

Fig. 16.24 Pressure readings at the transition from the oscillating to the superflow regime for the HDPE at 170 °C. *D* designates the reduced output rates according to Eq. 15.32. (Reprinted from [33] by permission from The Society of Rheology)

Fig. 16.25 Velocity profiles of the HDPE in the regime of superflow at different positions x along the slit axis. The lines are drawn to guide the eye. (Reprinted from [33] by permission from The Society of Rheology)

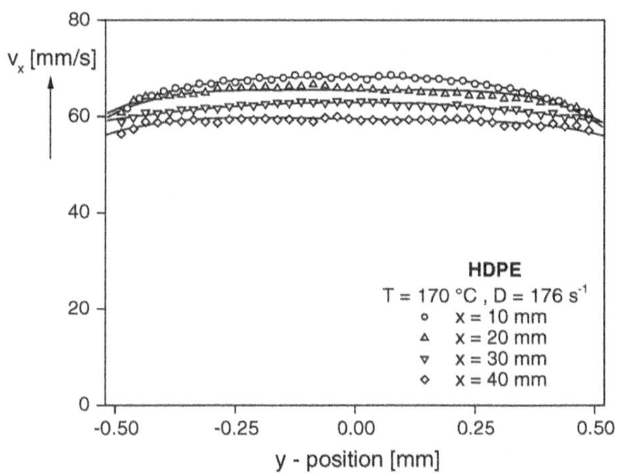

There is a slight dependence of the shape of the velocity curve on the position along the slit axis. Approaching the slit exit, the velocities become smaller and more constant across the slit height.

The deficiency to calculate a viscosity function in region II is evident, since pressure and output rates oscillate and, therefore, a material function cannot be obtained. A determination of accurate viscosity functions for the HDPE within region III remains illusive, too, as the flow is dominated by wall slip effects.

At still higher output rates the velocities and pressures become unstable again as it is known from experiments with high pressure capillary rheometers. This region could not be reached, however, with the LDV-setup used.

16.3.4 Flow at the Die Exit

Quantitative investigations of the velocity fields at the die exit during the extrusion of a polymer melt are of interest as two phenomena are observed there which are of relevance for processing, namely extrudate swell and melt fracture. Whereas the extrudate swell can be related to elastic properties of a polymer, the various appearances of melt fracture are not fully understood up to now. Widely accepted is the assumption that "sharkskin" is caused by stress concentrations at the die exit which were found in [30], for example, using birefringence. By this method, insights into the localized distribution of stresses are not possible, however, and, therefore, the determination of flow fields which can be measured by LDV with high spatial resolution bears the potential to get more detailed insights into the processes taking place at the die exit.

In Fig. 16.26 some velocity profiles of LDPE 2 near the slit exit are presented. The polymer was extruded at a temperature of 135 °C and an apparent shear rate of $D = 12.5\ s^{-1}$. The origin of the coordinate system is located at the slit entrance (cf. Fig. 16.2). Due to the length of the slit of 50 mm positions smaller than 50 mm lie within the die and those larger than this value are outside. The relatively high scatter of the velocities measured for the extruded strand (cf. Fig. 16.26) is due to the experimental fact that outside the flow channel the melt is not guided any more by walls and tends to wobble around the center axis of flow. Nevertheless, clear tendencies can be observed. The layer of the melt sticking to the wall within the capillary, i.e., having the velocity zero there, attains the finite velocity of the extruded strand at some distance from the exit. As a consequence, the outer layer of the extruded sample is elongated. The maximum velocities in the middle of the strand decrease with the distance from the capillary exit and gradually become equal to those at the rim, i.e., these parts are compressed.

The velocities v_y perpendicular to the extrusion direction at different distances from the slit exit can be measured, too, as Fig. 16.27 demonstrates for the LDPE 2. They describe how the strand expands and give an insight into the kinematics of the extrudate swell. Obviously, $v_y(y)$ shows a point symmetry. As v_y is found to be at least partially a linear function of y, the gradients and therewith the elongational rates in the direction perpendicular to the flow can easily be assessed. The highest values are obtained close to the die exit. They become smaller with growing distance from the exit due a superposition of two effects: The tendency of the extrudate swell to approach its equilibrium and the cooling of the strand leading to longer retardation times.

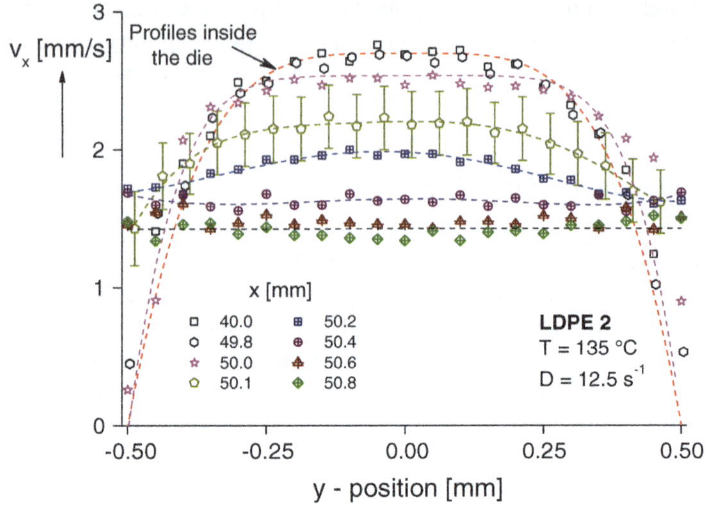

Fig. 16.26 Velocity profiles v_x in extrusion direction across the extruded samples (y-position) for the LDPE 2 at different positions x near the slit exit. The origin of the coordinate system is located at the entrance of the slit 50 mm in length. The error bars at $x = 50.1$ mm give an idea of the scatter of velocity measurements outside the slit. $T = 135$ °C relates to the temperature of the melt within the slit. (Reprinted from [31] by permission from Elsevier)

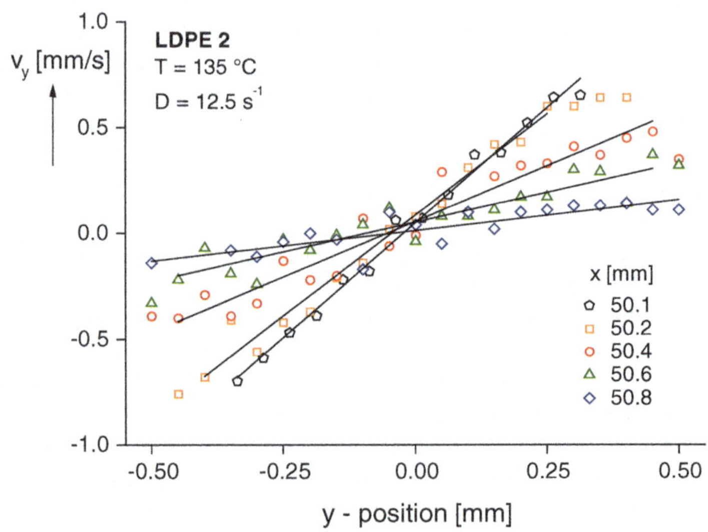

Fig. 16.27 Velocity profiles perpendicular to the extrusion direction for the LDPE 2 at different positions outside the slit. The temperature of the melt within the slit is $T = 135$ °C (Griess HJ Unpublished results). The lines are drawn to guide the eye

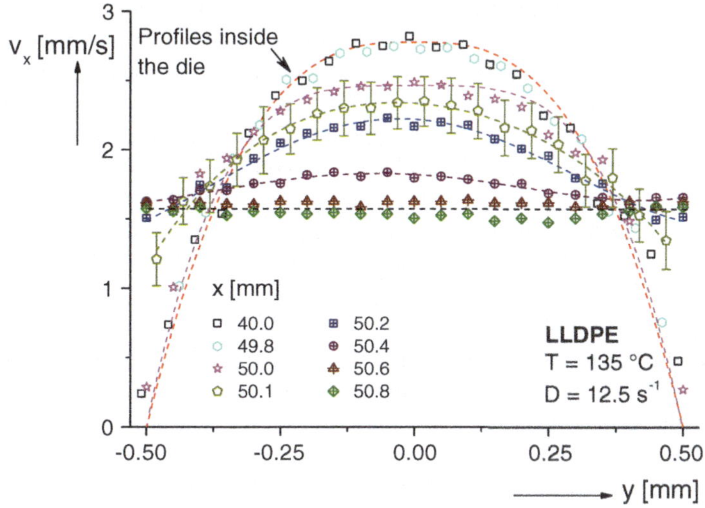

Fig. 16.28 Velocity profiles in extrusion direction for an LLDPE near the exit of the slit. $x = 50$ mm marks the slit exit, $T = 135$ °C is the temperature of the melt within the slit (Griess HJ Unpublished results)

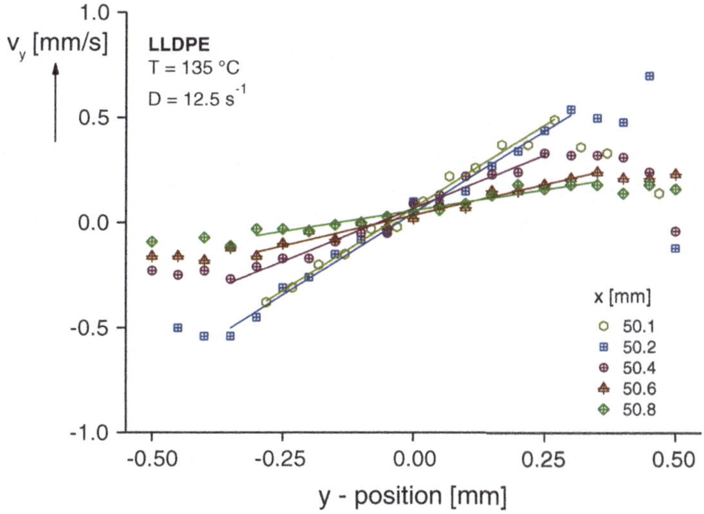

Fig. 16.29 Velocity profiles perpendicular to the extrusion direction for the LLDPE at different distances x from the slit exit. $x = 50$ mm marks the slit exit, the temperature of the melt within the slit is $T = 135$ °C (Griess HJ Unpublished results).The lines are drawn to guide the eye

The flow profiles of an LLDPE near the exit inside and outside the slit which was extruded under the same conditions as the LDPE are presented for the x-direction in Fig. 16.28 and for the y-direction in Fig. 16.29. They are qualitatively similar to those of the LDPE.

For both melts adhesion to the slit wall can be assumed and a distinct velocity maximum occurs along the slit axis. The strands of the LLDPE and the LDPE reach a constant velocity of 1.5 mm s^{-1} in extrusion direction at some distance from the slit exit. The velocity perpendicular to it, which describes the time dependence of the swell of the extrudate, obviously approaches zero within the accuracy of the measurements when the extrusion velocity has equilibrated. The flatter flow profile of the LDPE within the slit in comparison to the LLDPE goes back to its more pronounced shear thinning as discussed in Sect. 17.2. It is interesting to note, that this effect is still visible 0.4 mm from the exit ($x = 50.4$ mm) demonstrating the memory effect of the molecules.

Close to the slit exit the gradient of v_y which corresponds to the elongational rate perpendicular to the extrusion direction is distinctly higher for the LDPE than for the LLDPE which goes along with the larger extrudate swell observed for the LDPE. With further distance from the exit the elongation rates approach each other.

The results presented show the complexity of the flow at the exit. Besides the obviously non isothermal state, the elongational deformation is of a biaxial nature. Moreover, the elongational rates in extrusion direction and perpendicular to it change along the extruded strand from their largest values close to the slit exit to zero in the totally recovered or solidified state.

16.3.4.1 Relation Between Exit Flow and *"Shark Skin"*

The local flow profiles in the exit region are taken as the base for a discussion of "shark skin". As "shark skin" is a surface defect, the flow fields and the elongational rates at the rim, i.e., at $|y| = 0.5$ mm are of special interest. For a qualitative discussion the elongational rate in extrusion direction $\dot{\varepsilon}_{xx}$ is assessed by

$$\dot{\varepsilon}_{xx} \approx \Delta v_x / \Delta x \tag{16.8}$$

and that one perpendicular to it by

$$\dot{\varepsilon}_{yy} \approx \Delta v_y / \Delta y \tag{16.9}$$

Within the capillary, the velocity v_x is zero at $|y| = 0.5$ mm for the LDPE and the LLDPE investigated as the melts stick to the wall. Obviously the elongational rates and following from them the tensile stresses are the highest close to the exit and decrease along the extruded strand. Therefore, it can be assumed that the flow conditions existing around the exit are decisive for the formation of "shark skin". As $x = 50.1$ mm is the position closest to the slit exit for measurements by the LDV setup used, the velocities there were taken for an assessment of the elongational rates. From the velocity profiles shown in the Figs. 16.26 and 16.28

averaged elongational rates of about 13 s^{-1} for the LDPE and 11 s^{-1} for the LLDPE follow in flow direction. The corresponding elongational rates perpendicular to the flow direction are 2.3 s^{-1} and 1.8 s^{-1} at $x = 50.1$ mm, i.e., in a first approximation the exit flow may be regarded as mainly uniaxial. Knowing the viscosity functions in uniaxial elongation, the stresses could be calculated. That is not possible, however, as reliable viscosity data at the rather high elongational rates and short times around 0.1 s^{-1} are not available.

"Shark skin" can neither be observed for the LDPE nor the LLDPE under these conditions [31]. Due to experimental limitations, a larger output rate could not be reached at 135 °C. This situation changes at $T = 220$ °C and the higher apparent shear rate of 70 s^{-1} at which the LLDPE does exhibit visually observable "shark skin", but the LDPE not [31]. From measurements of the velocity profiles it follows that the melts stick to the walls. Even for the experiments on the LLDPE showing "shark skin" no indications of velocity fluctuations within the die are observable indicating that it is an effect developing outside the slit (Griess HJ Unpublished results).

At a distance of 0.1 mm from the exit the elongational rate $\dot{\varepsilon}_{xx}$ determined according to Eq. (16.8) reaches a value of 85 s^{-1} for the LDPE, and 65 s^{-1} for the LLDPE, i.e., the velocity fields are not very different. This finding means that differences of the failure behavior of the two polyethylenes have to be taken into account. A quantitative discussion of the relations between flow fields, material properties and the occurrence of "shark skin" is hampered by the following conditions, however:

The time scales to be regarded lie at around 0.01 s. Stress-strain curves in the molten state are not able to be measured under these conditions. Even if the time-temperature superposition principle would find application for the stress-strain curves, it is an open question whether it could be assumed to be valid for failure properties which obviously govern the appearance of "shark skin".

In spite of a lack of a quantitative understanding of "shark skin" it can be concluded that

1. "Shark skin" is an effect originating from flow conditions in the exit region of a die.
2. "Shark skin" goes back to a failure of the surface layer of the extruded strand.
3. "Shark skin", therefore, is related to both, the abrupt change of the flow field exerted in the exit region and the resistance to failure of the extruded melt.
4. "Shark skin", is governed by extrusion conditions and material properties.

The findings in [32] and Sect. 17.3 that "shark skin" can be reduced by generating slip at the wall of a die support the conclusions made above.

References

1. Yeh Y, Cummins HZ (1964) Appl Phys Lett 4:176
2. Ramamurthy AV, McAdam J (1980) J Rheol 24:167
3. Lawler JV, Muller SJ, Brown RA, Armstrong RC (1986) J Nonnewton Fluid Mech 20:51
4. Wunderlich AM, Brunn PO, Durst FJ (1988) J Nonnewton Fluid Mech 40:267
5. Quinzani LM, Armstrong RC, Brown RA (1994) J Nonnewton Fluid Mech 52:1
6. Kramer H, Meissner J (1980) Proceedings of the 8th International Congress Rheology. Plenum, New York 2:463
7. Mackley MR, Moore IPT (1986) J Nonnewton Fluid Mech 21:337
8. Ahmed R, Mackley MR (1995) J Nonnewton Fluid Mech 56:127
9. Piau JM, El-Kissi N, Mezghani A (1995) J Nonnewton Fluid Mech 59:11
10. Wassner E, Schmidt M, Münstedt H (1999) J Rheol 43:1339
11. Schmidt M, Wassner E, Münstedt H (1999) Mech Time-Depend Mat 3:371
12. Durst F, Melling A, Whitelaw JH (1981) Principle and practice of laser-doppler anemometry. Academic Press, New York
13. Griess HJ, Burghelea T, Münstedt H (2012) Polymer Engng Sci 52:615
14. Bagley EB, Birks AM (1960) J Appl Phys 31:556
15. Ballenger TF, White JL (1971) J Appl Phys 15:1949
16. White JL, Kondo AJ (1977) J Nonnewton Fluid Mech 3:41
17. Debbaut B, Marchal JM, Crochet MJ (1988) J Nonnewton Fluid Mech 29:119
18. Feigl K, Öttinger CH (1996) J Rheol 40:21
19. Hertel D, Valette R, Münstedt H (2008) J Nonnewton Fluid Mech 153:82
20. Wassner E (1998) Doctoral Thesis, University Erlangen-Nürnberg, Shaker Verlag Aachen, ISBN 3-8265-4498-6
21. Den Otter JL, Wales JLS, Schijf J (1967) Rheol Acta 6:205
22. Münstedt H, Schwetz M, Heindl M, Schmidt M (2001) Rheol Acta 40:384
23. White SA, Gotsis AD, Baird DG (1987) J Nonnewton Fluid Mech 24:121
24. Schwetz M, Münstedt H, Heindl M, Merten A (2002) J Rheol 46:797
25. Münstedt H (1977) J Macrom Sci Phys 14:195
26. Uhland E (1979) Rheol Acta 18:1
27. Hatzikiriakos SG, Dealy JM (1992) J Rheol 36:703
28. Hatzikiriakos SG, Dealy JM (1992) J Rheol 36:845
29. Mooney M (1931) J Rheol 2:210
30. Arda DR, Mackley MR (2005) J Nonnewton Fluid Mech 126:47
31. Burghelea TI, Griess HJ, Münstedt H (2010) J Nonnewton Fluid Mech 165:1093
32. Kissi NE, Piau JM (1994) J Nonnewton Fluid Mech. 52:249
33. Münstedt H, Schmidt M, Wassner E (2000) J Rheol 44:413
34. Mooney M (1931) Trans Soc Rheol 2:210
35. Yoshimura A, Prud'homme RK (1988) J Rheol 32:53

Chapter 17
Rheological Properties and Processing

Rheological properties of a polymer can be very useful to assess some aspects of processing. For a successful application of rheological data it has to be taken into account, however, that they strongly depend on temperature, time, and stress or strain rate, respectively. Therefore, the rheological behavior measured in laboratory tests can be of relevance for processing only, if the regimes of the parameters mentioned above are comparable to each other. While matching the temperatures normally is not a big problem due to the validity of the time-temperature shift for most of the polymer melts, particular attention has to be paid to the time scale, which in many cases is distinctly shorter in processing operations than in rheological measurements. Additionally, it has to be taken into consideration that pronounced nonlinear deformations are dominant, the range of which sometimes is not easy to cover by rheological experiments.

In spite of these restrictions, experiments on a laboratory scale have the potential to show directions to go for the optimization of processing operations, particularly with respect to selecting suitable materials. Some examples are given in the following.

17.1 Melt Flow Rate

The rheological quantity most frequently used today in practice is the melt flow rate (MFR). It is easy to determine using commercial equipments based on a capillary rheometer with fixed geometries of the barrel and the die laid down in various standards, e.g., [1]. The melt heated up to a prescribed temperature is extruded under different loads generated by a piston with definite weights. They are given in the standards and chosen in order to get output rates conveniently to measure. The mass of the exiting strand is determined and registered in the somewhat unusual, but prescribed unit g/10 min. For this quantity the expression melt flow index (MFI) can be found in the technical literature, too, which may be misleading insofar as an index normally stands for a dimensionless quantity. Besides the melt flow rate (MFR) measured in g/10 min, the melt volume rate (MVR) in cm^3/10 min is

H. Münstedt and F. R. Schwarzl, *Deformation and Flow of Polymeric Materials*,
DOI: 10.1007/978-3-642-55409-4_17,

sometimes determined. This quantity is of advantage, whenever a device is used which continuously measures the extruded volume by monitoring the displacement of the piston. Although the MFR reflects only one point of the viscosity curve of a material, i.e., the viscosity at one single shear rate around 1 s^{-1}, it has become well established over the years. One reason is that data are relatively easy to acquire with standardized equipments available on the market.

Besides the very narrow shear rate range the MFR is restricted to, there is another fact which limits its quantitative significance. As the radius of the capillary is $R = 1.048$ mm and its length $L = 8$ mm, the relatively small value of L/R may imply a remarkable contribution of entrance effects which have to be taken into consideration if the material property viscosity is determined (cf. Sect. 15.1.3). All in all the importance of the MFR has to be seen more in the field of a simple quality control than in an assessment of the flow behavior of a polymer melt for processing.

17.2 Role of Viscosity Functions

The viscosity function in shear, i.e., the dependence of the shear viscosity on the shear rate, has to be considered if a deeper insight into flow properties is required. It is widely used today as the base for modeling extrusion processes or designing screws and tools. An example of the significance of viscosity functions is shown in the following.

An LDPE and an LLDPE were processed by a single screw extruder with a screw diameter of $D = 30$ mm and a screw length of 20 D which feeds an annular die with a diameter of 36 mm and a gap width of 0.6 mm. The pressure was measured just in front of the die. The throughput was 2 kg/h corresponding to an apparent shear rate of $D = 31$ s^{-1} in the die. Details of the equipment are given in [2].

The viscosity functions of these two polyethylenes are shown in Fig. 17.1. They exhibit the typical feature already discussed in Sect. 11.5.1. At smaller shear rates the viscosity of the LDPE is higher than that of the LLDPE reflecting the melt flow rate which is lower by about 10 % for the LDPE. As the shear thinning of the LDPE is much more pronounced than that of the LLDPE, the viscosity of the LDPE comes to lie significantly below the curve of the LLDPE in the range of shear rates typical of extrusion processes. The pressures corresponding to an apparent shear rate of $D = 31$ s^{-1} measured just in front of the die were distinctly lower for the LDPE than for the LLDPE as expected from the viscosity functions in Fig. 17.1. This result is just contrary to what one would have expected from the melt flow rate (MFR) and it exemplarily demonstrates the very limited value of the MFR if the knowledge of the flow behavior of a melt relevant for processing is of importance. In fact, viscosity functions in dependence on shear rate and temperature are essential for an assessment of extrusion processes.

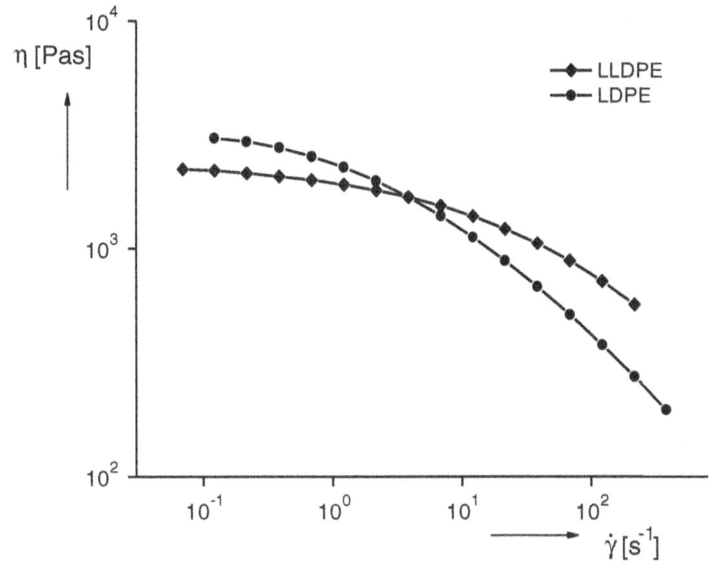

Fig. 17.1 Viscosity functions of an LDPE and an LLDPE at a temperature of 190 °C. The data were obtained from dynamic-mechanical experiments applying the Cox-Merz rule

17.3 Influence of Additives on the Surface Defect "Shark Skin"

An assessment of the performance of polymer melts with respect to processing is not only based on viscosity functions, however, as the following examples demonstrate.

A phenomenon which plays an important role for the processing of polymer melts is the surface defect "shark skin" (cf. Sect. 16.3.4.1). It is evident that it limits the extrusion velocity and following from that the efficiency of processing in all the cases in which a smooth surface is needed. Such a requirement holds for the wide field of films for packaging applications and a great number of various extruded items as well. In practice, additives are used to enhance the critical output rate for the occurrence of "shark skin". The additives generate slip of a melt in a die which normally sticks to the walls.

In Fig. 17.2, examples for the surface roughness of strands of a commercial linear low density polyethylene are presented. The LLDPE was extruded under various conditions through the slit die of the LDV-setup described in Sect. 16.2.1. At the small mass throughput of 0.07 g/s, the surface of the sample is smooth (Fig. 17.2 a), at the higher throughput of 0.23 g/s, a distinct roughness of several μm is measurable (Fig. 17.2b). If a fluoropolymer is added, the strand becomes smooth again, even at the somewhat higher output rate of 0.25 g/s (Fig. 17.2c). But this effect was reached after a continuous extrusion with the additive over 2 h,

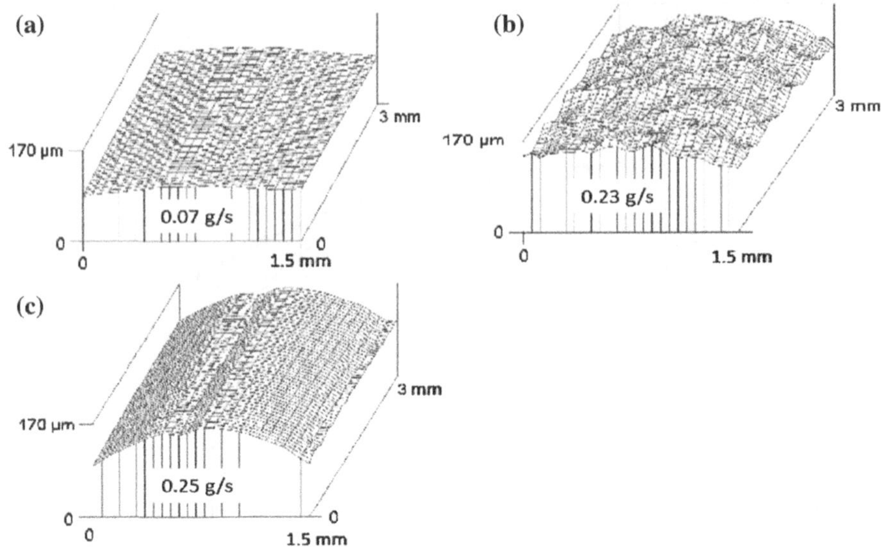

Fig. 17.2 Surface roughness of an LLDPE after extrusion at $T = 220$ °C and different mass throughputs. **a** and **b** unmodified material, **c** with 1000 ppm of the fluoropolymer additive Dynamar™ [3]. The profiles were obtained by scanning electron microscopy

only. That means some time is needed before the additive shows its full effectiveness. This finding provokes the question how the flow profiles change under the influence of the additive. Investigations of this kind offer the chance to better understand the mechanisms on which the effectiveness of additives is based that reduce "shark skin". Laser-Doppler velocimetry as discussed in Chap. 16 was applied to elucidate the effect of the additive. The results are presented in the following.

From the flow profiles of the LLDPE in Fig. 17.3, it can be concluded that up to the highest output rate of 0.23 g/s at which "shark skin" is pronounced, the flow velocities can be extrapolated to zero at the wall of the slit. Slippage is not obvious and irregularities like oscillations of the melt, for example, are not observed. This experimental finding underlines the previous conclusion that the origin of "shark skin" has not to be sought in flow irregularities within the die, but is an exit effect (cf. Sect. 16.3.4.1).

A totally different picture is obtained, if the fluoropolymer is added. Its concentration was set to 0.1 wt. % using a master batch of 5 wt. % fluoropolymer in LLDPE. At an output rate of 0.23 g/s the surface of the extruded strand was still rough at the beginning, but after 2 h of extrusion it became smooth and was comparable to that without additive at the smaller output rate (cf. Fig. 17.2c). The shear stress at the wall decreased from 0.22 MPa to 0.11 MPa during 2 h of extrusion and to 0.09 MPa after 3 h. These data are distinctly lower than the value of 0.18 MPa at which a pronounced "shark skin" comparable to that in Fig. 17.2 b

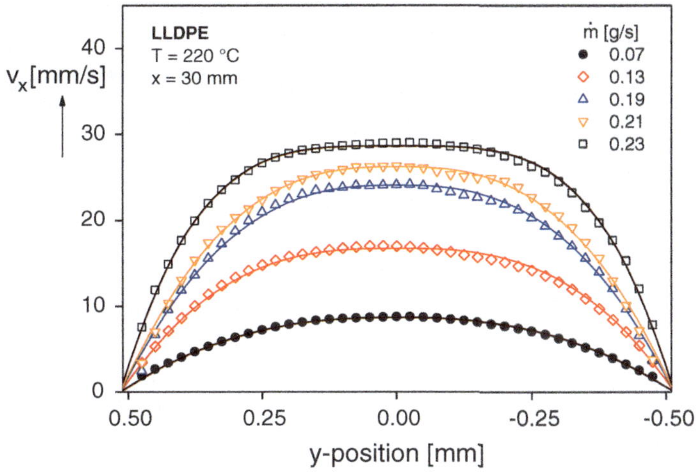

Fig. 17.3 Velocity v_x in flow direction for the unmodified LLDPE within a slit die at various mass flow rates $\dot{m}$ [3]. The flow profiles were measured at $x = 30$ mm from the slit entrance

Fig. 17.4 Velocity profile of the LLDPE mixed with 0.1 wt. % of a fluoropolymer after 2 h of extrusion at $x = 30$ mm from the slit entrance and 2 mm from the central plane in the direction of the glass window [3] (cf. Fig. 16.2)

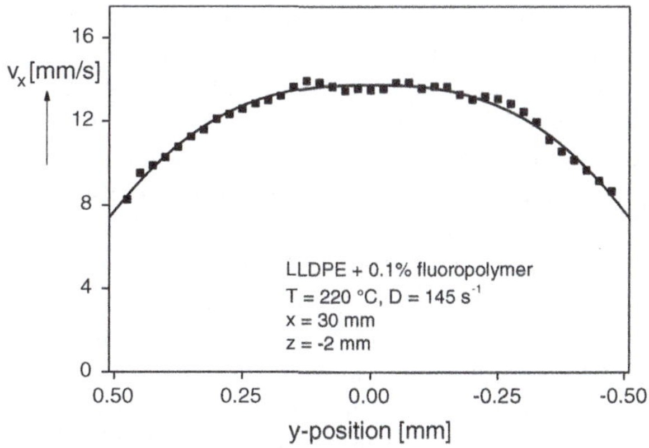

sets in. These observations are in agreement with results from the literature which give a range between 0.14 and 0.26 MPa as critical shear stresses for "shark skin" observed for various LLDPE in steel dies [4, 5].

The velocity profile of the LLDPE with 0.1 wt. % fluoropolymer after an extrusion time of 2 h is shown in Fig. 17.4. Due to the turbidity of the melt modified by the additive, the measurements were performed at $z = -2$ mm, i.e., two mm closer to the glass window than in the case of the profiles of the melt without additive which are usually determined in the center plane of the capillary. The turbidity and following from it the attenuation of the laser beam originates from the known nonmiscibility of the fluoropolymer and the polyethylene. The

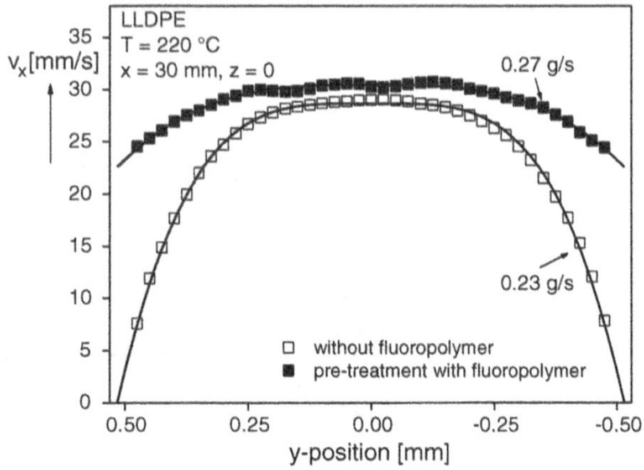

Fig. 17.5 Velocity profiles of the unmodified LLDPE across the slit after different extrusion histories. Filled squares: previous extrusion of 3 h with the LLDPE modified by the fluoropolymer followed by 2 h extrusion of the neat material. Open squares: no previous extrusion of LLDPE with fluoropolymer [3]

extrapolation of the measured velocity data to the walls reveals a distinct slippage. This finding explains the observed drop of the shear stress in comparison to the nonmodified LLDPE. In addition, from the remarkable slip velocities it becomes evident that the change in elongational velocity of the outer layers of the strands when exiting the slit is smaller than in the case of adhesion to the wall. Correspondingly, the stress acting on the sample surface is lower.

The slippage can be related to a layer of the fluoropolymer on the wall which is incompatible with the LLDPE. Obviously, the slippage occurs at stresses lower than those needed for shearing a melt sticking to the wall. Remarkable is the fact that it takes some time to build up the fluoropolymer layer. This finding is interesting from a fundamental and a practical point of view, as well. From detailed measurements a lot more could be learnt about the basic mechanisms of the layer formation, e.g., which role the diffusivity of the fluoropolymer within the polyethylene melt plays and how it could be optimized.

An example for the relevance of flow profiles for the practice of extrusion can be obtained from measurements related to the durability of the slip effect. Figure 17.5 shows the velocity profile of the LLDPE without any additive in the carefully cleaned slit die. Sticking to the wall is evident at the output rate chosen. The other curve in this figure represents the velocity distribution of the same unmodified LLDPE after 2 h of extrusion through the capillary which had previously been exposed to a 3 h extrusion of the melt with the fluoropolymer. The fact that a pronounced slip at the wall is still found, demonstrates the long-term efficiency of the preceding treatment. The conclusion is that even after 2 h at the extrusion conditions chosen the formed fluoropolymer layer is still effective in the slit die used.

The efficacy of the additive, although surprisingly durable, is finite, however, as the results of Fig. 17.6 demonstrate. After 12 h of extrusion with the neat LLDPE following the 3 h of pre-extrusion with the modified product as described before, slip is still observable but its feature has changed. From Fig. 17.6, it is obvious at a first glance that the flow profiles have become asymmetrical. The material

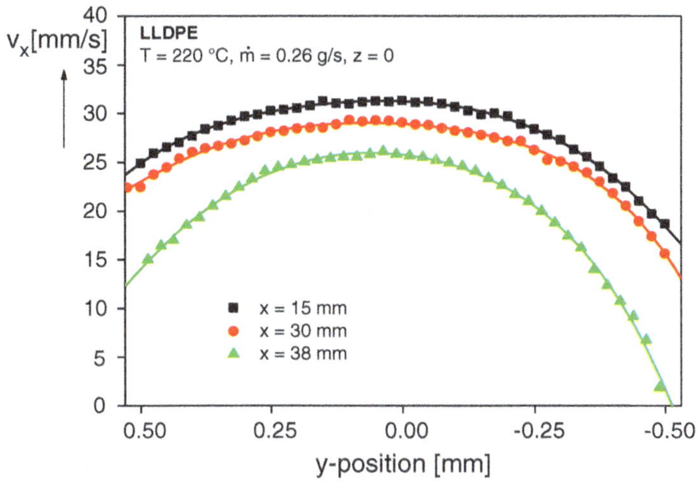

Fig. 17.6 Velocity profiles across the slit at different distances x from the slit entrance after 3 h extrusion with the modified LLDPE followed by 12 h extrusion of the neat material at $T = 220$ °C and a mass throughput of 0.26 g/s [3]

measured at $x = 38$ mm, i.e., relatively close to the capillary exit sticks to the wall again, at least on one side of the slit. At the positions further away from the exit, slip is still found, but the slip velocities on this wall are lower than after 3 h of extrusion (cf. Fig. 17.5). In parallel to the decrease of wall slip, the formation of "shark skin" is observed which gradually spreads across the surface of the extruded strand and mirrors the asymmetry of the change of the slip velocities.

The results described show that investigations of flow profiles within a slit die using laser-Doppler velocimetry are an interesting method to get an insight into the efficiency of additives for the reduction of "shark skin." Measurements of the slip velocities bear the potential to tailor additives for different polymers and wall materials in laboratory experiments.

17.4 Flow Profiles Inside a Flat Die for Film Casting

17.4.1 Motivation

Many processing operations comprise complicated flow patterns of polymer melts. Therefore, it is difficult to model them comprehensively using adequate equations of state and suitable rheological material functions. Laser-Doppler velocimetry is a method to quantitatively investigate flow fields in a tool and draw at least qualitative conclusions with respect to an optimization. An example is given for film casting using a flat die.

There exist two methods to produce thin films on an industrial scale: film blowing and film casting. The film blowing process is the more economical one. It will shortly be addressed in Sect. 17.5. The film casting process is usually preferred when films of good thickness uniformity are required which are favorable for many applications. Therefore, getting a deeper understanding of the basics of film casting is a challenging task.

There is quite a number of numerical simulations published in the literature, e.g., [6–8] based on one-, two-, or three-dimensional models which have been of only limited success, however, regarding their applicability. The number of experimental studies on film casting is much smaller, particularly, with respect to the investigation of velocity fields. In [9] LDV-measurements of flow profiles are presented for a polyethylene terephthalate (PET) in machine direction as well as in transverse direction. Convincing conclusions from these observations could not be drawn, however, due to the limited resolution of the experiments.

The laser-Doppler setup described in Sect. 16.2.1 was used to measure the flow fields along a flat die of 300 mm in width. Some few results are presented as examples of what can be learnt from a quantitative investigation of the flow in such an equipment. More comprehensive investigations can be found in [10] and [11].

17.4.2 Experimental Device

A principal view of the die used is shown in Fig. 17.7. The flat die is fed through the inlet I by a gear pump coupled to an extruder. The melt is distributed then by the manifold M, which has the shape similar to a coat hanger. The exit region of

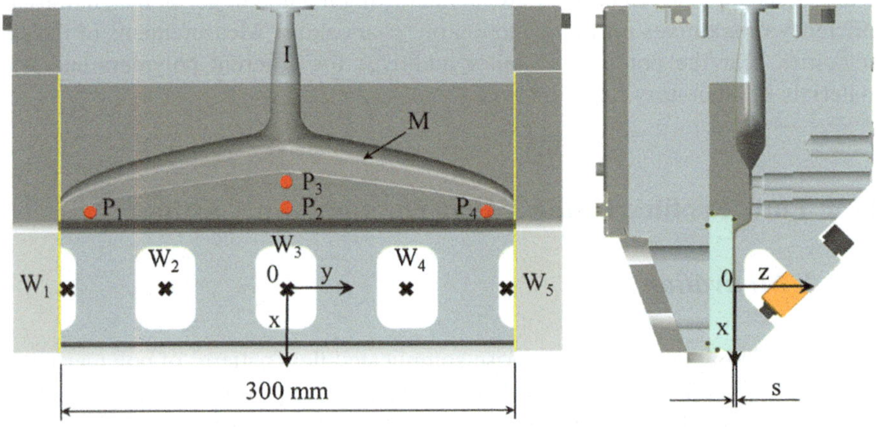

Fig. 17.7 Schematic view of the experimental setup of the flat die. *I* inlet, *M* manifold, $W_1 - W_5$ windows, $P_1 - P_4$ pressure transducers. *Left:* front view of the die. The crosses within the windows indicate the LDV measuring positions. *Right:* lateral view of the die, x is the extrusion direction and s the slit height. (Reprinted from [10] by permission from John Wiley and Sons)

the die is accessible for the laser light by the five glass windows W1–W5. A three-dimensional translation stage is applied to position the measuring volume with a spatial resolution of around 10 μm. The slit width was 300 mm, the slit height was chosen as 2 mm for all the measurements discussed. The extruded films were drawn by a special unit allowing a variety of drawing ratios.

The coordinate system is located in the center of window W3, its distance from the rim of the die is 60 mm. The *x*-coordinate points into the direction of flow, the *y*-coordinate marks the positions along the die width and the *z*-coordinate stretches across the slit height.

17.4.3 Flow Profiles

The polypropylene investigated had a weight average molar mass of $M_w = 530$ kg/mol and a polydispersity factor of $M_w/M_n = 5.3$. The flow profiles across the slit (*z*-direction) measured in the windows W_1, W_3, and W_5 are presented in Fig. 17.8 under the conditions given in the caption.

As expected, the linear polypropylene PP1 exhibits profiles principally similar to those of the LDPE in the slit rheometer (cf. Fig. 16.18). They are symmetrical with respect to the center of the die. Their extrapolations by making use of Eq. 16.4 give the velocity zero at the walls and, therefore, it can be assumed that the polypropylene melt sticks to the wall. Within the accuracy of the measurements the maximum velocity reached in the center of the slit is the same in the outer windows W_1 and W_5, but somewhat lower in the middle window W_3.

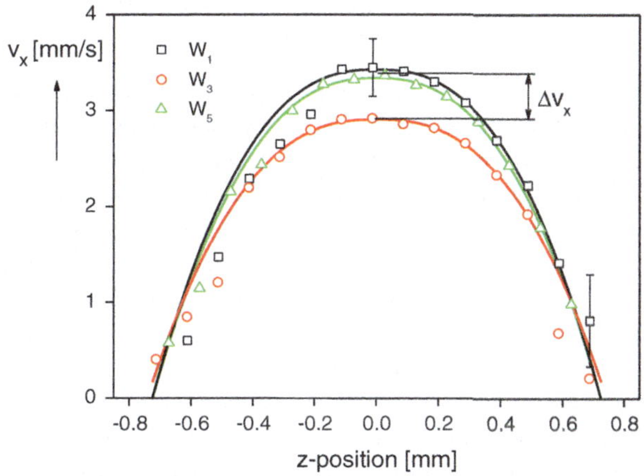

Fig. 17.8 Velocity profiles across the slit for a polypropylene of $M_w = 570$ kg/mol and $M_w/M_n = 5.3$ (PP 1) at a throughput of $\dot{V} = 90$ cm^3 min^{-1} and a temperature of $T = 230$ °C The *z*-coordinates represent apparent positions across the slit as they were not corrected for the different refractive indices between air and glass. The *solid lines* are fitted according to Eq. (16.4). (Reprinted from [10] by permission from John Wiley and Sons)

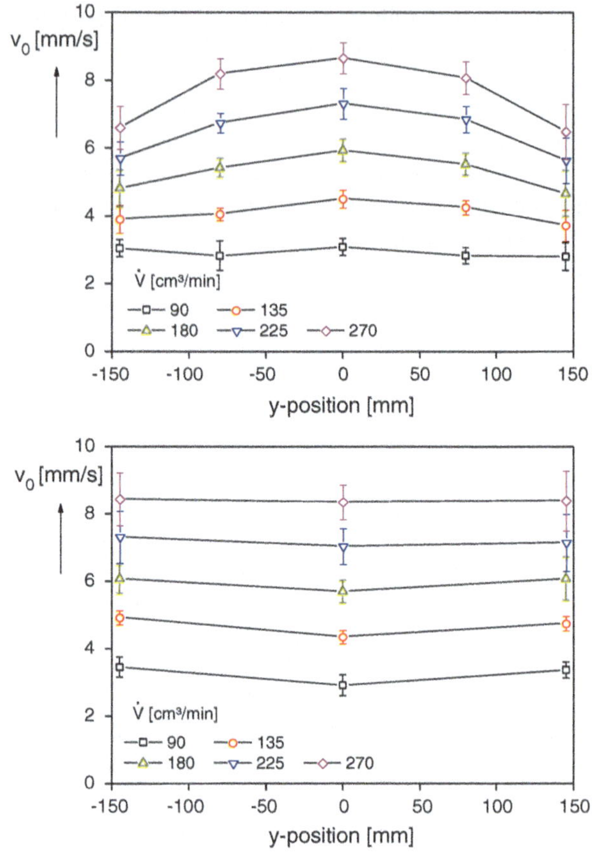

Fig. 17.9 Velocity distributions along the die for the polypropylene PP 1 at various volume rates $\dot{V}$ for 200 °C (upper part) and 230 °C (lower part). The error bars are obtained from five measurements each. (Reprinted from [10] by permission from John Wiley and Sons)

As a uniform velocity distribution along the slit can be regarded as one condition for a uniform film thickness, velocity measurements along the die present the focus of the following investigations with LDV. For the matter of clarity, only the maximum velocities v_0 are discussed instead of the whole profiles. In Fig. 17.9, v_0 obtained in the various windows is shown at the two temperatures of 200 and 230 °C for various throughputs. At 200 °C (upper part) the velocities in the middle of the die at $y = 0$ mm are significantly higher than at the rims 150 mm away. This effect becomes the more pronounced the higher the volume rate. With respect to applications the finding of a uniformity of the flow field decreasing with increasing output rate is of disadvantage as high extrusion rates are required in practice. The situation becomes more convenient at 230 °C as the lower part of Fig. 17.9 demonstrates. At this temperature the velocity distribution along the slit is not much different, particularly, at higher output rates.

As parameters like the temperature and the volume rate obviously have a measurable influence on the velocity distribution along the flat die the question arises which effects can be expected from variations of the molecular structure of

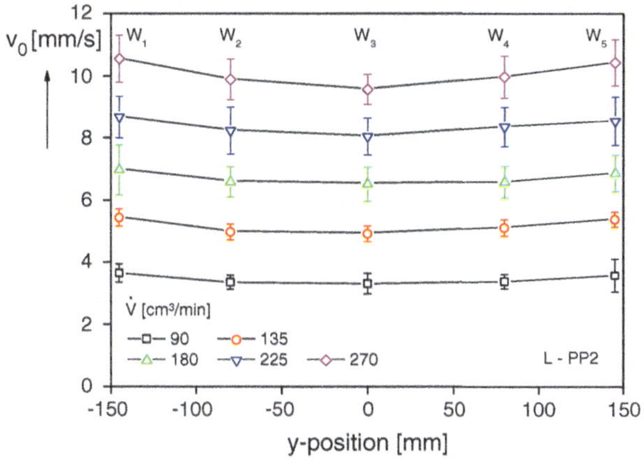

Fig. 17.10 Velocity distributions $v_0(y)$ along the die (windows $W_1 - W_5$) for the linear polypropylene PP2 at a temperature of 220 °C and various volume rates $\dot{V}$ The scatter *bars* give an idea of the reproducibility of the velocity measurements. (Reprinted from [11] by permission from John Wiley and Sons)

the extruded material. Of particular interest in this aspect are long-chain branched materials which are favorable with respect to the uniformity of sample deformation if elongation comes into play. For this purpose another linear PP with $M_w = 360$ kg/mol and $M_w/M_n = 6.9$ (PP 2) and a long-chain branched, distinctly strain-hardening sample with $M_w = 600$ kg/mol and $M_w/M_n = 9.6$ (LCB-PP) were chosen.

Figure 17.10 clearly demonstrates that under comparable conditions, PP 2 does not exhibit a constant velocity distribution along the die comparable to that of PP 1 in Fig. 17.9. particularly at the higher output rates. The velocities in the middle of the die are smaller than at the rims.

According to [12] the power-law index n related to the slope of the viscosity function by Eqs. (11.13) and (11.14) plays the decisive role for the velocity distribution along the die, whereas the zero-shear viscosity reflecting M_w is not of importance. The modeling described in [13] predicts a tendency to lower velocities in the center of the die for power-law indices higher than that one used for the die design. The die was optimized for PP 1 with the power-law index $n = 0.42$. The experimental results for PP 2 with $n = 0.53$ in Fig. 17.10 follow these predictions if compared with those in Fig. 17.9.

In the case of the LCB-PP, a power-law index similar to that of the PP 2 at 220 °C is obtained if the temperature is set to 200 °C. Figure 17.11 exhibits the results on the velocity distributions. The velocity profiles are rather similar. From this experimental finding it can be concluded that strain hardening does not influence the velocity distribution along the flat die chosen, but that the power-law index seems to be the decisive material parameter which allows the optimization of the uniformity of the flow at the die exit. This result supports the role of the power-law index as an important material-specific parameter for the modeling of extrusion through a flat die.

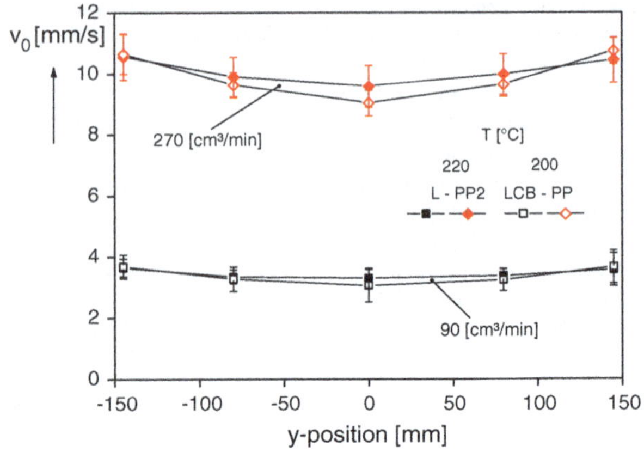

Fig. 17.11 Comparison of the velocity distributions $v_0(y)$ of the PP2 and the long-chain branched LCB–PP at throughputs of 90 cm^3 min^{-1} and 270 cm^3 min^{-1} and temperatures leading to rather similar power-law indices. (Reprinted from [11] by permission from John Wiley and Sons)

17.5 Role of Elongational Viscosity for Processing

17.5.1 General Considerations

The most outstanding feature of elongational flow is the occurrence of strain hardening for some types of polymers (cf. Sect. 12.5.2). This property should have a favorable influence on the uniformity of the geometry of a sample during extension as it becomes obvious from an experiment of thought presented in Fig. 17.12, which shows in a simplified form the elongational viscosity as a function of the Hencky strain for a non strain-hardening and a strain-hardening material. Let us regard the extension of a sample with a localized thinning of its diameter the reason of which can be manifold in practice. If stretched, the elongation of the thinner part becomes higher than that of the thicker one under conditions presented by the lower curve in Fig. 17.12. For a constant viscosity, the portion with the smaller diameter elongates more and more and the sample may fail, at least. In the case of strain hardening represented by the increasing viscosity curve in Fig. 17.12 the larger deformation would lead to a comparatively higher elongational viscosity which naturally results in a more pronounced resistance to deformation. A kind of self-healing mechanism occurs which should effect a more uniform extension of the sample than in the case of the absence of strain hardening.

17.5.2 Uniformity of the Elongation of Samples with Different Strain Hardening

The relevance of this experiment of thought was checked by investigations of the uniformity of sample deformation using the MTR tensile rheometer described in Sect. 12.3.5. The samples were video-recorded and then their diameters manually

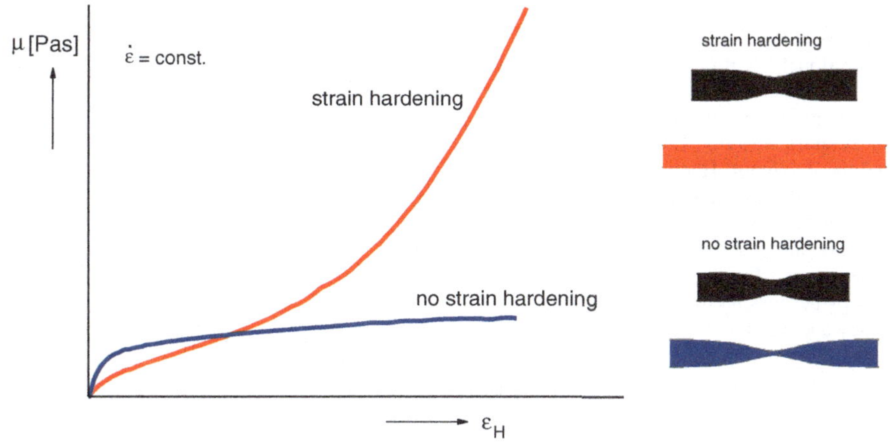

Fig. 17.12 The role of strain hardening for the uniformity of the sample geometry during extension

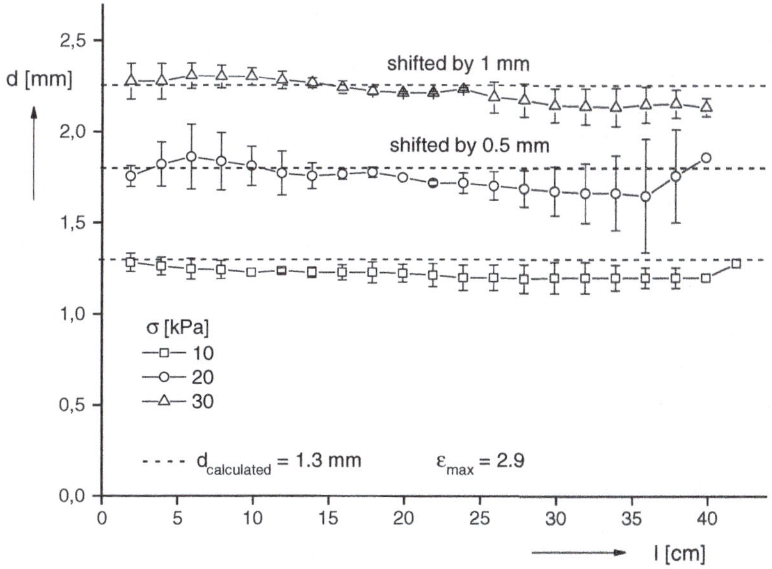

Fig. 17.13 Diameters along the samples of a commercial LDPE stretched at low, medium, and high constant stresses σ. The temperature was 150 °C. The initial length of the sample was 25 mm, its diameter 5.5 mm. The *curves* at the medium and high stress are shifted for the matter of a clear presentation. (Reprinted from [14] by permission from John Wiley and Sons)

measured at various positions along their length. As an example, results at a low, medium, and high tensile stress are depicted in Fig. 17.13 for a commercial LDPE. The initial sample of 25 mm in length and 5.5 mm in diameter was stretched 18 fold which corresponds to a Hencky strain of 2.9.

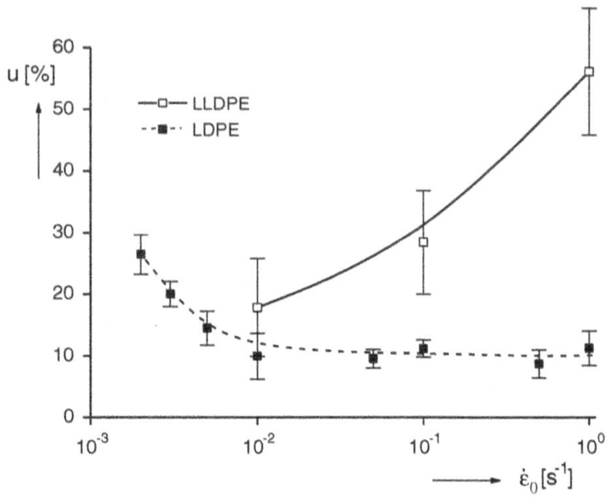

Fig. 17.14 Nonuniformity index as a function of elongational rate for an LDPE and an LLDPE stretched in a stressing experiment. The temperature was $T = 150$ °C and the maximum Hencky strain $\varepsilon_H = 2.9$. The scatter *bars* give an idea of the reproducibility of the experiments. (Reprinted from [14] by permission from John Wiley and Sons)

The diameter after deformation calculated under the assumption of a constant sample volume is shown in Fig. 17.13 as a broken line. The fact that the measured diameters are generally smaller than those expected from a uniform extension is due to the fastening of the samples to metal clamps by gluing which inhibits any deformation at their ends.

In order to simplify the assessment of the performance of an elongational experiment with respect to the uniformity of deformation, a so-called nonuniformity index u is defined as

$$u = a/d_c \tag{17.1}$$

with

$$a = \frac{1}{n}\sum_i^n |d_i - d_c| \tag{17.2}$$

d_c is the diameter calculated for a uniform elongation at constant volume, d_i the diameter of the sample at the position i along the sample and n the number of positions at which the diameters were determined.

Figure 17.14 compares the nonuniformity indices as a function of the elongational rate for samples of an LDPE and an LLDPE with geometries very similar to those in Fig. 17.13. The specimens were elongated using the MTR. Two features are evident. The LLDPE deforms much less uniform than the LDPE, particularly, at higher elongational rates and the scatter of the diameters obtained from several measurements is distinctly larger. The nonuniformity index u of the LDPE becomes smaller with increasing rate and stays nearly constant between 10^{-2} and 1 s^{-1}. However, the nonuniformity significantly grows with $\dot{\varepsilon}_0$ for the LLDPE.

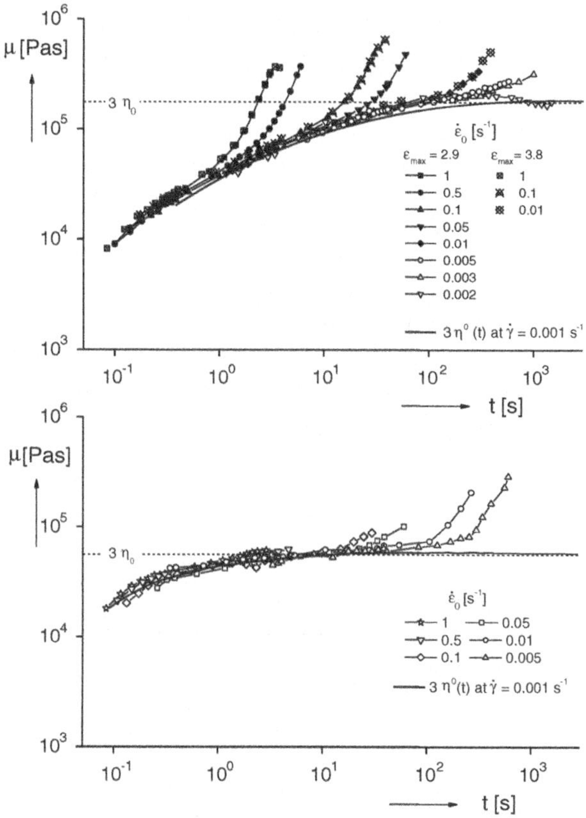

Fig. 17.15 Elongational viscosities μ of the two polyethylenes of Fig. 17.14 measured at $T = 150$ °C and different constant elongational rates $\dot{\varepsilon}_0$ (*upper figure* LDPE, *lower figure* LLDPE). The LDPE samples were stretched up to a maximum Hencky strain of 3.8, the LLDPE samples up to 3.0 [15]

The elongational viscosities as a function of time for the two polyethylenes throw some light on these results. They are shown in Fig. 17.15. The LDPE exhibits the well-known pronounced strain hardening at higher rates, which becomes smaller at lower ones and is hardly existent any more at $\dot{\varepsilon}_0 = 0.002\ \text{s}^{-1}$. The LLDPE shows a strain hardening behavior just the other way round. It is not significant up to $\dot{\varepsilon}_0 = 0.05\ \text{s}^{-1}$, but becomes pronounced at lower rates.

The nonuniformity indices of Fig. 17.14 correlate with these strain hardening features surprisingly well. In all the regimes in which the strain hardening is small the uniformity of the sample deformation is distinctly worse than in those with higher strain hardening. At the lowest elongational rates measured, at which a pronounced strain hardening for the LLDPE is found the nonuniformity indices of the two polyethylenes even approach each other.

17.5.3 Uniformity of Films Blown from Two Polyethylenes with Different Strain Hardening

Film blowing is a widely used technique, which is governed by distinct extensional flows. Over the years, it was tried by several authors to quantitatively describe this rather complicated technical process using rheological data, e.g., [16–18], but a comprehensive model is still missing. Particularly, the uniformity of the film thickness has not found much theoretical attention in the literature, although it is of crucial importance in practice. Therefore, in the following this topic is addressed from an experimental point of view.

The two polyethylenes whose uniformity during uniaxial elongation in the MTR device was discussed in the foregoing section were blown into films using an equipment of a laboratory scale the technical data of which are given in [15]. The uniformity of the films blown at various processing conditions is presented in Fig. 17.16, which shows the nonuniformity index u as a function of the take-up ratio TUR at a blow-up ratio BUR = 2. The ratios are defined as

$$\mathrm{TUR} = v/v_0 \tag{17.3}$$

with v being the velocity of the take-up device and v_0 that of the parison at the exit of the die and

$$\mathrm{BUR} = d/d_0. \tag{17.4}$$

d stands for the diameter of the film bubble in its equilibrium state and d_0 for that of the circular die extruding the parison.

The nonuniformity index u was determined according to Eqs. (17.1) and (17.2) replacing the diameter of the elongated cylindrical sample by the thickness of the extruded film. The data necessary for the calculation of the nonuniformity index were obtained in the following way. Over a length of 25 cm in extrusion direction,

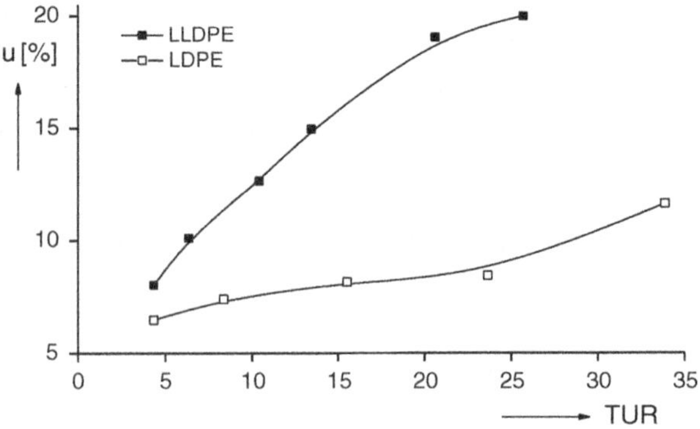

Fig. 17.16 Nonuniformity index u as a function of the take-up ratio TUR for films blown from the two polyethylenes characterized in the Figs. 17.14 and 17.15. The blow-up ratio was BUR = 2 and the processing temperature T = 205 °C. (Reprinted from [19] by permission from John Wiley and Sons)

the thickness of the solidified film was measured at distances of 1 cm. Five different positions across the film were chosen and averaged to get a mean thickness to which all the other thickness data were related according to Eq. (17.1). The nonuniformity indices u obtained in this way are plotted in Fig. 17.16. It is obvious that the LDPE films are more uniform than those from the LLDPE. At the smallest take-up ratio the two indices lie very close together. The index for the LDPE does not change much as a function of the take-up ratio whereas that of the LLDPE increases significantly. This behavior reflects the general favorable effect of strain-hardening on the uniformity of deformation as presented by the Figs. 17.12 and 17.14.

The qualitative agreement of the investigations on the uniformity of the blown film with those obtained by the well-defined elongational experiments is not necessarily conclusive and needs some further discussion as the experimental conditions are rather different. The most striking difference is that the laboratory experiment was performed in the uniaxial mode while in film blowing the deformation is biaxial. As shown in Fig. 12.19, however, strain hardening occurring in uniaxial elongation is reflected in a biaxial deformation, too, and, therefore, the beneficial effect of strain hardening with respect to the uniformity of deformation can be assumed for the film blowing of the LDPE.

Furthermore, the elongational rates are not at all constant along the blown film. They are obviously zero just at the die exit and beyond the frost line and attain a maximum value of 0.8 s^{-1} in case of the lowest take-up ratio and 2.2 s^{-1} at the highest displayed in Fig. 17.16 according to [15]. These elongational rates are in the range in which a distinct strain hardening is found for the LDPE, but not for the LLDPE (cf. Fig. 17.15). Insofar, the qualitative agreement between the laboratory and film blowing experiments with respect to the uniformity of deformation is not surprising.

The other big difference between the parameters of uniaxial extension and film blowing is that the elongational experiments with the rheometer were performed under isothermal conditions, whereas the film blowing is strongly nonisothermal. However, as discussed before (cf. Sect. 12.5.4), the strain hardening itself is existent in a wide range of temperatures, but shifts with respect to time and elongational rate if the temperature is changed. Therefore, strain hardening can be assumed to be effective under the nonisothermal conditions of the film blowing experiment, too.

The results presented convincingly demonstrate that uniaxial elongational experiments performed on a laboratory scale bear the potential for predictions regarding the uniformity of films blown in a technical process. Furthermore, from established relationships between molecular parameters and strain hardening in uniaxial elongation (cf. Sect. 13.3) conclusions can be drawn into which direction the molecular structure of polymers should be changed if the uniformity of blown films is intended to be improved.

17.5.4 Elongational Viscosity and Foaming

Another processing operation of great practical importance dominated by elongational flow is foaming. Particularly, the difficulties to generate polypropylene foams of high quality led to investigations on the relationship between rheological properties and foaming. Evidently, the elongational behavior became the focus of the research in this area and it was shown that the "melt strength," i.e., the maximum force an extruded melt strand can be loaded with before it breaks is a decisive property [20–22]. This failure quantity is usually determined with the so-called "Rheotens" [23] by drawing down the polymer melt between two counter rotating wheels and measuring the force. The value of the melt strength depends on quite a number of experimental conditions and, therefore, is not a well-defined material function. Strain hardening is experimentally more clearly defined and thus it may be a more suitable quantity to describe the foaming behavior of a polymeric material.

To answer the question whether such a relationship exists, a linear and a long-chain branched polypropylene and blends from them were investigated with respect to their foaming performance. The foaming was carried out by a laboratory equipment based on a capillary rheometer which is described in detail in [24] and [25]. As a chemical blowing agent azodicarbamide was applied. The temperature for the decomposition of the azodicarbamide was chosen as 200 °C, the foaming temperature for the investigations presented in the following was 160 °C.

The elongational viscosities of the various blends are shown in Fig. 17.17. For the linear PP (L-PP) the curves obtained at elongational rates between 0.01 and

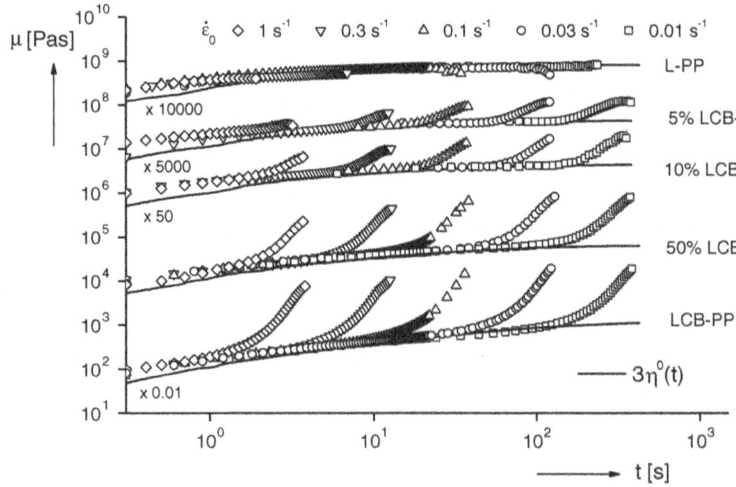

Fig. 17.17 Elongational viscosities μ of blends of a linear polypropylene (L-PP) and a long-chain branched polypropylene (LCB-PP) used for foaming. The *curves* are shifted to each other according to the factors indicated [24]

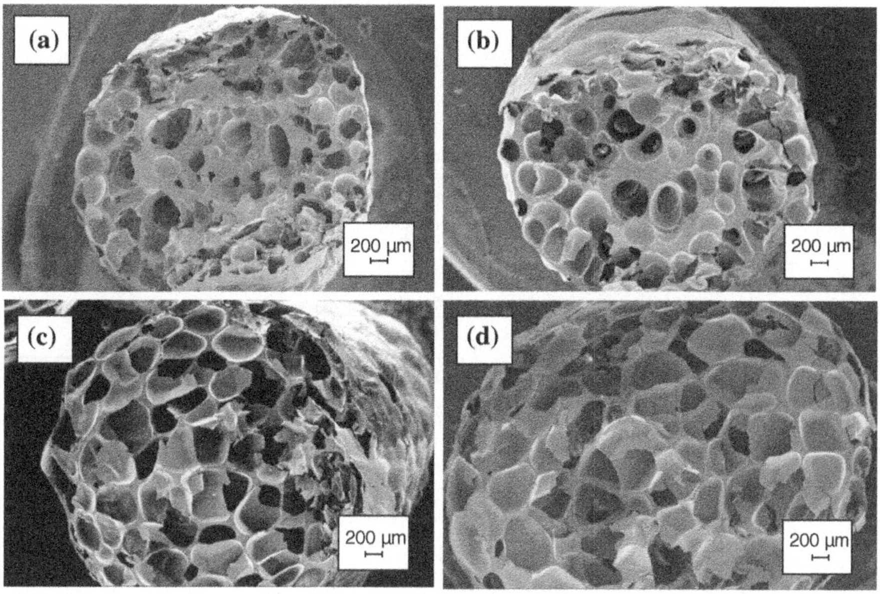

Fig. 17.18 Electron micrographs of foams from **a** the linear PP (L-PP), **d** the long-chain branched PP (LCB-PP) and two blends with **b** 5 and **c** 50 wt. % LCB-PP, respectively. The foaming temperature was 160 °C and the apparent shear rate in the die $D = 360\ s^{-1}$ [24]

$1\ s^{-1}$ fall together and agree very well with the linear relationship $3\eta^0\ (t)$. No strain hardening is observable up to a total strain of about 3 as expected for a linear polypropylene (cf. Fig. 13.19). The slight decreases at lower elongational rates are due to nonuniform sample deformations. The long-chain branched material (LCB-PP) exhibits significant increases of the elongational viscosity which are approximately the same for the elongational rates applied. The strain hardening of the blends shows a distinct dependence on the concentration of the LCB-PP. At 5 wt. % of the LCB-PP a strain hardening is visible at lower elongational rates, which becomes more pronounced at higher concentrations. At 50 wt. % LCB-PP the strain hardening has nearly reached the level of the neat LCB-PP and is approximately similar for all the elongational rates applied.

Electron micrographs of strands foamed from the two blend components L-PP and LCB-PP and blends of 5 and 50 wt. % of the branched material are displayed in Fig. 17.18. They were extruded at a temperature of 160 °C and an apparent shear rate in the capillary of $D = 360\ s^{-1}$.

It is obvious that the samples foamed from the LCB-PP and the material with 50 wt. % LCB-PP do show a distinctly more homogeneous cell structure and thinner cell walls than those of the other two products. Particularly, the foam from the L-PP contains areas at which foaming cannot be seen. An addition of 5 wt. % of the branched PP does not much change the morphology of the foam.

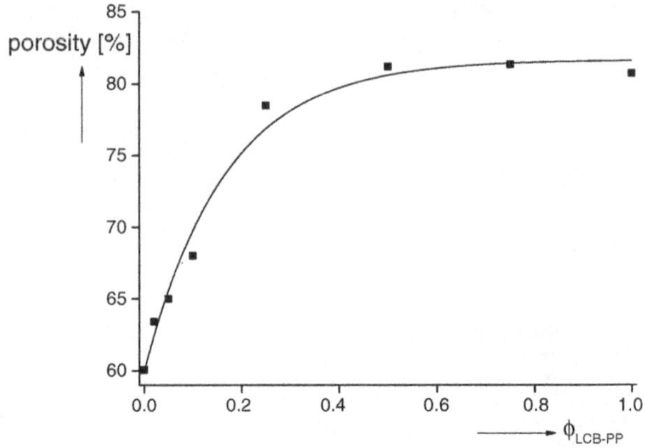

Fig. 17.19 Porosity of foams of PP blends as a function of the weight content Φ_{LCB-PP} of the long-chain branched component at a foaming temperature of 160 °C and an apparent shear rate of $D = 360\ s^{-1}$ [24]

A characteristic property of a foam is its porosity defined as

$$P = 1 - \frac{\rho}{\rho_0} \tag{17.5}$$

with ρ being the density of the foam and ρ_0 that of the bulk material. This quantity is plottedin Fig. 17.19 as a function of the content of the branched PP in the blend. The porosity of the foam increases up to around 80 % with growing content of LCB-PP. Up to about 25 wt. % LCB-PP the dependence is very pronounced, but for higher concentrations the curve levels off and approaches a plateau at about 50 wt. %. As the strain hardening significantly increases with the content of LCB-PP up to about 50 % and then remains nearly constant (cf. Fig. 17.17), a qualitative correlation between strain hardening in uniaxial extension and foam morphology is obvious, although foaming is a biaxial deformation. In Sect. 12.6 it was shown, however, that if strain hardening occurs in uniaxial extension it can be observed in the biaxial mode, too. Therefore, the "self-healing" effect of strain-hardening materials with respect to nonuniformities developing during deformation (cf. Sect. 17.5.1) can be assumed to be the reason for the better foamability of the samples containing long-chain branched PP. The cell walls can be stretched to a higher extent than in the case of a non strain-hardening material before they break.

Another interesting finding from Fig. 17.19 is the nearly constant degree of foaming for LCB-PP contents higher than 50 % which corresponds to the strain hardening not changing much in this range of LCB-PP addition.

These results demonstrate that elongational experiments and strain hardening in particular bear the potential for an assessment of the suitability of polypropylenes for foaming. The findings may be generalized with respect to other polymers, too.

References

1. International standard ISO 1183
2. Steffl T (2004) Doctoral thesis. University Erlangen-Nürnberg, Shaker Verlag, Aachen, ISBN 3-8322-2737-7
3. Schwetz M (2002) Doctoral thesis. Shaker Verlag, Aachen, ISBN 3-8322-0935-2
4. Ramamurthy AV (1986) J Rheol 30:337
5. Kalika DS, Denn MM (1987) J Rheol 31:815
6. Agassant JF, Avenas P, Sergent JP, Carreau PJ (1977) Polymer processing principles and modelling. Hanser Publishing, Munich
7. D'Halewyu S, Agassant JF, Demay Y (1990) Polym Eng Sci 30:335
8. Sakaki K, Katsumoto R, Kajiwara T, Funatsu K (1996) Polym Eng Sci 36:1821
9. Seyfzadeh B, Harrison GM, Carlson CD (2005) Polym Eng Sci 45:443
10. Griess HJ, Burghelea TI, Münstedt H (2012) Polym Eng Sci 52:615
11. Griess HJ, Münstedt H (2012) Polym Eng Sci 52:2253
12. Winter HH, Fritz HG (1986) Polym Eng Sci 26:543
13. Reid JD, Campanella OH, Corvalan CM, Okos MR (2003) Poly Eng Sci 45:693
14. Münstedt H, Kurzbeck S, Stange J (2006) Polymer Eng Sci 46:1190
15. Kurzbeck S (1999) Doctoral Thesis. University Erlangen-Nürnberg, Erlangen
16. Pearson J, Petrie C (1970) Plast Polym 38:85
17. Han CD, Park J (1975) J Appl Polym Sci 19:3257
18. Wagner M (1976) Doctoral thesis. University Stuttgart, Germany
19. Münstedt H, Kurzbeck S, Stange J (2006) Macromol Symp 245:181
20. Bradley MB, Phillips EM (1991) Plast Eng 47:82
21. Park CB, Cheung LK (1997) Polym Eng Sci 37:1
22. Den Doelder CFJ, Sammler RL, Koopmans RJ, Paquet AN (2002) Cell Polym 21:99
23. Meissner J (1971) Rheol Acta 10:230
24. Stange J (2006) Doctoral thesis. University Erlangen-Nürnberg, Shaker Verlag, Aachen, ISBN 3-8322-5521-4
25. Stange J, Münstedt H (2006) Cell Plas 42:445

Subject Index

H. Münstedt and F. R. Schwarzl, *Deformation and Flow of Polymeric Materials*, DOI: 10.1007/978-3-642-55409-4, © Springer-Verlag Berlin Heidelberg 2014

D

N

T

The manufacturer's authorised representative in the EU is Springer Nature Customer Service Centre GmbH, Europaplatz 3, 69115 Heidelberg, Germany. If you have any concerns regarding our products, please contact ProductSafety@springernature.com

Printed and bound by CPI Group (UK) Ltd, Croydon, CR0 4YY
22/09/2026
02220956-0015